Machining technology for composite materials

Related titles:

Composite joints and connections
(ISBN 978-1-84569-990-1)
The growing use of composites for structural applications in sectors such as aerospace demands that engineers have a thorough understanding of how composite joints behave under various loads. Gaining this understanding is not straightforward because joints and connections in composites present an entirely different set of problems to engineers when compared to metals. This book will address these differences, looking at the design, modelling and testing of joints and connections, using both bonded and mechanical joining techniques.

Composite reinforcements for optimum performance
(ISBN 978-1-84569-965-9)
This new book is concerned with improving the properties and performance of composites by modelling the reinforcements. This is a novel approach as it allows designers to optimise the performance of composite parts by modelling and therefore manipulating the reinforcements before the composite is manufactured. Reinforcements are an integral part of all composites, but often they are simply used during the moulding process with no thought about how their performance can be optimised in the finished product. The book will allow manufacturers to apply modelling techniques to improve the quality and performance of their products.

Non-crimp fabric composites
(ISBN 978-1-84569-762-4)
Non-crimp fabric (NCF) composites are composites that are reinforced with woven mats of straight (non-crimped) fibres. Straight fibres deform much less under tension. NCF composites are being used in applications in the aerospace, automotive, civil engineering and wind turbine sector where strength is important. *Non-crimp fabric composites* reviews production, properties and applications of this important class of composites.

Details of these and other Woodhead Publishing materials books can be obtained by:

- visiting our web site at www.woodheadpublishing.com
- contacting Customer Services (e-mail: sales@woodheadpublishing.com; fax: +44 (0) 1223 832819; tel.: +44 (0) 1223 499140 ext. 130; address: Woodhead Publishing Limited, 80 High Street, Sawston, Cambridge CB22 3HJ, UK)
- contacting our US office (e-mail: usmarketing@woodheadpublishing.com; tel.: (215) 928 9112; address: Woodhead Publishing, 1518 Walnut Street, Suite 1100, Philadelphia, PA 19102-3406, USA)

If you would like to receive information on forthcoming titles, please send your address details to: Francis Dodds (address, tel. and fax as above; e-mail: francis.dodds@woodheadpublishing.com). Please confirm which subject areas you are interested in.

Machining technology for composite materials

Principles and practice

Edited by
H. Hocheng

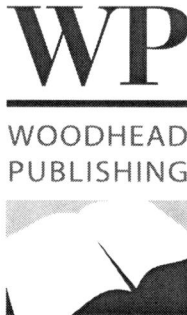

WP

WOODHEAD
PUBLISHING

Oxford   Cambridge   Philadelphia   New Delhi

Published by Woodhead Publishing Limited,
80 High Street, Sawston, Cambridge CB22 3HJ, UK
www.woodheadpublishing.com

Woodhead Publishing, 1518 Walnut Street, Suite 1100, Philadelphia, PA 19102-3406, USA

Woodhead Publishing India Private Limited, G-2, Vardaan House, 7/28 Ansari Road, Daryaganj, New Delhi – 110002, India
www.woodheadpublishingindia.com

First published 2012, Woodhead Publishing Limited
© Woodhead Publishing Limited, 2012
The authors have asserted their moral rights.

British Library Cataloguing in Publication Data
A catalogue record for this book is available from the British Library.

Library of Congress Control Number: 2011938828

ISBN 978-0-85709-030-0 (print)
ISBN 978-0-85709-514-5 (online)

The publisher's policy is to use permanent paper from mills that operate a sustainable forestry policy, and which has been manufactured from pulp which is processed using acid-free and elemental chlorine-free practices. Furthermore, the publisher ensures that the text paper and cover board used have met acceptable environmental accreditation standards.

Typeset by Toppan Best-set Premedia Limited, Hong Kong
Printed by TJI Digital, Padstow, Cornwall, UK

Contents

S. S. Joshi, Indian Institute of Technology Bombay, India

Contributor contact details

(* = main contact)

Editor

H. Hocheng
Department of Power Mechanical
    Engineering
National Tsing Hua University
Hsinchu 30013
Taiwan
E-mail: hocheng@pme.nthu.edu.tw

Chapter 1

H. A. Kishawy
Machining Research Laboratory
Faculty of Engineering and
    Applied Science
University of Ontario Institute of
    Technology (UOIT)
2000 Simcoe Street North
Oshawa
Ontario L1H 7K4
Canada
E-mail: hossam.kishawy@uoit.ca

Chapter 2

C. C. Tsao
Department of Mechatronics
    Engineering
Tahua Institute of Technology
Hsinchu 30740
Taiwan
E-mail: aetcc@thit.edu.tw

Chapter 3

S. D. El Wakil
Chancellor Professor of
    Mechanical Engineering
The University of Massachusetts
    Dartmouth
Dartmouth
MA 02747
USA
E-mail: selwakil@umassd.edu

Chapter 4

G. Caprino* and A. Langella
Department of Materials and
    Production Engineering
University of Naples Federico II
Piazzale Tecchio 80
80125 Naples
Italy
E-mail: caprino@unina.it; antgella@
    unina.it

Chapter 5

J. Sheikh-Ahmad*
Department of Mechanical
    Engineering
The Petroleum Institute
Abu Dhabi
UAE
E-mail: jahmad@pi.ac.ae

J. P. Davim
Department of Mechanical
    Engineering
University of Aveiro
Campus Santiago
3810-913 Aveiro
Portugal
E-mail: pdavim@ua.pt

Chapter 6

K. Palanikumar
Department of Mechanical
    Engineering
Sri Sairam Institute of Technology
Sai Leo Nagar
Chennai 600 044
India
E-mail: palanikumar_k@yahoo.com

Chapter 7

Q. Feng and C. Z. Ren
School of Mechanical Engineering
Tianjin University
Tianjin 300072
China
E-mail: fq141@163.com; renchz@
    tju.edu.cn

Z. J. Pei*
Department of Industrial and
    Manufacturing Systems
    Engineering
Kansas State University
Manhattan
KS 66506
USA
E-mail: zpei@ksu.edu

Chapter 8

B. Lauwers*, O. Malek, K. Brans
    and K. Liu
Department of Mechanical
    Engineering (MECH)
Katholieke Universiteit Leuven
Celestijnenlaan 300b – box 2420
3001 Heverlee
Leuven
Belgium
E-mail: bert.lauwers@mech.
    kuleuven.be

J. Vleugels
Department of Metallurgy and
    Materials Engineering (MTM)
Katholieke Universiteit Leuven
Kasteelpark Arenberg 44 – box
    2450
B-3301 Heverlee
Leuven
Belgium

Chapter 9

J. W. Liu
State Key Laboratory of Pulp and
  Paper Engineering
South China University of
  Technology
Guangzhou
China 510640
E-mail: fejwliu@scut.edu.cn

T. M. Yue*
The Advanced Manufacturing
  Technology Research Centre
Department of Industrial and
  Systems Engineering
The Hong Kong Polytechnic
  University
Hung Hom
Hong Kong
E-mail: mftmyue@inet.polyu.edu.
  hk

Chapter 10

G. Chryssolouris* and K. Salonitis
Laboratory for Manufacturing
  Systems and Automation
Department of Mechanical
  Engineering and Aeronautics
University of Patras
26500 Patras
Greece
E-mail: xrisol@lms.mech.upatras.gr

Chapter 11

R. Negarestani and L. Li*
Laser Processing Research Centre
School of Mechanical, Aerospace
  and Civil Engineering
The University of Manchester
Manchester
M13 9PL
UK
E-mail: Negarestani@manchester.
  ac.uk; lin.li@manchester.ac.uk

Chapter 12

F. Fischer*
Former address:
Laser Zentrum Hannover e.V.
Department of Technologies for
  Non-Metals
Head of Composite Group
Hollerithallee 8
30419 Hannover
Germany

Current address:
Technische Universität
  Braunschweig
Institut für Fuge- und
  Schweißtechnik
Langer Kamp 8
38106 Braunschweig
Germany
E-mail: fabian.fischer@
  tu-braunschweig.de

L. Romoli
Department of Mechanical,
    Nuclear and Production
    Engineering
University of Pisa
Pisa
Italy
E-mail: l.romoli@ing.unipi.it

R. Kling and D. Kracht
Laser Zentrum Hannover e.V.
Hollerithallee 8
30419 Hannover
Germany

Chapter 13

H. Attia* and M. Meshreki
Aerospace Manufacturing
    Technology Centre
Institute for Aerospace Research
National Research Council of
    Canada
P.O. Box 40, Station
    Côte-des-Neiges
Montreal
Quebec H3S 2S4
Canada
E-mail: helmi.attia@nrc-cnrc.gc.ca

A. Sadek
Department of Mechanical
    Engineering
Room 270, Macdonald Engineering
    Building
817 Sherbrooke Street West
Montreal
Quebec H3A 2K6
Canada

Chapter 14

Y. Yildiz*
Technology Faculty
Department of Manufacturing
    Engineering
Dumlupinar University
Kutahya
Turkey
E-mail: ykpyldz@hotmail.com

M. M. Sundaram
Micro and Nano Manufacturing
    Laboratory
Department of Mechnical
    Engineering
School of Dynamic Systems
University of Cincinnati
Cincinnati
OH 45221-0072
USA
E-mail: murali.sundaram@uc.edu

Chapter 15

M. Balazinski*
Department of Mechanical
    Engineering
École Polytechnique de Montréal
C. P. 6079, Succ. Centre-Ville
Montreal
Quebec H3C 3A7
Canada
E-mail: Marek.balazinski@
    polymtl.ca

V. Songmene
Department of Mechanical
    Engineering
École de Technologie Supérieure
    – ETS
Université du Québec
1100 Notre-Dame West
Montreal
Quebec H3C 1K3
Canada
E-mail: Victor.songmene@etsmtl.ca

H. A. Kishawy
Machining Research Laboratory
Faculty of Engineering and
    Applied Science
University of Ontario Institute of
    Technology (UOIT)
2000 Simcoe Street North
Oshawa
Ontario L1H 7K4
Canada
E-mail: hossam.kishawy@uoit.ca

Chapter 16

G. Kowaluk
Warsaw University of Life Sciences
    – SGGW
Department of Technology,
    Organization and Management
    in Wood Industry
159 Nowoursynowska Street
02-776 Warsaw
Poland
E-mail: gkowaluk@gmail.com

Chapter 17

S. S. Joshi
Department of Mechanical
    Engineering
Indian Institute of Technology
    Bombay
Powai
Mumbai 400 076
India
E-mail: ssjoshi@iitb.ac.in

Part I
Traditional methods for machining composite materials

1
Turning processes for metal matrix composites

H. A. KISHAWY, University of Ontario Institute of Technology (UOIT), Canada

Abstract: Metal matrix composite materials (MMCs) offer various mechanical properties that are not offered by conventional unreinforced monolithic metal counterparts; specifically, high temperature stability, specific strength, and wear resistance. As a result, these composite materials have different applications in several industries including automotive and aerospace. However, machining of MMCs still remains a challenge. Understanding the manufacturing methods, strengthening mechanisms and hence mechanical properties of MMCs is crucial to comprehension of their deformation behavior during machining and the resulting workpiece surface integrity and tool wear. This chapter describes the types of composites and their unique physical properties. In addition, the cutting performance of some composite materials is discussed.

Key words: MMC tool material, wear, self-propelled tool.

1.1 Introduction

The rise in industrial development has generated a steady demand for improving the properties of available materials. These properties are sometimes very hard to obtain by conventional alloying methods. Continuous and extensive research and development has broadened our material knowledge to aid in the improvement of mechanical properties and the generation of a new class of materials known as composite materials. Because of their unique mechanical properties, including high specific strength and stiffness, high damping ratio, and low coefficient of thermal expansion, applications of composite materials are growing increasingly, from primary applications such as in the interiors of automobiles to very advanced applications in automotive, aerospace, marine and off-shore industries.

A composite material is a combination of at least two chemically distinct materials, with a distinct interface separating the constituents. It is usually designed and formed to obtain properties which would not otherwise be achieved by any of the individual constituents. The distinction of composites

3

from other alloys with two or more phases comes from the processing of the composites where the different phases are mixed together. Composites, in general, consist of at least two components namely *matrix* and the *reinforcement.*

1.1.1 The matrix

In general, alloys are used for the matrix components of MMCs. The most commonly used matrix materials are aluminum-based alloys, magnesium alloys, titanium alloys, copper, and nickel. Among these, aluminum based alloys (2024, 2124, 5156, 6061, 7075 and 7090) are the most commonly used matrices, due to their low density and high thermal conductivity. Within the composite material system, the matrix provides the ductility and load trans-fer between the reinforcements. It also provides transverse strength in the case of fiber-reinforced composites.

1.1.2 The reinforcements

Inter-metallic compounds, oxides, carbides or nitrides are used as reinforce-ment materials. The most dominant reinforcements are SiC, graphite, carbon, Al_2O_3, boron, B_4C, tungsten, Si_3N_4 and TiB_2.

In processing, reinforcements are used either in continuous forms (long fibers) or in discontinuous forms (whiskers, particulates, chopped fibers and platelets) in the case of metal matrix composites. Based on the type of reinforcement, composite materials are typically classified in the following three categories; long fiber reinforced, short fiber (whiskers) reinforced, and particulate reinforced. Regardless of the type of reinforcement, the obtained physical and mechanical properties are always better than those offererd by the individual monolithic materials.

In the case of particulate reinforced composites, particle morphology, particle size distribution, dispersion uniformity, surface chemistry, volume fraction, particle shape, and particle wetability are some of the factors that affect the properties of the composite. The most desired advantages of MMCs are their resistance to severe environments, toughness, high elastic modulus, and retention of strength at high temperatures. Because it is pos-sible to obtain the required mechanical strength and stiffness from the reinforcements, the development and selection of matrix materials for a composite structure can be characterized based on their environmental stability, such as oxidation and corrosion resistance at elevated temperature. For long fiber reinforced composites, knowledge of the shear strength requirements of the matrix are essential since the matrix serves only to transfer load into the filaments. However, the strength of the matrix dominates the yielding behavior in short fiber and particulate reinforced MMCs. Also, practical experience indicates that the improved physical properties offered by short fibers (whiskers) or particulate composites are modest compared to those with long fibers. This would explain why most of the early research has focused on the development of continuous filament composites. Initially, the continuous reinforcement was developed for aero-space applications. Owing to the high production cost of this type of com-posite, its non-aerospace applications are minimal despite the numerous attractive mechanical and high temperature properties.

The adoptation of discontinuous reinforcements leads to significant cost reduction and greater flexibility in fabrication which makes this attractive and feasible for various applications. With the introduction of SiC whiskers, extensive research has been conducted to process discontinuous reinforced composites composed of ceramic particulate reinforcements, such as SiC, Al_2O_3 and B_4C in aluminum based matrixes. Adding to to the inexpensive techniques used for the fabrication of particulate MMCs, another very important advantage of discontinuously reinforced MMCs is that they can be formed by conventional processes such as extrusion, forging, and rolling. However, these forming processes are usually followed by a finishing machining operation. It should be mentioned here that these conventional forming techniques can be applied to composites with up to 40% particulate reinforcements. The characteristics of the reinforcement dictate the optimal properties of the composite.

1.1.3   Manufacturing methods

The methods employed to produce metal matrix composites can be divided into two steps: primary material production and secondary con-solidation operations. Primary material production encompasses the oper-ations by which the composite is fabricated from its raw materials. Secondary operations include additional procedures such as extrusion, rolling and forging processes, which transform the primary composite into a usable shape.

Most of the primary operations are employed to produce long-fiber rein-forced composites that are near net shape components. This step is used to avoid damage to fibers which is commonly caused by secondary operations or machining.

Most of the discontinuous reinforced metal matrix composites can be processed by secondary operations in order to convert billets into useful shapes. Conventional extrusion techniques are usually employed to produce aluminum based SiC and Al_2O_3 based MMCs with only minor modification to operations. Forging and rolling can also be used to produce discontinuous reinforced metal matrix composites.

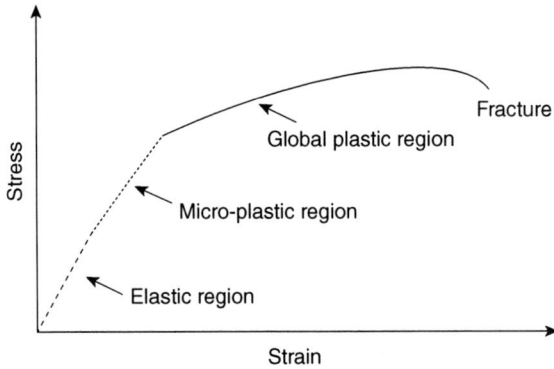

1.1 Schematic stress–strain curve of MMCs under tensile loading.

1.1.4   Strengthening mechanisms

Generally, the monotonic strength and stiffness of MMCs is much higher than that of the unreinforced alloys. Figure 1.1 shows a general schematic of the evolution of damage in a MMC during monotonic loading.[1]

Since the reinforcing phase is typically much stiffer than the matrix, a significant fraction of the stress is initially carried by the reinforcement. As a result, micro-plasticity occurs at fairly low values of stress and strain, corresponding to the original deviation from elastic linearity in the stress–strain curve. This point is termed the *proportional limit stress*. Micro-plasticity in the composites has been attributed to stress concentrations in the matrix at the sharp ends of fibers, whiskers, and particles, as well as at the poles of the reinforcement. With increase in strain, micro-plasticity increases in magnitude; therefore, the plasticity occurs in a global sense throughout the matrix. Compared with the unreinforced matrix, the rate of work hardening of MMCs is increased as a result of the integration of reinforcement.

1.2   Turning of metal matrix composites (MMCs)

A simple schematic diagram of the two-dimensional cutting operation of MMC is illustrated in Fig. 1.2. In the case of a perfectly sharp tool and the as a result of the relative motion between the tool and the workpiece, the material is sheared along the shear plain, forming a chip. However, a typical cutting tool usually has an edge radius, either by design or as a result of the manufacturing processes. The edge radius alters the effective undeformed chip thickness and moves the cutting action to the dotted line, as shown in Fig. 1.2, and the effective undeformed chip thickness is reduced to $(t - \Delta)$. The part of the chip material of thickness $(t - \Delta)$ will undergo a combination of elastic and plastic deformation, and the particles within this small layer

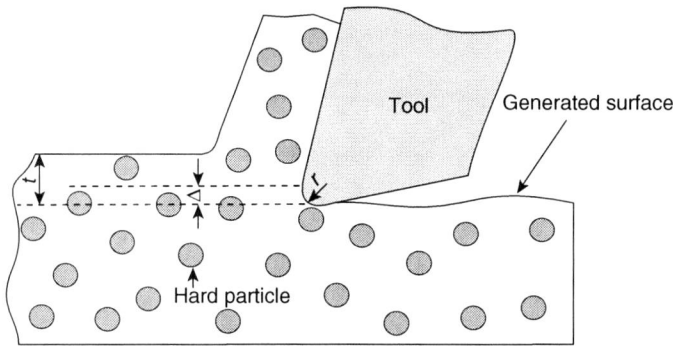

1.2 Schematic of MMC cutting process.

may be pushed into the soft matrix material or crushed under the compressive stress generated between this layer and the tool. The exact mechanism encountered by the particles, within this small layer, will largely depend on the deformation formed during cutting. At higher values of strain rate, the material matrix is more difficult to deform and thus particles tend to be sheared by the tool. At lower speeds and due to low strain, the matrix is easier to deform and particles are easily pulled out or pushed into the material.[2,3] After the material passes beneath the tool, some elastic recovery may occur.

Just prior to the shearing action, and depending on the location of the reinforcement particles through the undeformed chip thickness, the reinforcement particles encounter one of three different scenarios. If the particle is (i) above or (ii) below the cutting line, it will be pushed into the chip or below the machined surface. This action leads to generating more dislocations along the interface between the particle and the matrix as a result of the difference in material properties and difference in the thermal expansion. In the case where the particle is located (iii) along the cutting line, it will be either sheared or pulled up, leaving behind a cavity on the generated surface. The exact deformation mechanism depends on the matrix deformation rate as stated above. The sheared particle results in a better surface finish while the pulled out particle has mixed consequences; on one hand, in surface-to-surface contact applications, these cavities are filled with oil and provide the needed lubricant to reduce the friction, while on the other hand, the pulled particle represents a severe source of wear when it rolls freely along the flank surface between the tool and workpiece/chip causing three-body abrasion. The scanning electron microscopy (SEM) image in Fig. 1.3 shows a typical case of different deformation scenarios encountered by the reinforcement particles. Figure 1.4 shows a simplified illustration of different deformation scenarios encountered by the particle in front of the

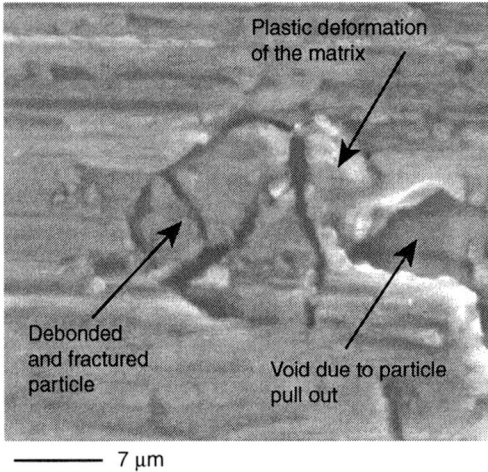

1.3 Scanning electron microscopy (SEM) image showing different deformation mechanisms in a MMC.

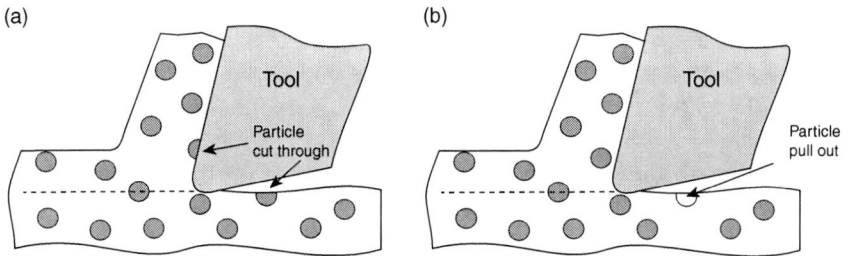

1.4 Schematic diagram for cutting MMCs: (a) particle cut through; (b) particle pull out.

tool tip during the MMC cutting operation. When the particle is sheared, a better surface finish is obtained, whereas poor surface finish is usually observed when particles are pulled out, due to the formation of cracks and pits on the machined surface.[3]

1.3   Cutting tools for turning Al/SiC based MMCs

Published literature on the machinability of particulate MMCs indicates that only cutting tools harder than the reinforcements have acceptable performance. Therefore, poly-crystalline diamond (PCD) tools, given their higher hardness compared with most of the common reinforcements, provide longer tool life. However, because of the high cost of PCD tools, other tools such as cemented carbides and ceramics have also been utilized to machine particulate MMCs.[4] Ceramic tools were found to be unsatisfactory, while the carbides were preferred over other types when machining at low cutting speeds and high feed rates.[4] Several researchers have investigated the effect of different coatings on the performance of carbide cutting tools during machining of MMCs.[5-7] It was suggested that coatings with less hardness than those of the reinforcement do not improve tool performance.

The properties of the composites mainly depend upon the reinforcement constituents and volume fraction used. It has been documented by many researchers that the change in the average particulate size and volume fraction of reinforcement, drastically affects the machining characteristics of MMCs[8-13] and the particulate reinforcement size and volume fraction together with the cutting speed are the dominant factors affecting the tool life.[14-16]

Analysis of modes of tool failure revealed that tool wear was severe when the volume fraction of reinforcements exceeded a critical value.[15] Analyzing the modes of tool failure showed that two-body and three-body abrasive wear are the dominant wear mechanism during machining of MMCs. This observation was confirmed by several researchers.[8,12,15,17] In previous research work by the author and his co-workers presented a novel approach for the prediction of tool flank wear progression during three dimensional (3D) bar turning of aluminum based particulate-reinforced metal matrix composites.[12] The flank wear rate was quantified by considering the tool geometry in 3D bar turning. Seeman et al.[18] investigated the effect of cutting speed and feed rate on the tool flank wear and surface integrity when machining LM 25 aluminum alloy reinforced with green-bonded silicon carbide particles using uncoated carbide tools. By examining the surface topographies of the worn tool, abrasive and adhesive wear were reported as the main mechanisms of tool wear on the cemented carbide tool. Better tool performance was observed at low cutting speed and feed rate, and the reason was attributed to the formation of a built-up edge which protects the cutting edge from further wear.

Aluminum alloys are frequently used as the matrix phase in metal matrix composites. The reinforcement can be in the form of fibers, whiskers and particles. The most common materials for reinforcements are silicon carbide (SiC) and aluminum oxide (Al_2O_3).[19] Table 1.1 shows the properties of typical discontinuous reinforcements.[20]

Aluminum alloy reinforced with discontinuous ceramic reinforcements (Al/SiC-MMC) is one of the more significant composites among MMCs that is harder than tungsten carbide (WC). Al/SiC-MMCs have a very light weight and low density, with high stiffness and hardness, high fatigue strength, high temperature resistance and wear resistance. Applications of all MMCs in industry are limited for several reasons including:

Table 1.1 Properties of typical discontinuous reinforcements for aluminum

	Al_2O_3	SiC particle	Al_2O_3 particle
Crystal structure	—	Hexagonal	Hexagonal
Density (g/cm³)	3.3	3.2	3.9
Average diameter (μm)	3	Variable	Variable
Strength (MPa)	2000	—	—
Young's modulus (GPa)	300	200–300	380

(i)   High machining cost due to their poor machinability. The tool life is very short during machining of MMCs and it is necessary to use advanced cutting tools (such as PCD tools) to obtain a reasonable tool life. HSS tools work for only a few seconds while conventional and coated carbides last only a few minutes.

(ii) Application of advanced cutting tools such as PCD tools is not as common as other tool materials and also they are very expensive. These types of tool materials are not readily available in complex cutting tools geometries such as taps, or very small diameter drills and reamers.

1.3.1   Conventional cutting tools

It can be concluded from the above discussion that machining is one of the major problems that limit the widespread use of MMCs.[11] Machining of MMCs is very costly because of short tool life. Common cutting tool materials for machining of MMCs are: PCD, CBN, PCBN, TiN-coated carbide, Al_2O_3-TiC ceramic, and CVD.

Existing coated tools with coatings such as titanium nitride and titanium carbide, perform well cutting steel, but show very poor performance in cutting MMCs. Furthermore, TiAlN coated tools show mediocre results in machining of Al + 20% SiC.[21] Diamond coating is a financially viable and environmentally friendly solution for machining MMCs. New coating methods such as hot filament and DC plasma jet CVD have to be applied to produce thin diamond layers with good adherence. Diamond inserts and diamond-coated tools can be used without cooling fluid, making the process environmentally friendly. This also has a positive effect on the health of operators and workers.[20]

Several studies on tool wear and surface integrity have found that tool wear is primarily due to abrasion by the hard reinforcement particles in the MMCs.[8,15,19,20,22–29] The grain size is an important factor which has a significant effect on the tool wear during machining of MMCs.[7] Li and Seah[15] showed that tool wear increases rapidly when the percentage of reinforcement particles in the MMC exceeds a critical value. This critical value is determined by the size and density of the reinforcement particles. El-Gallab and Skald[20] demonstrated that the built-up edge (BUE) can protect the cutting tool from abrasion wear. However, the unstable BUE may induce tool chipping and negatively affect the quality of the machined surface.[20,30]

The heat generated during machining can soften the metal matrix and enable the reinforcing particles to be embedded into the machined work-piece, which eliminates/reduces the tool particle interaction and in turn provides improved tool performance.[4] This promotes laser-assisted machining as a remedy for poor tool performance. The laser-assisted machining will provide enough heating to induce localized softening to facilitate particle squeezing into the machined surface. However, it should be noted here that care is needed to control the laser-induced heat to eliminate its negative impact on the generated surface integrity. In addition, an acceptable surface quality is produced when the particles are cut instead of being pulled from the matrix material.[29,31] Studies also suggest that dislocation pile-ups in the matrix material around the particles might result in the development of cracks and voids.[29,30] Polycrystalline diamond (PCD) tools show better wear resistance and they can produce smoother surfaces in comparison to carbide or alumina tools.[20,24,32,34] This is due to the higher hardness and lower chemical affinity with the MMC material.

1.4   Cutting with rotary tools

1.4.1   Rotary tools

During machining with a conventional single-point cutting tool, a small part of the cutting edge is continuously subjected to extremely high temperatures and cutting forces. This leads to excessive wear along the area of contact. In pioneer work, Shaw et al.[35] presented a study of a lathe-type cutting tool in the form of a disk that rotates around its center. The continuous spinning of the tool around its center allows for the use of the entire circumference of the insert. As a result of the tool spinning, a fresh portion of the cutting edge is provided and therefore a better distribution of tool flank wear over the entire cutting edge is generated. The spinning action of the tool also provides a way for carrying the cutting fluid to the tool point, as in the case of a journal bearing. In addition, the period when any portion of the tool is not cutting (engaged with the workpiece) is an opportunity used to dissipate the heat generated during machining and offers a self-cooling feature by which the heat is continuously carried away from the cutting zone. Rotary tools are found in two forms: driven or self-propelled. Kishawy and Gerber[36] have presented a model for heat transfer when using rotary tools and showed the self-cooling feature of this tool.

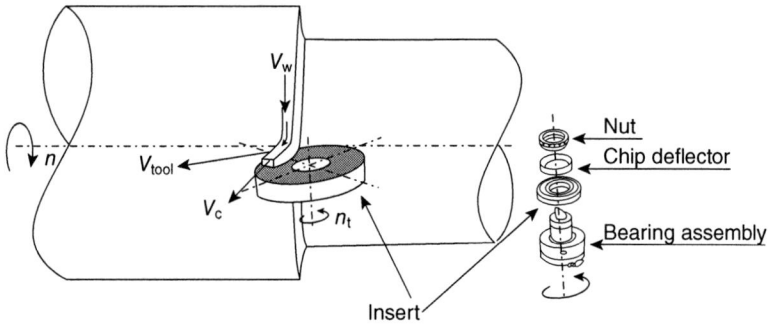

1.5 Schematic diagram showing self-propelled rotary tool during bar turning.

The tool spinning action in a driven tool is supplied by an independent external source. In the self-propelled tool, the spinning action is achieved by the interaction between the tool and the chip. The driven tool can be either orthogonal or oblique to the cutting direction, while the self-propelled tool requires the cutting edge to be oblique to the cutting direction. The rotational speed of the driven tool is independent of the process parameters. In the self-propelled tool, the rotational speed is a function of the cutting velocity and the angle between the workpiece and the rotary cutting edge velocity vectors. Generally, driven rotary tools provide more control over the rotational speed. Figure 1.5 shows a typical machining process set-up when a rotary tool is used, and indicates the main motions encountered in rotary tools.

Experimental investigation has shown low temperature generation with rotary tools. In addition to the self-cooling feature of this type of tool, the tool wear is evenly distributed along the cutting edge.[37] This is attributed to the nature of the rotating tool that leads to an even distribution of both mechanical and thermal loading along the cutting edge. Analysing chips formed by this tool showed similarity to those obtained in transient cutting. Previous investigations by the author have shown that the application of ceramic and carbide self-propelled rotary tools in machining of hardened steels leads to a great improvement in tool life and better wear resistance.

1.4.2   Rotary tool performance when turning MMCs

Chen[38] evaluated the cutting performance of rotary tools during turning of MMCs. Composites consisting of an AC8A cast aluminum matrix and silicon carbide whisker reinforcements of 18% volume fraction were used as workpieces for testing. Different tools were used, including rotary circular inserts and traditional single point tools. The tool performance was measured in terms of tool life, cutting forces and finished surface quality. A study of the progress of tool wear showed that the performance of circular rotary tools is comparable to that generated when using diamond tools.

The study compared the tool life and metal removal rate of a fixed circular insert and a rotary one. The influence of insert rotation on the overall tool performance was easily recognized. The analysis showed that while the feed and depth of cut have negligible influence on the fixed tool performance, their influence on the rotary tool is clear. Also, increasing the feed from 0.4 mm/rev. to 0.8 mm/rev. reduces the tool life by 15%; however, this increases the metal removal rate by almost 40%. Increasing the feed beyond 0.8 mm/rev. did not show any significant improvement on the metal removal rate.

More recently, Manna and Bhattacharyya[39] presented a comprehensive study on the machinability of MMC using self-propelled tools. Dry cutting tests were performed on Al-SiC MMC with 10% volume fraction. To demonstrate the effectiveness of the rotary tool during machining of MMCs, corresponding data were also collected by using fixed tools with the same radii. Various tools were employed in this investigation, namely self-propelled rotary circular tool (RCT), fixed circular tool (FCT), fixed square tool (FST), and fixed rhombic tool (FRT). It was found that, while the tool life of the rotary tool was 8 min, the tool life of the other tested tools was less than 3 min. An improvement of the obtained surface quality was also demonstrated by comparing the measured machined surface roughness. For all the tools, surface roughness was higher at lower cutting speeds and improved as the cutting speed increased. The improvement of surface roughness at higher speeds was more pronounced when using rotary tools.

1.5   Conclusions

Composites have high potential to replace traditional materials and alloys in several applications due to their unique properties which are not offered by traditional materials and alloys. Widespread application is hindered by their low machinability index which is attributed to the abrasive nature of the reinforcement particles. Several researchers have conducted turning tests and compared the performance of several cutting tool materials and configurations. The studies illustrated that PCD tools are the best option for acceptable tool life when machining MMC. The improved machinability performance by PCD is attributed to its high hardness and chemical stability. Self-propelled rotary tools were also used and their performance was compared with non-rotating ones. Compared to conventional cutting tools, self-propelled rotary cutting tools exhibit excellent performance in terms of progression of tool wear and tool life. Even carbide tools, which usually have a low machinability index, have shown an acceptable performance when used in self-propelled tools. Generally, the rotary tools showed superior wear resistance at high feed and cutting speeds due to the continuous spinning that provides efficient heat dissipation. Therefore, rotary tools are an alternative candidate to remarkably improve the production rate. Considering machining quality, generally the surface roughness generated by a rotary tool is much lower than that of a conventional tool, due to the large insert diameter. Therefore, rotary carbide tools present a good alternative for the high-performance cutting of MMCs and an economical substitute for the more expensive conventional diamond tools.

1.6   References

1 Chawala, N. and Chawla, K. K. *Metal Matrix Composites*, Springer, 2006.
2 Kishawy, H. A., Kannan, S. and Balazinski, M. An Energy-based Analytical Force Model for Orthogonal Cutting of Metal Matrix Composites. *CIRP Annals – Manufacturing Technology*, 2004, **53**(1), pp. 91–94.
3 Chan, K. C., Cheung, C. F., Ramesh, M. V., Lee, W. B. and To, S. A Theoretical and Experimental Investigation of Surface Generation in Diamond Turning of an Al6061/SicP metal matrix composite, *International Journal of Mechnical Sciences*, 2001, **43**, pp. 2047–2068.
4 Tomac, N., Tannessen, K. and Rasch, F. O. Machinability of Particulate Aluminium Matrix Composites. *CIRP Annals – Manufacturing Technology*, 1992, **41**(1), pp. 55–58.
5 Lane, C. The Effect of Different Reinforcements on PCD Tool Life for Aluminium Composites, in: *Proceedings of the Machining of Composite Materials Symposium. ASM Material Week.* Chicago, IL, 1992, pp. 17–27.
6 Quigley, O., Monaghan, J. and O'Reilly, P. Factors Affecting the Machinability of an Al/SiC Metal-matrix composite. *Journal of Materials Processing Technology*, 1994, **43**(1), pp. 21–36.
7 Weinert, K. and König, W. A Consideration of Tool Wear Mechanism when Machining Metal Matrix Composites (MMC). *CIRP Annals – Manufacturing Technology*, 1993, **42**(1), pp. 95–98.
8 Hung, N. P., Loh, N. L. and Xu, Z. M. Cumulative Tool Wear in Machining Metal Matrix Composites. Part II: Machinability. *Journal of Materials Processing Technology*, 1996, **58**(1), pp. 114–120.
9 Hung, N. P., Yeo, S. H., Lee, K. K. and Ng, K. J. Chip Formation in Machining Particle-reinforced Metal Matrix Composites. *Materials and Manufacturing Processes*, 1998, **13**(1), pp. 85–100.
10 Songmene, V. and Balazinski, M. Machinability of Graphitic Metal Matrix Composites as a Function of Reinforcing Particles. *CIRP Annals – Manufacturing Technology*, 1999, **48**(1), pp. 77–80.
11 Cronjäger, L. and Meister D. Machining of Fibre and Particle-reinforced Aluminium. *CIRP Annals – Manufacturing Technology*, 1992, **41**(1), pp. 63–66.
12 Kishawy, H. A., Kannan, S. and Balazinski, M. Analytical Modeling of Tool Wear Progression During Turning Particulate Reinforced Metal Matrix Composites. *CIRP Annals – Manufacturing Technology*, 2005, **54**(1), pp. 55–58.

13 Kishawy, H. A., Kannan, S. and Balazinski, M. An Energy Based Analytical Force Model for Orthogonal Cutting of Metal Matrix Composites. *CIRP Annals – Manufacturing Technology*, 2004, pp. 91–94.

14 Ciftci, I., Turker, M. and Seker, U. Evaluation of tool wear when machining SiCp-reinforced Al-2014 alloy matrix composites. *Materials & Design*, 2004, **25**(3), pp. 251–255.

15 Li, X. and Seah, W. K. H. Tool wear acceleration in relation to workpiece reinforcement percentage in cutting of metal matrix composites. *Wear*, 2001, **247**(2), pp. 161–171.

16 Sahin, Y. and Sur, G. The effect of Al_2O_3, TiN and Ti (C,N) based CVD Coatings on Tool Wear in Machining Metal Matrix Composites. *Surface and Coatings Technology*, 2004, **179**, pp. 349–355.

17 Sahin, Y., Kok, M. and Celik, H. Tool Wear and Surface Roughness of Al_2O_3 Particle-reinforced Aluminium Alloy Composites. *Journal of Materials Processing Technology*, 2002, **128**(1–3), pp. 280–291.

18 Seeman, M., Ganesan, G., Karthikeyan, R. and Velayudham, A. Study on Tool Wear and Surface Roughness in Machining of Particulate Aluminum Metal Matrix Composite-response Surface Methodology Approach. *The International Journal of Advanced Manufacturing Technology*, 2010, **48**(5–8), pp. 613–624.

19 D'Errico, G. E. and Calzavarini, R. Turning of Metal Matrix Composites. *Journal of Materials Processing Technology*, 2001, **119**(1–3), pp. 257–260.

20 El-Gallab, M. and Skald, M. Machining of Al/SiC Particulate Metal-matrix Composites. Part I: Tool Performance. *Journal of Materials Processing Technology*, 1998, **83**(1–3), pp. 151–158.

21 Ding, X., Liew, W. Y. H. and Liu, X. D. Evaluation of Machining Performance of MMC with PCBN and PCD Tools. *Wear*, 2005, **259**(7–12), pp. 1225–1234.

22 Iuliano, L., Settineri, L. and Gatto, A. High-speed Turning Experiments on Metal Matrix Composites. *Composites Part A: Applied Science and Manufacturing*, 1998, **29**(12), pp. 1501–1509.

23 Coelho, R. T., Yamada, S., Aspinwall, D. K. and Wise, M. L. H. The Application of Polycrystalline Diamond (PCD) Tool Materials when Drilling and Reaming Aluminium-based Alloys Including MMC. *International Journal of Machine Tools and Manufacture*, 1995, (5), pp. 761–774.

24 Hooper, R. M., Henshall, J. L. and Klopfer, A. The Wear of Polycrystalline Diamond Tools used in the Cutting of Metal Matrix Composites. *International Journal of Refractory Metals and Hard Materials*, 1999, **71**(1–3), pp. 103–109.

25 Andrewes, C. J. E., Feng, H.-Y. and Lau, W. M. Machining of an Aluminum/SiC Composite using Diamond Inserts. *Journal of Materials Processing Technology*, 2000, **102**, pp. 25–29.

26 Ferreira, J. R., Coppini, N. L. and Miranda, G. W. A. Machining Optimisation in Carbon Fibre Reinforced Composite Materials. *Journal of Materials Processing Technology*, 1999, **92–93**, pp. 135–140.

27 Hung, N. P., Tan, Z. W. and Yeow G. W. Ductile-regime Machining of Particle-reinforced Metal Matrix Composites. *Machining Science and Technology: An International Journal*, 1999, **3**(2), pp. 255–271.

28 Paulo Davim, J. and Monteiro Baptista, A. Relationship between Cutting Force and PCD Cutting Tool Wear in Machining Silicon Carbide Reinforced Aluminium. *Journal of Materials Processing Technology*, 2000, **103**(3), pp. 417–423.

29 Yuan, Z. J., Geng, L. and Dong, S. Ultraprecision Machining of SiC$_w$/Al Composites. *CIRP Annals – Manufacturing Technology*, 1993, **42**(1), pp. 107–109.
30 El-Gallab, M. and Sklad, M. Machining of Al/SiC Particulate Metal Matrix Composites: Part II: Workpiece Surface Integrity. *Journal of Materials Processing Technology*, 1998, **83**, pp. 277–285.
31 Cheung, C. F., Chan, K. C., To, S. and Lee, W. B. Effect of Reinforcement in Ultra-precision Machining of Al6061/SiC Metal Matrix Composites. *Scripta Materialia*, 2002, **47**(2), pp. 77–82.
32 Chambers, A. R. The Machinability of Light Alloy MMCs. *Composites. Part A: Applied Science and Manufacturing*, 1996, 27(2), pp. 143–147.
33 Chen, P. High-performance Machining of SiC Whisker-reinforced Aluminium Composite by Self-propelled Rotary Tools. *CIRP Annals – Manufacturing Technology*, 1992, **41**(1), pp. 59–62.
34 Durant, S., Rutelli, G. and Rabezzana, F. Aluminum-based MMC Machining with Diamond-coated Cutting Tools, *Surface and Coatings Technology*, 1997, **94–95**, pp. 632–640.
35 Shaw, M. C., Smith, P. A. and Cook, N. A. The Rotary Cutting Tool, *Transactions of the ASME*, 1952, **74**, pp.1065–1076.
36 Kishawy, H. A. and Gerber, A. G. A Model for the Tool Temperature During Machining With a Rotary Tool. *International Mechanical Engineering Congress and Exposition Symposium ECE2001/MED-23312*, 2001, pp. 1–10.
37 Kishawy, H. A. and Wilcox, J. Tool Wear and Chip Formation During Hard Turning with Self-propelled Rotary Tool, *International Journal of Machine Tool and Manufacture*, 2003, **43**, pp. 433–439.
38 Chen, P. and Hoshi, T. High-performance Machining of SiC Whisker-reinforced Aluminium Composite by Self-propelled Rotary Tools, *Annals of the CIRP*, 1992, **41**(1), pp. 59–62.
39 Manna, A. and Battacharyya, B. A study on Different Tooling System During Machining of Al/SiC-MMC. *Journal of Material Processing Technology*, 2002, **123**, pp. 476–482.

2
Drilling processes for composites

C. C. TSAO, Tahua Institute of Technology, Taiwan

Abstract: Since composites are neither homogeneous nor isotropic, drilling raises specific problems that can be related to subsequent damage in the region around the holes. The chapter first discusses theoretical models of delamination for various drill bits which explain the correlation between drilling-induced delamination and thrust force. Delamination measurement and assessment, and the influence of drilling parameters on drilling-induced delamination are next considered. Producing a consistent delamination-free hole still provides a vigorous challenge for the future.

Key words: delamination, thrust force, special drills, composite materials.

2.1 Introduction

2.1.1 Background and major issues

Composite materials have increased applications in many industries because of their excellent mechanical characteristics, such as strength-to-weight, stiffness-to-weight, corrosion resistance, fatigue and thermal expansion compared with metals. Furthermore, by deliberately designing the ratios and constituents of composite materials, and the direction of each ply in laminates, one can tailor material properties to fit specific needs. In fact, composite materials are composed of two constituent materials, namely reinforcement and matrix. The reinforcing material provides the key structural properties of the entity. The reinforcement is in the form of particulates, flakes, whiskers, lamina, or fibers, according to the selected materials and manufacturing processes. The most frequently used fibrous reinforcements are carbon, glass, graphite, aramid, silicon carbide and boron. The fibers can be further processed into chopped strand mat, unidirectional tape, or woven cloth. Generally, the presence of flaws decides the fiber strength because the fibers are brittle. Handling during processing is liable to induce and grow surface flaws that can weaken the fiber substantially. Most composite materials for structural use are laminates. Polymers, metals and ceramics can be used as the matrix. Polymer matrix composites have gained attention thanks to their improved toughness and processing convenience over some conventional materials. Epoxy resin is one of the most common matrix materials for structural composites. It has the advantages of non-volatility, good thermal and dimensional stability, and high bond

17

strength. Compared with polymer matrix composites (PMC), metal matrix composites (MMC) and ceramic matrix composite (CMC) have better high temperature properties, providing unique engineering applications.

Due to their anisotropic and abrasive nature, the process of machining composite materials has been identified as being different from that for homogeneous metal removal in the need to avoid creating splintering, delamination or burning. The various properties of the fiber and the matrix combined with fiber orientation have a significant effect on the machining process. The proper choice of cutting conditions becomes difficult due to the presence of both hard and abrasive fibers and soft matrix. Based on experimental observations, little plastic deformation of composite materials occurs during cutting, and the fracture resistance is ten to one hundred times lower than that of common steels.

In recent years, customer requirements have led to greater emphasis on the development of better machining techniques with vigorous challenges to manufacturers. One of the main advantages of composites is the near-net-shape in a structure in accordance with design requirements. Therefore, a sophisticated technique for machining of composites becomes a must in order to achieve high efficiency and low cost. Machining is indispensable for different stages of production, such as laminate preparation, parts manu-facturing, and assembly. Poor machined quality will result in poor assembly tolerance and long-term structural performance deterioration. Hole-making is one of the most common processes in secondary machining of composites due to the need for riveting and fastening in mechanical parts and structures. Several non-traditional machining processes, such as laser-beam drilling (Yung *et al.*, 2002, 2007; Dubey and Yadava, 2008), water-jet drilling (with or without abrasives) (Hocheng, 1990), ultrasonic drilling (Hocheng *et al.*, 2000; Zhang *et al.*, 2003; Azarhoushang and Akbari, 2007), and electrical discharge machining (Hocheng *et al.*, 1997; Singh *et al.*, 2004a,b), have been reported as alternatives. Nevertheless, conventional drilling using twist drills is still the most economic and convenient operation for composites. However, the drilling process has not received the same attention though as much as 40% of the machining time is devoted to hole-making as revealed by a survey of medium-sized industry (Widia, 1985). Poor hole quality accounts for an estimated 60% of all part rejection (Wong, 1982), and since holes are drilled in finished products, part rejections due to poor hole quality prove very costly. In fact, the twist drill has quite a complicated tool geometry in comparison to a straight edge tool. The effi-ciency of the cutting action varies, being most efficient at the outer diameter of the drill and least efficient at the center. The chisel edge and the lips near the center of the twist drill have a negative rake angle. The effect of a large negative rake angle is to fortify this action and make chip formation more difficult. The relative velocity decreases linearly toward the center of the drill, approaching zero, which limits its performance in generating the hole. As a result, the materials under the chisel edge of the drill point that penetrate into the hole are more likely to be extruded than cut. The thrust force for pushing the twist drill through the work is therefore high and this and the heat generated make the chisel edge of the drill point wear. Owing to the uncut thickness (last lamina) withstanding the drilling thrust force as the chisel edge approach as the exit plane, delamination can occur.

2.1.2   Literature review of drilling-induced delamination

Unlike metals, composites are made of two or more phases with dissimilar strength, hardness and thermal conductivity. Due to anisotropy and local inhomogeneity of composites, drilling-induced damage often originates from the fiber/matrix interface. Drilling-induced damage, such as delamination, burrs, swelling, splintering and fiber pullout, is a characteristic in machining composites. However, delamination damage is one of the major concerns because of its serious threat to structural reliability when the part is placed in service. Numerous studies have been examined of the drilled quality versus tool geometry, materials, and drilling parameters in drilling composites (Galloway, 1957; Haggerty and Ernst, 1958; Doran and Maikish, 1973; Friedrich *et al.*, 1979; Wu *et al.*, 1982; Doerr *et al.*, 1982; Koenig *et al.*, 1985; Miller, 1987; Komanduri *et al.*, 1991; Caprino and Tagliaferri, 1995). Kobayashi (1967) stated that a rapid increase in feed rate at the end of drilling causes cracking around the exit edge of the hole. He also found that the larger the feeding load, the more serious the cracking. Koenig *et al.* (1984, 1985) investigated the effect of processing variables on drilling damage.

In general, drilling-induced delamination occurs both at the entrance and the exit planes of the workpiece. Investigators have studied analytically and experimentally cases in which delamination in drilling have been correlated to the thrust force during exit of the drill. It is believed that there is a 'critical thrust force' below which no damage occurs (Koenig *et al.*, 1985). However, reducing thrust force depends on the geometry and materials of the tool, the materials of the workpiece, the drilling parameters, and coolant use. In addition, pre-drilling and back-up plates can reduce delamination when drilling composite materials (Won and Dharan, 2002; Tsao and Hocheng, 2003, 2005a; Tsao, 2006). Some important research issues have been given thorough treatment to solve the troublesome problems for drilling of composite materials (Hocheng and Tsao, 2009).

2.2   Delamination analysis

Hocheng and Dharan (1990) employed linear elastic fracture mechanics (LEFM) method and determined the critical thrust force that relates the delamination of composite laminates to drilling parameters and composite material properties. Chandrasekharan *et al.* (1995) have proposed a mechanistic approach and found that the thrust force is primarily a function of feed rate and tool geometry.

2.2.1 Physical model

The reference Hocheng and Tsao (2009) covers the work in this section.

In drilling composite laminates, the uncut thickness withstanding the drilling thrust force decreases as the drill approaches the exit plane. The laminate at the bottom may get separated from the interlaminar bond around the hole. At some point the loading exceeds the interlaminar bond strength and delamination occurs. Figure 2.1 depicts the model of drilling in composite materials. At the propagation of delamination, the drill movement of distance dX is associated with the work done by the thrust force F_A, which is used to deflect the plate as well as to propagate the interlaminar crack.

The energy balance equation gives

$$G_{IC}dA = F_A dX - dU \qquad [2.1]$$

where dU is the infinitesimal strain energy, dA is the increase in the area of the delamination crack, and G_{IC} is the critical crack propagation energy per unit area in mode I. The value of G_{IC} is assumed to be a constant and

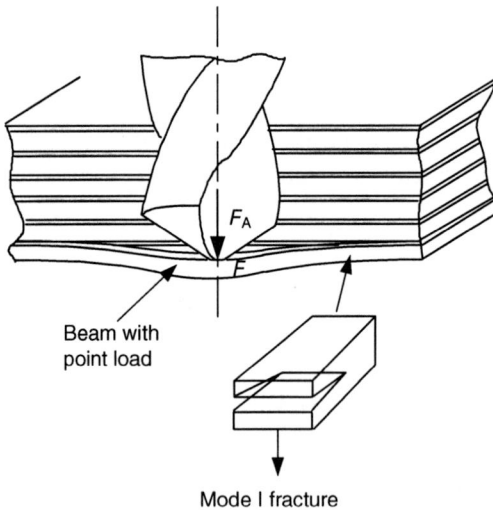

2.1 Schematics of drilling in composite materials (Hocheng and Tsao, 2009).

to be a mild function of strain rate by Saghizadeh and Dhahran (1986): the applicability of fracture mechanics to conventional isotropic materials has been demonstrated in an early reference (Dharan, 1978).

2.2.2 Twist drill without peripheral moment

The twist drill is a commonly used tool for hole-making in industrial practice. Twist drills have quite a complicated tool geometry and they appeared less than 100 years ago. The chisel edge of the twist drill plays an important role in this ensemble. The chisel edge of the drill point pushes the uncut thickness (last lamina) at the center as it approaches the exit plane in drilling composites. Therefore, the thrust of the twist drill can be regarded as a concentration force by proposing an analytical model for the chisel edge. Figure 2.2 depicts a twist drill and the induced delamination. In Fig. 2.2, the center of the circular plate is loaded by a twist drill of diameter d. F_A is the thrust force, X is the displacement, H is the workpiece thickness, h is the uncut depth under the tool, and a is the radius of delamination.

Isotropic behavior and pure bending of the laminate are assumed in the model. The thrust force at the onset of crack propagation can be calculated (Hocheng and Dharan, 1990):

$$F_A = \pi\sqrt{32G_{IC}M}$$
$$= \pi\left[\frac{8G_{IC}Eh^3}{3(1-v^2)}\right]^{1/2} \tag{2.2}$$

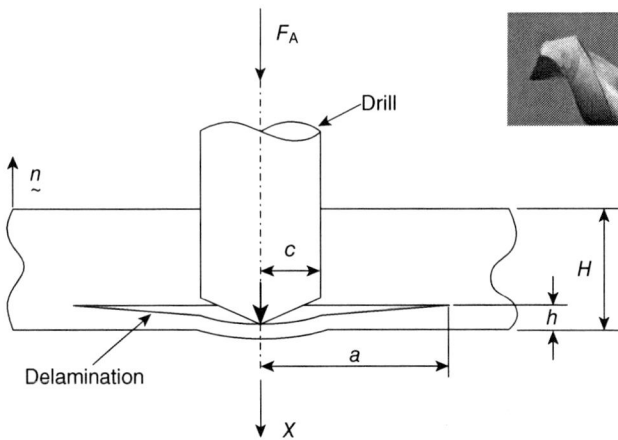

2.2 Circular plate model for delamination analysis (twist drill) (Hocheng and Tsao, 2009).

where $M = \dfrac{Eh^3}{12(1-v^2)}$ is the stiffness per unit width of the fiber reinforced
material, E is Young's Modulus and v is Poisson's ratio for the material.

To avoid drilling-induced delamination, the applied thrust force should not exceed this value, which is a function of the material properties and the uncut thickness. The thrust force can be correlated with the feed rate. When the uncut thickness progressively decreases, the strategy is to drill as fast as practically permissible in the beginning, and to gradually reduce the feed rate as the tool approaches the exit.

2.3   Delamination analysis of special drills

Drilling-induced delamination correlates closely with drilling thrust force. Reducing the thrust to avoid delamination in drilling composites is an effective solution method. Teti has pointed out that standard tool geometries are not suggested for machining composites because the individual fibers can be separated in a clean cut only under simultaneous pre-stress (Teti, 2002). During the past four decades, researchers have developed many types of drills, including multifacet drills, saw drills, candlestick drills, core drills, step drills and trepanning drills, all aimed at making better holes. Figures 2.3–2.6 show photographs of video footage using special drills. The twelve-frame sequence shows the drilling process from the point in which the first signs of puncture (or bulge) in the laminate are seen to the time when the drill leaves delamination around the hole exit. The sequence was observed using a 16 mm drill with a feed rate of 10 mm/min.

With increasing demand for advanced composite materials, not only new geometries and concepts of tooling but also different realms of cutting conditions are needed when machining them. Hocheng and Tsao (2005) have developed a series of analytical models, based on LEFM, for special drills to correlate the thrust force with the onset of delamination. According

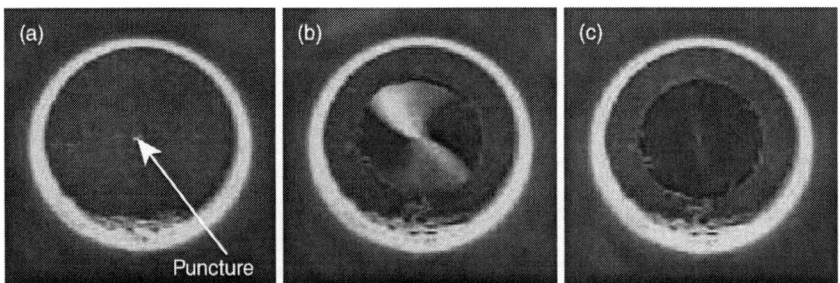

2.3 Photographs of video footage of twist drill after drilling exit: (a) drill puncture; (b) drill exit; (c) delamination after drilling.

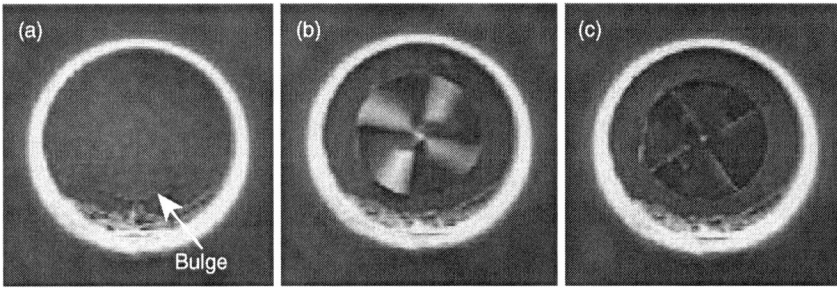

2.4 Photographs of video footage of saw drill after drilling exit:
(a) drill bulge; (b) drill exit; (c) delamination after drilling.

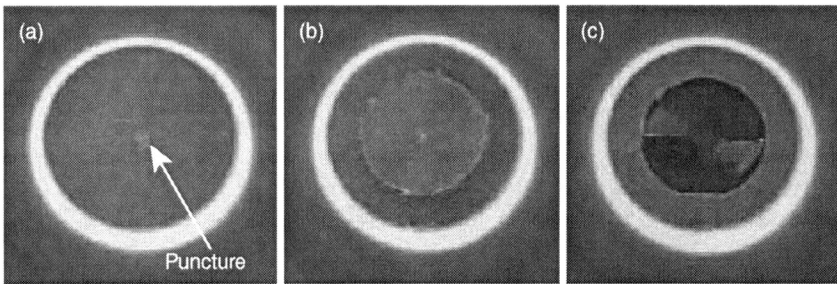

2.5 Photographs of video footage of candlestick drill after drilling exit:
(a) drill puncture; (b) drill exit; (c) delamination after drilling.

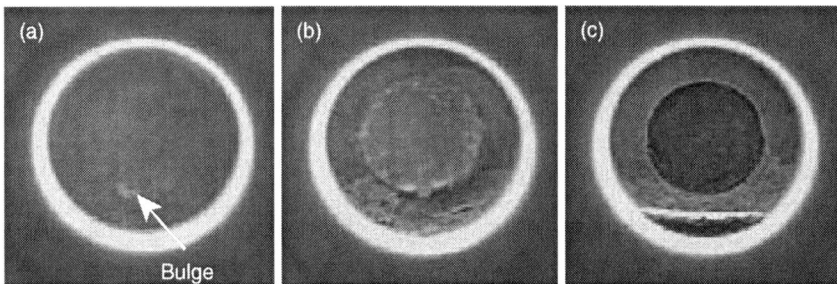

2.6 Photographs of video footage of core drill after drilling exit:
(a) drill bulge; (b) drill exit; (c) delamination after drilling.

to the functions of the tool, the drill bits can be classified as (i) conventional drills, (ii) special drills and (iii) compound drills, as shown in Fig. 2.7. Special tools can be classified into candlestick drills, saw drills and core drills. Compound tools are composed of a combination of conventional and special tools.

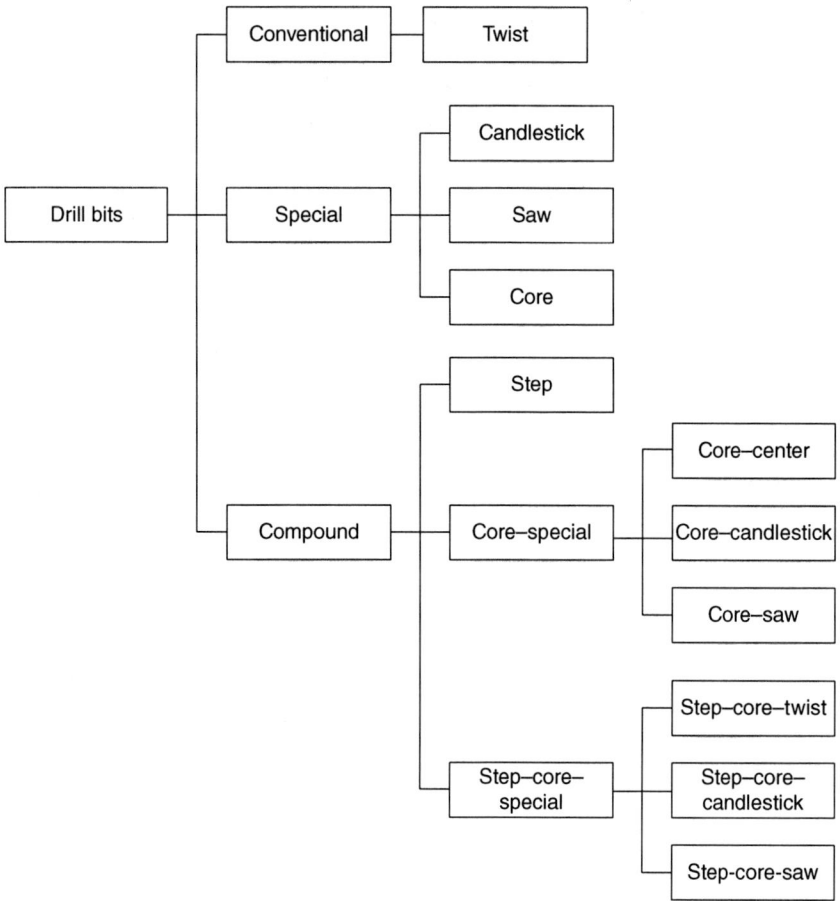

2.7 Classifications of drill bits.

Mathew *et al.* (1999a) have reported a trepanning tool to reduce the thrust force and torque during drilling of glass fiber reinforced plastic (GFRP) laminates. The cutting action of a trepanning tool starts from the periphery of the cutting edge that puts the fibers in tension during the entire cutting operation (Mathew *et al.*, 1999b). Chandrasekharan *et al.* (1995) have proposed a mechanistic approach to develop cutting force models to predict the thrust and the torque in drilling. Armarego and Wright (1984) developed a model that can be used to estimate thrust and torque for three different drill flank configurations. Mechanical models to predict the thrust and torque in vibration-assisted drilling of fiber-reinforced composite materials were constructed by Wang and Wang (1998) and Zhang *et al.* (2001). Under the same cutting conditions, the thrust and the torque by this method are reduced by 20–30 percent compared with conventional drilling. Langella *et al.* (2005) presented a mechanistic model for predicting the thrust and torque during composite material drilling. They specified that the action of the chisel edge of the twist drill on the thrust increases with the feed rate and may account for over 80% of the total force needed to drill a hole. The torque-associated crack propagation in Mode III is therefore considered of secondary significance in the analysis of drilling-induced delamination.

Although significant efforts have been made to realize drilling-induced delamination for saw drills, candlestick drills and core drills, there have been few papers reporting the effect of distributed peripheral moment on delamination in drilling composite materials based on mechanical and energy analysis.

2.3.1 Saw drills

Compared with twist drills, saw drills are more complicated in drill geometry and manufacture. The saw drill has in fact an equally spaced discrete loading during drilling. Therefore, the thrust of the saw drill can be regarded as a uniform circular load by proposing an analytical model for its discrete cutting edges. Saw drills can provide better machining quality in drilling composite laminates. One reason is that the saw drill utilizes the peripheral distribution of thrust for drilling. However, the cutting edges of the saw drill are prone to rapid wear in drilling composites because the cutting edges are very sharp. For graphite–epoxy or glass–epoxy, suitable tool materials are suggested as polycrystalline diamond (PCD) or solid tungsten carbide.

Without peripheral moment

The reference Hocheng and Tsao (2005) covers the work in this section. Figure 2.8 depicts a saw drill and the induced delamination, where c is the radius of the saw drill. Saw drills can acquire better machining quality in drilling composite laminates. As previously stated, one reason is that they utilize the peripheral distribution of thrust for drilling the laminates.

The critical thrust force of a saw drill (F_S) at the onset of crack propagation can be calculated:

$$F_S = \pi \sqrt{\frac{32 G_{IC} M}{(1 - 2s^2 + s^4)}}$$ [2.3]

where $s = c/a$. A comparison of F_S and F_A in Eq. [2.3] and Eq. [2.2] gives

$$\frac{F_S}{F_A} = \frac{1}{\sqrt{(1 - 2s^2 + s^4)}}$$ [2.4]

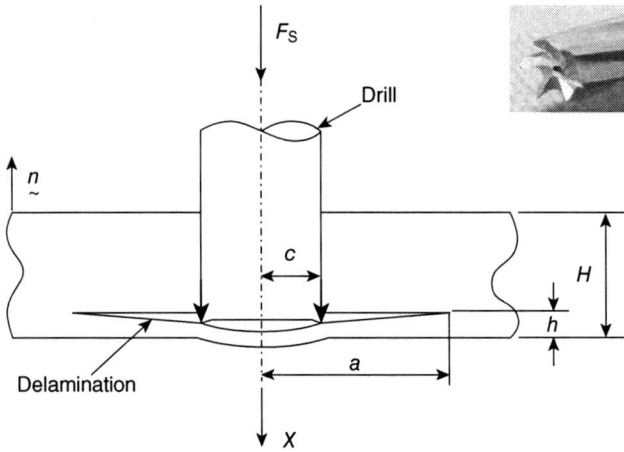

2.8 Circular plate model for delamination analysis (saw drill) (Hocheng and Tsao, 2005).

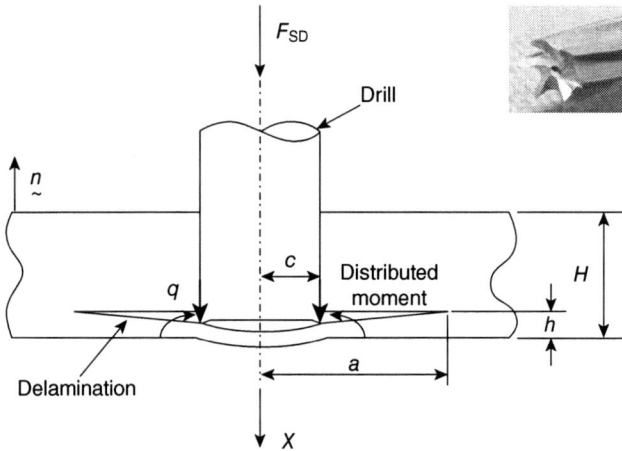

2.9 Circular plate model of delamination for saw drill with peripheral movement (Tsao and Hocheng, 2008a).

With peripheral moment

The reference Tsao and Hocheng (2008a) covers the work in this section. Figure 2.9 depicts a saw drill with distributed peripheral moment and the induced delamination, where c is the radius of saw drill. F_{SD} is the thrust force of the saw drill with distributed peripheral moment.

The critical thrust force of the saw drill with distributed peripheral moment (F_{SD}) at the onset of crack propagation can be calculated:

$$F_{SD} = \pi \sqrt{\frac{32 G_{IC} M}{C_1 + v C_2}} \qquad [2.5]$$

where

$C_1 = 1 - (3 + 2\ln s)s^2 + (3 + 2\ln s)s^4 - s^6$, $C_2 = (1 + 2\ln s)s^2 - (2 + 2\ln s)s^4 + s^6$
and $s = c/a$.

A comparison of F_{SD} and F_A in Eq. [2.5] and Eq. [2.2] gives

$$\frac{F_{SD}}{F_A} = \sqrt{\frac{1}{C_1 + v C_2}} \qquad [2.6]$$

Effect of saw drill on threshold thrust force

The reference Tsao and Hocheng (2008a) covers the work in this section. Figure 2.10 depicts the theoretical critical thrust ratio for a saw drill with and without peripheral moment. It is seen that the thrust ratio for the saw drill without peripheral moment is slightly larger than the thrust ratio for the saw drill with peripheral moment up to $s \approx 0.85$. This means that the peripheral moment can contribute to the delamination in addition to the thrust force in drilling composite materials. As pointed out by DiPaolo *et al.* (1996), spalling occurred via Mode I (opening) and Mode III (tearing) damage mechanisms. The Mode III was subjected to the drilling torque and

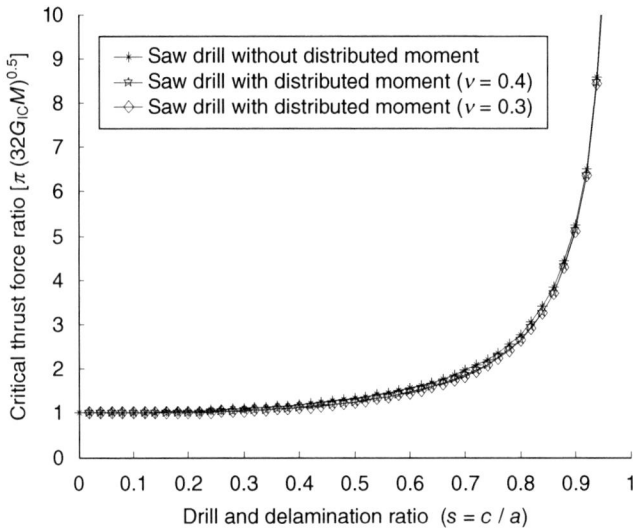

2.10 Theoretical critical thrust ratio for saw drill with and without peripheral moment (Tsao and Hocheng, 2008a)

a twisting due to the combination of the downward thrust force and the back rake angle along the cutting lips. Figure 2.10 also shows that increasing Poisson's ratio (large v) has a mild effect on the growth of the critical thrust force for the saw drill with peripheral moment.

2.3.2 Candlestick drills

Candlestick drills can be regard as being composed of a twist drill and a saw drill, and they are extensively used for drilling composite materials. The thrust force of the candlestick drill can be considered as a concentrated center load a the distributed circular load.

Without peripheral moment

The reference Hocheng and Tsao (2005) covers the work in this section. Figure 2.11 depicts the schematics of a candlestick drill and the induced delamination. Candlestick drills are extensively used for drilling composite materials. As stated, the thrust force of the candlestick drill can be considered as a concentrated center load (p_1) plus the distributed circular load (p_2).

The thrust force of the candlestick drill (F_C) at the onset of crack propagation can be calculated:

$$F_C = \pi(1+\alpha_c)\sqrt{\frac{32 G_{IC} M}{1+\alpha_c^2(1-2s^2+s^4)}}$$

[2.7]

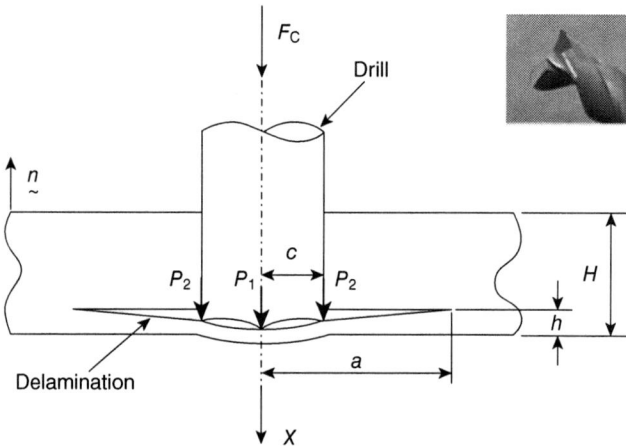

2.11 Circular plate model for delamination analysis (candlestick drill) (Hocheng and Tsao, 2009).

where α_c is the ratio of the distributed circular load and the concentrated center load.

A comparison of F_C and F_A in Eq. [2.7] and Eq. [2.2] gives

$$\frac{F_C}{F_A} = \frac{(1+\alpha_c)}{\sqrt{1+\alpha_c^2(1-2s^2+s^4)}} \qquad [2.8]$$

With peripheral moment

The reference Tsao and Hocheng (2008a) covers the work in this section. Figure 2.12 depicts the schematics of a candlestick drill with distributed peripheral moment and the induced delamination. The thrust force of the candlestick drill with distributed peripheral moment (F_{CD}) can be expressed as

$$F_{CD} = \pi(1+\alpha_c)\sqrt{\frac{32G_{IC}M}{1+\alpha_c^2(C_1+vC_2)}} \qquad [2.9]$$

A comparison of F_{CD} and F_A in Eq. [2.9] and Eq. [2.2] gives

$$\frac{F_{CD}}{F_A} = (1+\alpha_c)\sqrt{\frac{1}{1+\alpha_c^2(C_1+vC_2)}} \qquad [2.10]$$

Effect of candlestick drills on threshold thrust force

The reference Tsao and Hocheng (2008a) covers the work in this section. The results of the theoretical critical thrust ratio calculation for a

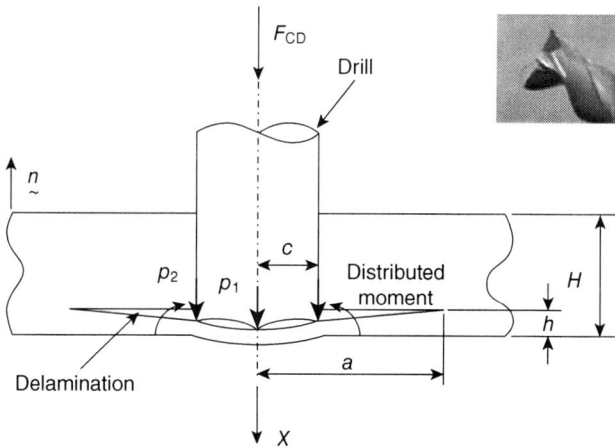

2.12 Circular plate model of delamination for candlestick drill with peripheral moment (Hocheng and Tsao, 2009)

(a)

(b)

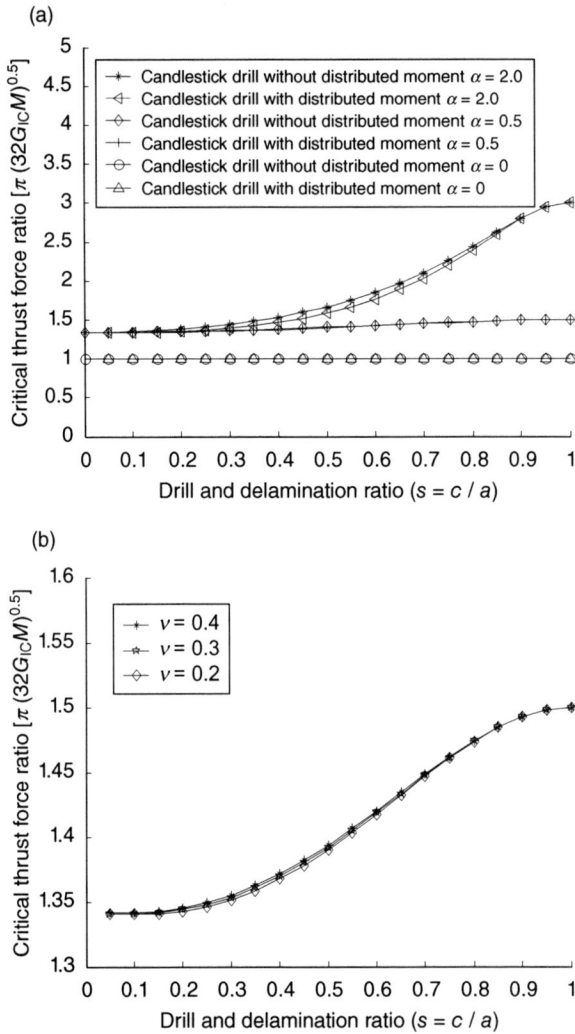

2.13 Theoretical critical thrust ratio for candlestick drill (a) with and without peripheral moment; (b) with peripheral moment at varying Poisson's ratio ($\alpha = 0.5$) (Tsao and Hocheng, 2008a).

candlestick drill with and without peripheral moment are presented in Fig. 2.13a. Since the total thrust force of a candlestick drill is distributed toward the periphery at ratio α, the drill is expected to be advantageous in allowing a higher threshold thrust force at the onset of delamination. Figure 2.13a illustrates that the more the thrust force is distributed toward the periphery (larger α), the higher becomes the critical threshold. In general, α is related to the geometry of the candlestick drill. From this figure, it can be seen that the thrust force varies within a narrow range as s increases at $\alpha \geq 0.85$. Figure 2.13b shows that the increasing Poisson's ratio (large v) has a mild effect on the increase of the critical thrust force for a candlestick drill with peripheral moment.

2.3.3 Core drills

A core drill is a hollow grinding drill with bonded diamonds and a limited thickness. This tool results in a much smaller thrust and much better hole quality when compared with a twist drill in the drilling process (Jain and Yang, 1992; Hocheng and Tsao, 2006). In general, the core drill is used for drilling hard, brittle materials, as in civil engineering structures, jewels and glass. However, the saw drill is a special type of core drill in drilling appli-cations, where the thickness of the core drill approaches zero.

Without peripheral moment

The reference Hocheng and Tsao (2005) covers the work in this section. Figure 2.14 depicts a schematic of a core drill and the induced delamination. The outer and inner deflection of a circular plate of radius a, which is clamped and subjected to an annular distributed load over a round area of radius c, is given. c^* and c are the inner and outer radius of the core drill, respectively, t is the thickness of the core drill, and β is the ratio between the thickness and the radius of the core drill (namely, $\beta = t/c$).

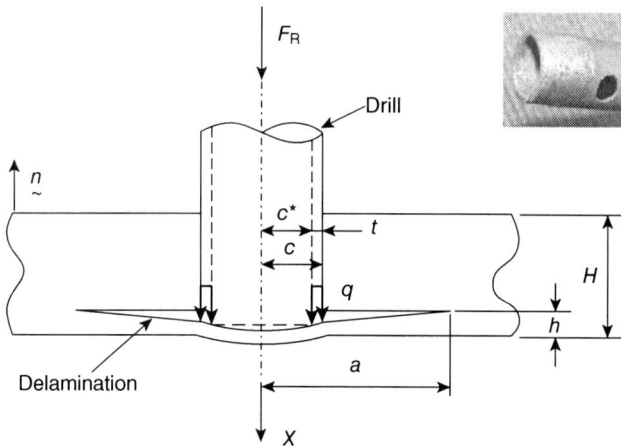

2.14 Circular plate model for delamination analysis (core drill) (Hocheng and Tsao, 2005).

One obtains the thrust force of the core drill (F_R) at the onset of crack propagation as

$$F_R = \pi \left\{ \frac{32 G_{IC} M}{1 - K_1 s^2 + K_2 s^4} \right\}^{1/2}$$ [2.11]

where

$$K_1 = \left(2 - 2\beta + \frac{3\beta^2}{2} \right) + \frac{4(1-\beta)^2}{\beta(2-\beta)} \ln(1-\beta)$$

$$K_2 = \frac{(2 - 4\beta + 5\beta^2 - 3\beta^3 + \beta^4)}{2} + \frac{2(1-\beta)^2(2 - 2\beta + \beta^2)}{\beta(2-\beta)} \ln(1-\beta)$$

A comparison of F_R in Eq. [2.11] and F_A in Eq. [2.2] gives

$$\frac{F_R}{F_A} = \left\{ \frac{1}{1 - K_1 s^2 + K_2 s^4} \right\}^{1/2}$$ [2.12]

With peripheral moment

The reference Tsao and Hocheng (2008a) covers the work in this section. Figure 2.15 depicts the schematics of a core drill with distributed peripheral moment and the induced delamination.

The critical thrust force of the core drill with distributed peripheral moment (F_{RD}) at the onset of crack propagation can be calculated:

$$F_{RD} = \pi \sqrt{\frac{32 G_{IC} M}{C_3 + v C_4}}$$ [2.13]

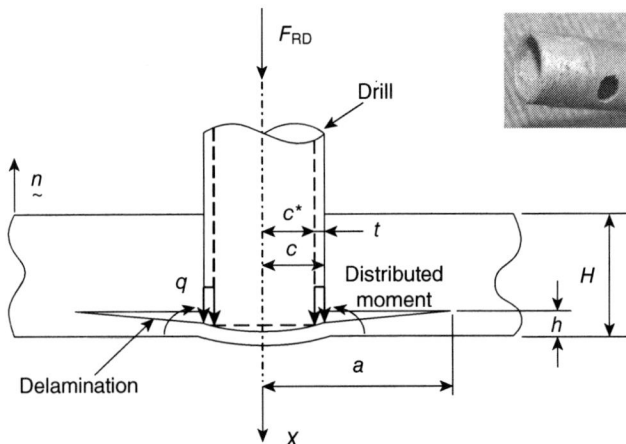

2.15 Circular plate model of delamination for core drill with peripheral moment (Tsao and Hocheng, 2008a).

$$C_3 = 1 - \left[\left(2 - 2\beta + \frac{3}{2}\beta^2 \right) + 2\ln s + \frac{2(1-\beta)^2}{\beta(2-\beta)} \ln(1-\beta) \right] s^2$$
$$+ \left\{ (2 - 2\beta + \beta^2) \left[\frac{(2-\beta+\beta^2)}{2} + \ln s + \frac{(1-\beta)^2}{\beta(2-\beta)} \ln(1-\beta) \right] \right\} s^4$$
$$- \frac{(2 - 2\beta + \beta^2)^2}{4} s^6$$

$$C_4 = \left[2\ln s - \frac{2(1-\beta)^2}{\beta(2-\beta)} \ln(1-\beta) \right] s^2$$
$$+ \left\{ (2 - 2\beta + \beta^2) \left[-\frac{1}{2} - \ln s + \frac{(1-\beta)^2}{\beta(2-\beta)} \ln(1-\beta) \right] \right\} s^4$$
$$+ \frac{(2 - 2\beta + \beta^2)^2}{4} s^6$$

A comparison of F_{RD} and F_A in Eq. [2.13] and Eq. [2.2] gives

$$\frac{F_{RD}}{F_A} = \sqrt{\frac{1}{C_3 + vC_4}} \qquad [2.14]$$

Effect of core drills on threshold thrust force

The reference Tsao and Hocheng (2008a) covers the work in this section. Figure 2.16a compares the theoretical critical thrust ratio of a core drill with and without peripheral moment with varying s. In this figure, the critical thrust ratio of the core drill with and without peripheral moment increases with a reduction of β. The critical thrust ratio of the core drill without peripheral moment is much higher than that of the core drill with peripheral moment. Figure 2.16b shows that increasing Poisson's ratio (large v) has little effect on the growth of the critical thrust force for a core drill with peripheral moment. However, the critical thrust ratio of a core drill with peripheral moment is below 1 at s values between 0 and 0.6.

2.3.4 Comparison and reducible relationship among drill bits

The reference Hocheng and Tsao (2005) covers the work in this section. Some reducible relationships of critical thrust force among the special drills are shown in Table 2.1: these are associated with the mathematical expressions as follows.

(i)   The saw drill (○) when s (which is c/a) = 0, reduces to the twist drill case. With increasing s (approaching to 1), the threshold of the thrust

(a)

(b)

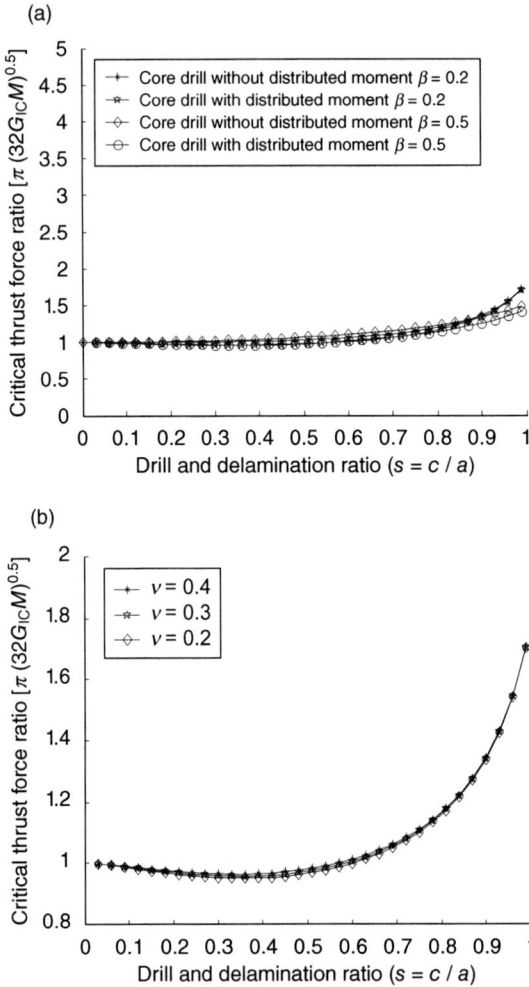

2.16 Theoretical critical thrust ratio for core drill (a) with and without peripheral moment; (b) with peripheral moment at varying Poisson's ratio ($\beta = 0.2$) (Tsao and Hocheng, 2008a).

Table 2.1 Reducible relationships of critical thrust force of special drills (Hocheng and Tsao, 2005)

	Twist drill •	Saw drill ○
Saw drill ○	$s = 0$	—
Candlestick drill ☉	$\alpha = 0$	$\alpha = \infty$
Core drill ○	$\beta = 0, s = 0$	$\beta = 0$

force becomes very high, showing the advantageous effects of the saw drill.

(ii)   For the candlestick drill ($\odot$), α (which is the ratio of circular load to centered load) = 0 reduces to the case of the twist drill with concentrated central load only, while $\alpha = \infty$ approaches to the use of the saw drill with circular load exclusively.

(iii) The core drill ($\bigcirc$) when β (which is t/c) = 0 and s = 0, reduces to the twist drill case, while $\beta = 0$ approaches to the use of the saw drill.

2.4   Delamination analysis of compound drills

2.4.1   Step drills

The reference Hocheng and Tsao (2005) covers the work in this section. The step drill can be considered to be composed of a primary stage (of diameter $2b$) and a secondary stage (of diameter $2c$), as shown in Fig. 2.17. DiPaolo *et al.* (1996) used an experimental setup to view the crack growth as the drill emerged from the bottom side of the workpiece. The eventual crack growth was due to the force of the cutting lips after the chisel edge has exited the laminate, similar to the case of Hocheng and Tsao (2005) considered here.

Figure 2.17 depicts a schematic of a step drill and the induced delamination, where F_T is the thrust force and Q is the circular load exerted by the secondary cutting lips. Isotropic behavior and pure bending of the laminate

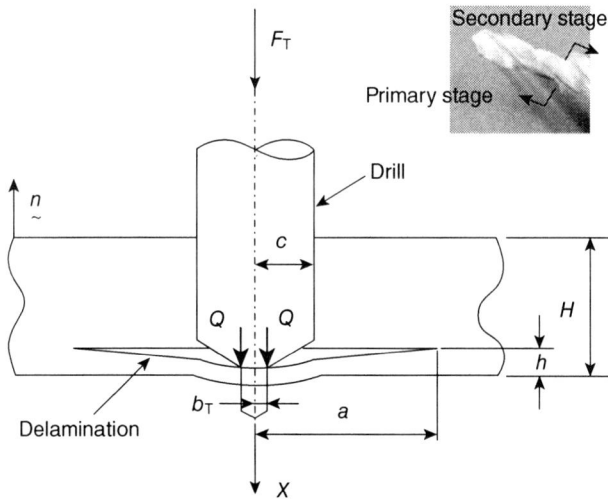

2.17 Circular plate model for delamination analysis (step drill) (Hocheng and Tsao, 2005).

are assumed in the model. A mathematical model of a plate subjected to symmetrical bending by the force Q along the circular edge of a hole is shown in Fig. 2.17.

The critical thrust force at the onset of crack propagation with a step drill when the final (secondary) stage of drilling proceeds is given by

$$F_T = \frac{\sqrt{2}\pi}{1-v} \left[\frac{32 G_{IC} M\{(1-v)+2(1+v)\kappa^2\}^2}{\begin{array}{c}(1+v)\{2(1-v)(1+2v^2) \\ -(12-4v+3v^2+3v^3)\kappa^2-8(1+3v)\kappa^2 \ln \kappa\}\end{array}} \right]^{1/2} \qquad [2.15]$$

where $\kappa = b_T/c$. The ratio between Eq. [2.15] and Eq. [2.2] is

$$\frac{F_T}{F_A} = \frac{\sqrt{2}}{1-v} \left[\frac{\{(1-v)+2(1+v)\kappa^2\}^2}{\begin{array}{c}(1+v)\{2(1-v)(1+2v^2) \\ -(12-4v+3v^2+3v^3)\kappa^2-8(1+3v)\kappa^2 \ln \kappa\}\end{array}} \right]^{1/2} \qquad [2.16]$$

Figure 2.18 depicts the ratio between the critical thrust force of the step drill and the twist drill varying with the drill diameter ratio. It is seen that the value of the thrust force ratio increases with increasing κ, which means the circular thrust force is distributed more outwardly. It also shows that increasing Poisson's ratio (large v) has a mild effect on the reduction of the critical thrust force.

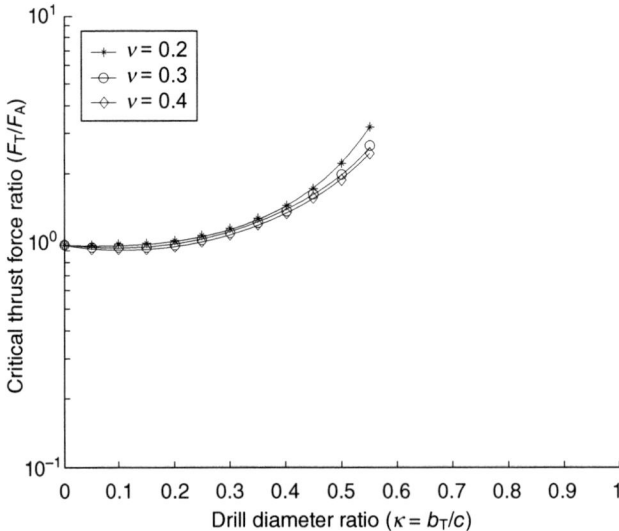

2.18 Critical thrust ratio between step drill and twist drill (Hocheng and Tsao, 2005).

2.4.2 Core–special drills

The references Tsao (2008a,b), Tsao and Hocheng (2008b) and Tsao et al. (2009) cover the work in this section. For conventional and special drill bits, thrust forces were sensitive to variations in the feed. Among the four, the core drill offers the highest critical thrust force, followed by the saw drill and the candlestick drill, while the traditional twist drill allows for the lowest thrust force. The saw drill, candlestick drill and core drill allow for a larger critical thrust force to be operated at a larger feed rate without the delamination damage that is encountered with the twist drill. Drill design can be developed based on the proposed models, especially when the thrust force is distributed toward the drill periphery, such as with the saw drill, the candlestick drill and the core drill. However, the removal of chips poses problems when using the core drill. To resolve this, compound core–special drills and step–core–special drills are designed to reduce chip removal in drilling.

Core–center drills

Figure 2.19 depicts a schematic of a core–center drill and the induced delamination. The center of the circular plate is loaded by a twist drill of radius c. F_{CC} is the thrust force of the core–center drill. The thrust load of the drill is simulated by a composition of a twist drill and a core drill; namely, the summation of the concentrated center load (l_2) and the annular area load (l_1).

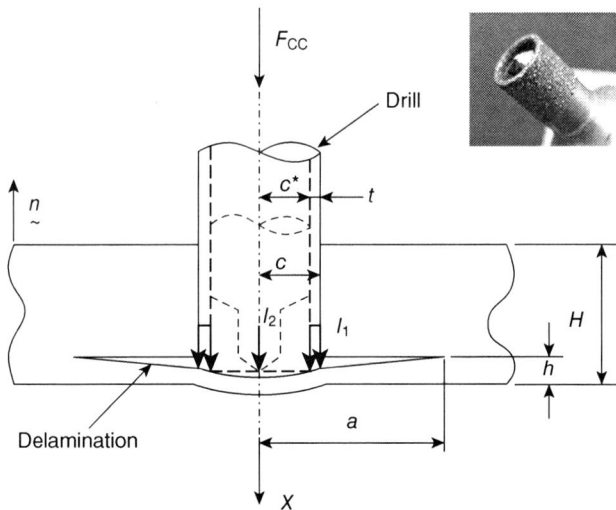

2.19 Circular plate model for delamination analysis (core–center drill) (Hocheng and Tsao, 2009).

The thrust force of the core–center drill (F_{CC}) at the onset of crack propagation can be obtained as:

$$F_{CC} = \pi(1+\gamma)\left\{\frac{32G_{IC}M}{1+\gamma^2(1-K_1s^2+K_2s^4)}\right\}^{1/2} \qquad [2.17]$$

where γ is the ratio of the central concentrated force and the annular area force.

A comparison of Eq. [2.17] and Eq. [2.2] gives

$$\frac{F_{CC}}{F_A} = \frac{(1+\gamma)}{\sqrt{1+\gamma^2(1-K_1s^2+K_2s^4)}} \qquad [2.18]$$

Results for the critical thrust force predicted by the core–center drills are presented in Fig. 2.20. The core–center drill exerts a thrust force on the laminate that is composed of the concentrated central force and the annular area force. Since the total thrust force is distributed towards the periphery

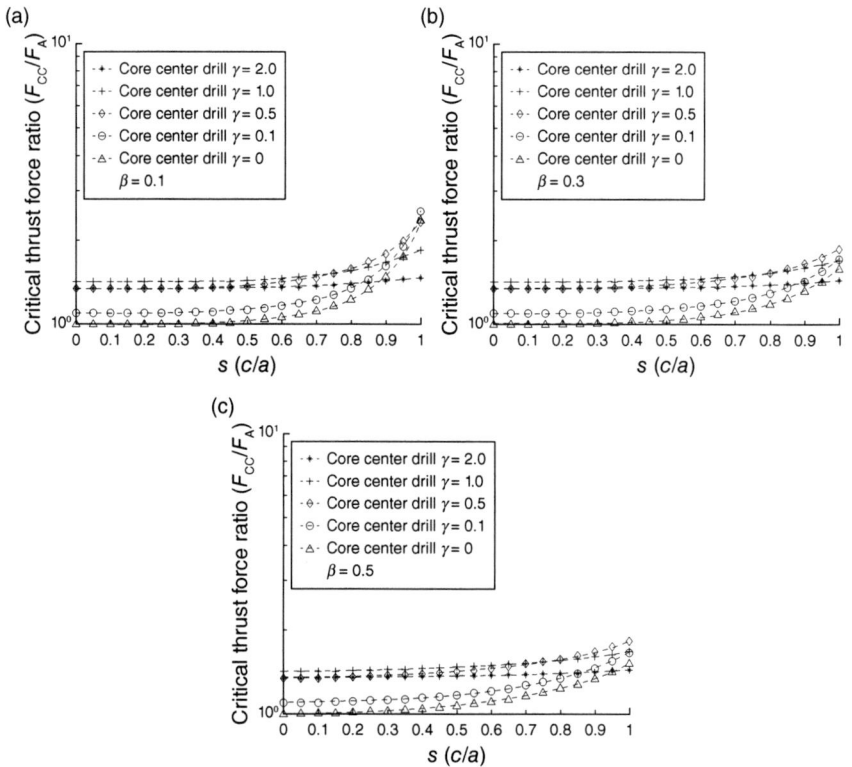

(a)

(b)

(c)

2.20 Critical thrust ratio between core–center drill and twist drill: (a) $\beta = 0.1$; (b) $\beta = 0.3$; (c) $\beta = 0.5$ (Hocheng and Tsao, 2009).

at a ratio of γ, the drill is expected to be advantageous in allowing a larger critical thrust force before the onset of delamination, similar to the effect of a core drill. Figure 2.20 illustrates that the more the thrust force is distributed toward the periphery (smaller β), the larger becomes the critical threshold. A comparison of Fig. 2.16a with Fig. 2.20 in β shows that the core–center drill performs much better than the core drill apparently. The more γ, the more critical thrust force. In fact, the core–center drill is physically an intermediate between the twist drill and the core drill, and mathematically the general solution is reducible to the particular cases of both the twist drill (completely concentrated force) and the core drill (completely circular area force). One notices in Fig. 2.20, that $\gamma = 0$ represents the twist drill case. While the value of γ is over 1, the critical thrust force varies in a narrow range.

Core–candlestick drills

Figure 2.21 depicts the schematics of a core–candlestick drill and the induced delamination. In Fig. 2.21, c^* is the inner radius of the core drill, b_C is the radius of the candlestick drill, t^* is the distance between c^* and b_C, and ϕ is the ratio between t^* and radius of the core drill (namely, $t^* = \phi c$). The thrust force of the core–candlestick drill can be considered as a sum of concentrated central load, periphery circular load and the annular area load. Using the method of superposition, the thrust force F_{CCS} can be expressed as follows

$$F_{CCS} = f_1 + f_2 + f_3 \qquad\qquad [2.19]$$

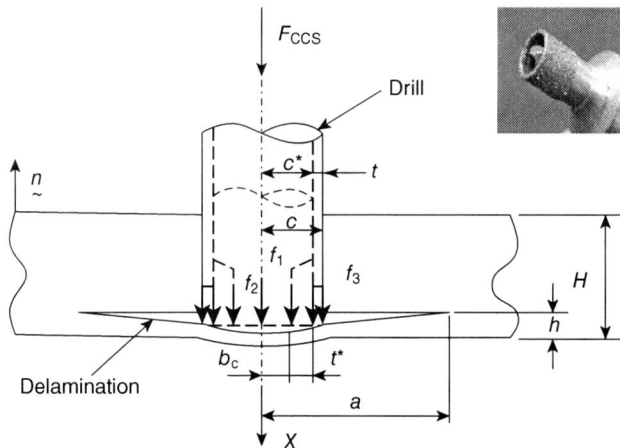

2.21 Circular plate model for delamination analysis (core–candlestick drill).

where f_1 is the central concentrated force, f_2 is the peripheral circular force and f_3 is the annular area force, respectively. Let

$$f_2 = \eta_C f_1 \qquad\qquad [2.20]$$

$$f_3 = \gamma_C f_2 \qquad\qquad [2.21]$$

One obtains the thrust force of the core–candlestick drill at the onset of crack propagation:

$$F_{CCS} = \pi[1+\eta_C(1+\gamma_C)]\left\{\frac{32G_{IC}M}{\begin{array}{c}1+\eta_C^2\{[(1+\gamma_C)-[2(1-\beta-\phi)^2+\gamma_C^2 K_1]s^2\\+[(1-\beta-\phi)^4+\gamma_C^2 K_2]s^4\}\end{array}}\right\}^{1/2}$$

$$[2.22]$$

The comparison of Eq. [2.22] and Eq. [2.2] gives

$$\frac{F_{CCS}}{F_A} = \frac{[1+\eta_C(1+\gamma_C)]}{\sqrt{\begin{array}{c}1+\eta_C^2\{[(1+\gamma_C)-[2(1-\beta-\phi)^2+\gamma_C^2 K_1]s^2\\+[(1-\beta-\phi)^4+\gamma_C^2 K_2]s^4\}\end{array}}} \qquad [2.23]$$

Results for the critical thrust force predicted by the core–candlestick drill are presented in Fig. 2.22. The core–candlestick drill exerts a thrust force on the laminate that is composed of the concentrated force, the periphery circular force and the circular area force. Since the total thrust force is distributed toward the periphery at a ratio of γ_C, η_C and ϕ, the drill can be expected to be advantageous in allowing a larger critical thrust force at the onset of delamination, similar to the effect of the core drill.

Figure 2.22 illustrates that the more the thrust force is distributed toward the periphery (smaller β and ϕ, larger γ_C and η_C), the larger becomes the critical threshold. Compare η_C and β in Fig. 2.22; the core–candlestick drill is much better than the core drill apparently.

Core–saw drills

Figure 2.23 depicts the schematics of a core–saw drill and the induced delamination. In Fig. 2.23, c^* is the inner radius of the core drill, b_S is the radius of the saw drill, t^* is the distance between c^* and b_S, and φ is the ratio between t^* and the radius of the core drill (namely, $t^* = \phi c$). The thrust force of the core–saw drill can be considered as a periphery circular load plus the annular area load.

The thrust force of the core–saw drill (F_{CS}) at the onset of crack propagation can be calculated:

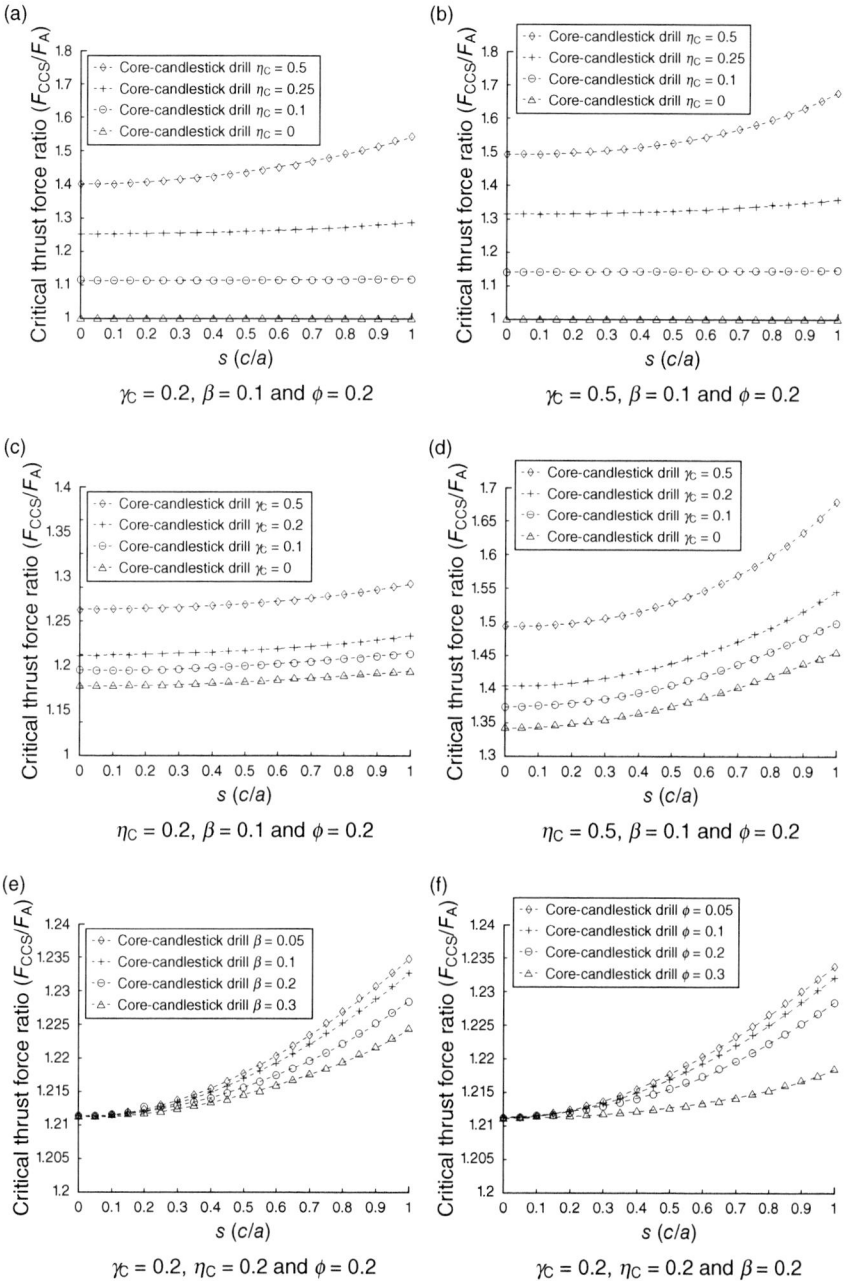

(a)

$\gamma_C = 0.2$, $\beta = 0.1$ and $\phi = 0.2$

(b)

$\gamma_C = 0.5$, $\beta = 0.1$ and $\phi = 0.2$

(c)

$\eta_C = 0.2$, $\beta = 0.1$ and $\phi = 0.2$

(d)

$\eta_C = 0.5$, $\beta = 0.1$ and $\phi = 0.2$

(e)

$\gamma_C = 0.2$, $\eta_C = 0.2$ and $\phi = 0.2$

(f)

$\gamma_C = 0.2$, $\eta_C = 0.2$ and $\beta = 0.2$

2.22 Critical thrust ratio between core–candlestick drill and twist drill.

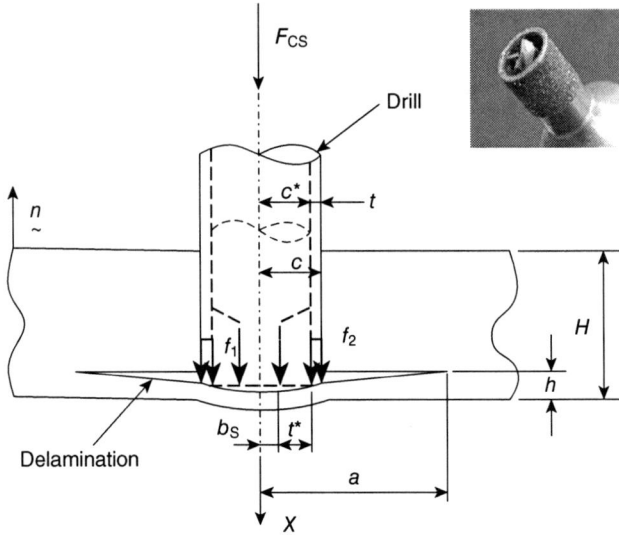

2.23 Circular plate model for delamination analysis (core–saw drill) (Hocheng and Tsao, 2009).

$$F_{CS} = \pi(1+\lambda)\left\{\frac{32G_{IC}M}{\lambda^2 + \{1-[K_1+\lambda^2(1-\beta-\phi)^2]s^2 + [K_2+\lambda^2(1-\beta-\phi)^4]s^4\}}\right\}^{1/2}$$

[2.24]

where λ is the ratio of the periphery circular force and the annular area force.

A comparison of Eq. [2.24] and Eq. [2.2] gives

$$\frac{F_{CS}}{F_A} = \frac{(1+\lambda)}{\sqrt{\lambda^2 + \{1-[K_1+\lambda^2(1-\beta-\phi)^2]s^2 + [K_2+\lambda^2(1-\beta-\phi)^4]s^4\}}}$$

[2.25]

Results for the critical thrust force predicted by the core–saw drill are presented in Fig. 2.24. The core–saw drill exerts a thrust force on the laminate that is composed of the periphery circular force and the circular area force. Since the total thrust force is distributed toward the periphery at a ratio of η and ϕ, the drill is expected to be advantageous in allowing a larger critical thrust force at the onset of delamination, similar to the effect of the core drill. Figure 2.24 illustrates that the more the thrust force is distributed toward the periphery (smaller β and ϕ, larger η), the larger becomes the critical threshold. Comparing Fig. 2.16a and Fig. 2.24 in η and β, the core–saw drill is apparently much better than the core drill.

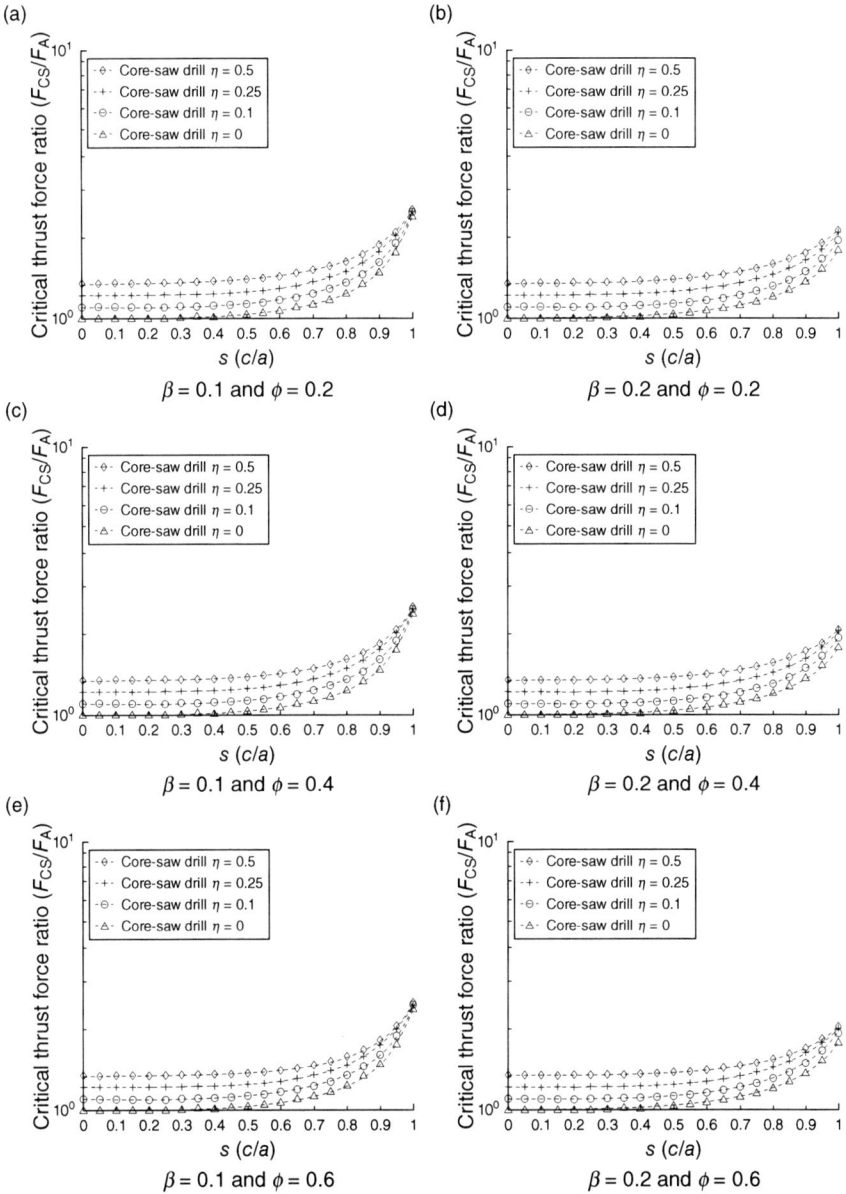

2.24 Critical thrust ratio between core–saw drill and twist drill (Hocheng and Tsao, 2009).

2.4.3   Step–core–special drills (step–core–twist drills, step–core–candlestick drills and step–core–saw drills)

The reference Tsao (2008c) covers the work in this section. Usually, the function of a step–core–special drill in the drilling of composite materials is similar to that of the pre-drilled pilot hole or a step drill that can reduce the delamination. The step–core–special drill is composed of a core drill and special drill (twist drill, saw drill and candlestick drill). Different step–core–special drills are shown in Fig. 2.25. Although significant efforts have been devoted to understanding drilling-induced delamination, there have been few papers reporting the effect of step–core–special drills in drilling composite materials. Hence, this section aims to examine experimentally the drilling-induced thrust force in composite materials when using step–core special drills.

The special drills performs front-edge cutting before step–core edge cutting. Delamination formed in the primary stage cutting by the front cutting edge can be removed in secondary stage cutting by the step–core edge. In particular, new delamination is formed through secondary-stage cutting. Figure 2.26 compares thrust-time characteristics for various step–core–special drills. Owing to the different drill geometries, the thrust force curves differ significantly from each other. During the first drilling cycle, a conventional twist drill can be drilled with a higher thrust force for the same drilling conditions when compared with a saw drill and a candlestick drill. The ultrasonic C-scan images at the exit plane of a workpiece for drills of different diameters and geometries are shown in Figure 2.27. As can be seen, the saw drill and the candlestick drill produced good-quality holes at the exit plane of the workpiece. This reason for this difference is that the cutting speed will be zero at the chisel edge of the twist drill, which easily pulls out the fiber at the exit plane of the workpiece. In other words, the candlestick drill and the saw drill produce less delamination damage compared with the twist drill in drilling. Due to the cutting mechanism and

2.25 Various step–core–special drills (Tsao, 2008c).

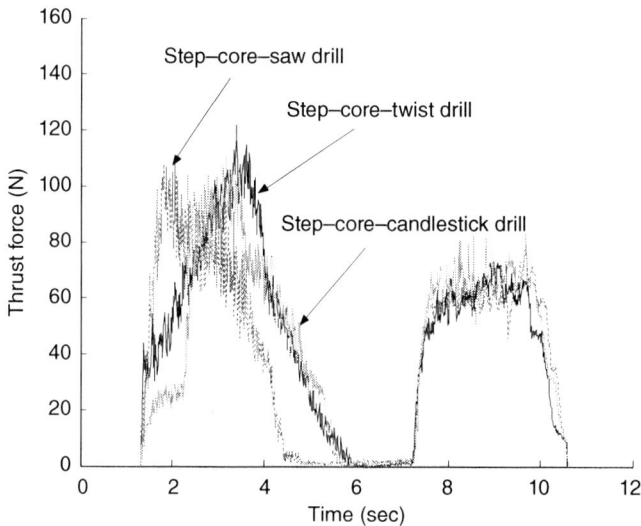

2.26 Thrust–time characteristic for various compound core drills (diameter ratio ($\beta = d/D$) = 0.74, feed rate = 16 mm/min and spindle speed = 800 rpm) (Tsao, 2008c).

geometry of drills such as the saw drill and the candlestick drill, the peripheral circular force at the exit plane are alike, as shown in Fig. 2.26. During the second drilling cycle, the saw drill and the candlestick drill show no significant difference of exit drilling thrust, whereas the twist drill causes a reduction in drilling thrust force of about 27% at the exit.

Figure 2.28 shows the thrust forces of various step–core–special drills with various diameter ratios and geometries (twist drill, candlestick drill and saw drill). As can be seen, the thrust force increases significantly with increase in drill diameter. Hence, a drill bit with a large diameter (d = 7.4 mm) should produce a large induced delamination in the inside of the step–core–special drill. However, for a small diameter (d = 5.5 mm), the thrust force of a step–core–special drill will increase, which may produce larger delamination at the outside of the drill. The thrust force of step–core–special drills with various diameter ratios is shown in Fig. 2.29. As seen in the figure, the thrust force of the drills increases with decreasing diameter ratio.

The correlation between the thrust force and feed rate for various step–core–special drills is shown in Fig. 2.30. The trends of thrust forces for various step–core special drills in drilling CFRP laminates are identical to each other. With increasing feed rate, the thrust force also increases. It is concluded that the step–core–saw drill has the highest thrust force, followed

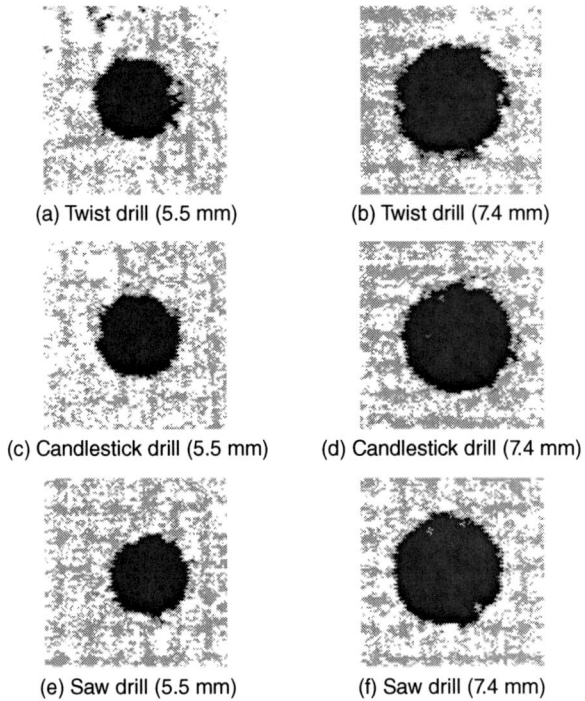

(a) Twist drill (5.5 mm)   (b) Twist drill (7.4 mm)

(c) Candlestick drill (5.5 mm)   (d) Candlestick drill (7.4 mm)

(e) Saw drill (5.5 mm)   (f) Saw drill (7.4 mm)

2.27 Ultrasonic C-scan images at the exit plane of workpiece for drills of various diameters and geometries (feed rate = 16 mm/min and spindle speed = 800 rpm) (Tsao, 2008c).

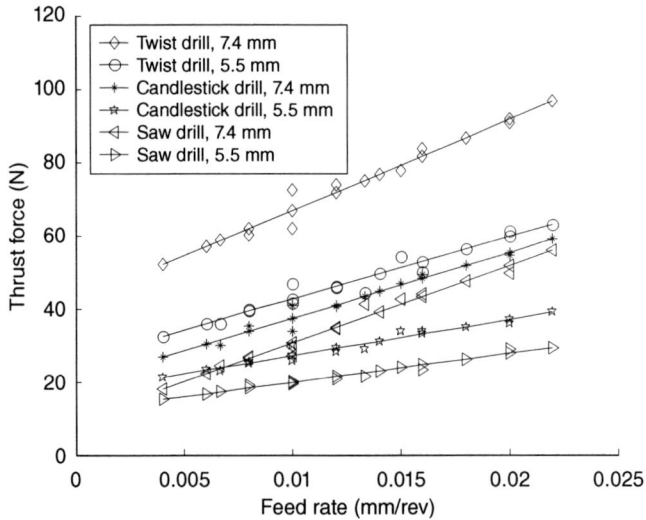

2.28 Thrust forces of various drill bits (twist drill, candlestick drill and saw drill) with various drill diameters (Tsao, 2008c).

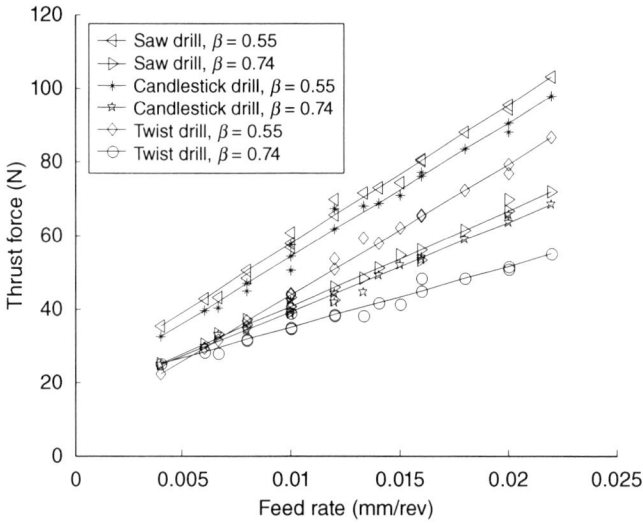

2.29 Thrust forces of various step–core–special drills with various diameter ratios (Tsao, 2008c).

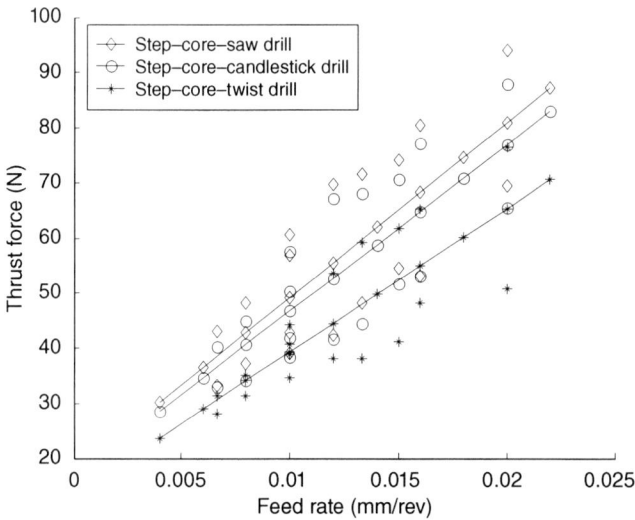

2.30 Correlation between thrust force and feed rate for various step–core–special drills (Tsao, 2008c).

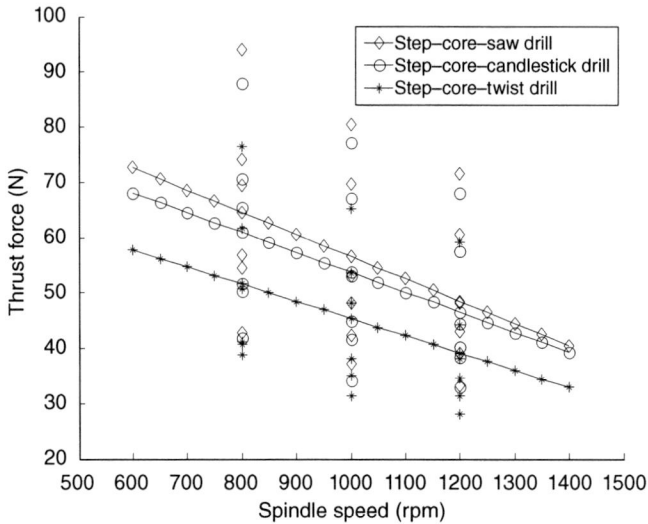

2.31 Effect of spindle speed on thrust force for various step–core–special drills (Tsao, 2008c).

by the step–core–candlestick drill and step–core–twist drill. It can be seen that the step–core–saw drill and step–core–candlestick drill generated very close results. Slight differences leads to higher thrust force at higher feed rates between the two step–core–special drills.

Figure 2.31 shows the effect of spindle speed on the thrust force for various step–core–special drills. The thrust force decreases significantly with an increase in spindle speed. As a result, the generated heat at the tool edge may soften the polymer matrix, which decreases the thrust force in drilling. As is seen from Fig. 2.31, the drill geometry, with obviously different effects, is associated with thrust force at the selected cutting conditions. However, when a step–core–special drill is used for drilling, the effect of thrust force on spindle speed must be considered since the operating spindle speed may cause drastic tool wear.

Shaw and Oxford found that the thrust force (F) was related to the feed rate (f) and diameter of the drill (d) using the following equations (Shaw and Oxford, 1957; Shaw, 1984):

$$F = K_{S1}(df)^{1-\alpha_S} + K_{S2}d \qquad [2.26]$$

where K_{S1}, K_{S2} and α_S are the empirical constants. The constants, K_{S1}, K_{S2} and α_S, determined the magnitudes of the drilling forces. With Eq. [2.26] developed by Shaw and Oxford, the semi-empirical equation for thrust force of various compound core drills was derived as follows:

(i) Thrust force of step–core–twist drill (F_W^E)

$$F_W^E = 72.83(\beta f)^{0.26} - 118.24\beta \qquad [2.27]$$

(ii) Thrust force of step–core–candlestick drill (F_C^E)

$$F_C^E = 93.22(\beta f)^{0.24} - 150.14\beta \qquad [2.28]$$

(iii) Thrust force of step–core–saw drill (F_S^E)

$$F_S^E = 102.09(\beta f)^{0.23} - 162.40\beta \qquad [2.29]$$

Among the three drills, the step–core–saw drill offers the highest drilling thrust force followed by the step–core–candlestick drill and the step–core–twist drill due to their higher K_{S1}, K_{S2} and lower α_S values.

2.5   Delamination measurement and assessment

The references Hocheng and Tsao (2002) and Tsao and Hocheng (2005b) cover the work in this section. Drilling-induced delamination occurs both at the entrance and the exit planes of the workpiece. Visualization and assessment of internal delamination is a difficult and challenging task. Interrogation of the composite materials to obtain a comprehensive knowledge of the size, shape and location of delamination nondestructively is highly desirable. Carbon fiber-based composites are particularly optically disadvantaged for conventional visual inspection.

2.5.1 Delamination measurement

The references Hocheng and Tsao (2002) and Tsao and Hocheng (2005b) cover the work in this section. To determine the extent of the drilling delamination area produced by various drill bits, two NDE methods were studied. The specimens were examined using ultrasonic C-scan and computerized tomography (CT). A schematic of the X-ray computed tomography technique used for the attenuation and detection of radiation is shown in Fig. 2.32. The image reconstruction was performed by a FUJIX Medical Image Processor MF-300S, which ensured fast reconstruction within 6–14 seconds. The reconstructed images were displayed on a high definition monitor, from which they were copied onto a film. All the images presented in this chapter are positive copies on photographic paper obtained from the negatives.

Ultrasonic C-scan is the preferred technique to detect defects and damage in composite structures caused during manufacture and in service conditions. Fracture mechanisms in composites are quite complex, involving the interaction of matrix cracking, fiber matrix debonding, fiber pullout, delamination and fiber breakage. Basically, higher feed rates produce not only

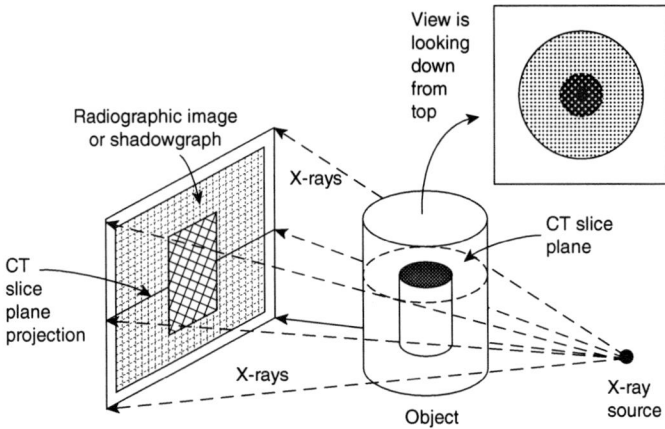

2.32 Schematic of X-ray computed tomography techniques for the attenuation and detection of radiation (Hocheng and Tsao, 2002).

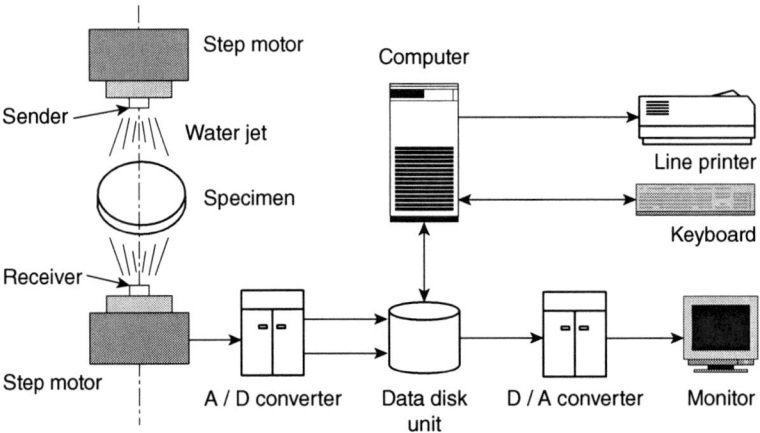

2.33 Schematic of ultrasonic C-scan (Tsao and Hocheng, 2005b).

larger delamination, but also more irregular shape of the delamination. Digital images of the drilling delamination area were produced from carbon fiber-reinforced composite material sections obtained by ultrasonic C-scan. The ultrasonic C-scan equipment was an AIT-5112 (Automated Inspection Technologies Inc.) unit. The specimen was placed between the sender and the receiver. When the sender emits an ultrasonic wave, the attenuation through the receiver is recorded on a computer. A schematic of the ultrasonic C-scan is shown in Fig. 2.33.

The drilling-induced delamination due to using a twist drill, from the ultrasonic C-scan and CT, is shown in Fig. 2.34. Around the hole in the

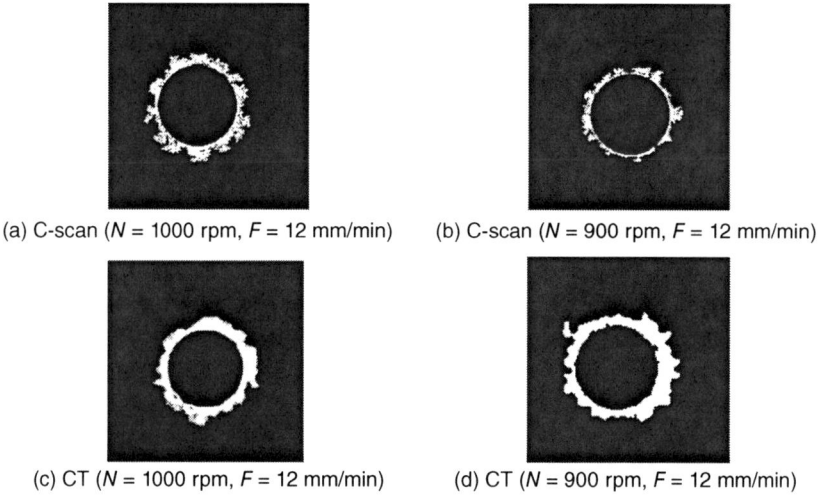

(a) C-scan (N = 1000 rpm, F = 12 mm/min)   (b) C-scan (N = 900 rpm, F = 12 mm/min)

(c) CT (N = 1000 rpm, F = 12 mm/min)   (d) CT (N = 900 rpm, F = 12 mm/min)

2.34 Ultrasonic C-scan and CT showing the extent of drilling defect (delamination) (Hocheng and Tsao, 2002).

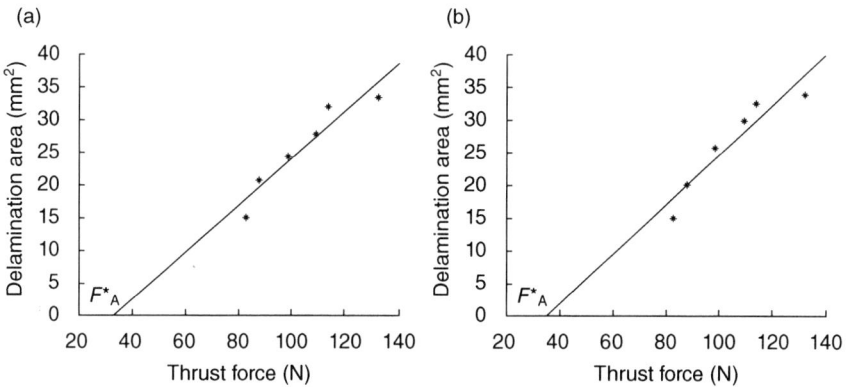

2.35 Experimental correlation between thrust force and delamination by (a) C-scan and (b) CT (Hocheng and Tsao, 2002).

specimen, the damage was evident from the edge of the hole to an extent. Note that the defects occur primarily at the hole exit. At a low feed rate, the area of delamination is small. The area of delamination enlarges with increasing feed rate. The shape of the delamination between two cutting conditions was found to be slightly different. Higher feed rates produce not only larger delaminations, but also more irregular shapes of delamination. Figure 2.35 illustrates that the same relationship exists for the ultrasonic C-scan and the CT. Both show the same scattering (within 6%) and reveal experimentally the critical thrust force for the onset of delamination. Hence, CT is demonstrated as a feasible and effective tool for the evaluation of drilling-induced delamination.

2.5.2 Delamination assessment

The reference Tsao et al. (2010) covers the work in this section. In general, the quality control and evaluation of drilling-induced delamination during the drilling of fiber-reinforced composite materials is rather difficult. Many investigators have reported X-ray (Chen, 1997; De Albuquerque, 2010), optical microscope (Davim and Reis, 2003a,b), ultrasonic C-scan (Tsao and Hocheng, 2004, 2005b) and digital photography (Piquet et al., 2000; Tsao and Hocheng, 2005b) techniques, which have been commonly used to acquire the size, shape and location of delamination in composite laminates. Chen (1997) first proposed the concept of the 'delamination factor' (F_a) (i.e. the ratio of the maximum diameter D_{max} in the damage zone to the hole diameter D_o), to easily analyze and compare the degree of delamination in the drilling of CFRP composite laminates. The equation of the conventional delamination factor can be expressed as follows:

$$F_a = \frac{D_{max}}{D_o} \qquad [2.30]$$

The advanced technology of digital image processing to measure drilling-induced delamination has been widely used because it emphasizes the improvement of process efficiency, stability of image quality and cost-saving. Davim et al. (2007) proposed the adjusted 'delamination factor' (F_{da}) to evaluate the delamination zone by digital image processing after drilling composite laminates. The advantage of this measurement technology is that it incorporates the novel approach of area function. F_{da} gave better results of measurement compared with F_a for different drilling damage where D_{max} was identical. So, the F_{da} can provide a more effective and actual measure for hole defects. The equation of the adjusted delamination factor can be expressed as follows

$$F_{da} = F_a + \frac{A_d}{A_{max} - A_o}(F_a^2 - F_a) \qquad [2.31]$$

where A_{max} is the delamination area related to the D_{max}, A_o is the drilled area of the D_o, and A_d is the delamination area in the vicinity of the drilled hole. The first part of Eq. [2.31] represents the size of the crack contribution, and the second part represents the damage area contribution (Davim et al., 2007). From Eq. [2.31], the higher the damage on A_d is, the higher the effect on F_{da} is. However, the delamination area is minimal or maximal, F_{da} is not

(a)   (b)

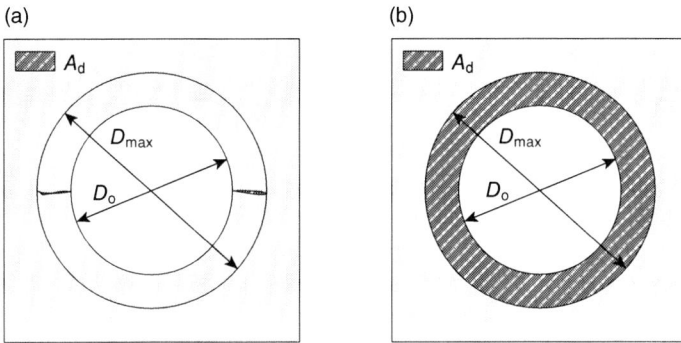

2.36 Critical cases in drilling composite laminate: (a) minimal delamination area and (b) maximal delamination area (Tsao *et al.*, 2010).

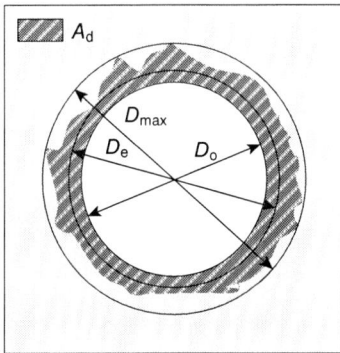

2.37 Scheme of the F_{ed} in drilling a composite laminate with a drill bit (Tsao *et al.*, 2010).

equal to F_a. The schemes of two critical cases in the drilling composite laminate for the drill bit are shown in Fig. 2.36.

A novel approach was proposed to characterize the delamination factor by Tsao *et al.*, namely the 'equivalent delamination factor' (F_{ed}), calculated through Eq. [2.32] (Tsao *et al.*, 2010). The scheme of the F_{ed} in drilling a composite laminate with the drill bit is shown in Fig. 2.37.

$$F_{ed} = \frac{D_e}{D_o} \qquad [2.32]$$

where D_e is the equivalent delamination diameter and can be expressed as

$$D_e = \left[\frac{4(A_d + A_o)}{\pi} \right]^{0.5} \qquad [2.33]$$

This section presents a comprehensive analysis of delamination with various delamination factor models. In this analysis, the equivalent delamination factor is compared with the adjusted delamination factor and the conventional delamination factor.

Table 2.2 shows the effect of the delamination parameters on various delamination factor models. It is seen that F_{da} and F_{ed} gave better discrimination of delamination damage compared with F_a, which had the same delamination factor for Test Nos. 2 to 4. In addition, Table 2.2 shows that the trend for F_{da} and F_{ed} is similar for Test Nos. 2 to 4. Increasing A_d causes higher delamination factors for F_{da} and F_{ed}. However, F_{da} is clearly larger than F_{ed}. The ultrasonic C-scan shows identical F_a with various tests as shown in Fig. 2.38. This figure shows that the delamination area in Test No. 4 possesses a regular distribution in the vicinity of the drilled hole. However, for Test Nos. 2 and 3 the delamination area presents an irregular form, containing long and fine breaks and cracks at the hole exit. When there was

Table 2.2 Effect of delamination parameters on various delamination factor models (Tsao *et al.*, 2010)

Test No.	Delamination parameter				Delamination factor model		
	D_0 (mm)	D_{max} (mm)	A_d (mm^2)	D_e (mm)	F_a	F_{da}	F_{ed}
1	10	10.000	0	10.000	1.0000	1.4375	1.0000
2	10	14.375	6.099	10.381	1.4375	1.4839	1.0381
3	10	14.375	14.977	10.911	1.4375	1.5513	1.0911
4	10	14.375	31.102	11.815	1.4375	1.6739	1.1815
5	10	14.375	83.756	14.375	1.4375	2.0664	1.4375

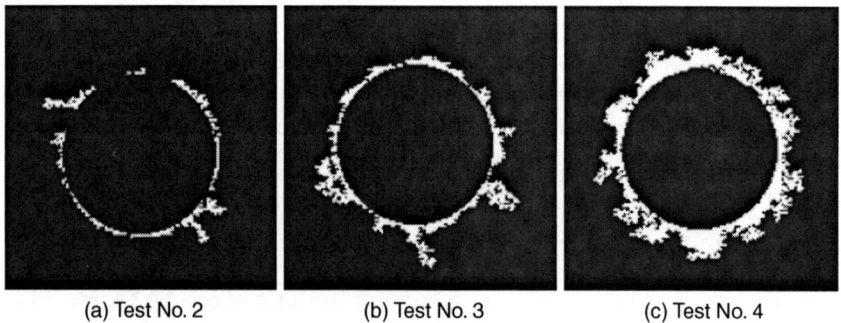

(a) Test No. 2   (b) Test No. 3   (c) Test No. 4

2.38 Ultrasonic C-scan showing F_a with various tests (Tsao *et al.*, 2010).

a larger delamination area occurring at the hole exit, then the variance of the measured delamination factor F_{da}.was clearer. Additionally, the delamination factor between F_{da} and F_{ed} for Test No. 1 (delamination-free) and Test No. 5 (full uniform delamination area) in Table 2.2 was compared. However, F_{da} is null for two critical cases (minimal or maximal delamination area). In other words, F_{da} is comparatively suitable for the regular delamination area. Table 2.2 shows that F_{ed} is the same as F_a in the two critical cases. However, the difference between F_{ed} and F_a increases with the delamination area.

2.6   Influence of drilling parameters on drilling-induced delamination

The drilling parameters are now considered to obtain desirable performance, such as good surface finish, dimensional accuracy of the component, minimum tool wear, easy chip removal and so on. In addition, they must satisfy economic criteria, such as minimum production cost or maximum production rate. DiPaolo *et al.* (1996) conducted an experiment to investigate the size of the delamination region during drilling, which is confirmed to be related to the thrust force developed at the drill exit. Most of the published papers and manufacturers' literature correlates the drill materials, drill geometries, drill wear and drilling variables to the delamination produced by twist drills (Galloway, 1957; Russell, 1962; Jain and Yang, 1993; Lin and Chen, 1996; Chen, 1997; Okafor and Birdsong, 1999; Davim and Reis, 2003a; Arul *et al.*, 2006a; Tsao and Hocheng, 2007a; Hocheng and Tsao, 2009). The factors affecting drilling quality when drilling composites are shown in Fig. 2.39.

2.6.1 Drill materials

Successful tool material selection for operating a drill depends on the required application. Some pioneering works found that high speed steel (HSS) suffers extreme wear and should not be used for composites removal. Tungsten carbide (WC) and PCD tooling instead can provide a good compromise between tool life and production costs. Teti (2002) has presented a thorough review on the effectiveness of tool materials for various machining processes of composites. In the case of GFRP and CFRP, it is the cutting tool material that dominates the tool selection. In the case of aramid fiber-reinforced plastics, it is the tool geometry that is the most significant factor in the choice of cutting tool. The hardness of glass and, more especially, of carbon fibers, results in a high rate of tool wear (Teti, 2002).

A coated layer on the surface of drill bits provides lubricity, heat resistance, as well as corrosion resistance during the drilling process This results

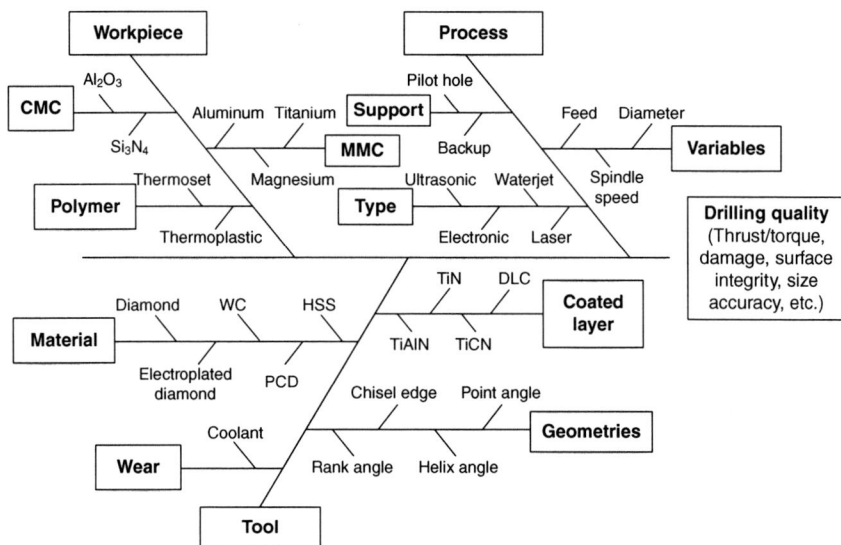

2.39 Factors affecting drilling quality when drilling composites.

in a longer drill life than is obtained with typical uncoated drill bits. Ti-based coatings (titanium nitride (TiN), titanium carbon nitride (TiCN) and titanium aluminum nitride (TiAlN)) and diamond-like-carbon (DLC) have been commonly used to reduce the thrust force and to extend the cutting life when drilling composite materials (Murphy *et al.*, 2002; Teti, 2002). TiCN and TiAlN are superior to TiN and can extend tool life five or more times. Diamond powder is used as an abrasive, most often for cutting brittle and hard materials, such as stone, glass, MMC and CMC. Large amounts of heat are generated, and diamond-coated drill bits often have to be water cooled to prevent damage to the drill bit or the workpiece.

2.6.2 Drill geometries

In general, drill geometry is one of the most frequently considered parameters to improve the drilled quality in drilling composite materials. The twist drill is indeed an important tool in industrial applications such as hole-making. However, the thrust of a twist drill, which can be regarded as a concentration force, pushes uncut thickness at the center as it approach the exit plane when drilling composites. It is well known that drilling-induced damage correlates with the drilling thrust. The consequence of a larger point angle for the twist drill is an enlargement of the area responsible for the extrusion effect on the uncut plies of the laminate and consequently a higher thrust force (De Albuquerque *et al.*, 2010). Some researchers point out that not only for new concepts of tooling but also for different realms of cutting conditions to solve the drilling-induced damages in drilling composite materials (Miner, 1980; Mackey, 1982). Piquet *et al.* (2000) completed an experimental analysis of drilling damage in thin carbon/epoxy plates using special drills. They concluded that the use of a small rake angle, a great number of cutting edges, and a point angle of 118° for the main cutting edges can reduce the damage involved. In addition, a reduction of the chisel edge dimension can prevent delamination onset (Jain and Yang, 1993; Won and Dharan, 2002; Tsao and Hocheng, 2003). Hocheng and Tsao also found that the advantage of the special drills compared with twist drills is that the thrust force exerted is distributed toward the drill periphery when drilling (Hocheng and Tsao, 2005, 2006).

2.6.3 Drill wear

Conventional drilling can usually be applied to composite materials, but there is often a problem of excessive tool wear. A high rate of tool wear is usually associated with fiber-reinforced plastic laminate machining, due to the high abrasiveness of the reinforcement fiber. The effect of cutting speed is evidently attributable to the tool encountering more frictional contact, while fibers resist excavation more strongly and hence more stress and heat are generated in the tooling. When a drill bit becomes dull, it increases the drilling thrust and decreases further the tool life. Chen (1997) found that the delamination factor increases with increasing flank wear of the drill as a result of increasing the thrust force. Lin and Chen (1996) showed that increasing cutting speed accelerates tool wear, and the thrust force increases as drill wear increases. Arul *et al.* (2006b) showed that, during early stages of vibratory drilling (1 to 10 holes), there is a progressive rise in the thrust force which is usually associated with run-in wear of the cutting wedge. Tsao and Hocheng (2007a) employed LEFM methods to solve the drill run-out that causes the thrust force which relates the delamination of composite laminates to machining parameters and composite material properties. It is noted that tool wear increases with feed, demanding a greater thrust force with increasing delamination damage. Rawat and Attia (2009) presented an experimental investigation of the wear mechanisms of WC drills during dry high speed drilling of quasi-isotropic woven graphite fiber epoxy composites. They found that chipping and abrasion were the main mechanisms controlling deterioration of the WC drills. Iliescu *et al.* (2010) have pointed out that the wear of uncoated and diamond-coated carbide tools when drilling CFRP is strongly dependent on the axial force applied to the cutting edge by the contact length between the cutting edge and the composite material.

2.6.4   Drilling variables

Owing to the anisotropy and inhomogeneity of composites, the selection of drilling parameters plays a very important role. Armarego and Wright (1984) found that the effects of feed rates, spindle speeds and the geometrical characteristics of three drills on the resulting torque and thrust values were comparable, regardless of the three different drill flank configurations used. Feed rate is one of the cutting parameters of greatest interest in drilling composite materials. Generally, as the feed rate becomes greater, an improvement in productivity can be achieved. Numerous studies have recognized that thrust force (or delamination) increases with feed rate (Koenig *et al.*, 1985; Hocheng and Dharan, 1990; Caprino and Tagliaferri, 1995; Davim and Reis, 2003a; De Albuquerque *et al.*, 2010). Low values of feed rate ensure minimum thrust force in order to reduce the drilling-induced delamination. Response surface methodology (RSM) is a very useful tool for quality and productivity improvement in industry. Many researchers have used RSM for correlating machining parameters (Mohan *et al.*, 2005; Palanikumar, 2007; Tsao, 2008a; Krishnamoorthy *et al.*, 2009). From these studies, it is understood that drilling-induced delamination can be improved by the careful selection of drilling parameters with selected tool geometries. From the viewpoint of cost and productivity, optimization of drilling processes is extremely important for the manufacturing industry. Taguchi's method of off-line quality control has been successfully used in the design and selection of optimal process parameters in many areas of manufacturing processes, and is a powerful tool in the design of a high quality system. Kackar (1985) reported that Taguchi had developed more than 60 performance statistics that could be used, depending on the problem being investigated. Without a systematic approach, there may occur some difficulties in solving problems or developing projects in any area. A number of researchers show that using Taguchi's method leads to significant improvements in the thrust and hole quality when drilling composites (Enemuoh *et al.*, 2001; Davim and Reis, 2003b; Tsao and Hocheng, 2004; Tsao, 2007; Tsao and Hocheng, 2007b, 2008c). However, a single performance characteristic (Taguchi method) may not be straightforward in assessing multiple performance characteristics, especially under highly conflicting conditions. Therefore, the Grey–Taguchi method and artificial neural networks have been suggested to solve the multiple performance characteristics and complicated problems in manufacturing processes (Pan *et al.*, 2007; Fung, 2003; Yang *et al.*, 2003; Sing *et al.*, 2004; Sardiñas *et al.*, 2006; Tzeng *et al.*, 2009; Karnik *et al.*, 2008; Panda *et al.*, 2008; Huang *et al.*, 2010; Kilickap, 2010).

2.7   Conclusions

Hole-making is one of the most commonly used processes in the secondary machining of composites, due to the need for riveting and fastening in mechanical parts and structures. Poor machined quality will result in poor assembly tolerances and long-term structural performance deterioration. In the machining of composite materials, however, a sound theoretical analysis for innovative tools is still evolving. Composites are locally anisotropic, inhomogeneous and abrasive in nature. Such characteristics have a significant effect on the machining process and result in the complexity of the mechanical modeling of the material removal process. Having established the arguments above, this chapter has dealt with typical drilling processes, that are frequently adopted to make holes for screws, rivets and bolts. Interlaminar cracking (or delamination) around the hole exit has been the primary concern. Some analytical models using classic plate theory and LEFM have been developed. Delamination is interpreted as resulting from the action of the thrust force and distributed peripheral moment. From the equations, the critical thrust force at the onset of delamination can explicitly be determined as a function of material properties and specimen configuration, and tool geometry. The higher the interlaminar Mode I fracture toughness or the modulus of elasticity, or the thinner the ply thickness for a twist drill, the larger is the thrust that can be allowed. However, special and compound drills utilize the characteristic of distributed peripheral thrust to acquire higher critical thrust force and better machining quality in drilling composite laminates. Experiments show satisfactory agreement between these results and the models. According to these approaches, a control scheme to provide the optimal thrust force can be employed in the future. On the other hand, feed governs the thrust while the cutting speed has only a mild influence. The diameter ratio for compound drills leads to a significant effect on the thrust and hole quality in drilling composites.

The advanced technology of digital image processing to measure drilling-induced delamination has been widely used because it emphasizes the improvement of process efficiency, stability of image quality and cost-saving. A comparative analysis on various delamination factor models of drilled composite materials has been presented in this chapter. The adjusted delamination factor F_{da} and the equivalent delamination factor F_{ed} had better discrimination of delamination damage compared with the conventional delamination factor F_a, which was identical at the hole exit. However, F_{da} is null for minimal or delamination-free. F_{ed} obtained through digital image processing is considered suitable for characterizing delamination at the exit of the drilled hole.

2.8   References

Armarego EJA and Wright JD (1984), 'Predictive models for drilling thrust and torque – a comparison of three flank configurations', *Ann CIRP*, **33**(1), 5–10.

Arul S, Vijayaraghavan L, Malhotra SK and Krishnamurthy R (2006a), 'Influence of tool material on dynamics of drilling of GFRP composites', *Int J Adv Manuf Technol*, **29**, 655–662.

Arul S, Vijayaraghavan L, Malhotra SK and Krishnamurthy R (2006b), 'The effect of vibratory drilling on hole quality in polymeric composites', *Int J Mach Tools Manuf*, **46**, 252–259.

Azarhoushang B and Akbari J (2007), 'Ultrasonic-assisted drilling of Inconel 738-LC', *Int J Mach Tools Manuf*, **47**, 1027–1033.

Caprino G and Tagliaferri V (1995), 'Damage development in drilling glass fiber reinforced plastics', *Int J Mach Tools Manuf*, **35**(6), 817–829.

Chandrasekharan V, Kappor SG and DeVor RE (1995), 'A mechanistic approach to predicting the cutting forces in drilling: with application to fiber-reinforced composite materials', *Trans ASME J Eng Ind*, **117**, 559–570.

Chen WC (1997), 'Some experimental investigations in the drilling of carbon fiber-reinforced plastic (CFRP) composite laminates', *Int J Mach Tools Manuf*, **37**(8), 1097–1108.

Davim JP and Reis P (2003a), 'Drilling carbon fibre reinforced plastics manufactured by autoclave – experimental and statistical study', *Mater Des*, **24**, 315–324.

Davim JP and Reis P (2003b), 'Study of delamination in drilling carbon fiber reinforced plastics (CFRP) using design experiments', *Compos Struc*, **59**, 481–487.

Davim JP, Rubio JC and Abrao AM (2007), 'A novel approach based on digital image analysis to evaluate the delamination factor after drilling composite laminates', *Compos Sci Technol*, **67**, 1939–1945.

De Albuquerque VHC, Tavares JMRS and Durão LMP (2010), 'Evaluation of delamination damage on composite plates using an artificial neural network for the radiographic image analysis', *J Compos Mater*, **44**(9), 1139–1159.

Dharan CKH (1978), 'Fracture mechanics of composite materials', *Trans ASME J Eng Mater Technol*, **100**, 233–247.

DiPaolo G, Kapoor SG and DeVor RE (1996), 'An experimental investigation of the crack growth phenomenon for drilling of fiber-reinforced composite materials', *Trans ASME J Eng Ind*, **118**, 104–110.

Doerr R, Greene E, Lyon B and Taha S (1982), *Development of Effective Machining and Tooling Techniques for Kevlar Composites: Technical Report, No. AD-A117853*, Du Pont, Wilmington, Delaware.

Doran JH and Maikish CR (1973), 'Machining boron composite', in Noton BR, *Composite Materials in Engineering Design*, ASM, 242–250.

Dubey AK and Yadava V (2008), 'Experimental study of Nd:YAG laser beam machining – an overview', *J Mater Process Technol*, **195**, 15–26.

Enemuoh EU, El-Gizawy AS and Okafor AC (2001), 'An approach for development of damage-free drilling of carbon fiber reinforced thermosets', *Int J Mach Tools Manuf*, **41**(12), 1795–1814.

Friedrich MO, Burant RO and McGinty MJ (1979), 'Cutting tools/drills: Part 5 – Point styles and applications', *Manuf Eng*, **83**, 29–31.

Fung CP (2003), 'Manufacturing process optimization for wear property of fiber-reinforced polybutylene terephthalate composites with Grey relational analysis', *Wear*, **254**, 298–306.

Galloway DF (1957), 'Some experiments on the influence of various factors on drill performance', *Trans ASME*, **79**, 191–237.

Haggerty WA and Ernst H (1958), 'The spiral point drill – self-centering drill point geometry', *ASTE Paper No. 101*, 58.

Hocheng H and Dharan CKH (1990), 'Delamination during drilling in composite laminates', *Trans ASME J Eng Ind*, **112**, 236–239.

Hocheng H (1990), 'A failure analysis of water jet drilling in composite laminates', *Int J Mach Tools Manuf*, **30**(3), 423–429.

Hocheng H, Tai NH and Liu CS (2000), 'Assessment of ultrasonic drilling of C/SiC composite material', *Compos Part A Appl Sci Manuf*, **31**(2), 133–142.

Hocheng H, Lei WT and Hsu HH (1997), 'Preliminary study of material removal in electrical-discharge machining of SiC/Al', *J Mater Process Technol*, **63**, 813–818.

Hocheng H and Tsao CC (2002), 'Assessment of delamination by computerized tomography and C-scan in drilling of composite materials', *Proc 6th Int Conf Prog Mach Technol (ICPMT 6)*, Xian, China.

Hocheng H and Tsao CC (2005), 'The path towards delamination-free drilling of composite materials', *J Mater Process Technol*, **167**, 251–264.

Hocheng H and Tsao CC (2006), 'Effects of special drill bits on drilling-induced delamination of composite materials', *Int J Mach Tools Manuf*, **46**, 1403–1416.

Hocheng H and Tsao CC (2009), 'A treatment of drilling-induced delamination of composite materials', in Davim JP, *Drilling of Composite Materials*, Nova Science Publishers, New York, 1–43.

Huang CH, Tsao CC, Wang SS and Hsu CY (2010), 'Optimization of the sputtering process parameters of GZO films using the Grey-Taguchi method', *Ceram Int*, **36**(3), 979–988.

Iliescu D, Gehin D, Gutierrez ME and Girot F (2010), 'Modeling and tool wear in drilling of CFRP', *Int J Mach Tools Manuf*, **50**(2), 204–213.

Jain S and Yang DCH (1992), 'Delamination-free drilling of composite laminates', *Trans ASME Mater Div Publ MD*, Anaheim, CA, USA, 45–59.

Jain S and Yang DCH (1993), 'Effects of feed rate and chisel edge on delamination in composite drilling', *Trans ASME J Eng Ind*, **115**, 398–405.

Kackar RN (1985), 'Off-line quality control, parameter design and Taguchi method', *J Qual Eng*, **17**, 176–209.

Karnik SR, Gaitonde VN, Campos Rubio J, Esteves Correia A, Abrão AM and Davim JP (2008), 'Delamination analysis in high speed drilling of carbon fiber reinforced plastics (CFRP) using artificial neural network model', *Mater Des*, **29**(9), 1768–1776.

Kilickap E (2010), 'Optimization of cutting parameters on delamination based on Taguchi method during drilling of GFRP composite', *Expert Sys Appl*, **37**(8), 6116–6122.

Kobayashi A (1967), *Machining of Plastics*, McGraw-Hill, New York.

Koenig W, Grass P, Heintze A, Okcu F and Schmitz-Justin C (1984), 'Developments in drilling, contouring composities containing Kevlar', *Prod Eng*, 56–61.

Koenig W, Wulf C, Grass P and Willerscheid H (1985), 'Machining of fiber reinforced plastics', *Ann CIRP*, **34**(2), 538–548.

Komanduri R, Zhang B and Vissa CM (1991), 'Machining of fiber reinforced composites', *Proc Manuf Compos Mater*, **49/27**, 1–36.

Krishnamoorthy A, Rajendra Boopathy S and Palanikumar K (2009), 'Delamination analysis in drilling of CFRP composites using response surface methodology', *J Compos Mater*, **43**(24), 2885–2902.

Langella A, Nele L and Maio A (2005), 'A torque and thrust prediction model for drilling of composite materials', *Compos Part A Appl Sci Manuf*, **36**, 83–93.

Lin SC and Chen IK (1996), 'Drilling carbon fiber-reinforced composite material at high speed', *Wear*, **194**, 156–162.

Mackey BA (1982), 'A practical solution for the machining of accurate, clean, burr free holes in Kevlar composites', *Proc 37th Conf Soc Plast Ind*, Reinf Plast/Compos Inst, 1–5 (24D).

Mathew J, Ramakrishnan N and Naik NK (1999a), 'Investigations into the effect of geometry of a trepanning tool on thrust and torque during drilling of GFRP composites', *J Mater Process Technol*, **91**, 1–11.

Mathew J, Ramakrishnan N and Naik NK (1999b), 'Trepanning on unidirectional composites: Delamination studies', *Compos Part A Appl Sci Manuf*, **30**(8), 951–959.

Miller JA (1987), 'Drilling graphite/epoxy at Lockheed', *Am Mach Auto Manuf*, 70–71.

Miner LH (1980), 'Cutting and machining Kevlar aramid composites', *Kevlar Compos Symp*, El Segundo, California, Soc Plast Eng, 85–93.

Mohan NS, Ramachandra A and Kulkarni SM (2005), 'Influence of process parameters on cutting force and torque during drilling of glass fiber polyester reinforced composites', *Compos Struc*, **71**, 407–413.

Murphy C, Byrne G and Gilchrist MD (2002), 'The performance of coated tungsten carbide drills when machining carbon fiber-reinforced epoxy composite materials', *Proc Inst Mech Eng Part B*, **216**(2), 143–152.

Okafor AC and Birdsong SR (1999), 'Effect of drilling conditions, drill material and point angle on acoustic emission and hole exit delamination in drilling advanced fiber reinforced composites', *Proc. SPIE – Inter Soc Opt Eng*, **3589**, 101–114.

Palanikumar K (2007), 'Modeling and analysis for surface roughness in machining glass fibre reinforced plastics using response surface methodology', *Mater Des*, **28**, 2611–2618.

Pan LK, Wang CC, Wei SL and Sher HF (2007), 'Optimizing multiple quality characteristics via Taguchi method-based Grey analysis', *J Mater Process Technol*, **182**, 107–116.

Panda SS, Chakraborty D and Pal SK (2008), 'Flank wear prediction in drilling using back propagation neural network and radial basis function network', *Appl Soft Comput*, **8**(2), 858–871.

Piquet R, Ferret B, Lachaud F and Swider P (2000), 'Experimental analysis of drilling damage in thin carbon/epoxy plate using special drills', *Compos Part A Appl Sci Manuf*, **31**(10), 1107–1115.

Rawat S and Attia H (2009), 'Wear mechanisms and tool life management of WC-Co drills during dry high speed drilling of woven carbon fibre composites', *Wear*, **267**, 1022–1030.

Russell WR (1962), 'Drill design and drilling conditions for improved efficiency', *ASTME Paper No. 397*, 62.

Saghizadeh H and Dharan CKH (1986), 'Delamination fracture toughness of graphite and aramid epoxy composites', *Trans ASME J Eng Mater Technol*, **108**, 290–295.

Sardiñas RQ, Reis P and Davim JP (2006), 'Multi-objective optimization of cutting parameters for drilling laminate composite materials by using genetic algorithms', *Compos Sci Technol*, **66**(15), 3083–3088.

Shaw MC (1984), *Metal Cutting Principles*, Oxford University Press, New York.

Shaw MC and Oxford Jr. JC (1957), 'On the drilling of metals. Part II: The torque and thrust force in drilling', *Trans ASME*, **79**, 139–148.

Singh NP, Raghukandan K, Rathinasabapathi M and Pai BC (2004a), 'Electric discharge machining of Al-10%SiCP as-cast metal matrix composites', *J Mater Process Technol*, **155–156**, 1653–1657.

Singh NP, Raghukandan K and Pai BC (2004b), 'Optimization by Grey relational analysis of EDM parameters on machining Al-10% SiCP composites', *J Mater Process Technol*, **155–156**, 1658–1661.

Teti R (2002), 'Machining of composite materials', *Ann CIRP*, **51**, 611–634.

Tsao CC and Hocheng H (2003), 'The effect of chisel length and associated pilot hole on delamination when drilling composite materials', *Int J Mach Tools Manuf*, **43**(11), 1087–1092.

Tsao CC and Hocheng H (2004), 'Taguchi analysis of delamination associated with various drill bits in drilling of composite material', *Int J Mach Tools Manuf*, **44**(10), 1085–1090.

Tsao CC and Hocheng H (2005a), 'Effects of exit back-up on delamination in drilling composite materials using a saw drill and a core drill', *Int J Mach Tools Manuf*, **45**(11), 1282–1287.

Tsao CC and Hocheng H (2005b), 'Computerized tomography and C-scan for measuring delamination in drilling of composite material using various drills', *Int J Mach Tools Manuf*, **45**, 1282–1287.

Tsao CC (2006), 'The effect of pilot hole on delamination when core drill drilling composite materials', *Int J Mach Tools Manuf*, **46**, 1653–1661.

Tsao CC and Hocheng H (2007a), 'Effect of tool wear on delamination in drilling composite materials', *Inter J Mech Sci*, **49**, 983–988.

Tsao CC and Hocheng H (2007b), 'Parametric study on thrust force of core drill', *J Mater Process Technol*, **192–193**, 37–40.

Tsao CC (2007), 'Taguchi analysis of drilling quality associated with core drill in drilling of composite material', *Int J Adv Manuf Technol*, **32**, 877–884.

Tsao CC and Hocheng H (2008a), 'Effects of peripheral drilling moment on delamination using special drill bits', *J Mater Process Technol*, **201**, 471–476.

Tsao CC and Hocheng H (2008b), 'Analysis of delamination in drilling of composite materials by core–saw drill', *Int J Mater Prod Technol*, **32**, 188–201.

Tsao CC and Hocheng H (2008c), 'Evaluation of thrust force and surface roughness in drilling composite material using Taguchi analysis and neural network', *J Mater Process Technol*, **203**, 342–348.

Tsao CC (2008a), 'Comparison between response surface methodology and radial basis function network for core–center drill in drilling composite materials', *Int J Adv Manuf Technol*, **37**, 1061–1068.

Tsao CC (2008b), 'Thrust force and delamination of core–saw drill during drilling carbon fiber reinforced plastics (CFRP)', *Inter J Adv Manuf Technol*, **37**, 23–28.

Tsao CC (2008c), 'Experimental study of drilling composite materials with step–core drill', *Mater Des*, **29**(9), 1740–1744.

Tsao CC, Kuo KL, Hsu IC and Chern GT (2009), 'Analysis of core–candlestick drill in drilling composite materials', *Key Eng Mater*, **419–420**, 337–340.

Tsao CC, Kuo KL and Hsu IC (2010), 'Evaluation of novel approach on delamination factor after drilling composite laminates', *Key Eng Mater*, **443**, 626–630.

Tzeng CJ, Lin YH, Yang YK and Jeng MC (2009), 'Optimization of turning operations with multiple performance characteristics using the Taguchi method and Grey relational analysis', *J Mater Process Technol*, **209**(6), 2753–2759.

Wang LP and Wang LJ (1998), 'Prediction and computer simulation of dynamic thrust and torque in vibration drilling', *Proc Inst Mech Eng Part B*, **212**(6), 489–497.

Widia (1985), 'Hole-making with carbides', *Technical Information Publication No. WI-T-129-00-85*, The Widia Company, Latrobe, PA, USA.

Won MS and Dharan CKH (2002), 'Chisel edge and pilot hole effects in drilling composite laminates', *Trans ASME, J Manuf Sci Eng*, **124**, 242–247.

Wong TL (1982), 'An analysis of delamination in drilling composite materials', *Proc 14th SAMPE Tech Conf*, Atlanta, GA, 471–483.

Wu SM, Shen JM and Chen LH (1982), 'Multifacet drills', *Proc 14th SAMPE Tech Conf*, Atlanta, GA, 456–463.

Yang SY, Girivasan V, Singh NR, Tansel IN and Kropas-Hughes CV (2003), 'Selection of optimal material and operating conditions in composite manufacturing. Part II: Complexity, representation of characteristics and decision making', *Int J Mach Tools Manuf*, **43**(2), 175–184.

Yung KC, Mei SM and Yue TM (2002), 'A study of the heat-affected zone in the UV YAG laser drilling of GFRP materials', *J Mater Process Technol*, **122**, 278–285.

Yung WKC, Wu J, Yue TM, Zhu BL and Lee CP (2007), 'Nd:YAG laser drilling in epoxy resin/AlN composites material', *Compos Part A Appl Sci Manuf*, **38**(9), 2055–2064.

Zhang LB, Wang LJ, Liu XY, Zhao HW, Wang X and Luo HY (2001), 'Mechanical model for predicting thrust and torque in vibration drilling fiber-reinforced composite materials', *Int J Mach Tools Manuf*, **41**(5), 641–657.

Zhang LB, Wang LJ and Wang X (2003), 'Study on vibration drilling of fiber reinforced plastics with hybrid variation parameters method', *Compos Part A Appl Sci Manuf*, **34**(3), 237–244.

3
Grinding processes for polymer matrix composites

S. D. EL WAKIL, The University of Massachusetts Dartmouth, USA

Abstract: The chapter begins by providing some of the current applications of grinding of fiber-reinforced polymeric composites, and explains the reasons for the difficulty of using composites as machine components. The problems encountered during the grinding of composites are explained, and the different approaches for addressing these problems are discussed. The various variables affecting the operation are explained in detail. References and further sources of information and advice are also given.

Key words: grinding of composites, surface quality of composites, dimensional accuracy of ground composites.

3.1   Introduction

There has recently been an ever growing necessity for the use of precision machining, such as grinding, in order to produce structural machine components made of fiber-reinforced polymeric (FRP) composites. The reason is that, unlike with other common applications of FRP composites (e.g. boats, skiing gear, and water-surfing boards), an adequate level of dimensional accuracy has to be achieved so that those composite parts function properly as machine components. Unfortunately, the grinding of composites is a far more difficult and complicated procedure than the grinding of metals. While the later process is well-understood and can be easily controlled and optimized to enable trouble-free operations, the nature as well as the physical and mechanical properties of composites make it very difficult to achieve that goal. In this chapter, two important questions are discussed; namely, why and when composite parts should be subjected to grinding. Then a detailed account of the problems encountered when grinding composites is given, together with the reasons for such problems. Next, a thorough coverage of the variables that affect the process is provided, together with methods to achieve trouble-free operations. For completion, future trends for improving the process of grinding of composites are covered and information for further reading and advice is provided

3.2   Applications of grinding processes for composites

Fiber-reinforced polymeric composites are currently very popular in fabricating a diverse group of products. Those include boats, skiing gear, water surfing boards, masts, construction panels and the like. Such applications do not require that such parts be produced with tight tolerances. On the other hand, design and manufacturing engineers are continuously trying to employ graphite–epoxy composites to replace lightweight, high strength alloys as structural machine components. This process has been quite limited due to the inability to achieve adequate levels of dimensional accuracy and good surface quality in the precision machining operations of FRP composites. There has been, however, some success in a few industries where graphite–epoxy composites are used in the manufacture of machine components. Examples include rollers for the processing of packaging film, rollers in the printing industry, and power transmission shafts of ships. Evidently, the light weight (i.e. low density) of the composites is a vital factor that makes them very attractive as a material for the manufacturing of torque transmitting shafts and rollers. The low density of the composite material results in a low mass moment of inertia of a shaft. Consequently, less energy is required to start the rotation of the shaft, and better control of its angular velocity during operation is an additional benefit. Needless to say, whether the composite product is a printing roller or a torque transmitting shaft of a boat, its cylindrical surface has to be extremely smooth, and it must also have very tight tolerances in order to enable those shafts to be supported in journal bearings. It is clear that the kind of surface quality and tolerance necessary in those applications can be achieved only by employing the process of cylindrical grinding. This process also ensures a high degree of concentricity between the inner and the outer surfaces of the roller or the shaft. Another emerging application for FRP composites is replacing steel or cast iron as a material for manufacturing the high-pressure pipes used in city water systems and in petroleum pipe lines. Their light weight and ability to undergo large elastic deformation, make the process of laying down the composite pipeline far easier and more advantageous than installing steel pipe lines. While only one mile of a steel pipe line can be completed in a day, several miles per day can be laid down when using composite pipes. Moreover, the maintenance of composite pipe lines is much cheaper than that of steel pipe lines because of the corrosion resistance of the composites. While steel pipes have to be welded together, composite pipes are joined together by adhesive bonding. Each two successive pipes are joined together by a lap joint using a third short external pipe. The ends of each pipe are subjected to grinding to ensure concentricity and yield the required clearance between the outer connecting pipe and the pipes to be joined. That clearance is filled with appropriate adhesive in order to produce the joint.

3.3   Problems associated with the grinding of composites

If composite materials are to be used for manufacturing machine components, they must successfully undergo trouble-free grinding operations. Unfortunately, this has always been extremely difficult and has held up the production of machine components from composites. Many problems are encountered when grinding FRP composites such as the graphite–epoxy that can be used to fabricate machine components. Examples of such problems include degradation and thermal damage of the ground surface, excessive noise and vibrations during the grinding operation, and clogging of the grinding wheel. Since those problems are not encountered when grinding metals (if the parameters of the operation are correctly chosen), it is logical to conclude that they come as a result of the nature and the characteristics of FRP composites, which are different from those of metals. A good step to understand what causes those problems, and thus be able to eliminate them, is to compare the physical and mechanical properties of composites with those of a commonly used metal such as steel (see Table 3.1).

Table 3.1 Comparison between some properties of steel and those of a graphite–epoxy composite

Property	Steel (low carbon)	Graphite–epoxy composite 70/30
Properties, in general	In this respect, steel is isotropic. It has the same value of a property in all directions.	It is clearly anisotropic. The value of a property varies with the direction along which it is measured.
Coefficient of thermal conductivity (W/m.K)	51.9	4.963 Very poor
Coefficient of thermal expansion (°C^{-1})	11.7×10^{-6}	40×10^{-6} Transverse direction
Density (kg/m^2)	7.87×10^3	1.633×10^3
Specific heat (J/kg.K)	486	938
Volumetric specific heat (J/m^3.K)	3.825×10^6	1.532×10^6
Thermal diffusivity = thermal conductivity/volumetric specific heat (m^2/s)	13.57×10^{-6}	3.24×10^{-6}

Needless to say, the properties of graphite–epoxy composites depend upon the graphite fibers-to-epoxy matrix volumetric ratio. Again, not all the physical properties of the composites are direct functions or linearly vary with the above mentioned ratio. While the values of some physical properties may increase with increasing fiber-to-matrix ratio, the values of others may actually decrease.

The process of grinding of composites has been subjected to numerous experimental investigations, and its physics is fairly well understood. An explanation of the problems encountered during the process was therefore possible, although it was not enough to help optimize the process because of the absence of practical quantitative relationships or analytical models that could mathematically be optimized. Selection of the process variables was accordingly based on trial and error, and/or personal experience of machinists on the shop floor.

There have recently been two approaches for providing a quantitative description of the relationship between the various variables involved in the process of grinding of composites. Each has its advantages and disadvantages. In the first approach, a mathematical model is established, based on an analysis of the thermal energy generated as a result of grinding, and the dissipation of that heat into the surroundings, i.e. the work piece and the cutting tool. The goal is to identify the conditions that result in thermal damage of the surface of the composite workpiece. The rate of heat generation is dependent upon the material removal rate (MRR) – in other words the depth of cut and cutting speed – as well as on a characteristic of the material called the specific energy (also called the unit horse power in the US), which is the amount of energy generated per unit volume of removed material per unit time. Thermal energy is therefore generated when grinding FRP composites, and its rate of generation is dependent on the specific energy of the composite and the grinding process parameters. Now, the thermal conductivity of composites is much less than that of metals, as previously shown in Table 3.1, where the thermal conductivity of the 70/30 graphite–epoxy composite is one tenth that of steel. Therefore, the heat cannot be transferred quickly away from the spot being machined and is accordingly retained, resulting in a localized increase in the temperature of the surface of the workpiece. Again, since the coefficient of thermal expansion of polymeric composites is much higher than that of metals (four times that of steel as shown in Table 3.1), appreciable radial expansion of the workpiece occurs, resulting in noise, excessive vibrations, thermal damage of the surface of the workpiece, and even clogging of the grinding wheel (because the workpiece expands toward the grinding wheel). While this approach is definitely sound, the resulting equations are complicated and cumbersome to handle, and render themselves impractical for shop floor applications.

Since the mechanics of the process are very complex, and depend on several process variables as well as the physical and mechanical properties of the workpiece and on the characteristics of the grinding wheel, the second approach is based on reducing the number of variables investigated by employing the Buckingham theorem for dimensional analysis. Three dimensionless numbers can be obtained, which quantitatively describe the cylindrical grinding operation of FRP composites. They are as follows:

$$Pi1 = t / D \qquad\qquad\qquad [3.1]$$

$$Pi2 = V / U \qquad\qquad\qquad [3.2]$$

$$Pi3 = UD / \alpha \qquad\qquad\qquad [3.3]$$

where t = depth of cut

D = diameter of the workpiece

U = axial feed rate

V = peripheral velocity of the workpiece

α = thermal diffusivity = K / C_v

K = thermal conductivity

C_v = volumetric specific heat = (density) × (specific heat)

Either the SI or English system of units can be used without affecting the values of those numbers since the latter are dimensionless. Now, the physical meaning of $Pi2$ can be understood because the tool path on the surface of the workpiece is actually a helix. The angle which the tangent to that helix makes with the axis of the workpiece is called the helix angle, ϕ. Let us rewrite $Pi2$ in the following form:

$$Pi2 = \pi DN / fN \qquad\qquad\qquad [3.4]$$

where N is the rev/min of the workpiece and f is the axial feed in meters or inches/revolution. When further simplifying the above given equation, it will take the form:

$$Pi2 = \pi D / f = \tan \phi \qquad\qquad\qquad [3.5]$$

In other words, $Pi2$ is the dimensionless number that indicates the geometry of the tool path. Again, we have to bear in mind that the direction of the peripheral velocity of the grinding wheel relative to the workpiece (which is actually the direction of cutting) is always normal to the helix of the tool path. Accordingly, if the direction of the cutting is to be normal to the reinforcing fibers, the latter must coincide with the helix of the tool path. As we will see later, that condition is necessary to minimize the amount of energy generated and thus produce the best surface finish.

When mathematically manipulating $Pi1$ and $Pi3$, we can obtain a new dimensionless number $Pi4$, as follows:

$$Pi4 = Pi3 \,/\, Pi1$$
$$= (UD \,/\, \alpha) \cdot (t/D) = Ut \,/\, D \qquad [3.6]$$

Let us examine $Pi4$ in order to understand its physical meaning. The numerator is an indication of the material removal rate when the diameter of the workpiece is kept constant. It is also an indication of the rate of the thermal energy generated due to grinding of a known material (i.e. the material is kept the same in all operations), and depends on the specific energy of the workpiece material. Meanwhile, the denominator is the ability of the material to diffuse thermal energy. Evidently, for a specific material and when the diameter of the workpiece is kept constant, $Pi4$ is the ratio between the rate of thermal energy generated and the amount of thermal energy which the workpiece can diffuse per unit time. The value of the actual ratio (not just $Pi4$) should be kept very low by reducing the material removal rate (or changing the workpiece material), and must not be allowed to exceed unity.

In addition to the above-mentioned fundamental dimensionless numbers, there is an additional one (say $Pi5$) which indicates the ratio between the mechanical and the thermal strains, as follows:

$$Pi5 = \varepsilon_m \,/\, \varepsilon_T \qquad [3.7]$$

where: ε_m is the mechanical strain and ε_T is the thermal strain

Evidently, all those dimensionless relations have to be validated by experimental work, in order to determine the variables that have a sensible effect on the grinding problem, and also to identify the values of those variables that will yield an excellent surface finish and a trouble-free grinding operation.

3.4 Various factors affecting the grinding of composites

For surface grinding operations, the process parameters, i.e. the linear feed, the cross feed, and the depth of cut, have a direct influence on the roughness of the ground surface. Higher values of any of them would result in a higher material removal rate, and therefore, a higher rate of generation of thermal energy. Such a condition would, as previously mentioned, lead to a rough surface finish and possibly thermal damage. It has also to be borne in mind that the direction of cutting must be normal to the direction of the fibers in the surface layer in order to obtain best surface finish. This is because the specific energy of the composite is dependent upon the direction, and

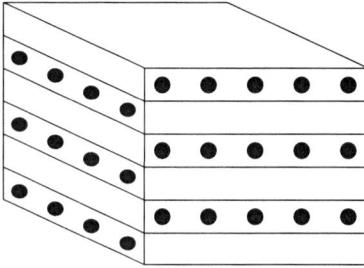

3.1 Block of composite material.

is at a minimum when the fibers are cut normal to their axis. The depth of cut plays another important role if it is large enough to make more than one layer of the reinforcing fibers be cut at the same time. Since each layer of fibers has a direction different from the subsequent one, as can be seen in Fig. 3.1, the cutting direction can be normal to one but not to both of them. Therefore, if the depth of cut is large enough to enable cutting of both of them at the same time, the cutting conditions will not be optimal for one of them, thus leading to increased roughness of the ground surface. Accordingly, it is always better to limit the depth of cut to remove only one layer at a time, if the surface quality (and not the material removal rate) is the major objective.

In the case of cylindrical grinding, the tool path, and hence the direction of cutting, is determined by the magnitudes of the axial feed rate and the preferal velocity of the workpiece, as previously explained. The direction of cutting (linear preferal velocity of the grinding wheel) is always normal to the tool path, as can be seen in Fig. 3.2. It should also be normal to the direction of the fibers in the outer layer, in order to minimize the amount of energy generated due to grinding, and accordingly produce a ground surface with low surface roughness. Again, larger depths of cut would remove more than one layer of fibers at the same time, and would therefore detrimentally influence the quality of the ground surface, because the direction of cutting cannot be normal to the fibers in both layers at the same time, as previously mentioned.

The type of the grinding wheel used is another factor that influences the quality of the ground surface of the composite. When comparing the performance of two types of grinding wheel, namely AZ46I8V32A and AZ60J8V32A (type numbers are according to the American Standard Marking System), the results consistently indicated that the wheel with a softer grade (AZ46I8V32A) was superior to the one with a harder grade (AZ60J8V32A), even though the latter had finer cutting edges (i.e. smaller size of abrasive particles). This can be explained by the fact that a composite

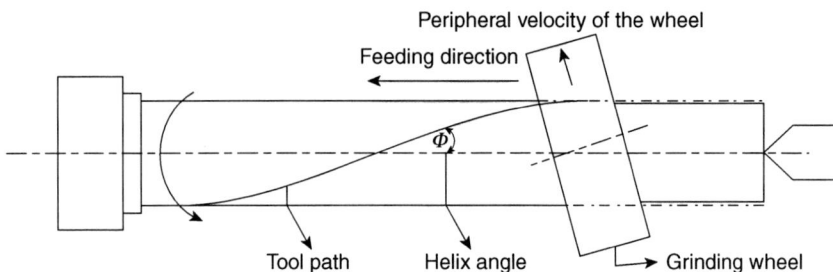

3.2 Tool path and grinding direction in cylindrical grinding.

is a heterogeneous system of materials. When the abrasive particles plow into the composite during the grinding operation, the high strength fibers act as obstacles in the way of those particles. The abrasive particles of the wheel with the softer grade would preferentially be pulled off rather than rupturing the surface and causing clogging of the grinding wheel. This is, in effect, a continuous process of self-dressing of the grinding wheel, thus ensuring that the surface of the grinding wheel always has sharp, fresh abrasive particles.

The physical and mechanical properties of the workpiece have a marked effect on the quality of the ground surface, and are, in turn, dependent upon the volumetric ratio of the fibers compared with the polymeric matrix. Higher fiber content would evidently result in higher strength of the composite, and accordingly higher specific energy. This would increase the rate of energy generated during the grinding operation, yielding thermal damage of the ground surface and/or higher values of surface roughness. Experiments have confirmed this trend when the fiber content increased from 0% (i.e. pure epoxy) to 50% during grinding of graphite–epoxy composites. The trend is, however, reversed when the fiber content is increased from 50% to 70%; in other words, the surface roughness decreased for the higher fiber content composite, under otherwise identical conditions of grinding. This can be explained by the fact that the 70–30 composite has a higher thermal diffusivity than the 50–50 or the 0–100 composites. This physical property enables the generated thermal energy to be diffused away from the spot being machined, thus preventing thermal damage and yielding better surface quality. This trend becomes more evident for larger depths of cut. On the other hand, it is not a good idea to decrease the polymer content to less than 30% (e.g. have a 90–10 composite with 90% fibers) because this would result in a weaker bond between the fibers, making it difficult to grind such a composite or even to manufacture it. We have to bear in mind, however, that the above mentioned optimal fiber-to-matrix ratio is only valid for graphite–epoxy composites.

It can be different for other composites, depending on the physical and mechanical properties of both reinforcing fibers and the matrix polymer in the composite.

3.5   Future trends

There have been some efforts to improve the process and eliminate thermal damage. These were based upon enhancing the thermal diffusivity of the composites. The approach adopted involved having some very thin metal wires side-by-side with the reinforcing fibers in each layer. Early results indicated effectiveness of the approach. The function of the metal wires was not to reinforce the polymeric matrix, but to transfer the heat generated during grinding quickly and efficiently away from the spot undergoing grinding.

Another approach proposed by some professional engineers in the composites industry involved using nickel-coated graphite fibers to reinforce the epoxy matrix. This had the effect of enhancing the diffusivity of the composite without causing any shortcomings. The only factor that has to be taken into consideration for this approach is the cost of the nickel-coated fibers as compared with that of ordinary graphite fibers.

3.6   Sources of further information

- SME Composite Manufacturing Tech Group
  http://www.sme.org/cgi-bin/getsmepg.pl?/communities/cma/cmahome.htm&&&sme&
- Society for the Advancement of Material and Process Engineering (SAMPE)
  http://www.sampe.org/
- American Composites Manufacturers Association (ACMA)
  http://www.acmanet.org/
- American Society for Composites
  http://www.ase-composites.org
- Canadian Association for Composite Structures and Materials
  http://www.cacsma.ca/
- Composites Division of the Society of Plastics Engineers (SPE)
  http://www.compositeshelp.com/
- Composites Europe, a special interest group of the Society of Plastics Engineers (SPE)
  http://www.4spe.org/technical-groups/special-interest-groups/composites-europe
- Automotive Composites Consortium of United States Council for Automotive Research (USCAR)

http://www.uscar.org/guest/teams/25/automotive-composites-consortium
- International AVK (Association of Reinforced Plastics)
  http://avk-tv.de (in German with English translation)
- European Composites Industry Association
  http://www.eucia.org
- SWEREA/SICOMP Swedish Institute of Composites
  http://www.sicomp@swerea.se

3.7   Bibliography

Desai, R. (2007), *Application of DOE Method to Study the Grinding of Graphite–Epoxy Composites*, Master's Thesis, University of Massachusetts Dartmouth, USA.

El Wakil, S.D. and Fares, G. (2006), 'The Grinding of Epoxy–Graphite Composites.' *The 9th International AVK Conference*, Essen, Germany, Vol. 1. A10, pp. 1–6.

El Wakil, S.D., Azab, N.A. (2008), 'Effect of Process Parameters on the Grinding of Polymeric Composites,' *13th European Conference on Composites Materials*, Stockholm, Sweden.

El Wakil, S.D. and Srinagesh, K. (2008), 'Effect of Physical and Mechanical Properties of Composites on their Grinding Characteristics.' In: DeWilde, W.P., Brebbia, C.A. (eds). High Performance Structures and Materials IV, 4th International Conference on Structures and Materials, Algarve, May 2008. *WIT Transactions on the Built Environment*, Vol. 97, pp. 149–155.

Eshgy, S. (1967), 'Thermal Aspects of the Abrasive Cut-off Operation.' *Journal of Engineering for Industry*, Vol. **66**, pp. 356–364.

Fares, G.F. (2006), *Grinding of Fiber-reinforced Polymeric Composites*, Master's Thesis, University of Massachusetts Dartmouth, USA.

Kim, J., and Lee, D.G. (2000), 'Grinding Characteristics of Carbon Fiber Epoxy Composite Hollow Shafts.' *Journal of Composite Materials*, Vol. **34**, No. 23, pp. 2016–2035.

Laoulache, R.N. and El Wakil, S.D. (2002), 'Grinding of Fiber-reinforced Polymeric Composites.' *Proceedings of the 6th International Conference on Production Engineering & Design for Development*, Vol. 1, pp. 134–142.

Lee, D.G. and Kim, P.J. (2000), 'Temperature Rise and Surface Roughness of Carbon Fiber Epoxy Composites During Cut-off Grinding.' *Journal of Composite Materials*, Vol. **34**, No. 23, pp. 2061–2080.

Park, K.Y., Lee, D.G., and Nakagawa, T. (1995), 'Mirror Surface Finish Grinding Characteristics and Mechanism of Carbon Fiber Reinforced Plastics.' *Journal of Materials Processing Technology*, Vol. **52**, pp. 386–398.

Snoeys, R., Lewen, K.U., Maris, M., and Peters, J. (1978), 'Thermally Induced Damage in Grinding.' *Annals of the CIRP*, Vol. **27**, No. 2, pp. 571–581.

4

Analysing cutting forces in machining processes for polymer-based composites

G. CAPRINO and A. LANGELLA,
University of Naples Federico II, Italy

Abstract: The cutting forces arising in machining fibre-reinforced plastics are considered. The mechanisms of chip formation and material damage in the orthogonal cutting of unidirectional composites are recalled; the trends of cutting forces as functions of the operating parameters, as highlighted by the experimental results available, are illustrated; a number of schemes for interpreting the cutting forces, taking into account or disregarding the tool flank action, are presented; and a number of analytical and numerical models, put forward in the literature, are critically reviewed. The topic of forces developing in drilling and milling, two of the most important machining operations in composite industry, is then examined. Here, certain tendencies identifying the main parameters affecting the cutting forces are discussed, along with models that aim to predict the cutting forces depending on tool geometry.

Key words: orthogonal cutting, drilling, milling, machining, composite materials.

4.1   Introduction

The use of fibre-reinforced plastic laminates, often containing graphite reinforcement, has increased tremendously over recent decades, driven, in particular, by the needs of the aeronautical industry. Although the production methods available for these materials mean that the final part can be created in its near-net shape, two basic machining operations, namely edge-trimming and drilling, are very often necessary to fulfil the design requirements.

In optimizing metal cutting, the primary objectives are to reduce working time and the amount of energy expended. The situation is very different in the case of polymer-based composites, the inhomogeneity, anisotropy, and laminar structure of which can give rise to extensive damage, impaired quality and unachieved dimensional tolerances, or even result in part rejection. The main goal for these materials becomes, therefore, to obtain a clean cut, by preserving tool sharpness and properly governing the cutting forces. To this end, the interaction phenomena that occur at the tool–workpiece interface, and lead to chip generation, must be thoroughly understood.

75

The simplest machining condition for revealing the basic mechanisms of material removal is orthogonal cutting, the results of which can be easily transferred to trimming, thereby providing fundamental information for any type of cutting operation based on conventional tools. Furthermore, a number of studies have shown that the objective of high cut quality is particularly hard to achieve where unidirectional composites are concerned,[1,2] since certain failure modes during cutting are suppressed in multidirectional laminates, due to the beneficial support provided by adjacent plies.[3,4] Consequently, most experimental research efforts have been devoted to orthogonal machining of unidirectional composites.[1-21] Their findings will be critically reviewed in Section 4.2 of this chapter, with reference to the mechanisms of chip generation, trends exhibited by cutting forces as a function of working parameters, and models proposed for their prediction.

Avoiding delamination is the fundamental objective when drilling composite laminates by conventional methods.[22,23] Since delamination, apart from the fact that it depends on interlaminar material properties, is mainly determined by thrust force,[24,25] the importance of theoretical models capable of correlating the cutting forces in drilling with tool geometry is apparent. This topic is dealt with in Section 4.3, where some of the models developed for the purpose are reviewed.

For many years, composite structures for industrial use were typically thin (just a few millimetres in thickness). With the application of this class of materials to primary aeronautical structures, the situation has changed rapidly, and thicknesses of up to 100 mm have gained practical interest. As a result, the dimensional tolerances necessary for connecting two mating surfaces have been tightened dramatically, and milling, which is examined in Section 4.4, has emerged as a critically important process for achieving the required level of precision.

Suggestions for future research and sources of further information are provided in Sections 4.5 and 4.6, respectively.

Finally, an appendix (Section 4.8) defines the symbols used in this chapter.

4.2   Orthogonal cutting of unidirectional composites

The structure of a unidirectional, continuous fibre composite lamina is schematically depicted in Fig. 4.1a. Apart from in the case of boron, the fibre diameter is too small (5–10 μm) to allow for a control of the filament location along the thickness during fabrication. Thus, with reference to the principal material axes in Fig. 4.1a (indicated as 1-, 2-, 3-axis), the fibre distribution in the 2–3 plane is random, and the lamina behaves as a transversely isotropic material.[26] In terms of machinability, this implies that the same cutting forces will be experienced for all the cutting planes passing along the 1-axis.

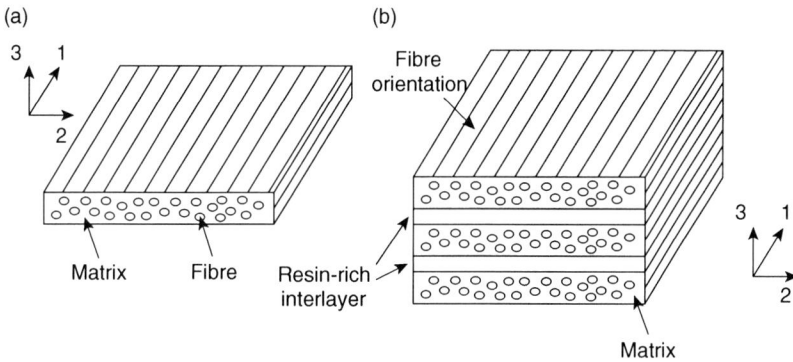

4.1 (a) A unidirectional lamina; (b) a unidirectional laminate.

A single lamina is too thin (about 0.13 mm) for most engineering applications. It is, therefore, common practice to stack and bond a suitable number of layers together to obtain a unidirectional laminate of appropriate thickness. In so doing, some differences in the mechanical properties between the 2- and 3-directions are often generated, due to the resin-rich interlayers (depicted on an exaggerated scale in Fig. 4.1b), meaning that transverse isotropy is lost, and the material becomes orthotropic. The presence of the resin-rich interlayers, being preferential sites for delamination, cannot be neglected in drilling, where avoiding workpiece damage is the main goal. Nevertheless, interlayers are not relevant when the cutting plane passes through the 3-axis. Since many of the laminates of engineering interest are typically thin, this condition is fulfilled by the major studies on orthogonal cutting, the results of which are discussed in this section.

4.2.1 Mechanisms of chip formation

Composites are brittle materials, exhibiting insignificant plastic deformation. Thus, machining them results in discontinuous chips, the thickness of which almost coincides with the depth of cut set. The chip formation mechanisms are highly dependent on fibre orientation (θ in Fig. 4.2), being negligibly affected, except for $\theta = 0°$, by other operating parameters, such as fibre and matrix type, tool angles, depth of cut, and cutting speed. Therefore, the evolution of chip development and detachment will be illustrated with reference to θ.

Fibre orientation $\theta = 0°$

Only for this angle, seemingly, is chip development influenced by the rake angle, γ. When $\gamma = 0°$, intense compression stresses are generated in the

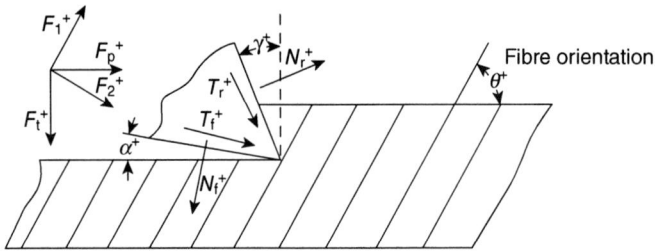

4.2 Force scheme and sign convention in the orthogonal cutting of a unidirectional composite.

work at the tool rake–material surface (Fig. 4.3a), resulting in brooming failures: multiple splits, randomly spaced within the chip thickness and the work width, form and propagate parallel to the fibre direction (Fig. 4.3b); the filaments formed are bent by the advancing tool, sliding along its rake face and occasionally following out-of-plane trajectories, until they are fractured (Fig. 4.3c).[20] This process is repeated to produce a new chip, which is typically irregular in shape and incoherent.[3,11,20]

The periodicity of fracture events in chip formation is also reflected in the signatures of the cutting forces. In Lopresto et al.,[20] where a precise correlation between the principal force and chip development was found, the maximum principal force was associated with brooming initiation, and the filament length was observed to be comparable to the depth of cut.

A chip formation mechanism completely different from that above is documented for positive rake angles (Fig. 4.4a). In this case, a Mode I load component that favours splitting along the cutting plane is provided. Under an appropriate cutting force, therefore, a longitudinal split originates and propagates some distance from the cutting edge (Fig. 4.4b); the forming chip, which takes on the approximate form of a cantilever beam, begins to slide along the tool rake, generating extensive fibre bridging, until it is broken in flexure (Fig. 4.4c). The advancing tool engages fresh material, and the process begins anew.

The width of a chip generated according to the mechanism depicted in Fig. 4.4 coincides with the work width, and its thickness is determined by the original longitudinal split (Fig. 4.4b), roughly following the cutting plane; the chip length is given by the distance between two subsequent bending failures undergone by the material removed. The latter increases as the depth of cut t increases: in Arola et al.,[3] where t was in the range 0.1 to 0.4 mm, chips 1–2 mm long were measured; a much greater chip length (about 7 mm) was recorded by Lopresto et al.,[20] setting $t = 0.8$ mm. The results obtained by Lopresto et al., also showed that, for large depths of cut, a second split, parallel to the original one and located approximately at the

(a) Fibre orientation (b) (c)

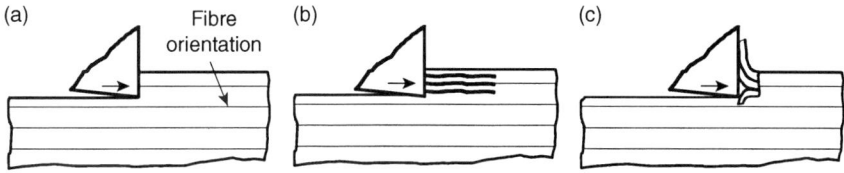

4.3 Chip formation in the case of fibre orientation $\theta = 0°$ and rake angle $\gamma = 0°$.

(a) Fibre orientation (b) (c)

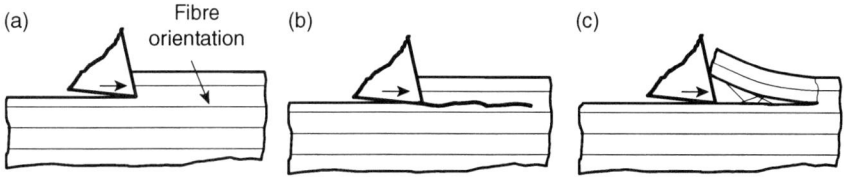

4.4 Chip formation in the case of fibre orientation $\theta = 0°$ and rake angle $\gamma > 0°$.

4.5 Chip formation in the case of fibre orientation $\theta = 0°$ and rake angle $\gamma > 0°$ (30°). Depth of cut $t = 0.8$ mm.[20] Material: CFRP.

mid-thickness of the forming chip (Fig. 4.5), can occur, probably because of the intense shear stresses arising during cantilever bending. It was found that the maximum principal force is experienced when the original split starts propagating.

Not enough experimental data are available to assess how chip formation progresses for a negative rake angle, a solution seldom used in composite machining. If $\gamma < 0°$, the load condition schematically depicted in Fig. 4.6, where F_{pr}, F_{tr} are the principal and thrust forces arising at the tool rake, is conceivably established. Therefore, the Mode I component, which is primarily responsible for the nucleation and propagation of the split in Fig. 4.4b when $\gamma > 0°$, is suppressed. On the other hand, a Mode II component which tends to propagate the longitudinal crack also exists, as well as the local

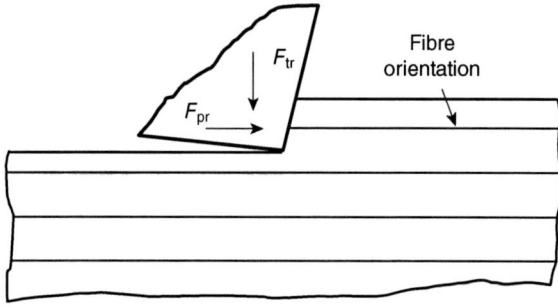

4.6 Force scheme in the case of a rake angle $\gamma < 0°$.

(a)   (b)   (c)

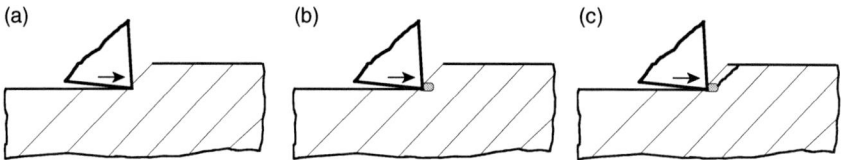

4.7 Chip formation in the case of fibre orientation $0° < \theta < 90°$.

compression stresses at the tool rake–material contact surface. Therefore, two possible failure modes, namely split propagation and brooming, can be foreseen. In both cases, the direction of the thrust force hinders chip sliding along the rake, so that the final chip detachment is probably provoked by buckling and subsequent failure in flexure of the entire chip, or of the individual filaments generated by brooming.

Fibre orientation $0° < \theta \leq 90°$

Irrespective of the fibre and matrix type, and of the machining parameters adopted, the chip formation proceeds according to the scheme shown in Fig. 4.7. The tool edge engages fresh material (Fig. 4.7a), creating an intense local compression stress state; because of this, a process zone develops (grey area in Fig. 4.7b), made of crushed fibres and microcracks in the matrix; from the process zone, a compression-induced shear crack (Fig. 4.7c) starts and quickly propagates along the fibre direction, giving rise to a blocky chip,[3,10,20] which is typically from a few microns up to 1 mm long. The maximum principal cutting force is found at the initiation of the shear crack.[20]

In Fig. 4.8a, a view of the forming chip during CFRP machining is shown:[20] the process zone and the shear crack resulting in a blocky chip are highlighted by an oval. The seemingly curled chip appearing in the figure is made

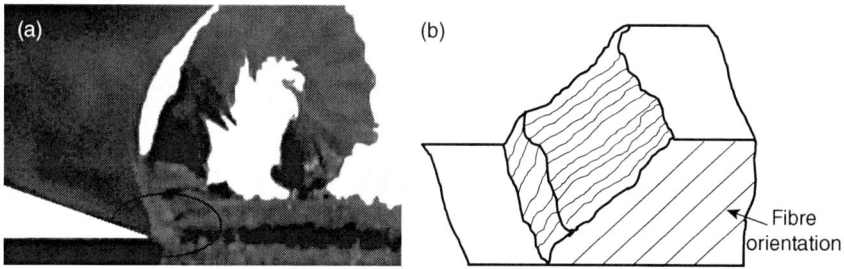

4.8 Chip formation for fibre orientation $0° < \theta < 90°$: (a) in-process chip formation[20] and (b) image of a freshly generated surface, obtained after a quick-stop experiment.[3] Material: CFRP.

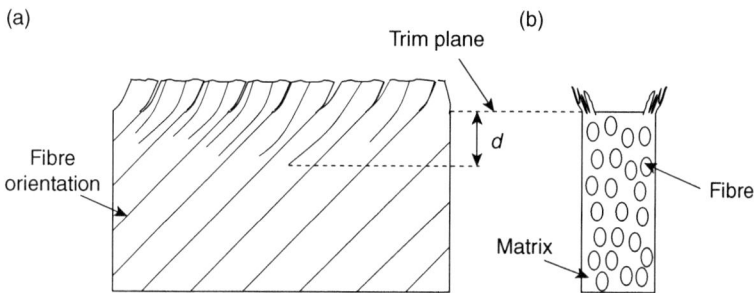

4.9 Main features of the damage observed when machining a unidirectional composite along a fibre direction approaching 90°: (a) side view; (b) cross view.

of many blocky chips, loosely adhering to each other.[2,20] The process zone and the surface created by the shear crack are schematically shown in Fig. 4.8b, as obtained by Arola et al.,[3] after a quick-stop experiment on CFRP.

With increasing θ, a new phenomenon occurs, regarding the work material below the trim plane:[16,19,20] cracks, possibly extending some millimetres into the workpiece, are formed along the fibre direction (Fig. 4.9a); at the same time, two thin layers, severely bending out-of-plane, passing uncut below the tool, and sliding against the flank are created at the unsupported edges of the work (Fig. 4.9b). Of course, this gives rise to a poor cut quality and also causes unacceptable sub-surface damage to the work. The portion of removed material in between the lateral layers is detached according to the laws described above, with the formation of shear-induced blocky chips.

Puw and Hocheng,[16] in studying orthogonal cutting perpendicular to the fibre axis, observed that chip formation can occur by bending: the material in front of the tool is bent until the resulting stress overcomes the material strength, causing failure. Due to this mechanism, the point of chip failure can be found below the cutting plane, and the chip thickness becomes larger than the depth of cut. The authors proposed a simple mechanical model to correlate cutting forces, chip length, and chip thickness.

Fibre orientation $90° < \theta < 180°$

Establishing the mechanisms of chip formation beyond $\theta = 90°$, where the tool moves against the reinforcing fibres, is very difficult. The nature of the chip generated, made of small dusty particles, renders it hard not only to identify the fracture sequences, but also to measure chip size, even when sophisticated methods of analysis, such as macrochip and quick-stop techniques, are used.[3] A further complication comes from the uncut layers generated at the unsupported sample edges, which undergo severe out-of-plane displacement, thereby causing sub-surface damage that penetrates deeply into the work. In Lopresto *et al.*,[20] the volume of uncut material was found to be dependent on both rake angle and fibre orientation; the worst working conditions were recorded for $\gamma = 0°$ and $\theta = 120°$, where almost all the material to be removed passed uncut under the cutting edge; furthermore, the length of the sub-surface damage (*d* in Fig. 4.9) increased greatly as the depth of cut increased, reaching about 9 mm[19] for $t = 0.3$ mm.

4.2.2 Experimental evidence

In an instrumented orthogonal cutting test, only the principal and transverse cutting forces (indicated by F_p and F_t in Fig. 4.2) are commonly measured, through a two-channel dynamometer. Both these components are affected by high-frequency oscillations, as anticipated from the fragmented nature of the chips generated during machining. In an attempt to correlate the force fluctuations with chip dimensions, Arola and co-workers[3] inspected the force signatures using a fast Fourier transform analysis. The results were inconclusive, since no dominant spectrum was revealed by the examination of either the principal or the thrust force. The authors attributed this to the rate of data acquisition of the dynamometry employed, which was too low compared with the force signal frequency. In fact, Lopresto *et al.*,[20] using a particularly low cutting speed (10 mm/min), were able to associate precisely the principal force evolution within a single fluctuation with the progression of chip detachment, at least in the range $0° \geq \theta < 90°$.

A comparison of the force data available in the literature on a quantitative basis is almost impossible, due to the multiplicity of operating and material parameters affecting them. Nevertheless, from their analysis, a number of trends suggesting the appropriate tool geometry and depth of cut can be identified.

Some authors have observed that cutting forces are quite insensitive to cutting speed V.[6,11,15] However, in carrying out orthogonal turning tests on CFRP discs 100 mm in diameter,[27] the cutting forces underwent a considerable decrease as V increased from 46 to 245 rev/min. Hocheng and Leu,[28] in commenting on the results of turning tests on CFRP tubes, hypothesized the existence of a critical level for V: below this threshold, the increase in strain rate results in a more brittle behaviour of the composite, and the cutting forces tend to decrease; beyond it, the effect of heating prevails, and more energy is required to machine the material.

In general, for a given fibre orientation, an increase in the tool relief angle results in a decrease in the thrust force, negligibly affecting the principal force.[11,12]

Using larger rake angles yields lower principal cutting forces.[2,6,11,13,21] However, quite recently Nayak et al.[2] found that, beyond $\gamma = 30°$, the principal cutting force, independently of fibre orientation, starts rising, so that 30° is the optimum tool rake angle, when the aim is to minimize F_p.

The effect of γ on the thrust force is quite controversial. Koplev and co-workers[6] did not find a clear trend of F_t as a function of the rake angle. Wang et al.[11] noted an increase in F_t with increasing γ up to $\theta = 60°$; in the range $\theta = 75°$ to 90°, the tool rake angle had less significant influence on cutting forces. Caprino et al.,[17] cutting CFRP specimens along the fibre direction, recorded a considerable influence of γ on the thrust force, which decreased with increasing rake angle. The rate of decrease was higher, the larger the depth of cut adopted.

In the authors' opinion, one of the topics requiring deeper investigation is the relationship occurring between cutting forces and depth of cut. According to many researchers,[6,11,17] the principal cutting force grows quite linearly with t. Should this be the case, the concept of 'unit cutting force' (also known as 'specific energy'), defined as the ratio of F_p to chip transverse area, could be useful for easy prediction of the principal cutting force. Unfortunately, evidence exists that, besides being affected by fibre orientation, the unit cutting force is also strongly sensitive to cutting depth[2,17] and rake angle,[17] decreasing as both t and γ increase. This phenomenon, which is also well established in metal machining, and is known as 'size effect', is graphically represented by the positive intercept of the straight line fitting the experimental points in the F_p-t diagram (Fig. 4.10). A physical interpretation of the size effect was provided by Nayak et al.,[2] who performed cutting tests at 0.1 and 0.2 mm depth of cut, carefully controlling the tool edge radius: both the principal and thrust forces increased with increasing edge radius, the effect of which diminished for the greater depth of cut. The authors commented that the size effect is likely to be correlated with a ploughing effect at the tool flank surface, leading to higher cutting forces. The size effect was also revealed by the cutting data generated by Santo

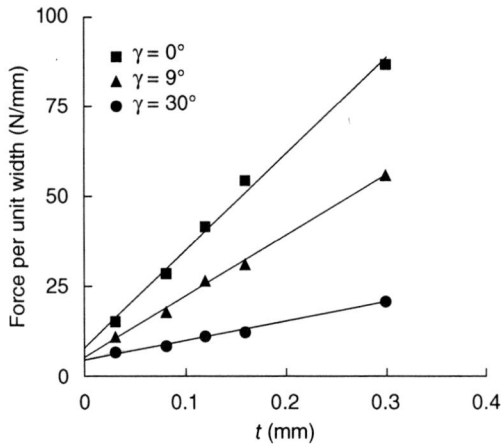

4.10 Principal cutting force per unit width versus depth of cut, t, for different rake angles γ. Fibre orientation θ = 0°. Material: CFRP. Data from reference 17.

et al.,[1] who trimmed unidirectional CFRP pultruded bars having a relatively low fibre volume fraction V_f (30%). In these experiments, the linear relationship correlating F_p and t was found only when $θ = 0°$; for all other orientations, the F_p-t curve was non-linear, exhibiting a decreasing slope with increasing depth of cut.

Further evidence of the importance of edge radius in determining cutting forces, albeit indirect, is provided by their rapid variation with tool wear. In references 6 and 14, where CFRP specimens were trimmed along the fibre direction using HSS tools, qualitatively similar results were obtained, showing that the thrust force is much more sensitive than the principal force to tool wear. In both papers, it was implied that the forces arising at the tool flank, rather than those occurring at the rake, are responsible for this behaviour.

Of course, the cutting forces are very sensitive to fibre orientation, reflecting the strong anisotropy of unidirectional composites. Many authors[1,2,7,11,18] found a maximum in the principal force at 90°, with local minima eventually occurring between 15° and 30°. However, depending on the work material, rake angle, and depth of cut used, the maximum F_p has sometimes been observed at 60°[1,7] or 120°.[1] Some experimental data selected from Santo et al.,[1] illustrating the general trend of cutting forces per unit work width with varying θ, are plotted in Fig. 4.11. Interestingly, with the exception of $θ = 0°$, the principal force (Fig. 4.11a) is lower than the thrust force (Fig. 4.11b) at low fibre orientations;[10,11,18] the contrary occurs for sufficiently large θ values.

(a)

(b)

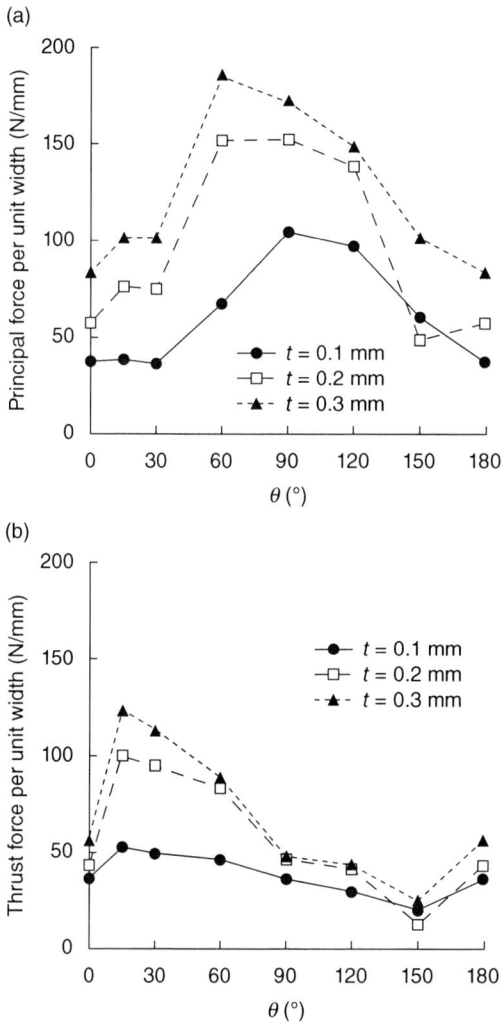

4.11 (a) Principal cutting force per unit specimen width and (b) thrust force per unit width against fibre orientation, θ. Material: CFRP. Data from reference 1.

Typical unit cutting forces experienced in cutting unidirectional composites cover the range from 500 to 1000 MPa for the most difficult fibre orientations, and 100 to 250 MPa for those that are easiest to machine. These values seem to be negligibly influenced by the degree of cure of the matrix.[21] Indeed, a parameter critically affecting them should be the fibre content, although no systematic data supporting this statement are yet available.

4.2.3 Schemes for the interpretation of cutting forces

In order to understand the tool-material interaction and chip formation mechanisms, there is a need to share the principal and thrust cutting forces between the rake face, which actually contributes to chip development and detachment, and flank, which is mainly responsible for parasitic effects. In principle, this task can be easily accomplished, if the validity of the Amontons' law of friction:[29]

$$T = \eta N \qquad [4.1]$$

correlating the force of friction T with the normal force N acting on two sliding surfaces, is accepted. In Eq. 4.1, η is the coefficient of dynamic friction, which is hypothesized to be a constant, uniquely dependent on the couple of bodies coming into contact.

Applying Eq. 4.1 to the normal and tangential forces developing at the tool rake and flank during machining yields:

$$T_r = \eta_r N_r \qquad [4.2]$$

$$T_f = \eta_f N_f \qquad [4.3]$$

where the indexes r and f refer to the rake and flank, respectively.

From simple force equilibrium along the horizontal and vertical directions, and adopting the symbology and sign convention in Fig. 4.2, the following relationships are obtained for N_r and N_f:

$$N_r = \frac{F_t(\eta_f \cos\alpha - \sin\alpha) - F_p(\eta_f \sin\alpha + \cos\alpha)}{(\eta_r \cos\gamma - \sin\gamma)(\eta_f \cos\alpha - \sin\alpha) - (\eta_r \sin\gamma + \cos\gamma)(\eta_f \sin\alpha + \cos\alpha)} \qquad [4.4]$$

$$N_f = \frac{F_p(\eta_r \cos\gamma - \sin\gamma) - F_t(\eta_r \sin\gamma + \cos\gamma)}{(\eta_r \cos\gamma - \sin\gamma)(\eta_f \cos\alpha - \sin\alpha) - (\eta_r \sin\gamma + \cos\gamma)(\eta_f \sin\alpha + \cos\alpha)} \qquad [4.5]$$

The application of Eqs. 4.2 to 4.5 seems to be straightforward in providing the forces developing at the tool rake and flank. However, some serious drawbacks, deriving from the actual values of the coefficients of friction to adopt, arise in practice.

Since a unidirectional composite is anisotropic, its coefficient of friction is expected to vary with fibre orientation. For this reason, two different η values have been assumed in Eqs. 4.2 and 4.3 for the rake–material and flank–material couples.

Direct evidence of the dependence of η on the reinforcement orientation has been given in references 30 and 31. In reference 30, CFRP specimens cut from graphite/vinyl ester pultruded bars at fibre angles $\theta = 0°–180°$ were characterized by sliding a flat-ended HSS punch against their surface under controlled closing forces. In reference 31, the coefficient of friction was measured in the $\theta = 0°–90°$ range by classical pin-on-disk tests on GFRP samples.

The results obtained in references 30 and 31 are plotted in Fig. 4.12, where η is shown against θ. From them, the coefficient of dynamic friction, irrespective of fibre type, monotonically increases up to $\theta = 90°–120°$, decreasing thereafter. Furthermore, it has been shown in reference 30 that the linear relationship postulated in Eq. 4.1 is violated beyond a given threshold of the closing pressure, p. The sensitivity of η to p was different for the different fibre orientations, and can be seen in Fig. 4.12, where the open circles refer to a vanishingly small pressure, and the full ones to $p = 30$ N/mm^2. It is important to realize that the latter value is a great deal lower than the pressure actually occurring in a cutting operation,[32] so that larger variations in the coefficient of dynamic friction might occur during machining.

A further difficulty in applying Eqs. 4.2 to 4.5 is the fact that, while the tool flank certainly slides along the newly generated work surface, the same is not obvious at the rake–material interface. As discussed previously (Fig. 4.6), no sliding at all between the forming chip and the tool is reasonably expected under appropriate working conditions, so the coefficient of static friction, which is generally higher than η_r, should more correctly be adopted in some calculations.

Of course, both forces at the rake and flank contribute to F_p and F_t (Fig. 4.2) according to the relationships:

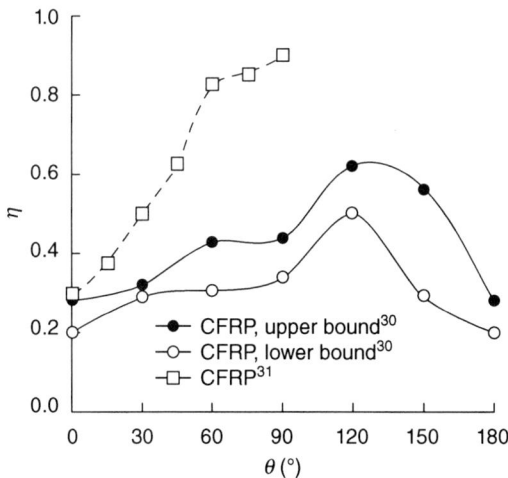

4.12 Coefficient of dynamic friction, η, against fibre orientation, θ

$$F_p = F_{pr} + F_{pf} \tag{4.6}$$

$$F_t = F_{tr} + F_{tf} \tag{4.7}$$

in which, from equilibrium:

$$F_{pr} = N_r \cos\gamma + T_r \sin\gamma \tag{4.8}$$

$$F_{pf} = T_f \cos\alpha - N_f \sin\alpha \tag{4.9}$$

$$F_{tr} = T_r \cos\gamma - N_r \sin\gamma \tag{4.10}$$

$$F_{tf} = N_f \cos\alpha + T_f \sin\alpha \tag{4.11}$$

As noted previously, the characterization of a unidirectional composite is usually performed along its principal material coordinates. It is useful, therefore, to refer also the cutting forces to the same directions as well, thereby obtaining (Fig. 4.2):

$$F_1 = F_p \cos\theta - F_t \sin\theta \tag{4.12}$$

$$F_2 = F_p \sin\theta + F_t \cos\theta \tag{4.13}$$

In metal cutting, many classical models (e.g. Ernst and Merchant, Lee and Shaffer, etc.[33]) have been put forward, all of them assuming that no tool–material interaction takes place at the tool flank, so that the forces measured during cutting develop only at the tool rake–chip interface. Of course, in this hypothesis $N_f = T_f = F_{pf} = F_{tf} = 0$, and Eq. 4.4 becomes:

$$N_r^* = \frac{F_p}{\eta_r^* \sin\gamma + \cos\gamma} = \frac{F_t}{\eta_r^* \cos\gamma - \sin\gamma} \tag{4.14}$$

From Eq. 4.14, we get:

$$\eta_r^* = \frac{F_p tg\gamma + F_t}{F_p - F_t tg\gamma} \tag{4.15}$$

which is often utilized in interpreting metal cutting data. In Eqs. 4.14 and 4.15, the asterisk has been adopted to distinguish the approximate N_r, η_r values from the exact ones.

The hypothesis of no contact forces at the tool flank has been used by many authors in composite machining too,[7,10,34,35] despite certain experimental evidence to the contrary: first of all, thrust force, in particular, has been found to be significantly influenced by the flank angle α, generally decreasing as the latter increases;[6,12] secondly, similarly to metals, intense flank wear is also found in composite cutting.[14,36] Both these observations suggest that a significant interaction between tool flank and work material occurs.

It is important to note that the previous force schemes hold in the ideal case of a perfectly sharp tool. The edge of an actual tool is always characterized by a round nose, which results in a deviation of the actual geometry from the ideal one. Of course, the effect of the edge radius is expected to be the more evident, the lower the depth of cut.[2]

4.2.4 Predictive models

The models proposed for predicting cutting forces can be divided into two categories: (i) analytical models based on chip formation mechanics; (ii) numerical simulations relying on finite element (FE) analyses.

To the authors' knowledge, all the closed-form models available at this time try to calculate the cutting forces in the range of fibre orientations between 0° and 90°; beyond 90°, the mechanisms of chip formation become too complex for any reasonable hope that simple analytical solutions will capture the major features of tool–material interaction.

An attempt to estimate cutting forces using a closed-form model was made by Takeyama and Iijima,[7] where unidirectional GFRP specimens were cut using various tools, characterized by different rake angles. The authors applied the classical Merchant's model, developed for metals, while recognizing that the anisotropy of composites renders their shear strength dependent on the orientation:

$$F_{\mathrm{p}} = \frac{w_{\mathrm{w}} \cdot t \cdot \tau(\theta') \cdot \cos(\beta - \gamma)}{\cos(\phi + \beta - \gamma)\sin\phi} \qquad [4.16]$$

$$F_{\mathrm{t}} = \frac{w_{\mathrm{w}} \cdot t \cdot \tau(\theta') \cdot \sin(\beta - \gamma)}{\cos(\phi + \beta - \gamma)\sin\phi} \qquad [4.17]$$

In Eqs. 4.16 and 4.17, w_{w} is the workpiece width, β the friction angle at the rake face, ϕ the shear angle, and $\theta' = \theta - \phi$ the angle between fibre orientation and shear plane, all quantities which are known, or can be easily calculated.[33] To find the shear strength $\tau(\theta')$ of the composite along the shear plane, Takeyama and Iijima[7] carried out appropriate characterization tests. Limiting their attention to a single depth of cut, the computed and the experimental data were found to agree to a fairly large extent.

Observing that, for $0° \le \theta \le 90°$, chip detachment occurs for compression-induced shear along the fibre direction, Bhatnagar *et al.*[10] postulated that the shear angle coincides with the fibre orientation, so that $\theta = \phi$, and Eqs. 4.16 and 4.17 become:

$$F_{\mathrm{p}} = \frac{w_{\mathrm{w}} \cdot t \cdot \tau(\theta) \cdot \cos(\beta - \gamma)}{\cos(\theta + \beta - \gamma)\sin\theta} \qquad [4.18]$$

$$F_{\mathrm{t}} = \frac{w_{\mathrm{w}} \cdot t \cdot \tau(\theta) \cdot \sin(\beta - \gamma)}{\cos(\theta + \beta - \gamma)\sin\theta} \qquad [4.19]$$

To assess their model, Bhatnagar *et al.*[10] performed Iosipescu shear tests on CFRP samples, determining $\tau(\theta)$. Cutting tests were then conducted at various θ values, setting $t = 0.25$ mm, and the cutting forces were recorded; from these, $\tau(\theta)$ was calculated, using Eqs. 4.18 and 4.19. Finally, the shear strengths directly measured from the Iosipescu tests and those drawn from the cutting tests were compared. Reasonable agreement was found up to 60° fibre orientation; however, the divergence was remarkable beyond this angle.

According to Eqs. 4.16–4.19, the trend of the cutting forces with increasing t should be represented by a straight line passing through the origin. As noted previously, this is likely not to be the case, due to the size effect. Furthermore, one of the basic assumptions in developing these solutions is the absence of forces at the tool flank, which can be misleading in interpreting cutting force data, as will be shown later on.

The principal and thrust forces per unit work width measured by Bhatnagar and co-workers[10] are plotted in Figs. 4.13a and 4.13b, respectively, against θ (full symbols). The open symbols were obtained by applying Eqs. 4.4 to 4.11 to predict F_{pr}, F_{tr}, adopting three different hypotheses for the coefficient of friction: (i) $\eta = 0.2$, independent of θ (open triangles); (ii) $\eta = 0.6$, independent of θ (open squares); (iii) η variable according to all the CFRP data in Fig. 4.12, fitted by a fourth-order polynomial (open circles). Clearly, disregarding the role played by the tool flank can result in sizeable inaccuracies, especially where the thrust force is concerned. Indeed, whatever the η value, a major portion of F_p is attributable to the flank action for some fibre orientations, and the rake action becomes negligible if η is low.

A force scheme accounting for the forces developing at the tool flank was proposed in reference 12. From the analysis of tool wear, the authors concluded that the tangential force at rake (T_r in Fig. 4.2) is negligible. Putting $T_r = 0$ in the force scheme, and using X to refer to the unit cutting force defined as:

$$X = \frac{F_{pr}}{t \cdot w_w} \qquad [4.20]$$

the following relationship was found from simple trigonometric considerations:

$$X^* = X \cdot (1 + \eta_f \cdot \tan \gamma) + \eta_f \cdot \frac{F_t}{t \cdot w_w} \qquad [4.21]$$

where $X^* = F_p/tw_w$ is the unit cutting force defined as usual in metal cutting.

The tests performed in references 12 and 17, where the tool angles and the depth of cut were varied, concerned only CFRP and GFRP cut along the fibre direction ($\theta = 0°$). Attributing the experimental cutting forces to

(a)

(b)

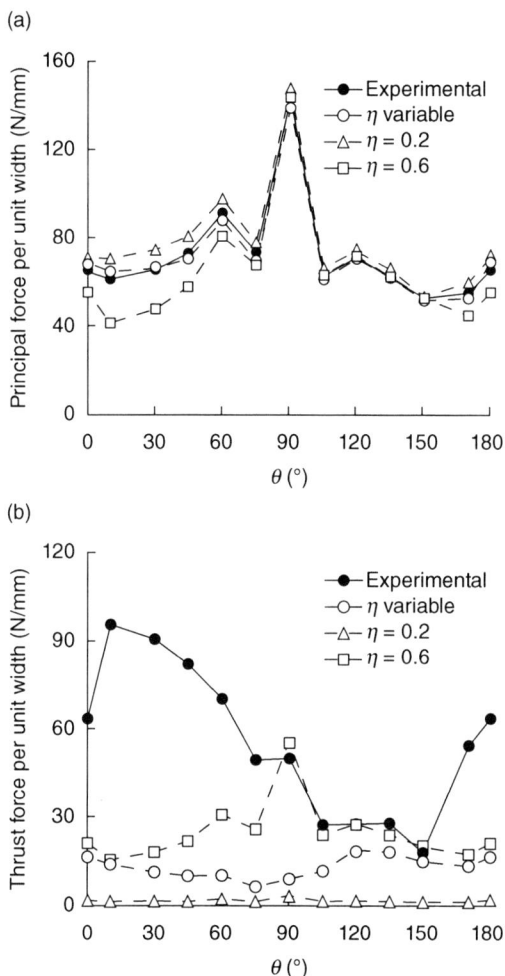

4.13 (a) Principal and (b) thrust force per unit specimen width against fibre orientation, θ. Material: CFRP. Experimental data adapted from reference 10.

the rake face only, the apparent coefficient of friction η_f^* as well as X^* were strongly dependent on t, and clearly decreased as the depth of cut increased. When Eq. 4.21 was used, η_f became a constant, independent of both the tool geometry and depth of cut, whereas X was affected only by the rake angle, as expected. Some indication that Eq. 4.21 might be also applicable for $\theta \neq 0°$ was given in reference 27, where the model was used to interpret cutting data deriving from CFRP machined over the entire range from 0° to 180° of fibre orientations

An interesting force prediction model was developed by Zhang *et al.*[18] The authors divided the tool into three distinct regions, consisting of rake (chipping region), nose (pressing region), and flank (bouncing region), and calculated the associated force portions separately. The forces arising at the chipping region were evaluated according to classical metal machining theory; the action of the tool nose was assimilated to a cylindrical indenter pressing on the work material; the contact forces at the tool flank were attributed to the bouncing back of the workpiece, and treated accordingly. The equations obtained are not given here for the sake of brevity. The model predictions were in reasonable agreement with experiments, carried out on two types of CFRP, in which tool geometry, depth of cut, and fibre orientation in the range 0° to 90° were varied. The maximum error in prediction, which concerned the thrust force, was 37%.

An initial attempt to predict cutting forces by FE analysis was performed by Arola and Ramulu.[37] To simplify the problem, a two-dimensional FE model, relying heavily on experimental observations of chip formation mechanisms, was constructed; on the basis of this, the tool was shaped as a rigid body consisting of rake face, flank, tool nose, and flank wear land. Meshing was carried out by predefining the primary fracture on the trim plane, which coincided with the flank surface of the cutting tool, as well as the secondary fracture, which occurred along the fibre direction. The cutting forces were evaluated from the analysis of the forces resulting in primary and secondary fractures. The agreement with experimental principal forces, generated while working on CFRP over the range $\theta = 0°–90°$, was satisfactory when the Tsai–Hill criterion was adopted to predict secondary fracture. However, thrust force predictions were an order of magnitude lower than the experimental values, probably due to inaccuracies implicit in the hypothesis of a primary fracture developing along the cutting plane.

Using 2-D elements, Nayak and co-workers[31] carried out two FE analyses within $\theta = 0°–90°$, with the aim of evaluating not only cutting forces, but also the extent of sub-surface damage occurring in GFRP specimens. In the first model (macro-model), the material was considered to be locally homogeneous and orthotropic; in the second (micro-model), fibre and matrix were meshed as distinct physical entities. The influence of fibre orientation on η_f was taken into account (Fig. 4.12) in the macro-model, where η_r was assumed to be nil, as suggested in reference 12; in the micro-model, the effect of friction was disregarded. Both the models quite satisfactorily predicted the principal forces, with a maximum error of about 10%. However, the macro-model was unable even to correctly follow the trend of the thrust forces. The latter were well represented by the micro-model, which also provided useful indications on the mechanisms of chip formation. By using the macro-model, the trend of sub-surface damage was captured, but the calculated values were much lower than the experimental ones when $\theta = 60°$ was exceeded.

Recently, other authors[32] have used the micro-mechanical approach to evaluate cutting forces in orthogonal cutting by FE. To reduce computational time, fibre and matrix were modelled separately only in the portion of work material adjacent to the cutting tool. The fibre material was assumed to be linearly elastic, and the matrix elasto-plastic, with a modulus degrading linearly to failure. Portions of the work away from the tool were modelled as an equivalent homogeneous material. Zero-thickness cohesive elements were adopted to implement damage initiation and evolution. The prediction of principal and thrust forces was quite satisfactory for both GFRP and CFRP, substantially confirming the results of Nayak *et al.*[31] A similar study, in which an explicit dynamic formulation with mass scaling was developed to ease contact problem, was carried out later,[38] and possible mechanisms driving chip formation and sub-surface damage were revealed. The micro-mechanical method, together with cohesive zone modelling, was used also by Dandekar and Shin,[39] where the main scope was predicting debonding at the fibre–matrix interface and fibre pull-out. From the comparison of simulation results with experimental measurements, the analysis was effective in predicting not only damage, but also cutting forces.

Lasri *et al.*[40] attempted to overcome the inability of macro-mechanical simulations to effectively predict thrust force by better modelling the bouncing back effect noted by Zhang *et al.* To this end, the fracture path was not predetermined, and a selective stiffness degradation concept was applied to the work material. According to the numerical results, fibre breakage generally occurred at a plane elevated from the flank plane. Despite this, only the computed principal force agreed reasonably with the experimental one, while the thrust force was an order of magnitude lower than in experiments. Nevertheless, in reference 41, macro-mechanics was able to predict sub-surface damage effectively, using an adaptive mesh technique coupled with dynamic explicit elements.

All the previous FE analyses were quasi-static, focusing on a very short machining time. Santiuste and co-workers,[42] using a dynamic explicit analysis and modelling the material as homogeneous, performed FE simulations following the cutting process until a steady state condition was reached. The work material properties were degraded according to suitable criteria, and differed for GFRP and CFRP. The principal cutting force underwent an increase, with evident oscillations, up to the detachment of the first chip; after that, the mean force decreased, until the steady state was achieved. The authors concluded that, if the analysis is limited to the formation of the first chip, a cutting force in excess of the experimental one (generally measured under steady state conditions) is predicted. From the comparison with experiments, limited to CFRP, the principal force was well calculated by the model; the thrust force was lower than that measured, but showed better accuracy when compared with numerical results in the literature.

Finally, in reference 43, the discrete element method (DEM) was proposed as a valid alternative for the simulation of CFRP cutting. DEM simulations were performed for $\theta = 0°, 45°, 90°, 135°$, and the major phenomena resulting in chip formation and sub-surface damage were reconstructed. In addition, the correlation between predicted and experimental cutting forces was encouraging.

Further evidence of the usefulness of FE in studying the machinability of composite materials is given in references 44–47. The authors considered a particular system made of polycarbonate charged with randomly arranged[44–46] or aligned[47] carbon nanotubes (CNTs) (interesting for microfluidic devices) and developed a microstructure-based FE model to simulate its response to machining operations. By means of this model, a number of different features, including cutting forces, chip morphology and thickness, as well as the possible formation of adiabatic shear bands in the matrix, were captured for various CNT contents.

4.3   Drilling

Traditionally, even in studying metals, the conditions occurring during drilling have been considered too complex to be approached by taking into account the actual point-by-point variation of the tool geometry. Thus, extensive experimental campaigns, revealing the trends of the forces involved in this operation, have been carried out on one side. On the other side, simplified approaches, based on 'mean values' of the angles characterizing the cutting edges, have been developed. Although these methods are useful for standard tool geometries and traditional materials, and provide information on the expected thrust force and torque, they are intrinsically weak when new materials have to be drilled, or the performances of completely different tool geometries must be evaluated.

Over recent decades, some models devoted to thrust force and torque calculation in drilling metals have been proposed to fill the existing gap.[48–54] Generally, they refer to conventional twist drills, but in principle their application to other tool geometries would be straightforward, since the edges of the drill are analysed by dividing them into elementary cutting edges.

In the first part of this section, some experimental results, which identify the main tendencies of drilling forces as a function of the operating parameters, are discussed. Subsequently, a model proposed for metals,[55,56] the philosophy of which was also adapted to composite materials, is recalled. Finally, models specifically developed for composites,[57,58] relying on mechanistic relationships, are presented.

4.3.1   Experimental evidence

Thrust force and torque have an important effect on the quality of a machined hole.[24,59] In particular, the thrust force is responsible for delamination, possibly occurring at both the entry and at the exit side of a composite laminate, so that lowering its value is of paramount importance for a successful drilling operation.

From the experimental results available, thrust force and torque increase with increasing feed rate,[60–69] and are negligibly influenced by the nature of the matrix material.[65] This evidence substantially agrees with the information drawn from orthogonal cutting, as discussed previously. Lower fibre volume ratios result in lower thrust forces and torques.[63] As for orthogonal cutting, the effect of cutting speed[63] is contradictory, since increasing this parameter was beneficial in reference 63, whereas it did not significantly alter the cutting forces in references 61 and 70.

The thrust force can be substantially influenced by the chisel edge.[57,58,64] The effect of this parameter was highlighted by using a pilot hole, which meant that chisel edge action during drilling was eliminated. The analyses performed demonstrated that, in some cases, the chisel edge can account for up to 80% of the total thrust force, and that its contribution increases with increasing feed rate. Abrão *et al.*[59] showed (Fig. 4.14) the influence of feed rate on the thrust force (F) and torque (C), using two specific cutting coefficients, k_F and k_C, defined as:

$$k_F = \frac{2F}{f \cdot D} \qquad\qquad [4.22]$$

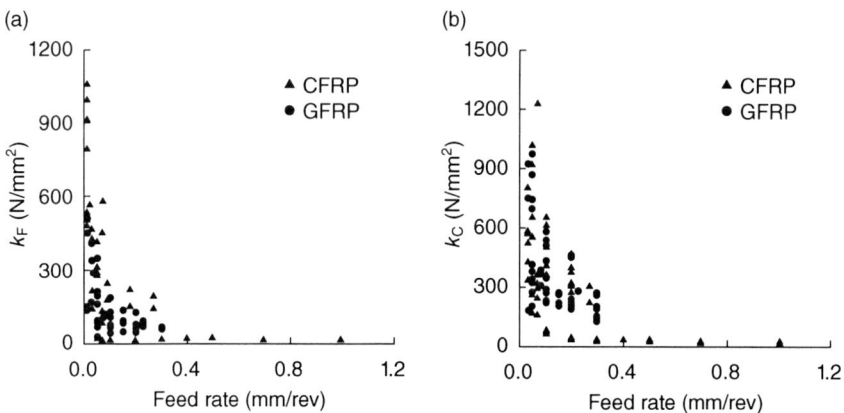

4.14 Influence of the feed rate on the specific cutting coefficients: (a) for thrust force, k_F, and (b) for torque, k_C. Data from reference 53.

$$k_C = \frac{8C}{f \cdot D^2}$$
[4.23]

where F is the thrust force, C the torque, f the feed rate, and D the diameter of drill.

Clearly, both coefficients, independently of the material considered, decrease notably as f increases, indicating that the machinability is improved, in the sense that lower energies are expended. Unfortunately, increasing the feed rate also results in greater damage and poorer residual properties of the workpiece.[71,72]

4.3.2 Cutting model for metal drilling

In order to predict thrust and torque in metal drilling, Elhachimi et al.[55] considered a generic element of the cutting lips (Fig. 4.15), located along the cutting edge at a distance r from the drill axis and having length dl, and the associated elemental thrust force and torque, indicated by dF_l and dC_l, respectively, hereafter. Of course, if dF_l and dC_l are known for each r value the contribution of the lips to total thrust force and torque can be calculated by integration along the entire cutting lip. Of course, the same reasoning applies to the chisel edge, so that

$$F = F_l + F_c$$
[4.24]

$$C = C_l + C_c$$
[4.25]

where the indexes l and c designate lips and chisel edge, respectively.

To find the elemental force and torque at the lips, the shear zone model in oblique cutting developed by Oxley[73] was incorporated into the analysis. The cutting forces at the chisel edge were modelled by orthogonal cutting, while the indentation occurring close to the tool axis was disregarded. Integrating over the cutting lips, the following formulae, which are valid when both the entire lip length and the chisel edge are engaged in the workpiece, were obtained:

$$F_l = 2 \int_{D'/2}^{D/2} k_{AB} \frac{f \sin p \cos \mu}{2 \sin \phi_n \cos \alpha_n} (\sin(\lambda_n - \gamma_s - \mu) \sin p - \cos p) \frac{r}{(r^2 - w^2)^{\frac{1}{2}}} dr$$
[4.26]

$$C_l = 2 \int_{D'/2}^{D/2} k_{AB} \frac{f \sin p \cos \mu}{2 \sin \phi_n \cos \alpha_n} (\sin(\theta_n - \gamma_s - i)) \frac{r^2}{(r^2 - w^2)^{\frac{1}{2}}} dr$$
[4.27]

$$F_c = 2 \int_{r_o}^{D'/2} \frac{\cos(\phi_d - \gamma_d)}{\cos(\phi_d + \lambda_d - \gamma_d)} \frac{f k_{AB} \cos \gamma_f}{2 \sin \phi_d} (\cos \gamma_f - \tan(\phi_d - \gamma_d) \sin \gamma_f) dr$$
[4.28]

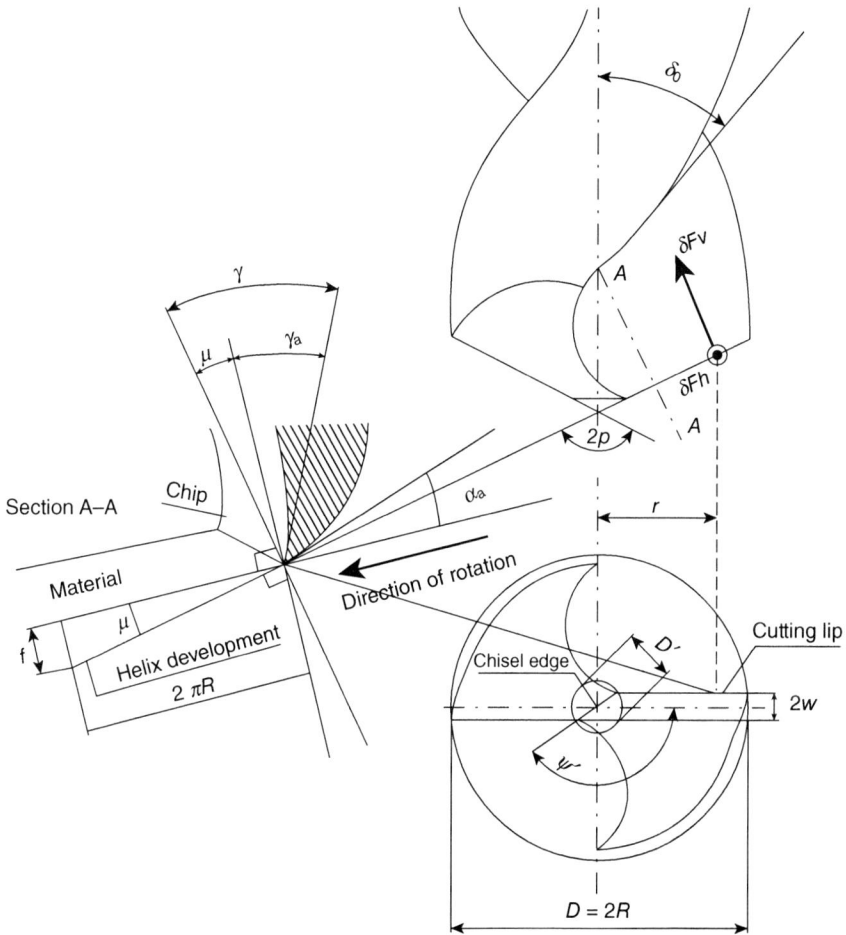

4.15 View of the drilling tool geometry and elemental forces acting on the cutting edges.

$$C_c = 2 \int_{r_0}^{D'/2} \frac{\cos(\phi_d - \gamma_d)}{\cos(\phi_d + \lambda_d - \gamma_d)} \frac{fk_{AB}\cos\gamma_f}{2\sin\phi_d} (\sin\gamma_f - \tan(\phi_d - \gamma_d)\cos\gamma_f)dr \quad [4.29]$$

where f is the feed rate, k_{AB} the shear flow stress along the shear plane, ϕ_n and ϕ_d the normal and dynamic shear angles, λ_n and λ_d the normal and dynamic friction angles, γ_d the dynamic cutting rake angle defined as $\gamma_d = \gamma_f - \gamma_w$, with γ_f and γ_w given by the expressions:

$$\tan\gamma_f = \frac{f}{2\pi r} \quad [4.30]$$

$$\tan\gamma_w = \tan p \sin\psi' \quad [4.31]$$

With reference to Fig. 4.15, the other terms in Eqs. 4.26 to 4.29 are defined as:

$$i = \sin^{-1}(\sin \omega \cdot \sin p) \tag{4.32}$$

$$\omega = \sin^{-1} \frac{w}{r} \tag{4.33}$$

$$\mu = \tan^{-1}(\tan \omega \cdot \cos p) \tag{4.34}$$

$$\gamma = \frac{\tan \delta(r) \cos \omega}{\sin p - \cos p \tan \delta \sin \omega} \tag{4.35}$$

$$\delta(r) = \frac{D}{2r} \tan \delta_0 \tag{4.36}$$

$$r_0 = \frac{f \tan p}{2\pi} \sin \psi' \tag{4.37}$$

$$\frac{D'}{2} = \frac{w}{\sin \psi'} \tag{4.38}$$

To refer back to some useful findings, Elhachimi et al.[55] used a numerical method of integration, together with the flow diagram shown in Fig. 4.16. The experimental tests were performed on steel, using a standard drill bit, whilst varying the feed and rotational speeds. Comparing the predicted and

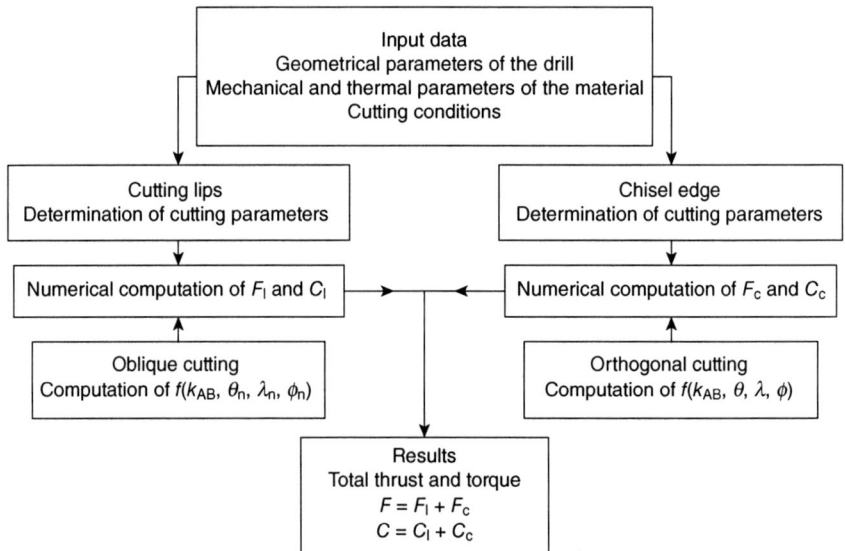

4.16 Flow diagram of thrust force and torque computation.[49]

measured values, the proposed model was able to follow with sufficient accuracy the trends of both thrust force and torque, although in general the thrust force was overestimated, and the torque underestimated.

4.3.3 Cutting models for composite drilling

As highlighted in Section 4.3.1, the mechanisms of material removal in composites are very different from those found in metals. Consequently, it is hard to accept that the Oxley model adopted by Elhachimi et al.[55] will also be valid for these materials. Recognizing this, Chandrasekharan et al.[57] assumed the following empirical relationships for the normal elemental force δF_v and the tangential elemental force δF_h occurring at the cutting edge:

$$\delta F_v(\rho) = K_v(\rho)\frac{f}{2}dx = K_v(\rho)\frac{f}{2}\cos i(\rho)Rd\rho \qquad [4.39]$$

$$\delta F_h(\rho) = K_t(\rho)\frac{f}{2}dx = K_t(\rho)\frac{f}{2}\cos i(\rho)Rd\rho \qquad [4.40]$$

where $\rho = r/R$, while $K_v(\rho)$ and $K_t(\rho)$ are the specific cutting pressures in the normal and tangential directions, defined as:

$$K_v(\rho) = C_1\rho^a \qquad [4.41]$$

$$K_v(\rho) = C_2\rho^b \qquad [4.42]$$

and $i(\rho)$ is the inclination angle at the generic point on the cutting lip, given by:

$$i(\rho) = \sin^{-1}\left(\frac{w}{R\cdot\rho}\sin p\right) \qquad [4.43]$$

The constants C_1, C_2, a, and b depend on machining conditions and drill geometry.

By integrating the elemental force and torque on the entire cutting line of the drill (Fig. 4.17a and b), the following closed-form expressions were obtained:

$$F_1 = \frac{C_1 Rf \sin p}{(a+1)}[1-\tau^{(a+1)}] - \frac{C_1 fw^2 \sin^3 p}{2R(a-1)}[1-\tau^{(a-1)}] \qquad [4.44]$$

$$C_1 = \frac{C_2 R^2 f}{(b+2)}[1-\tau^{(b+2)}] - \frac{C_2 fw^2 \sin^2 p}{2b}[1-\tau^b] \qquad [4.45]$$

where:

$$\tau = \frac{w}{R\sin\psi} \qquad [4.46]$$

(a)   (b)

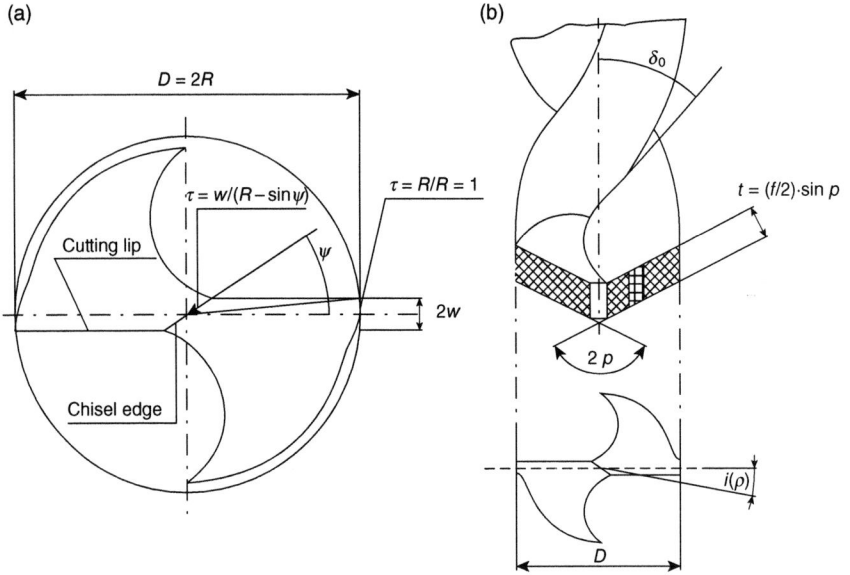

4.17 (a) Transverse and (b) side view of the drill bit.

The meaning of the angle ψ is shown in Fig. 4.17a.

The constants of the model were determined by suitable experiments, described in detail in reference 57. Since Eqs. 4.44 and 4.45 take into account only the action of the lips, the calibration tests were performed using a pilot hole (Fig. 4.18), to exclude the contribution of the chisel edge. For the latter, no particular solution dedicated to composites was proposed, but the application of the slip-line field method, developed by Kachanov[74] for perfectly plastic materials, was attempted.

Chandrasekharan *et al.*[57] verified the applicability of their model for metals and composites. The predictions agreed to a large extent with the experimental results in both cases as regards the force and torque at the cutting lips. The correlation was unsatisfactory when the effect of chisel edge in composite drilling was considered.

Langella *et al.*[58] executed preliminary orthogonal cutting tests on GFRP tubes having various fibre orientations and, with reference to the symbology in Fig. 4.2, reduced the force data according to empirical relationships proposed in references 12 and 13, obtaining:

$$F_{pu} = A + B \cdot 10^{-c\gamma} \cdot t \qquad\qquad [4.47]$$

$$F_{tu} = B \cdot 10^{-c\gamma} \cdot t^{0.5} \qquad\qquad [4.48]$$

The index u in Eqs. 4.47 and 4.48 indicates that the forces per unit specimen width were considered.

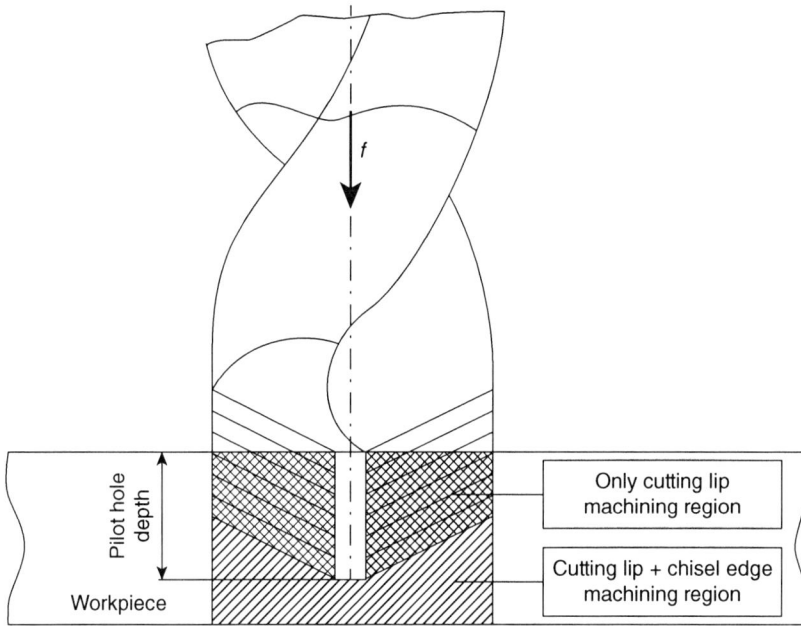

4.18 Drilling a workpiece with a pilot hole.

From the experimental data, the best values for the constants A, B, c appearing in Eqs 4.47 and 4.48 were calculated. The agreement between the empirical formulae and measured results was quite good, since a total average error of about 10% was found.

Following a procedure similar to that used by Chandrasekharan *et al.*, Langella and co-workers then used Eqs. 4.47 and 4.48 to express the elemental horizontal and vertical forces along the lips of a conventional drill bit. Since the analytic solution could not be integrated along the length of the lips without the aid of a numerical analysis, the authors also proposed a simplified version, thanks to which the thrust force and torque generated at the lips under stationary working conditions could be reduced to the closed form:

$$F_1 = 2 \cdot B \cdot 10^{cy} \left(\frac{f}{2}\right)^{0.5} G' \qquad\qquad [4.49]$$

$$C_1 = 2 \left[A + B \cdot 10^{cy} \left(\frac{f}{2}\right) \right] G \qquad\qquad [4.50]$$

with G, G' determined only by the drill geometry, and given by:

$$G = \int_\tau^1 \left(1 - \frac{w^2 \sin^2 p}{2\rho^2 R^2}\right) \rho R^2 d\rho = \frac{(1-\tau^2)R^2 + w^2 \sin^2 p \ln \tau}{2} \qquad [4.51]$$

$$G' = \int_\tau^1 \left(1 - \frac{w^2 \sin^2 p}{2\rho^2 R}\right) R \sin p\, d\rho$$

$$= \frac{\sin p(1-\tau)(2\tau R^2 - w^2 \sin p)}{2\tau R} \qquad [4.52]$$

In their analysis, the authors disregarded the contribution of the chisel edge to torque, but treated its contribution F_C to thrust force as an orthogonal cutting, using a relationship formally identical to Eq. 4.48:

$$F_C = 2wH \cdot 10^{-d\gamma_{ch}} \cdot f^{0.5} \qquad [4.53]$$

where γ_{ch} is the chisel rake angle, and H and d are two constants to be determined experimentally.

To assess their model, Langella and co-workers carried out drilling tests on three types of GFRP, i.e. mat/epoxy, unidirectional/epoxy, and fabric/epoxy. Significantly, the constant c was drawn directly from the preliminary orthogonal cutting tests. Consequently, the calibration stage necessary for obtaining the constants A and B could be substantially simplified, requiring a single drilling test with pilot hole (Fig. 4.18) for each of the materials tested.

A comparison of the experimental results and theoretical predictions, concerning tests performed adopting various drilling parameters, is shown in Fig. 4.19. The correlation of predicted values with measurements is very accurate for all the materials tested.

Figure 4.20 shows the evolution of experimental thrust force and torque recorded during a typical drilling test executed in reference 58. The thick lines are the theoretical predictions, obtained by progressively extending the limits of integration of Eqs. 4.51 and 4.52 to the portion of lip length actually engaged in the workpiece. The analysis was stopped at the point of maximum thrust force and torque, so no attempt was made to simulate the tool action during exit from the material.

A relevant aspect emerging from the results discussed in reference 58 is the direct correlation established between orthogonal cutting and drilling. It is hoped, therefore, that a complete characterization of composite machinability under simple cutting conditions will be useful in simulating more complex machining operations. Unfortunately, only limited knowledge is presently available on composite laminates cut along planes pertaining to axes other than the 3-axis in Fig. 4.1.

(a)

(b)

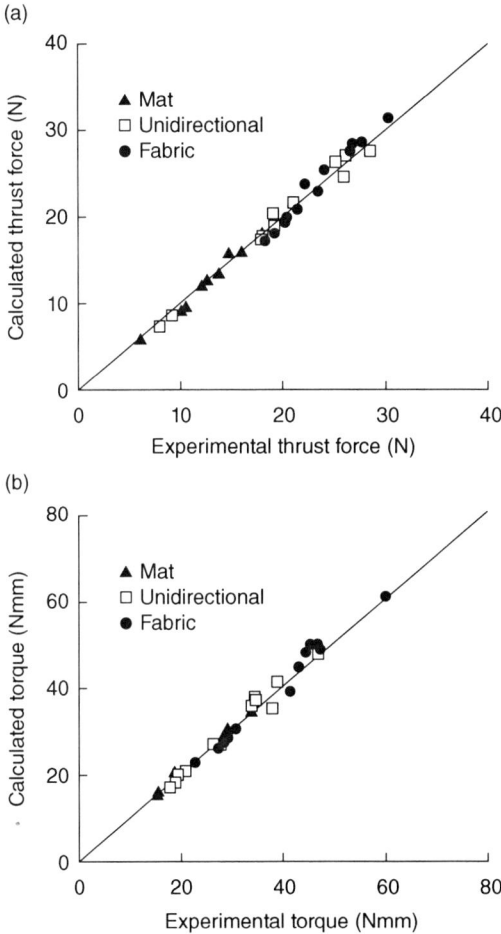

4.19 Comparison of experimental and theoretical values of: (a) thrust force and (b) torque. Data adapted from reference 52.

4.4   Milling

4.4.1   Experimental trends

The experimental results available show that the machining forces in milling increase with increasing feed rate.[75–78] Consequently, the delamination factor, defined as the ratio between maximum damage and nominal width of cut, increases too.[76,79] However, it has also been observed that lowering the machining forces does not necessarily result in improved cut quality: in reference 80, increasing cutting speed at low feed rates adversely affected

(a)

(b)

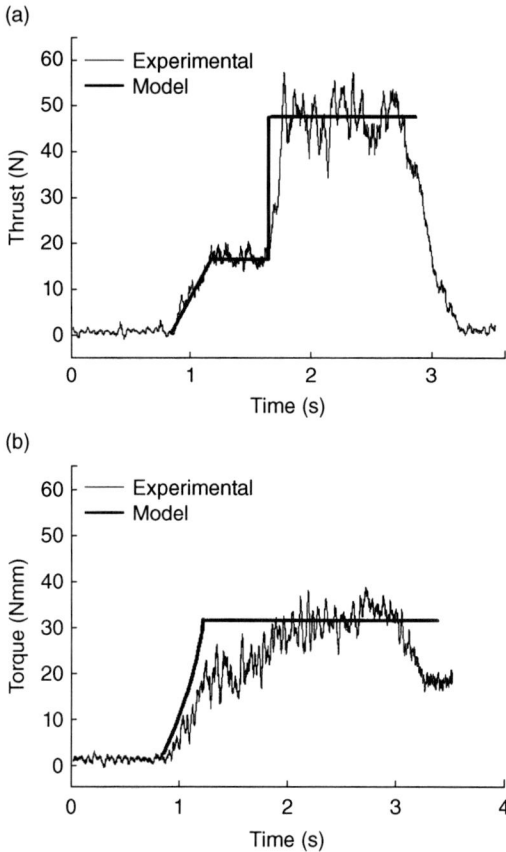

4.20 Comparison of theoretical and experimental: (a) thrust force and (b) torque. Material: mat/epoxy. Rotational speed $n = 1250$ rpm; feed rate $f = 0.25$ mm/rev.

roughness, because of the heat generated during the process, coupled with the low thermal conductivity of the material.[80] In fact, a 0.1 mm per tooth feed rate and a 50 m/min cutting speed were recommended for optimum surface roughness. Of course, the surface damage can significantly influence the mechanical properties, as found in reference 81, where the mechanical properties of unidirectional specimens machined by side milling were measured.

In helical milling, the axial force is not significantly influenced by the axial and tangential feed per tooth,[82] while the normal force increases as both the axial feed per tooth is increased and the tangential feed per tooth is decreased.

4.4.2 Predictive models

A number of theoretical models, taking into account the most important process parameters such as spindle speed rotation, feed per tooth, axial and radial depth of cut, and mill geometry, have been developed and verified for metals.[83–85]

For composites, only a few studies that try to predict the cutting forces in milling are available. Apparently, Puw and Hocheng[78] presented the first mechanistic model to evaluate the milling forces in a unidirectional FRP using single-insert end mills. To characterize the machinability of the material studied, the authors carried out planing tests parallel and perpendicular to the fibre direction of unidirectional reinforced plastics, varying the cutting speed, V, depth of cut t, and specimen width, w_w (coinciding with the laminate thickness). The experimental results were interpreted in terms of unit cutting forces, k, using the following empirical relationship:

$$k_{i,\theta} = \text{const} \cdot V^{\chi} w_w^{\xi} t^{\zeta} \qquad [4.54]$$

with the index $i = p,t$ individuating the principal (p) and thrust (t) force, and $\theta = 0°, 90°$. From the data generated, the set of constants const, χ, ξ, ζ appearing in Eq. 4.54 were calculated for each of the i,θ couples.

Knowing the unit cutting forces, and noting that the local chip thickness t_c in milling (Fig. 4.21) varies with angular position according to $t_c = w_w \sin\phi$, the local milling forces F_T, F_R as a function of ϕ were easily calculated, and from these, F_x, F_y were obtained using simple trigonometric considerations.

To assess their model, Puw and Hocheng[78] recorded the variation of F_x, F_y with ϕ in milling unidirectional carbon/PEEK and carbon/epoxy

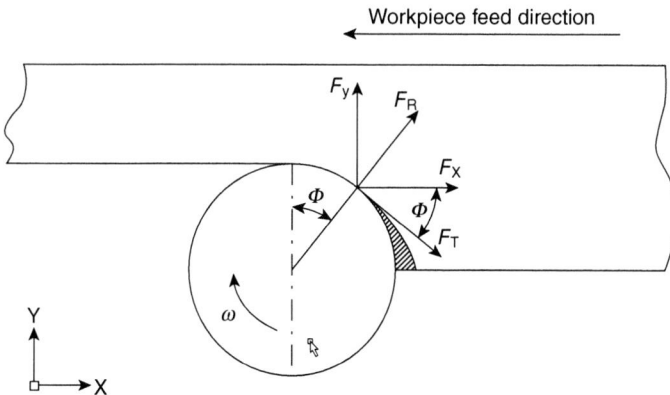

4.21 Force scheme in milling.

composites, cut parallel and perpendicular to fibre orientation. The agreement was satisfactory for carbon/thermoplastic, while discrepancies were found in the case of carbon/thermoset. The correlation between predictions and measured values improved considerably when the constants in Eq. 4.54 were considered to be dependent on θ, as well as on the damage generated during machining. Of course, this required an extensive experimental campaign, in which planing tests were conducted at various fibre angles.

A mechanistic approach, similar to that proposed by Puw and Hocheng, was developed by Sheikh-Ahmad.[86] From experimental tests, the dependence of the unit tangential and radial cutting forces (indicated by the symbols k_T and k_R in Fig. 4.22a and b, respectively) on fibre orientation and chip thickness were highlighted. Subsequently, regression analysis was used to determine appropriate functions for k_T and k_R, from which the milling forces were predicted. The agreement with the experimental data was reasonably accurate, not only for unidirectional, but also for multidirectional laminates. For the latter, the cutting forces were simply calculated as the sum of the cutting forces required to independently machine each unidirectional ply.

Recently, Kalla and co-workers[87] designed, trained, and used an artificial neural network (ANN) architecture to predict unit cutting forces. The training set consisted of an experimental data base, previously developed[88] by performing orthogonal up-milling tests on unidirectional CFRPs. The authors applied a procedure, previously proposed for metals,[89,90] to calculate the cutting forces in the helical end milling of composites: the helical cutting edge was divided into a number of discs of thickness dz along its z-axis (Fig. 4.23); the cutting action of an individual tooth within each disc was evaluated by transforming the unit cutting force data from orthogonal cutting to oblique cutting, to take into account the inclination angle i; finally, the total forces on the end mill were obtained from the sum of the elemental forces acting on each tooth segment at every disc. To assess the model, helical end milling tests were performed on unidirectional and multidirectional laminates. The agreement was reasonably good for unidirectional, and fair for multidirectional composites. According to the authors, a possible reason for the inaccuracies found was the mismatch between the unit cutting forces provided by ANN and the actual cutting forces, as the effect of the inclination angle was not accounted for in the ANN training set.

4.5   Conclusions and recommended future research

In this chapter, some fundamental features of composite materials machined by conventional cutting tools were recalled, and models aimed at the prediction of cutting forces in orthogonal cutting, drilling, and milling have been discussed.

(a)

(b)

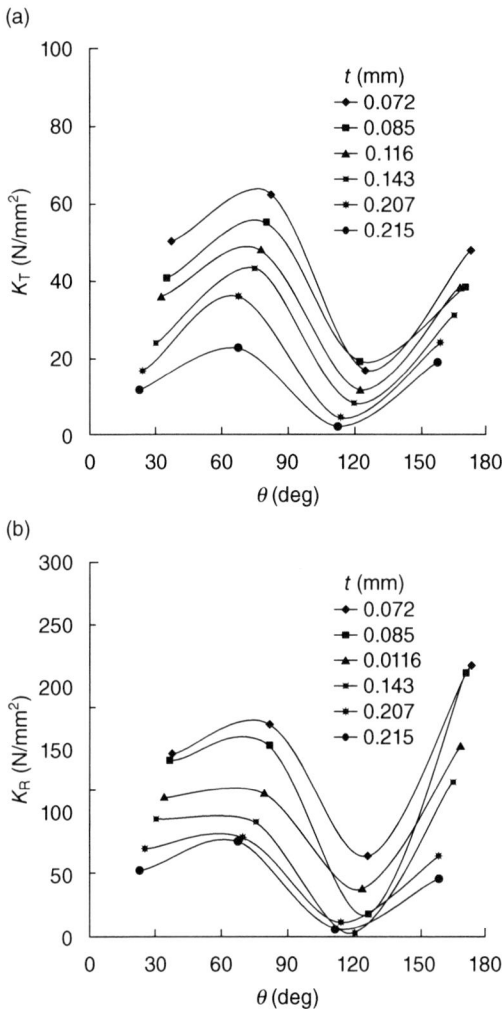

4.22 Dependence of specific cutting energy on fibre orientation: (a) k_T and (b) k_R. Data from reference 86.

Nowadays, the study of material removal in composite laminates by traditional techniques is still in its early stages, and requires substantial improvements in scientific and technical knowledge to fulfil industrial requirements that are becoming increasingly severe in terms of productivity and product quality. New challenges include machining of composite stacks (made of alternate layers of polymeric composites and metals), as well as laminates having very high thicknesses.

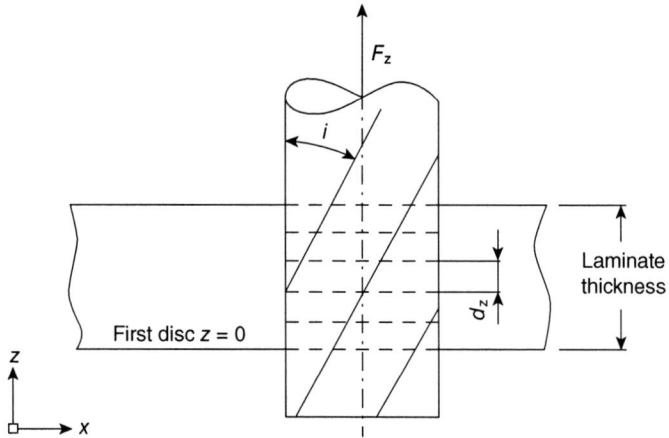

4.23 Scheme of helical end milling: tool segmentation in discs.

To satisfy industrial demand, different actions, as regards tools, machines, and working strategies, must be developed. As far as tools are concerned, minimizing cutting forces is of paramount importance, leading to cleaner cut surfaces with the least amount of tool wear. Of course, a pre-requisite for an engineered design of tool geometry would be a clear understanding of the mechanisms occurring at the tool–workpiece interface, which is lacking at the moment. For instance, as highlighted previously, most of the experimental works available in the literature on orthogonal cutting concern planes passing through the 3-axis. Consequently, our knowledge on the machinability of composites along a generic plane, which is necessary to approach drilling and milling, is very limited.

Another topical aspect in composite cutting, which could potentially increase productivity, is the effect of cutting speed on cut quality. Thanks to the relatively low forces required to machine them, composites could conceivably be machined at surface speeds even exceeding the possibilities of the present machine tools. Unfortunately, the heat generated during cutting can damage or melt the matrix, also affecting chip evacuation, and resulting in tool wear and clogging. How to transfer heat effectively and remove material in dry cutting, while still obtaining acceptable results, is an issue involving machine tool architecture, as well as new process strategies.

4.6   Sources of further information

Over recent years, several textbooks have been published on composite machining using conventional and non-conventional techniques. Among them, references 91 and 92 are recommended for further information as including the most appropriate parameters to set in typical cutting operations, as well as information on matters of health and safety.

A myriad of companies around the world today offer uncoated carbide, CVD diamond-coated and PCD cutting tools with continuously improved geometry, purposely designed for composite materials, or even custom-made products, developed for specific applications. Sandvik (https://www.sandvik.coromant.com), AMAMCO (http://www.amamcotool.com/composite.php), Lach Diamond (http://www.lach-diamond.com) and Kennametal (http://www.kennametal.com) are just some examples of these.

4.7   References

1 L Santo, G Caprino, I De Iorio, 'Cutting forces and cut quality in orthogonal cutting of unidirectional carbon fibre reinforced plastics', *Acts III Biennial World Conf. on Integrated Design & Process Technology*, Berlin, July 6–9, 1998.
2 D Nayak, N Bhatnagar, P Mahajan, 'Machining studies of UD-FRP composites. Part 1: Effect of geometrical and process parameters', *Mach Sci Technol*, **9**, 2005, 481–501.
3 D Arola, M Ramulu, DH Wang, 'Chip formation in orthogonal trimming of graphite/epoxy composite', *Composites Part A*, **27A**, 1996, 121–133.
4 M Ramulu, 'Machining and surface integrity of fibre-reinforced plastic composites', *Sadhana*, **22**, 1997, 449–472.
5 A Koplev, 'Cutting of CFRP with Single Edge Tool,' *Proc 3rd ICCM*, Paris, 1980, 1597–1605.
6 A Koplev, A Lystrup, T Vorm, 'The cutting process, chips, and cutting forces in machining CFRP', *Composites*, **14**, 1983, 371–376.
7 H Takeyama, N Iijima, 'Machinability of glassfiber reinforced plastics and application of ultrasonic machining', *Annals of CIRP*, **37**, 1988, 93–96.
8 T. Kaneeda, 1989, 'CFRP cutting mechanism,' *Proc 16th NAMRC*, Urbana, 1989, 216–221.
9 R Krishnamurthy, 'Machining of polymeric composites', *Proc Machining of Composite Materials Symposium*, Chicago, 1992, 139–148.
10 N Bhatnagar, N Ramakrishnan, NK Naik, R Komanduri, 'On the machining of fiber reinforced plastic (FRP) composite laminates', *Int J Mach Tools Manuf*, **35**, 1995, 701–716.
11 DH Wang, M Ramulu, D Arola, 'Orthogonal cutting mechanisms of graphite/epoxy composite. Part I: Unidirectional laminate', *Int J Mach Tools Manuf*, **35**, 1995, 1623–1638.
12 G Caprino, L Santo, L Nele, 'On the origin of cutting forces in machining unidirectional composites', *Proc III Biennial ESDA Conf*, Montpellier, July 1–4, 1996, 83–89.
13 G Caprino, L. Nele, 'Cutting forces in orthogonal cutting of unidirectional GFRP composites', *J Eng Mater Technol*, **118**, 1996, 419–425.
14 G Caprino, I De Iorio, L Nele, L Santo, 'Effect of tool wear on cutting forces in the orthogonal cutting of unidirectional glass fibre reinforced plastics', *Composites Part A*, **27A**, 1996, 409–415.

15 G Caprino, L Santo, I De Iorio, 'Chip formation mechanisms in machining uni-directional carbon fibre reinforced plastics', *Acts III AITEM Conf*, Salerno, Italy, Sept 17–19, 1997, 65–72.

16 HY Puw, H Hocheng, 'Chip formation model of cutting fiber-reinforced plastics perpendicular to fiber axis', *J Manufact Sci Eng*, **120**, 1998, 192–196.

17 G Caprino, L Santo, L. Nele, 'Interpretation of size effect in orthogonal machining of composite materials. Part I: Unidirectional glass fibre-reinforced plastics', *Composites Part A*, **29**, 1998, 887–892.

18 LC Zhang, HJ Zhang, XM Wang, 'A force prediction model for cutting unidirectional fibre-reinforced plastics', *Mach Sci Technol*, **5**, 2001, 293–305.

19 V Lopresto, L Santo, G Caprino, I De Iorio, 'Effect of fibre orientation on cutting forces and cut quality in machining unidirectional carbon fibre reinforced plastics', *Acts IV AITEM Conf*, Brescia, Italy, 1999, Sept.13–15, 451–458.

20 V Lopresto, L Santo, G Caprino, I De Iorio, 'Mechanisms of chip generation in orthogonal machining of unidirectional carbon fibre reinforced plastics', *Proc PRIME 2001*, Sestri Levante, Italy, June 20–22, 2001, 81–86.

21 XM Wang, LC Zang, 'An experimental investigation into the orthogonal cutting of unidirectional fibre reinforced plastics', *Int J Mach Tool Manuf*, **43**, 2003, 1015–1022.

22 H Hocheng, CC Tsao, 'The path towards delamination-free drilling of composite materials', *J Mater Proc Technol*, **167**, 2005, 251–264.

23 S Mohan, SM Kulkarni, A Ramachandra, 'Delamination analysis in drilling process of glass fiber reinforced plastic (GFRP) composite materials', *J Mater Proc Technol*, **186**, 2007, 265–271.

24 H Hocheng, CKH Dharan, 'Delamination during drilling in composite laminates', *J Eng Ind*, **112**, 236–239.

25 S Jain, DCH Yang, 'Effects of feed rate and chisel edge on delamination in composite drilling', *J Eng Ind*, **115**, 398–405.

26 RM Jones, *Mechanics of Composite Materials*, 2nd Ed., Taylor & Francis, Philadelphia, 1999.

27 L Santo, G Caprino, V Lopresto, I De Iorio, 'Machining of carbon fibre reinforced plastics. Part II: Analytical prediction of cutting forces', *Acts of ESDA 2000 Conf*, Montreaux, July 10–13, 2000, 39–43.

28 H Hocheng, SC Leu, 'Machining characteristics of carbon fiber-reinforced epoxy tube in turning', *Composites*, **32**, 1992, 136–140.

29 M Amontons, 'De la resistance causeé dans le machines', *Histoire de l'Académie Royale des Sciences*, 1699, 206–222.

30 G Caprino, L Nele, L Santo, I De Iorio, 'Dependence of the coefficient of friction on fibre orientation in the sliding of a HSS punch on a unidirectional CFRP surface', *Proc III Biennial ESDA Conf*, Montpellier, July 1–4, 1996, 33–39.

31 D Nayak, N Bhatnagar, P Mahajan, 'Machining studies of UD–FRP composites. Part 2: Finite element analysis', *Mach Sci Technol*, **9**, 2005, 503–528.

32 G Venu Gopala Rao, P Mahajan, N. Bhatnagar, 'Micro-mechanical modeling of machining of FRP composites – Cutting force analysis', *Compos Sci Technol*, **67**, 2007, 579–593.

33 M.C. Shaw, *Metal Cutting Principles*, 2nd Ed, Oxford University Press, Oxford, 2005.

34 MV Ramesh, KN Seetharamu, N Ganesan, MS Sivakumar, 'Analysis of machining of FRPs using FEM', *Int J Mach Tools Manuf*, **38**, 1998, 1531–1549.

35 PS Sreejith, R Krishnamurthy, SK Malhotra, K Narayanasamy, 'Evaluation of PCD tool performance during machining of carbon/phenolic ablative composites', *J Mater Proc Technol*, **104**, 2000, 53–58.

36 G Santhanakrishnan, R Krishnamurthy, SK Malhotra, 'Investigation into the machining of carbon-fibre-reinforced plastics with cemented carbides', *J Mater Proc Technol*, **30**, 1992, 263–275.

37 D Arola, M Ramulu, 'Orthogonal cutting of fiber-reinforced composites: A finite element analysis', *Int J Mech Sci*, **5**, 1997, 597–613.

38 G Venu Gopala Rao, P Mahajan, N. Bhatnagar, 'Machining of UD–GFRP composites chip formation mechanism', *Compos Sci Technol*, **67**, 2007, 2271–2281.

39 CR Dandekar, YC Shin, 'Multiphase finite element modeling of machining unidirectional composites: Prediction of debonding and fiber damage', *J Manuf Sci Eng*, **130**, 2008, 051016-1–051016-12.

40 L Lasri, M Nouari, M El Mansori, 'Modelling of chip separation in machining unidirectional FRP composites by stiffness degradation concept', *Compos Sci Technol*, **69**, 2009, 684–692.

41 A Mkaddem, M El Mansori, 'Finite element analysis when machining UGF-reinforced PMCs plates: Chip formation, crack propagation and induced damage', *Mater Des*, **30**, 2009, 3295–3302.

42 C Santiuste, X Soldani, MH Miguélez, 'Machining FEM model of long fiber composites for aeronautical components', *Compos Struct*, **92**, 2010, 691–698.

43 D Iliescu, D Gehin, I Iordanoff, F Girot, ME Gutiérrez, 'A discrete element method for the simulation of CFRP cutting', *Compos Sci Technol*, **70**, 2010, 73–80.

44 A Dikshit, J Samuel, RE DeVor, SG Kapoor, 'Microstructure-level machining simulation of carbon nanotube reinforced polymer composites – Part I: Model development and validation', *J Manuf Sci Eng*, **130**, 2008, 031114.

45 A Dikshit, J Samuel, RE DeVor, SG Kapoor, 'Microstructure-level machining simulation of carbon nanotube reinforced polymer composites – Part II: Model interpretation and application', *J Manuf Sci Eng*, **130**, 2008, 031115.

46 J Samuel, A Dikshit, RE DeVor, SG Kapoor, KJ Hsia, 'Effect of carbon nanotube (CNT) loading on the thermomechanical properties and the machinability of CNT-reinforced polymer composites', *J Manuf Sci Eng*, **131**, 2009, 031008.

47 J Samuel, SG Kapoor, RE DeVor, KJ Hsia, 'Effect of microstructural parameters on the machinability of aligned carbon nanotube composites', *J Manuf Sci Eng*, **132**, 2010, 051012.

48 EJA Armarego, CY Cheng, 'Drilling with flat rake face and conventional twist drills – 1. Theoretical investigation', *Int J Mach Tool Des Res*, **12**, 1972, 17–35.

49 S Wiriyacosol, EJA Armarego, 'Thrust and torque prediction in drilling from a cutting mechanics approach', *CIRP Ann*, **28**, 1979, 87–91.

50 AR Watson, 'Geometry of drill elements', *Int J Mach Tool Des Res*, **25**, 1985, 209–227.

51 AR Watson, 'Drilling model for cutting lip and chisel edge and comparison of experimental and predicted results. I – Initial cutting lip model', *Int J Mach Tool Des Res*, **25**, 1985, 347–365.

52 AR Watson, 'Drilling model for cutting lip and chisel edge and comparison of experimental and predicted results. II – Revised cutting lip model', *Int J Mach Tool Des Res*, **25**, 1985, 367–376.

53 AR Watson, 'Drilling model for cutting lip and chisel edge and comparison of experimental and predicted results. III – Drilling model for chisel edge', *Int J Mach Tool Des Res*, **25**, 1985, 377–392.

54 AR Watson, 'Drilling model for cutting lip and chisel edge and comparison of experimental and predicted results. IV – Drilling tests to determine chisel edge contribution to torque and thrust', *Int J Mach Tool Des Res*, **25**, 1985, 394–404.

55 M Elhachimi, S Torbatty, P Joyot, 'Mechanical modelling of high speed drilling. 1: Predicting torque and thrust', *Int J Mach Tools Manuf*, **39**, 1999, 553–568.

56 M Elhachimi, S Torbatty, P Joyot, 'Mechanical modelling of high speed drilling. 2: Predicted and experimental results', *Int J Mach Tools Manuf*, **39**, 1999, 569–581.

57 V Chandrasekharan, SG Kapoor, RE DeVor, 'A mechanistic approach to predicting the cutting forces in drilling with application to fiber-reinforced composite materials', *J Eng Ind*, **117**, 1995, 559–570.

58 A Langella, L Nele, A Maio, 'A torque and thrust prediction model for use in composite materials drilling', *Composites Part A*, **36**, 2004, 83 – 93.

59 AM Abrão, PE Faria, JC Campos Rubio, P Reis, JP Davim, 'Drilling of fiber reinforced plastics: A review', *J Mater Proc Technol*, **186**, 2007, 1–7.

60 JP Davim, P Reis, CC António, 'Experimental study of drilling glass fiber reinforced plastic (GFRP), manufactured by hand lay-up', *Compos Sci Technol*, **64**, 2004, 289–297.

61 I El-Sonbaty, UA Khashaba, T Machaly, 'Factors affecting the machinability of GFR/epoxy composites', *Compos Struct*, **63**, 2004, 329–338.

62 LB Zhang, LJ Wang, XY Liu, HW Zhao, X Wang, HY Luo, 'Mechanical model for predicting thrust and torque in vibration drilling fibre-reinforced composite materials', *Int J Machine Tools Manuf*, **41**, 2001, 641–657.

63 UA Khashaba, MA Seif, MA Elhamid, 'Drilling analysis of chopped composites', *Composites Part A*, **38**, 2007, 61–70.

64 K Ogawa, E Aoyama, H Inoue, T Hirogaki, H Nobe, Y Kitahara, T Katayama, M Gunjima, 'Investigation on cutting mechanism in small diameter drilling for GFRP (thrust force and surface roughness at drilled hole wall)', *Compos Struct*, **38**, 1997, 343–350.

65 UA Khashaba, 'Delamination in drilling GFR-thermoset composites', *Compos Struct*, **63**, 2004, 313–327.

66 LMP Durão, DJS Gonçalves, JMRS Tavares, VHC de Albuquerque, AA Vieira, A Torres Marques, 'Drilling tool geometry evaluation for reinforced composite laminates', *Compos Struct*, **92**, 2010, 1545–1550.

67 H Hocheng, CC Tsao, 'Effects of special drill bits on drilling-induced delamination of composite materials', *Int J Machine Tools Manuf*, **46**, 2006, 1403–1416.

68 CC Tsao, 'Experimental study of drilling composite materials with step-core drill', *Mater Des*, **29**, 2008, 1740–1744.

69 UA Khashaba, IA El-Sonbaty, AI Selmy, AA Megahed, 'Machinability analysis in drilling woven GFR/epoxy composites: Part I – Effect of machining parameters', *Composites Part A*, **41**, 2010, 391–400.

70 S Arula, L Vijayaraghavana, SK Malhotrab, R Krishnamurthya, 'The effect of vibratory drilling on hole quality in polymeric composites', *Int J Machine Tools Manuf*, **46**, 2006, 252–259.

71 V Tagliaferri, G Caprino, A Diterlizzi, 'Effect of drilling parameters on finish and mechanical properties of GFRP composites', *Int J Mach Tools Manuf*, **30**, 1990, 77–84.

72 G Caprino, V Tagliaferri, 'Damage development in drilling glass fibre reinforced plastics', *Int J Mach Tools Manuf*, **35** , 1995, 817–829.

73 PLB Oxley, 'Modeling machining processes with a view to their optimatization', *Rob Comp Integr Manuf*, **4**, 1988, 103–119.

74 LM Kachanov, *Foundation of the Theory of Plasticity*, North-Holland, Amsterdam, 1971.

75 R Rusinek, 'Cutting process of composite materials: An experimental study', *Int J Non Linear Mech*, **45**, 2010, 458–462.

76 JP Davim, P Reis, CC António, 'A study on milling of glass reinforced plastics manufactured by hand-lay up using statistical analysis (ANOVA)', *Compos Struct*, **64**, 2004, 493–500.

77 HY Puw, H Hocheng, 'Machinability test of carbon fiber-reinforced plastics in milling', *Mater Manufact Process*, **8**, 1993, 717–729.

78 HY Puw, H Hocheng, 'Milling of Polymer Composites', in *Machining of Ceramics and Composites*, Marcel Dekker, 1999.

79 K Colligan, M Ramulu, 'The effect of edge trimming on composites surface plies', *Manufact Rev*, **5**, 1992, 274–282.

80 H Hocheng, HY Puw, Y Huang, 'Preliminary study on milling of unidirectional carbon fiber reinforced plastics', *Compos Manufact*, **4**, 1993, 103–108.

81 P Ghidossi, ME Mansori, F Pierron, 'Influence of specimen preparation by machining on the failure polymer matrix off-axis tensile coupons', *Compos Sci Technol*, **66**, 2006, 1857–1872.

82 B Denkena, D Boehnke, JH Dege, 'Helical milling of CFRP–titanium layer compounds', *CIRP J Manufact Sci Technol*, **1**, 2008, 64–69.

83 KH Fuh, RM Hwang, 'A predicted milling force model for high-speed and milling operation', *Int J Machine Tools Manufact*, **37**, 1997, 969–979.

84 C Andersson, M Andersson, JE Stahl, 'Experimental studies of cutting force variation in face milling', *Int J Machine Tools Manufact*, **51**, 2011, 67–76.

85 HZ Li, WB Zhang, XP Li, 'Modelling of cutting forces in helical end milling using a predictive machining theory', *Int J Mech Sci*, **43**, 2001, 1711–1730.

86 J Sheikh-Ahmad, 'Model for predicting cutting forces in machining CFRP', *Int J Mater Prod Technol*, **32**, 2008, 152–167.

87 D Kalla, J Sheikh-Ahmad, J Twomey, 'Prediction of cutting forces in helical end milling fiber reinforced polymers', *Int J Machine Tools Manufact*, **50**, 2010, 882–891.

88 D Kalla, *Committee Neural Network Force Prediction Model in Milling of Fiber Reinforced Polymers*, PhD Thesis, Wichita State University, USA, 2008.

89 GCI Lin, P Mathew, PLB Oxley, AR Watson, 'Predicting cutting forces for oblique machining condition', *Proc Institut Mech Eng*, **196**, 1982, 141–148.

90 HZ Li, WB Zhang, XP Li, 'Modeling of cutting forces in helical end milling using a predictive machining theory', *Int J Mech Sci*, **43**, 2001, 1711–1730.

91 JY Sheikh-Ahmad, *Machining of Polymer Composites*, Springer, 2009. ISBN 978-0-387-35539-9.

92 JP Davim, *Machining Composite Materials*, Wiley, 2009. ISBN 978-1-848-21170-4.

4.8   Appendix: List of symbols used

C = torque in drilling
C_c = torque at drill chisel edge
C_l = torque at drill lips
D = drill bit diameter
D' = chisel edge diameter
f = feed rate
F = thrust force in drilling
F_1 = cutting force component along fibre direction
F_2 = cutting force component orthogonal to fibre direction
F_c = thrust force at drill chisel edge
F_h = tangential force at drill lips
F_l = thrust force at drill lips
F_n = normal force at drill lips
F_p = principal cutting force
F_{pf} = principal force component at tool flank
F_{pr} = principal force component at tool rake
F_R = radial cutting force in milling
F_T = tangential cutting force in milling
F_t = thrust cutting force
F_{tf} = thrust force component at tool flank
F_{tr} = thrust force component at tool rake
k_{AB} = shear flow stress along the shear plane
k_C = specific cutting coefficient for torque
k_F = specific cutting coefficient for thrust force
N = normal force in friction
N_f = normal force at the tool flank
N_r = normal force at the tool rake
N_r^* = apparent normal force at the tool rake
$2p$ = point angle
t = depth of cut
T = force of friction
T_f = tangential force at the tool flank
T_r = tangential force at the tool rake
V = cutting speed
V_f = fibre volume fraction
W = half-thickness of chisel edge
w_w = workpiece width
X = unit cutting force
X^* = apparent unit cutting force
α = flank angle
β = friction angle at the rake face

γ = rake angle
γ_d = dynamic cutting rake angle
δ_o = helix angle
ε = drill angle
λ_d = dynamic friction angle
λ_n = normal friction angle
η = coefficient of dynamic friction
η_r = coefficient of dynamic friction at tool rake
η_r^* = apparent coefficient of dynamic friction at tool rake
η_f = coefficient of dynamic friction at tool flank
θ = fibre orientation
$\tau(\theta)$ = shear strength
ϕ = shear angle
ϕ_d = dynamic shear angle
ϕ_n = normal shear angle

5

Tool wear in machining processes for composites

J. SHEIKH-AHMAD, The Petroleum Institute, UAE and
J. P. DAVIM, University of Aveiro, Portugal

Abstract: This chapter discusses the phenomena of tool wear in machining composite materials with various types of cutting tool materials. A discussion of viable cutting tool materials is first given and the important properties required for cutting composites are highlighted. This is followed by a discussion of specific wear mechanisms that arise in machining metal matrix and polymer matrix composites. The most common forms or types of tool wear are described. The effects of composite matrix, reinforcement phases and amount, and cutting process parameters on tool wear and tool life are also discussed.

Key words: metal matrix composites, polymer matrix composites, cutting tool materials, tool wear mechanisms, tool life.

5.1   Introduction

Composite materials are increasingly being used in the aerospace and automotive industries, because of their high specific strength and specific stiffness, and increased wear resistance over unreinforced materials. The matrix material is either metallic, polymeric or ceramic, and the reinforcement can be in the form of fibers, whiskers and particles that are generally of greater strength and hardness than the matrix. The resulting composite is therefore classified as metal matrix composite (MMC), polymer matrix composite (PMC) or ceramic matrix composite (CMC). The properties of the resulting composite are generally controlled by three critical components: the matrix, the reinforcement and the interface. Many of the considerations that arise with respect to composite fabrication, processing and service performance relate to processes that are taking place in the interfacial region between the matrix and reinforcement.

Today, the term metal matrix composite (MMC) covers a very wide range of materials, from relatively simple reinforcement of castings at low cost to complex continuous fiber lay-ups in exotic metallic alloys. MMC are finding increased application due to their very favorable properties, including high mechanical properties and good wear resistance. Silicon carbide (SiC) reinforced aluminum is among the most common and several compositions for

116

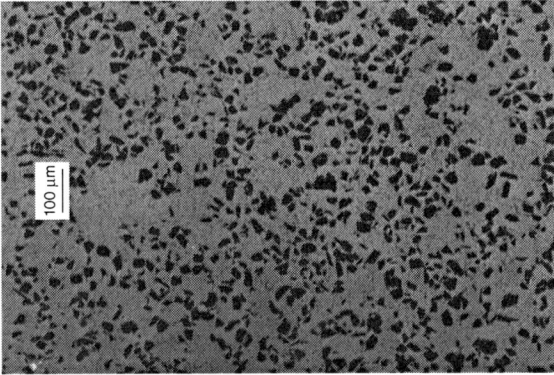

5.1 Typical microstructure of aluminum matrix composite A356-20%SiCp-T6.

the matrix are available commercially. MMCs have been applied in the aeronautic and automotive industries. A typical microstructure of Al/SiC particulate MMC with 20% vol. SiC particles is shown in Fig. 5.1.

Polymer matrix composites (PMC), sometimes termed fiber-reinforced plastics (FRPs), are characterized by high strength and stiffness at simultaneously low weight. So, FRPs have replaced conventional materials in various fields of application in mechanical engineering. Nowadays, carbon fiber-reinforced plastics (CFRPs), made by using carbon fibers for reinforcing a resin matrix such as epoxy or polyester, are characterized by having excellent properties such as light weight and high stiffness. In general, these properties make them especially attractive for aircraft and aerospace applications.

Composites are mostly produced near net shape. However, in other cases, most frequently in the application of MMC, machining is used to fabricate finished components from cast stock. Machining is also used for finishing parts to desired dimensional tolerances and for drilling holes for assembly operations. A continuing problem with composites manufacturing is that they are difficult to machine, due to their inhomogeneous structure and the high hardness and abrasive nature of the reinforcement phase. Machining challenges arise in the form of accelerated tool wear, poor surface finish, high cutting temperatures and poor surface integrity due to subsurface damage at the interfaces between the reinforcement and matrix. This chapter discusses the phenomena of tool wear in machining composites with different tool materials. It discusses the required properties of tool materials for machining composites and surveys available viable options. It also describes the various tool wear mechanisms that cause tool failure and how wear is manifested on the cutting edge. Tool life and the factors influencing it are also discussed.

5.2   Tool materials

As stated, composite materials are characterized by their high strength, inhomogeneous microstructure and high abrasiveness. Therefore, the cutting tools required for machining these materials should provide adequate levels of strength, toughness and hardness to combat the high cutting loads. Also, thermal conductivity and thermal shock resistance become of great importance when cutting intermittently or at high cutting speeds, where the thermal load applied is significant. From the wide range of cutting tool materials that are available for machining applications, a subset is capable of meeting these requirements. Materials in this subset are classified into two groups: hard materials and superhard materials. The hard materials group includes cemented carbides, coated carbides and ceramics. The superhard materials include cubic boron nitride, CVD diamond, polycrystalline diamond (PCD) and single crystal diamond.

Figure 5.2 shows a comparison of the relative hardness of materials in these two groups against single crystal diamond, which is known to be the hardest material to be found. Table 5.1 shows typical physical and mechanical properties of these hard and superhard materials. The properties of SiC and Al_2O_3, the main reinforcing phases in metal matrix composites are also included for reference.

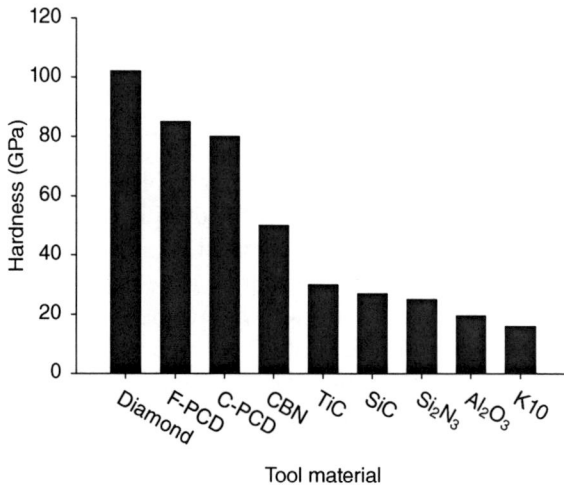

5.2 Typical hardness values for cutting tool materials. F-PCD and C-PCD, fine- and coarse-grain PCD, respectively.

Table 5.1 Representative properties of some cutting tool materials. *TRS* = transverse rupture strength, S_c = compressive strength, K_{IC} = fracture toughness, α = coefficient of thermal expansion, K = thermal conductivity

Tool material	TRS (MPa)	S_c (MPa)	K_{IC} (MPa m$^{-1/2}$)	HV (GPa)	α ($\times 10^{-6}/°C$)	K (W/m°C)
Al$_2$O$_3$-TiC	800	4500	4.5	16.7	8.0	16.7
Si$_3$N$_4$-Al$_2$O$_3$	900	—	—	15.7	3.0	—
WC-Co (6% Co)	1900	5380	12	17.2	4.3–5.6	80
PCBN	600–800	6900	—	41–49	3.2–4.2	110–200
PCD	600–2000	7700	3.0–9.0	54–88	1.5–3.8	543
CVD diamond film	1300	9000	5.5–8.5	50–100	3.84	500–2200
Single crystal diamond	1350	6900	3.4	60–102	0.8–4.8	600–2200

The data in Table 5.1 show a wide contrast in mechanical and thermal properties among the different materials within the same group and from one group to another. Noted are the high toughness of tungsten carbide, the high hardness, high strength and excellent thermal conductivity of diamond, and the poor thermal conductivity of ceramics and PCBN. All of these properties have tremendous influence on the wear behavior of the cutting tool during machining. Hardness describes the material's ability to resist abrasive wear, and toughness describes its ability to resist fracture under heavy and/or intermittent loads. It is evident from examination of the data in Fig. 5.2 and Table 5.1 that materials that are very hard tend to have very poor toughness, and *vice versa*. An ideal tool material, which is yet to be discovered, would have both high hardness and high toughness. It is also noted here that the hardness of SiC is higher than that of cemented carbides and ceramics tools, but is lower than that of CBN and PCD. Based on hardness alone, it becomes evident that CBN and PCD are the most appropriate cutting tool materials for machining SiC reinforced metal matrix composites. However, taking into account cyclic thermal and mechanical loading, this conclusion will be proven to be faulty, as performance of cutting tools is not determined by hardness alone. Under certain machining conditions, the lower hardness but tougher tools, such as carbides, may perform better than the harder CBN. The following sections explain the differences between the various groups of tool materials, their suitability for machining composites, and the processes by which they wear when machining.

5.2.1 Cemented carbides

The cemented carbide group of tool materials is mostly based on tungsten carbides (WC) as the hard phase, but also include other carbides such as

TiC and TaC. Cemented carbides based on Ti(C,N) as the hard phase are called Cermets. The tool blanks are produced near net shape by cold pressing of a suitable mixture of powders of carbides and a binder metal, such as cobalt or nickel, in weight fractions up to 20%. The tool blanks are then sintered at temperatures in the range from about 1350 to 1650 °C, to melt the binder and create bonding between the binder metal and the carbide grains. For decades, cemented carbides have been the work horse for the machining industry. Their processing technique allows them to be produced in different shapes and sizes in mass and at low cost. They can also be relatively easily ground, using diamond or CBN wheels. The technology of cemented carbides is well developed, which means significant improvements to the current state is not likely.

The properties of carbides are controlled, for the most part, by the binder content and grain size. Their hardness is imparted by the hard carbide phase, while the binder metal provides the necessary toughness and the even distribution of applied load to the harder and stronger phase. The hardness of the material is reduced and the toughness is increased as the binder metal weight fraction is increased. Hardness and toughness can be further improved by reducing grain size to the submicron and ultra fine levels.

The earliest cemented carbides consisted mainly of WC grains in Co binder. Until now, this hard material has constituted a large segment of the cutting tools produced. However, it was noticed that this carbide was not suitable for machining steel at high temperatures because of the rapid dissolution of WC in steel and the resulting drastic crater wear. This problem was later solved by adding TiC and TaC to the mixture, which considerably slowed down the diffusion of carbon from the carbide to the steel chip. Therefore, WC-Co based carbides are used mainly for cutting nonferrous materials and cast iron, while those based on WC-TiC-TaC-Co are used for cutting steel. The hardness and thermal stability of carbides are further improved by applying thin film coatings of hard ceramics on the cutting surfaces. The shortcomings of cemented carbides, however, are their inadequate hardness for machining advanced and highly abrasive composite materials.

5.2.2 Coated carbides

Hard ceramic coatings of several micrometers thick are deposited on cemented carbides in order to improve their wear resistance, particularly thermally driven wear, such as crater formation on the rake face. The coatings are formed at high temperatures (900–1050 °C) by chemical vapor deposition, CVD. Single layer and multi-layer coatings of TiC, TiN, TiCN and Al_2O_3 are utilized. These ceramic coatings possess superior high temperature hardness and are more thermally stable than the tungsten carbide substrate. Therefore, they act as a thermal barrier in machining at high cutting speeds and high chip loads. TiC is particularly used for its better hardness, Al_2O_3 for its chemical stability and TiN for its low coefficient of friction. Physical vapor deposition, PVD, was later developed to take advantage of low temperature vapor deposition (400–450 °C), which provides finer microstructures, higher toughness and less deterioration to the carbide substrate at high temperatures. In addition, the residual stresses in CVD coatings are tensile and in PVD coatings are compressive. Hence, PVD coatings have higher transverse rupture strength (TRS) and higher chipping resistance than CVD coatings. However, the adhesion strength and wear resistance are superior in CVD coatings. Because of these contrasting properties, CVD coatings are used in general turning and milling, while PVD coatings are applied where cutting forces are high and chipping resistance is required. PVD coatings are also more suitable for coating sharp cutting tools, such as solid carbide end mills and drills for machining composites.

5.2.3 Ceramics

Ceramics are sintered hard phases (mainly Al_2O_3 or Si_3N_4) with a sintering aid such as TiC, magnesium oxide or zirconia. They have the highest thermal stability among tool materials and exhibit excellent performance in high speed machining where the cutting temperatures are extremely high. On the other hand, their toughness is poor and they suffer from failure by chipping when used in heavy cuts or under interrupted loads. Ceramic tools are usually ground into negative rake angles (with cutting edge angle greater than 90°) in order to enhance their mechanical strength. Ceramics are also notorious for poor resistance to thermal shock because of their low thermal conductivity, another reason for their poor performance in interrupted cutting. Introducing TiC and ZrO_2 to the alumina base ceramics has been shown to significantly improve TRS and compressive strength. Also, reinforcement with SiC whiskers increases the toughness of alumina-based ceramics. This results in less chipping wear and better tool life. Silicon nitride-based ceramics are sintered with additions such as Al_2O_3 and Y_2O_3. They generally possess slightly higher toughness and higher hardness than alumina-based ceramics. Their low thermal expansion coefficient makes them less susceptible to thermal cracking than alumina-based ceramics.

5.2.4 Polycrystalline boron nitride (PCBN) and polycrystalline diamond (PCD)

These superhard tool materials were developed mainly for applications where long tool life and high productivity are of main concern. Similar to ceramics, these materials are also made by sintering of the hard phase, provided in the form of micrograin particles or polycrystals, with or without a binder phase. The hard phase is synthetic diamond or cubic boron nitride grits that are created at high temperature and pressure. The binder phase, which could be ceramic or metallic, improves the toughness and manufacturability of the cutting tools. Sintering is performed at extremely high temperatures and high pressures to help consolidation and bonding of the tool material. This makes their production cost very high. PCBN tools are generally much harder than alumina-based ceramics and possess excellent chemical wear resistance. But like ceramics, their toughness is poor. Therefore, they are more suited for finish turning of hardened materials and cast iron.

Polycrystalline diamond provides an impressive combination of mechanical and thermal properties, which makes it one of the most advanced cutting tool materials. Its hardness, toughness and strength are greater than those of PCBN and ceramics. It is also chemically stable, has low coefficient of friction and is an excellent heat conductor. The biggest disadvantage of sintered PCD tools is their cost. PCD tools typically cost ten times, or more, than conventional carbide and ceramic tools. The technology for high temperature, high pressure sintering is expensive, and cutting and grinding the cutting edge is very difficult. Therefore, production cost is the main factor affecting PCD tools cost. However, when used under proper cutting conditions, PCD tools are economically viable because they provide superior tool life and higher productivity. Another disadvantage of diamond cutting tools is their propensity to react with ferrous metals at high temperatures.

5.2.5 Chemical vapor deposition (CVD) diamond coating

Diamond crystals are grown from a carbon rich gas, mostly methane (CH_4), by low-pressure chemical vapor deposition (CVD) over a heated carbide substrate. Diamond films that are several μm thick can be grown using this process. The film is composed of pure polycrystalline diamond and thus should, in principle, provide better hardness and toughness characteristics than sintered diamond. The advantages of diamond coatings over PCD are large-scale production, lower production cost, and possibility for complex shaped tools such as drills and end mills.

The biggest obstacle to the widespread use of diamond coatings in machining is poor adhesion of the diamond film to the substrate. This, in part, is caused by large residual stresses in the film and substrate which are generated by the CVD process. Because of mismatches in thermal expansion coefficients of the diamond film and the substrate ($\alpha_D = 2.85 \times 10^{-6}\,°C^{-1}$ for the diamond coating and $\alpha_s = 5.0 \times 10^{-6}\,°C^{-1}$ for the substrate), and the high temperatures required for the CVD process (600–1000 °C), high thermal residual stresses are formed in the coating and substrate on cooling. These stresses affect the behavior of the coated tool in machining. Furthermore, additional stresses are introduced to the cutting edge by the external loads from machining and the peculiar shape of the cutting edge. A specific requirement for machining fiber reinforced composites is the availability of a sharp cutting edge with a small included angle, in order to cleanly sever the fibers and produce good surface quality. This, however, makes the edge mechanically weak, creates complex residual stresses and only accelerates edge wear.

A number of improvements have been effected, particularly in surface pretreatment technologies of the substrate. Some of these treatments include mechanical roughening of the surface, chemical etching of the cobalt binder to enhance diamond nucleation, etching of the tungsten carbide in order to increase surface roughness and mechanical interlocking, and the deposition of intermediate layers of hard coatings to act as chemical barriers. Some improvements have been reported and considerable tool life improvements have been realized in machining highly abrasive materials such as graphite, wood-based composites and FRPs. The failure mode of diamond coated carbides is by uniform wear of the diamond film and by film delamination. Therefore, the performance of diamond coated tools is critically dependent on the feed rate, workpiece material, cutting edge geometry, substrate type and preparation, and coating thickness (Köpf *et al.*, 2006).

5.3   Tool wear

Due to the severe conditions of pressure and temperature at the tool surfaces in contact with the workpiece and chip, wear of the cutting edge during machining is inevitable. Damage to the cutting edge may occur prematurely and on a large scale, or gradually, leading to the end of tool life. Premature and large-scale damage is usually associated with improper selection of cutting tool material, cutting edge geometry or cutting parameters. Under normal conditions, wear of the cutting edge progresses gradually until it reaches a stage when the cutting tool becomes much less effective in performing its principal functions, which are material removal and generating a good quality machined surface. This stage marks the end of tool life and a tool change has to be made. Tool wear leads to undesirable consequences, such as reduction in cutting edge strength, increased tool forces and power consumption, increased cutting temperatures, degradation in surface finish, loss of part dimensional accuracy, and eventually loss of productivity. Therefore, it is extremely desirable that tool wear is considerably minimized and controlled. Attempts are continuing to achieve this objective through the development of better wear resistant materials, better workpiece machining characteristics, and proper choices of machining conditions that promote long tool life.

Tool wear is a complex phenomenon and it occurs by several mechanisms or processes, which include abrasive wear, diffusion wear, erosive wear, corrosive wear and fracture (Shaw, 2005). Abrasive wear is associated with the presence of hard particles in the workpiece, which, under high cutting pressure between the tool face and workpiece, will indent into the tool and micro-cut tiny grooves in the tool surface. Diffusion wear is associated with migration of atoms from the tool to the workpiece and *vice versa*, under conditions of high temperatures and pressures. Erosive wear is similar to abrasive wear, in regard to the gouging or cutting action of loose abrasive particles carried in a fluid medium. Corrosive wear occurs by chemical attack of the tool surface and is mainly driven by oxidation under sufficiently high cutting temperatures. Fracture wear (chipping) and delamination wear occur in brittle materials by the initiation, propagation and coalescence of microcracks, transverse and parallel to the tool surface, leading to removal of particles or flakes from the tool surface. The formation of microcracks could be caused by mechanical or thermal loads. The above mentioned wear mechanisms do not behave in a similar manner under a given set of cutting conditions, and interactions between them may occur. The relative effects of these wear mechanisms are functions of cutting temperatures, cutting forces and workpiece machining properties. In the next sections, tool wear is described in detail in terms of its types (or forms), mechanisms and the influence of cutting parameters on tool wear. Tool life of different tool materials in machining composites are also be discussed.

5.3.1 Tool wear types and evolution

Tool wear and damage are demonstrated on the cutting edge in one or more of several types or forms, as shown in Fig. 5.3. Each one of these types of wear is associated with a wear mechanism or cause that is responsible for its appearance. Flank wear (1a) is generally caused by abrasion due to the friction between the underside of the tool (clearance face) and the workpiece. Flank wear is marked by tiny grooves that run parallel to the cutting direction. Notch wear (1b) is also caused by severe abrasion at the depth of cut mark. Crater wear is caused by diffusion and is generally associated with machining steels at high cutting speeds with WC-Co tools. Crater wear (2) appears on the rake face of the tool on which the chip flows and is usually positioned away from the cutting edge. This position coincides with the location of maximum temperature on the face of the tool. Plastic deformation (3) results from excessive cutting forces and is associated with high speed steels and the softer grades of WC-Co tools. Built-up edge wear (4) is caused by adhesion of the workpiece material to the phase of the tool.

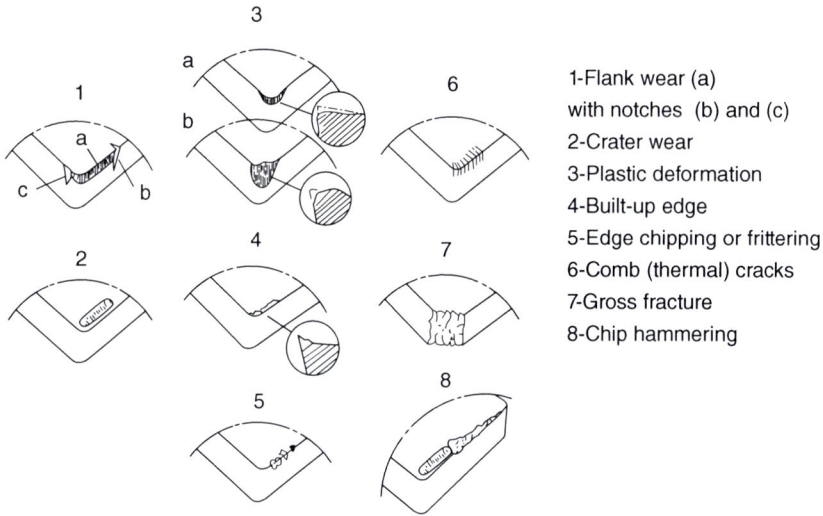

1-Flank wear (a)
with notches (b) and (c)
2-Crater wear
3-Plastic deformation
4-Built-up edge
5-Edge chipping or frittering
6-Comb (thermal) cracks
7-Gross fracture
8-Chip hammering

5.3 Types of failures and wear on cutting tools (adapted from Sandvik®).

This is more common when machining ductile metals such as aluminum and carbon steel.

Flank and crater wear are the most important measured forms of tool wear. Flank wear is most commonly used for wear monitoring because it is easily measured. According to the norm ISO 3685:1993 for wear measurements, the major cutting edge is divided into four regions, as shown in Fig. 5.4.

Region C contains the curved part of the cutting edge at the tool corner;

Region B is the remaining straight part of cutting edge between regions C and A;

Region A is the quarter of the worn cutting edge length *b* farthest away from the tool corner;

Region N extends beyond the area of mutual contact between the tool workpiece for approximately 1 to 2 mm along the major cutting edge. The wear in this region is of notch type.

The width of the flank wear land VB_B is measured within region B in the cutting edge plane *Ps* perpendicular to the major cutting edge. The width of the flank wear land is measured from the position of the original major cutting edge. The crater depth *KT* is measured as the maximum distance between the crater bottom and the original face in region B. Tool wear is most commonly measured using the toolmaker microscope (with video imaging systems and a resolution less than 0.01 mm) or a stylus instrument similar to a profilometer (with ground diamond styluses).

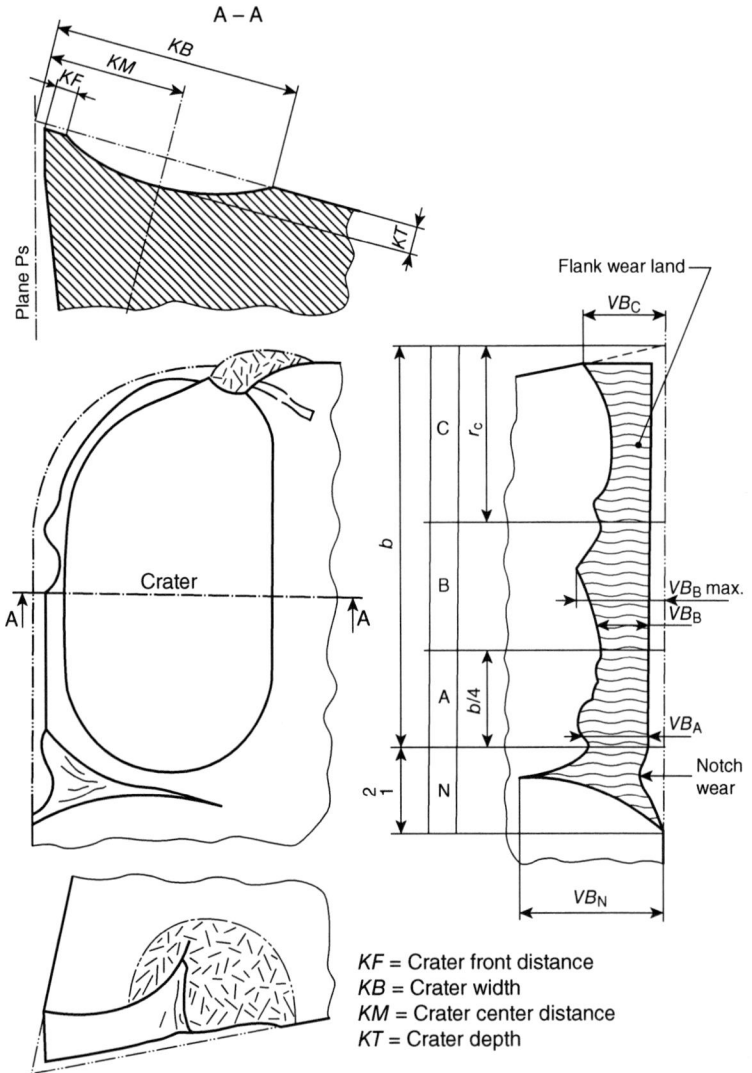

KF = Crater front distance
KB = Crater width
KM = Crater center distance
KT = Crater depth

5.4 Types of tool wear according to norm ISO 3685:1993.

In general, the failure and wear of cutting tools depend on tool material and geometry, workpiece material, cutting parameters (cutting speed, feed rate and depth of cut), cutting fluids and machine-tool characteristics. Under proper cutting conditions, the cumulative effect of the various wear mechanisms causes gradual degradation of the cutting edge with cutting time. Figure 5.5 shows a typical curve of the progress of flank wear land VB_B with cutting time for different cutting speeds.

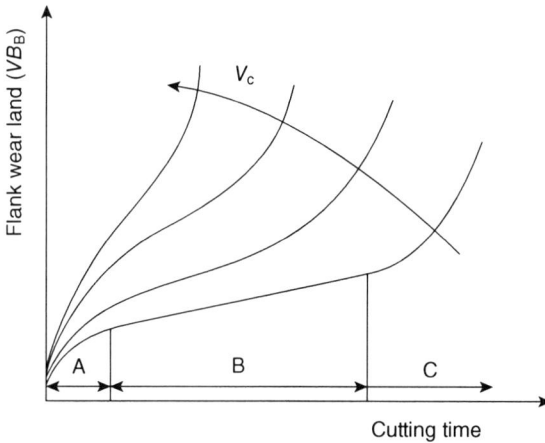

5.5 Evolution of flank wear land VB_B as a function of cutting time for different cutting speeds.

The wear curve in Fig. 5.5 can be divided into three zones. The wear of the cutting edge starts at a high rate (Region A), which is caused by the initial breakdown of the sharp corners and weak spots in the cutting edge. The cutting edge is then stabilized and wear proceeds by gradual removal of the tool material (Region B) up to a point when the cutting edge geometry becomes dysfunctional and rapid or catastrophic wear sets in (Region C). Region B, which corresponds to wear at a uniform rate, represents the useful life of the cutting edge. Wear in this zone is predictable and has been modeled mathematically. This has allowed successful prediction of tool life and scheduling of tool replacement. The influence of cutting speed on flank wear land is also shown in Fig. 5.5. It is evident that increasing the cutting speed causes a drastic increase in flank wear land VB_B for a given cutting time.

The criteria recommended by ISO 3685:1993 to define the effective tool life for cemented carbides tools, HSS (high-speed steels) and ceramics is listed below:

Cemented carbides

(i) $VB_B = 0.3$ mm, if the flank wear is uniform in Region B, or
(ii) $VB_{Bmax} = 0.6$ mm, if flank wear is non-uniform in Region B, or
(iii) $KT = 0.06 + 0.3f$, where f is the feed rate.

HSS and ceramics

(i)   Catastrophic failure, or
(ii) $VB_B = 0.3$ mm, if the flank is uniform in Region B, or
(iii) $VB_{Bmax} = 0.6$ mm, if flank is non-uniform in Region B

Table 5.2 Recommendations used in industrial practice for limit of flank wear
VB_B (mm) for several cutting tool materials

Operation	HSS	Cemented carbides	Coated carbides	Ceramics	
				Al_2O_3	Si_3N_4
Roughing	0.35–1.0	0.3–0.5	0.3–0.5	0.25–0.3	0.25–0.5
Finishing	0.2–0.3	0.1–0.25	0.1–0.25	0.1–0.2	0.1–0.2

General recommendations used in industrial practice for limit of flank wear
VB_B for several cutting materials are given in Table 5.2.

5.4   Tool wear in machining metal matrix composites

5.4.1 Wear of carbide tools

Carbides are the oldest hard tool materials in the machining trade. Their
hardness and toughness are tailored with great flexibility to fit the application by adjusting the weight fraction of the binder content and the grain
size. Most recently, the development of ultra-fine (less than 0.5 μm) and
layered carbide further enhanced their toughness and hardness. However,
the most obvious shortcoming of these hard materials is that their hardness
is not high enough to resist the abrading effects of reinforcement phases in
MMCs (see Fig. 5.2). Therefore, carbide tools fail miserably in machining
MMCs. Tool coatings provide only slight or no improvement to the wear
resistance of carbide tools, apparently for the same reasons of inadequate
levels of hardness. The adhesion of the tool coating to the substrate are not
an apparent problem as the abrasion grooves cut uniformly through the
coating and substrate (Tomac and Tonnessen, 1992; Sahin and Sur, 2004;
Sahin, 2005). The main mechanism of wear is abrasion on the flank face of
the tool, including two-body and three-body abrasion. Clearly visible
grooves, representing the micro-cutting path of reinforcing particles and
running parallel to the cutting direction, often mark the flank face.

The main factors affecting tool wear are volume fraction and type of the
reinforcement phase, particulate size, cutting speed and feed rate. A study
by Kilickap *et al.* (2005) compared the wear rate of uncoated and TiN
coated K10 carbide tools in machining Al-5%wt SiC MMC, with average
particle size of 24 μm. The variable process parameters in this study were
cutting speed, feed rate and depth of cut. Results from this study are shown
in Fig. 5.6 for a depth of cut of 1.0 mm. The figure clearly shows the profound effect of cutting speed on tool wear. The influence of feed rate and
depth of cut were found to be secondary. The figure also shows a slight

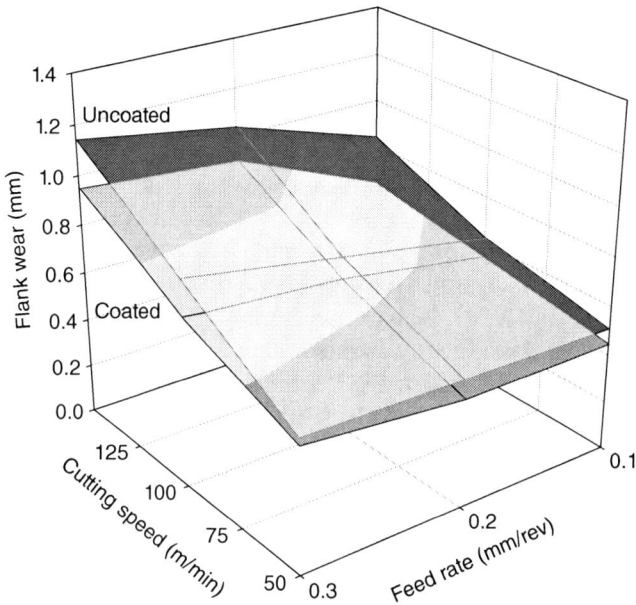

5.6 Effect of cutting speed and feed rate on tool wear in turning of Al-5%wt. SiC. Depth of cut = 1.0 mm, cutting time = 90 s, dry cutting conditions (Kilickap *et al.*, 2005).

improvement in tool wear imparted by the TiN coating. After 90 seconds of cutting, it is apparent that the tool life criterion of 0.3 mm flank wear has been exceeded by both tools at all cutting conditions. This demonstrates the unsuitability of carbide tools in machining MMCs. Tool wear was also shown to increase with an increase in reinforcement particle size and increased volume fraction (Sahin and Sur, 2004; Ciftci *et al.*, 2004b; Sahin, 2005).

5.4.2 Wear of CVD diamond coated carbides

CVD diamond is a more recent super hard tool material. It consists of a pure polycrystalline diamond coating over a cemented carbide substrate. Its hardness and toughness approaches that of single diamond crystal. However, its surface roughness is more than that of PCD and single crystal, due to the relatively large grain growth when thick diamond films are deposited. Despite some desirable properties, the advantages of CVD diamond are not fully realized due to the lack of adhesion of the diamond film to the carbide substrate. This often results in premature failure of the cutting edge by coating delamination and spalling. This problem is made worse by the high surface roughness of the diamond film. Asperities between the diamond crystals increase the tendency of the matrix material to adhere to the cutting tool, which leads to tool failure by seizure and delamination. Most of the work in the literature generally agrees that this failure mode is the main cause of tool failure. Bergman and Jacobson (1994) used a pendulum-type quick-stop device to study tool failure of diamond coated tools in machining aluminum matrix composites reinforced with alumina whiskers and SiC particles. It was shown for both composites that diamond coating failed mainly by adhering to the chip and peeling off the rake face at the moment of quick-stop. Andrewes *et al.* (2000) compared the performance of CVD and PCD tools in machining Al-20vol% SiC MMC at a cutting speed of 190 m/min. It was shown that after one third of the cutting time, wear on the CVD tool was 83% higher than that of the PCD tool. This faster wear rate was caused by both abrasion and adhesion wear. It was concluded that small crevices on the surface of the diamond film promoted adhesion of the aluminum matrix to the tool surface and accelerated the delamination of the diamond film and wear of the substrate.

Work presented by Chou and Liu (2005) shows that tool wear is sensitive to cutting speed and feed rate, the latter having the more profound effect. The dominant wear mechanism was catastrophic coating failure (peeling-off) caused by strong material adhesion and high cutting temperatures. High cutting temperatures will induce great interfacial stresses at the bonding surface due to different thermal expansions between the coating and substrate. Results presented by Davim (2002) show that PCD tools are important in the cutting of aluminum matrix composites with reduced machinability (see microstructure in Fig. 5.1). In this case, CVD diamond-coated tools show short life as tool wear evolution becomes very fast after coating failure.

Kremer and co-workers (2008) investigated the machinability of two Al/SiC particulate MMC with 5 and 15% vol. SiC particles. Dry turning tests were performed with three different CVD diamond-coated tools: a rough monolayer coating C1 (sharp crystals could be observed), a smooth 'cauliflower-like' coating C2, and a multilayered coating C3, which was a combination of the previous two. It was shown that the rough monolayer coating provided better performance than the smooth monolayer cauliflower-like. The smooth surface of the C2 tool induced a higher tendency for adhesion and greater stress on the diamond coating, resulting in stripping of the diamond particles from the tool top surface. The multilayer coating C3 (combination of smooth layer over a rough layer) gave results similar to C1 (tool life six times longer than C2). This work demonstrated the importance of coating morphology in determining friction conditions and adhesion tendency at the tool surface.

More recently, the performance of nanocrystalline CVD diamond coating in machining A359-20vol.% SiCp was investigated (Hu *et al.*, 2007;

Qin *et al.*, 2009). The nanocrystalline diamond coating (NCD) was compared with both microcrystalline coating (MCD) and PCD diamond. The grain size of MCD was in the range of 3–5 μm, while the NCD coating had ultrafine grains. NCD had greater hardness than both MCD and PCD (81, 57 and 50 GPa, respectively). However, its surface was rougher than PCD and smoother than MCD. Its elastic modulus was lower than both tools. It was shown that NCD performance in machining was comparable to that of PCD and much better than MCD. At a cutting speed of 240 m/min, a feed rate of 0.05 mm/rev and a depth of cut of 1 mm, both NCD and PCD tools reached a flank wear of 0.1 mm after 12.5 minutes while the MCD tool reached a flank wear of 0.8 mm after 2.8 minutes of cutting (Hu *et al.*, 2007). Both NCD and MCD tools failed by delamination of the coating. This delamination was also related to the adhesion on the matrix material on the flank face. The coated tools generally exhibited a slow increase in flank wear followed by an abrupt increase in wear land width in one single pass. The onset of this abrupt increase of wear was associated with coating delamination and was found to be dependent on cutting speed and feed rate. Feed rate was found to have the dominant effect, due to the increase in mechanical loads associated with the increase in feed rate. It is evident from this work that NCD tools represent a real opportunity to provide an economical solution to machining MMC. However, at the current state of development, PCD remains by far the most viable solution.

5.4.3 Wear of PCBN tools

PCBN is next only to diamond in hardness. As shown in Fig. 5.2, the hardness of PCBN is almost twice as much as that of SiC. Therefore, in principle it should be capable of resisting the aggressive abrasive wear of the reinforcing particles. However, PCBN tools are not widely used in machining MMCs due to their high cost and lack of toughness. The cutting tool geometry is often designed to increase the strength of the cutting edge. This is often done in the form of chamfering the edge (negative back rake angle) and increasing the nose radius. Both these conditions result in higher cutting forces. In a study conducted by Ciftci *et al.* (2004a), it was shown that the wear of PCBN is highly dependent on particulate size. Severe fracture of the cutting edge was the dominant wear mechanism when machining Al-16 wt% SiCp with a particle size of 110 μm. Reducing the particle size to 45 and 30 μm resulted in a significant reduction in tool wear by fracture. It was also shown that for small particulate size, the dominant wear mechanisms on the flank are abrasion and adhesion. Notch wear and flank wear caused by intergranular fracture were also found to be major types of wear occurring when machining Al-20 wt% SiC with CBN tools (Ding *et al.*, 2005).

Notch wear occurred on the flank face due to sliding contact with the feed mark ridges on the machined surface. This type of wear occurred more for PCBN grades with lower hardness and toughness. A binderless PCBN grade was found to be the most appropriate for combating this type of wear. It was also noted that notch wear was made more prevalent when a coolant was used. The detrimental effect of the coolant was more severe for the binderless PCBN grade. Material was removed from the flank face of the tool by the formation of intergranular cracks. Again, this form of wear was less severe for the binderless PCBN grade due to its high toughness.

Built-up edge formation was found to be common in machining aluminum matrix composites with PCBN tools (Looney *et al.*, 1992; Ciftci *et al.*, 2004a; Ding *et al.*, 2005). Because BUE is a transitional behavior, its breakage from the cutting edge often causes wear by removal of tool particles adhered to the cutting edge. The size of BUE was found to depend on the cutting speed. Looney *et al.* (1992) reported a significant decrease in the size of BUE as the cutting speed was increased from 17 m/min to 75 m/min in machining Al-25vol% SiC. However, it was shown by Ding *et al.* (2005) that increasing the cutting speed from 50 m/min to 400 m/min resulted in a marked increase in the size of BUE when machining Al-20%vol% SiC. Similar results were reported by Cifci *et al.* (2004a). This behavior was attributed to the softening of the matrix material as a result of the increase in cutting temperatures at higher cutting speeds. The softening of the matrix material increased the propensity for adhesion and resulted of the spread of BUE over a large area of the tool edge.

5.4.4 Wear of PCD tools

It is generally agreed that polycrystalline diamond is the most appropriate tool material for machining MMCs. Its hardness, strength and toughness surpass those of other superhard tool materials in its group. It also provides a lower coefficient of friction, less propensity for adhesion, high thermal conductivity and low coefficient of thermal expansion, which makes it more effective in managing thermal loads (see Table 5.1). In comparison to a coated carbide tool, it was shown that, when cutting Al-14wt% SiCp under the same conditions ($V = 60$ m/min, $doc = 0.4$ mm and $f = 0.1$ mm/rev), a PCD tool reaches a flank wear land of 0.4 mm after 62 minutes while a coated carbide tool reaches the same wear land after only two minutes (Tomac and Tonnessen, 1992). A similar finding was reported by Hung *et al.* (1996), where the PCD tool life was found to be 32 times more than that of K10 carbide and approximately 5 times more than that of a PCBN tool when facing Al-20wt% SiCp at a circumference speed of 6 m/min. As a matter of fact, most of the work in the literature suggests that PCD tools are best suited for machining Al-SiC composites at high cutting speeds, in

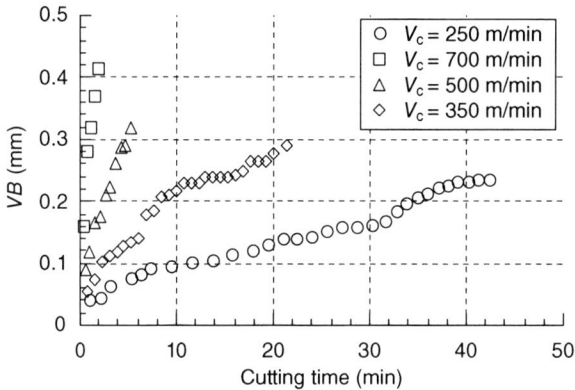

5.7 Wear curves for PCD insert obtained in turning MMC: $f = 0.1$ mm/ rev, depth of cut = 1 mm with cutting fluid (Davim and Baptista, 2000).

the range from 50 m/min up to 890 m/min, with high feed rates (El-Gallab and Sklad, 1998; Andrewes *et al.*, 2000; Heath, 2001; Ding *et al.*, 2005). The reason for this capability to perform at high cutting speeds may be attributed to a lowering of the cutting temperature as a result of the high thermal conductivity and the low coefficient of friction of the cutting tool (El-Gallab and Sklad, 2000). The inverse relationship between tool wear and feed rate is not well understood. Tomac and Tonnessen (1992) attributed this behavior to softening of the matrix material. This results in the reinforcing particles becoming pressed into the workpiece, causing less wear on the tool surface.

The progression in flank wear when turning aluminum matrix composites with PCD inserts at different cutting speeds is shown in Fig. 5.7. At a cutting speed of 250 m/min, it is possible to perform cutting for 45 min to reach an average flank wear $VB_B = 0.24$ mm. Also, at a cutting speed of 700 m/min, it is possible to machine for only 2 min for an average flank wear $VB_B = 0.42$ mm (Davim and Baptista, 2000). The effect of cutting speed on tool wear is also apparent in the figure. An increase in cutting speed results in a drastic increase in flank wear. This was attributed to the increase in kinetic energy of the abrading particles and degradation of the cutting tool material.

The predominant wear mechanisms in machining Al-SiCp MMC are two body abrasion and three body abrasion (Lane, 1992). Examples of this wear type in a turning insert are presented in Figs 5.8 and 5.9. The resistance to abrasion of cutting tool materials depends directly on the relative hardness of these materials. SiC particles used as reinforcement are harder than WC and ceramics, but less hard than PCBN and diamond. Therefore, abrasion of diamond cutting tools does not take place by microcutting of the diamond

5.8 The wear land on PCD insert used in turning aluminum matrix composite: V = 250 m/min; f = 0.1 mm/rev; depth of cut = 1 mm with cutting fluid.

crystals as in the conventional sense of abrasion of soft bodies with fixed or loose hard particles. Instead, abrasion takes place by removal of the diamond crystals from the surface of the tool. PCD consists of a thin layer of fine diamond particles, sintered together with a metallic binder and brazed onto a cemented carbide substrate (Fig. 5.8). The hardness and strength of the metallic binder, which is mostly cobalt, is lower than those of the SiC reinforcement particles (Davim, 2001). In such cases, the bonding between the cobalt and diamond crystals can be damaged by the hard particles degrading the tool. Abrasion takes place by the dislodging and removing of the diamond grit by the SiC particles. This results in pits and grooves on the surface of the tool, as shown in Fig. 5.9. The disintegration of the inserts material occurs next to the cutting edge as the maximum cutting temperature and maximum shear stress occur at this location (El-Gallab and Sklad, 1998). The mechanical damage of the cutting tool material comes from the kinetic energy transferred from the reinforcement particles of the cutting edge and depends essentially on the abrading particles size, cutting speed and diamond grit size (Lane, 1992). Tool wear increases with increase in cutting speed and increase in SiC particle size. Furthermore, increasing the size of diamond grit from 10 μm to 25 μm is shown to improve resistance to this abrasion because larger diamond particles are more difficult to dislodge. However, further increase in diamond grit size may also result in deteriorating tool performance due the decrease in toughness of the tool material.

Another important wear mechanism that occurs in machining Al-SiCp MMC is pull-out of the diamond grit by aluminum seizure or adhesion, both on the flank and rake faces. Aluminum film deposits on the flank face of the tool due to high pressures and reactivity of the newly formed aluminum

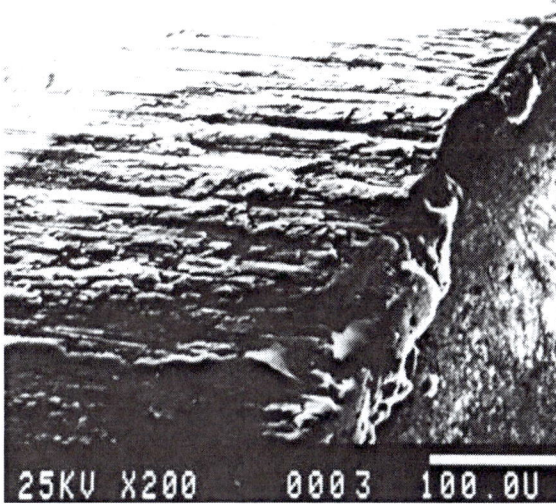

5.9 SEM image of abrasion flank wear on PCD insert used in turning aluminum matrix composite: V_c = 750 m/min; f = 0.1 mm/rev; depth of cut = 1 mm with cutting fluid.

surface. The formation of built-up edge has also been reported when machining with PCD at all cutting speeds. The deposited matrix material adheres to the diamond grit and subsequent removal of this material causes removal of diamond grit from the tool surfaces. This type of wear becomes evident after etching the cutting edge to remove aluminum deposition (El-Ghallab and Sklad, 1998; Andrewes *et al.*, 2000). Again, this type of wear could be reduced by increasing diamond grit size.

Effects of reinforcement phase on tool wear

The previous sections have indicated that the morphology of the reinforcement phases, its weight or volume fraction, and its hardness, are influential parameters on the progression of tool wear in machining MMCs. The main reinforcement phases, Al_2O_3 and SiC, are relatively hard phases as compared to other tool materials and therefore are capable of abrading these tool materials. SiC, being harder than Al_2O_3, has been shown to generate more wear or give shorter tool life (Sahin and Sur, 2004; Sahin *et al.*, 2002).

Particulate SiC phase was also found to cause more wear on the rake face of the tool than whisker-SiC (Bergman and Jacobson, 1994). This was attributed to the presence of weakly held fragments of the whiskers in the chip. It was postulated that the whiskers, being long, slender and brittle, break down into fragments as they pass through the primary deformation zone. The large plastic deformation of the matrix also weakens the interface between the reinforcement and the matrix. Therefore, the whiskers that slide on the rake face are smaller and less firmly held. They cause less wear on the rake face than the particulate reinforcement which, due to its shape, can pass through the shear zone intact.

Tool wear also increases with an increase in reinforcement weight fraction and an increase in particulate size (Li and Seah, 2001; Ciftci et al., 2004a,b). The larger the particle size, the more the momentum energy carried by the particle and the more abrasion it can cause. Ozben et al. (2008) investigated the effects on tool wear of weight fraction of SiC reinforcement particles, cutting speed, feed rate and depth of cut during turning of AlSiMg$_2$ matrix composite. Figure 5.10 shows results from this study for a feed rate of 0.2 mm/rev and depth of cut of 1.0 mm. It is apparent that tool wear increased with an increase in weight fraction of the reinforcement. However, the effects of cutting speed and feed rate on tool wear were more profound.

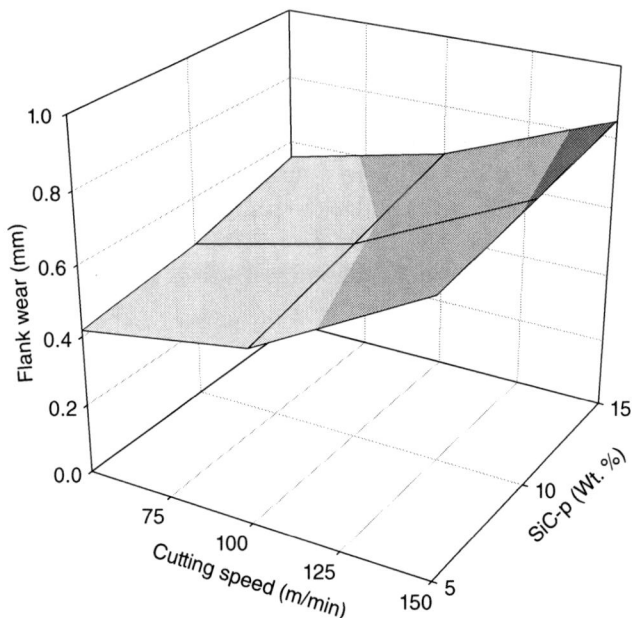

5.10 Effect of cutting speed and reinforcement weight fraction on tool wear in turning operation. Cutting tool: TiN coated K10 carbide, feed rate = 0.2 mm/rev, depth of cut = 1.0 mm, machining time = 90 s (Ozben et al., 2008).

5.5   Tool wear in machining polymeric matrix composites

Fiber reinforced polymer composites (PMC) differ in their interaction with the cutting tool from metal matrix composites. The matrix material in PMCs is not as strong or as ductile as metals (with the exception of thermoplastics which are more ductile). The fiber reinforcement is typically continuous and is utilized in greater volume fractions than in MMCs. This makes PMCs as a whole more brittle and thus their machining seldom produces a continuous chip. The material breaks ahead of the cutting tool in a series of fractures across the fibers and at the interfaces with the matrix. The absence of a chip significantly reduces the threat of wear on the rake face of the tool. However, the spring back of fibers on the newly machined surface increases the frictional conditions on the clearance face. In addition, loose fragments of the fibers entrapped between the machined surface and the clearance face give rise to three-body abrasion conditions. Thus, flank wear by two-body and three-body abrasion is more prevalent in machining PMCs.

The forces required for cutting PMCs are also lower than those required for cutting MMCs and thus the cutting temperatures are significantly lower. For example, a temperature of 260 °C was measured when machining carbon fiber reinforced polymers at a cutting speed of 200 m/min (Masuda et al., 1993). Similar temperatures were measured in the cutting of particleboard (Sheikh-Ahmad et al., 2003a), but higher temperatures between 260 to 400 °C were measured when machining glass fiber reinforced polymers at the same cutting speed, apparently due to the poor thermal conductivity of GFRP (Spur and Wunsch, 1988). Therefore, tool wear in machining PMCs is more or less controlled by mechanical rather than chemical processes and is affected by the fiber reinforcement type, architecture and volume fraction. The wear mechanisms that contribute to the overall wear of the cutting tool include gross fracture or chipping, abrasion, erosion and microfracture, or microchipping. While gross fracture results in a sudden or catastrophic failure of the cutting edge, usually in the early stages of cutting, the other wear mechanisms result in gradual or progressive wear.

Tool materials suitable for machining FRPs are those which possess high hardness and high thermal conductivity. Hardness is required for resisting abrasion, and thermal conductivity is required for dissipating heat – which is particularly important when machining low thermal conductivity FRPs such as aramid and glass FRPs. The most common tool materials used are cemented carbides, cubic boron nitride (CBN) and polycrystalline diamond (PCD). Diamond coated carbides are a promising class of materials, with many desirable properties. As diamond coating technology advances, and the problems with film adhesion and edge strength are resolved, this class of materials will probably provide an economically viable substitute for

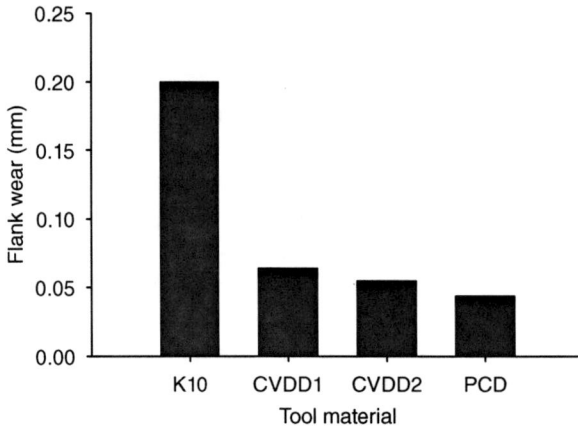

5.11 Flank wear of different tool materials in turning CFRP under wet cutting conditions. Cutting distance = 400 m, cutting speed = 100 m/min, feed rate = 0.1 mm/rev, depth of cut = 0.5 mm (Usuki *et al.*, 1991).

PCD tools. Figure 5.11 shows the flank wear of various tool materials in the turning of CFRP pipe (52 vol% carbon fiber). The cutting tool materials include a standard uncoated carbide grade K10, diamond coated silicon nitrides (CVDD1 with 10 μm film thickness and CVDD2 with 20 μm film thickness), and a PCD tool with 5 μm grain size diamond and cobalt as the binder. The figure shows that, after a cutting distance of 400 meters, the wear of the bare cemented carbide tool is considerably higher than that of the PCD (4.5 times) and the diamond coated silicon nitrides tools (3.5 and 3.1 times). This demonstrates the vulnerability of the carbide tools and the suitability of the harder diamond tooling. The higher wear resistance for the diamond coated tool CVDD2 is attributed to the thicker diamond film, which lasted longer before being abraded from the surface of the tool (Usuki *et al.*, 1991).

5.5.1 Wear of carbide tools

Depending on the type of PMC machined, volume fraction and cutting conditions, progressive wear of cemented carbide tools takes place by one or a combination of two wear mechanisms, namely soft abrasion and hard abrasion. Soft abrasion occurs when machining PMCs that are softer than the cemented carbides. Example of these are wood-based composites such as medium density fiberboard and particleboard (Sheikh-Ahmad and Bailey, 1999), sandstone and zirconia (Larsen-Basse and Devani, 1986). Hard abrasion occurs when machining carbon and glass fiber reinforced polymers and it takes place by two-body and three-body abrasion modes.

In both types of wear, the dominant forms of wear are flank wear and nose rounding. Crater wear or any other forms of rake face wear are almost absent.

Soft abrasion refers to a mechanism of wear whereby carbide grains, which are harder than the reinforcement, are removed from the tool edge by fracture and dislodging. This type of wear is more common for tool materials with a large mean free path of the binder (i.e. large grain size, large fraction of metallic binder). Soft abrasion increases with an increase in binder content and an increase in grain size. Figure 5.12 shows an illustration of this type of wear in machining particleboard (Sheikh-Ahmad and Bailey, 1999). Figure 5.12a shows an SEM micrograph of the cutting edge at small magnification. It is apparent from this figure that most of the wear occurs on the clearance face of the tool because of the rubbing action between the clearance face and the workpiece surface. This rubbing is facilitated by the elastic spring back of the reinforcement particles and matrix after cutting. Less wear is caused by flow of the chip dust and particles over the rake face. This is shown by the extent of removal of the original grinding marks from the rake face and the degree of roundness of the cutting edge. Flank wear may not be uniform and it may reflect the density variations across the thickness of the workpiece. Voids may also be seen on the tool nose, indicating material that was broken off the tool edge by microfracture. These voids are later smoothed out by wear. Figure 5.12b shows the microstructure of the wear surface in the flank wear region. Tungsten carbide grains are seen standing in relief on the wear surface, as

5.12 SEM pictures showing wear surfaces on WC-Co carbide cutting edge (97 and 3 wt.%, respectively) after edge trimming of fiberboard on a CNC router. Cutting speed = 1060 m/min, feed speed = 3.42 m/min, depth of cut = 1 mm, total cutting distance = 850 m. (a) Low magnification showing flank wear and edge roundness, (b) high magnification of wear microstructure showing preferential removal of binder and fractured grains.

well as cavities once occupied by carbide grains which have been removed. There also exists evidence of transgranular cracks in the WC grains. The presence of smaller than nominal grain size fragments of WC on the wear surface suggests that large grains have been fractured into smaller fragments. These fragments are later removed from the tool surface.

The characteristic microstructure of the wear surface in Fig. 5.12b indicates that wear of the tungsten carbide cutting edge occurs by preferential removal of the cobalt binder followed by fragmentation or fracture and uprooting of the WC grains. The binder phase is first partly removed from between the tungsten carbide grains by a combination of plastic deformation (extrusion) and micro-abrasion. The second stage of wear occurs when sufficient binder has been removed, and it involves removal of the carbide grains from the surface by fracture and uprooting. Moreover, loose micro-fragments of the hard and abrasive reinforcement phase at the interface between the tool and workpiece are able to penetrate, under cutting pressure, between the carbide grains, and preferentially erode the cobalt binder by micro-abrasion. High fluctuating forces that are generated by material inhomogeneity will then cause the tungsten carbide grains to rock slightly in their position in the WC–binder composite. This, in turn, results in the partial extrusion of the binder to the surface of the cutting tool where it is later removed by the workpiece. In addition, the relative motion of the brittle carbide grains results in developing cracks across the grains, which is followed by fracture and removal of parts of or the whole grain from the binder matrix, leaving large voids as shown in Fig. 5.12b.

The wear macro and micro appearance of hard abrasion of WC–Co tool materials is shown in Fig. 5.13. Figure 5.13a shows the flank surface of a C2 end mill after edge trimming of carbon fiber reinforced epoxy. Similar to soft abrasion, rubbing against the clearance face and the tool nose causes nose rounding and flank wear. However, the wear surfaces appear to be smoother and no visible pits or microfracture surfaces can be seen. Figure 5.13b shows the microstructure in the worn region. Examination of this microstructure reveals that the carbide grains have been polished flat by the abrading particles. No visible grooves can be seen on the carbide grains or the wear surface in general. This suggests that abrasion of the carbide grains takes place mostly by three-body and two-body abrasion mechanisms. The loose hard particles from the carbon fibers (generally in lengths of one or two orders of the fiber diameter) are dragged under pressure between the cutting tool surface and the workpiece surface. The fibers sticking out of the machined surface also act as a wear brush that smoothes out the tool surface. These combined actions cause a polishing-like action of the carbide grains, as shown in the figure. The figure also shows minimal removal of the binder from between the carbide grains. In addition to hard abrasion, shedding of the carbide grains due to brittle fracture has been

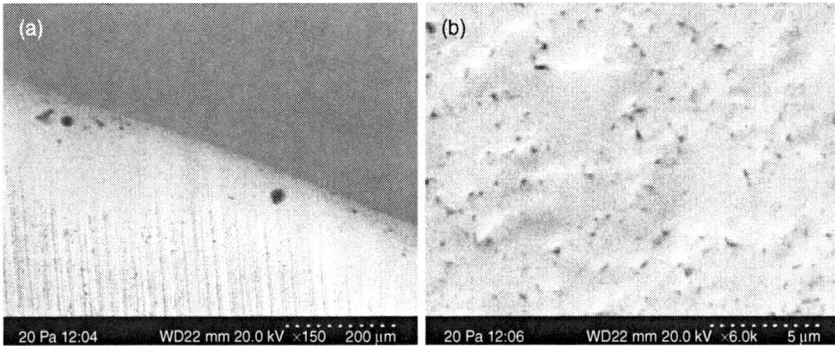

5.13 SEM pictures showing wear land on the flank face of C2 carbide end mill after cutting 31 m of CFRP at cutting speed of 150 m/min, feed rate of 1.27 m/min and depth of cut of 1 mm. (a) Low magnification showing flank wear and edge roundness, (b) high magnification of wear microstructure showing nearly polished tool surface and pits representing lost binder.

reported as a primary factor in machining CFRP and sintered carbon (Masuda *et al.*, 1993). Also in this study, the strong relationship between the mean free path of the binder and width of wear was revealed.

5.5.2 Wear of CVD diamond coated carbides

Figure 5.14 shows the performance of diamond coated and uncoated tools in terms of width of flank wear in milling CFRP at two feed rates and two cutting distances – feed rate of 1.27 m/min and lineal cutting distance of 30 m; and feed rate of 2.54 m/min and lineal cutting distance of 18 m. The cutting tool used was a general purpose C2 carbide, four fluted end mill with 30° helix angle, 15° rake angle and 65° cutting edge angle. The diamond coating was produced by hot filament CVD at two coating thicknesses, 10 μm (D10) and 20 μm (D20). Despite the higher cutting distance used for the low feed rate, it was shown that flank wear of both diamond coated tools was considerably less than that of the uncoated tool, and also less than that of the coated tools tested at the high feed rate. The lower feed rate produced a smaller chip per tooth and thus lower cutting forces and less severe impact on the cutting edge. This resulted in prolonging the diamond film life by reducing its chipping and delamination. Tool wear was also reduced by increasing the film thickness from 10 μm to 20 μm, also at the low feed rate. The thicker film thickness performed poorly at the high feed rate, possibly because of higher internal stresses. It is evident from this figure the significant role that feed rate plays in controlling the performance

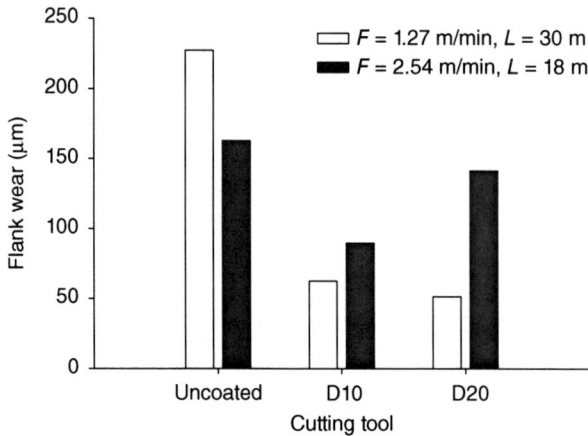

5.14 Wear of uncoated and diamond coated carbide tools in edge trimming of CFRP at cutting speed of 150 m/min and depth of cut of 1 mm. L is lineal cutting distance.

of diamond coated tools in machining PMCs. Similar findings regarding the influence of feed rate have also been reported in the machining of MMC (Hu *et al.*, 2007).

In the absence of premature failure of the cutting edge because of lack of mechanical strength, two wear mechanisms determine to the wear characteristics of diamond coated tools in machining PMCs. These are uniform abrasion and fracture-controlled delamination of the diamond film. When the film adhesion is sufficient, the diamond coated tools wear by uniform abrasion of the diamond film, as shown in Fig. 5.15a, and generally exhibit very long tool life in comparison with uncoated tools. This is attributed to the extreme hardness of the diamond film, its low coefficient of friction and its high thermal conductivity. The wear by abrasion proceeds by gradual smoothing and shedding of the diamond grains, leading to thinning of the diamond film. This in turn leads to fracture and localized removal of the film from the substrate. Accelerated wear by abrasion of both substrate and diamond film then follows. When the film adhesion is not sufficient or when conditions of residual stresses in the film are not favorable, localized fracture initiates at the nose of the cutting edge, in the form of a radial crack. The growth of this crack under loading causes large scale delamination of the diamond film, as shown in Fig. 5.15b. The substrate is then exposed, and accelerated wear by abrasion of the substrate and diamond film follows. High feed rates and small film thickness are found to promote wear by fracture and delamination of the diamond film (Sheikh-Ahmad *et al.*, 2003b; Köpf *et al.*, 2006).

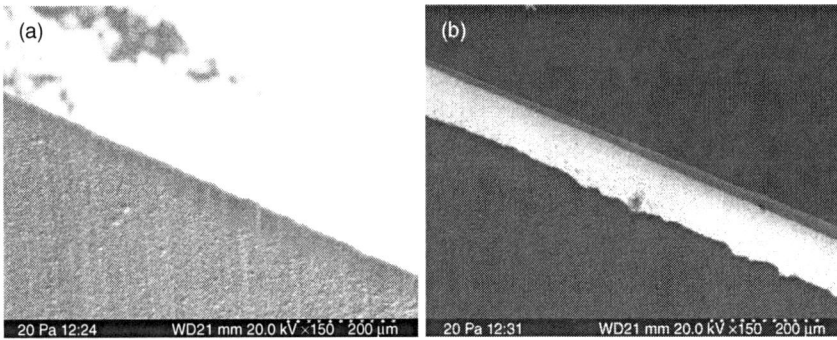

5.15 SEM pictures showing flank of diamond coated C2 carbide end mill after cutting 31 m of CFRP at cutting speed of 150 m/min, feed rate of 1.27 m/min and depth of cut of 1 mm. (a) 20 μm film thickness, (b) 10 μm film thickness.

5.5.3 Wear of PCD tools

Uniform wear by abrasion is the dominant wear mechanism when machining homogeneous composites with low feed rates. The wear mechanism appears to be similar to that present when machining with tungsten carbides (i.e. soft abrasion). Voids on the wear surface indicate locations where diamond grains were dislodged. Shallow abrasive grooves form on the flank face and run in the direction of cutting. The cobalt binder is first removed from between the diamond grains, which facilitates its dislodgment from the cutting edge. As the work material become less homogeneous, or when using high feed rates, especially in interrupted cutting, the oscillating forces on the cutting edge give rise to micro chipping and fracture as the dominant wear mechanism. Chipping can occur at a scale much larger than the diamond grain size, causing severe damage to the cutting edge. Rounding is also noticed on the rake face, but the wear land on the rake face is much narrower than that on the clearance face. Coarse PCD grades are found to perform better than the fine ones in the long run, as they provide smaller flank wear rate. On the other hand, fine grain PCD is less susceptible to micro chipping of the cutting edge than coarse grain PCD. Adhesive wear may also be present and is marked by the presence of carbonized or melted material deposits that settled on the tool surfaces (Ramulu *et al.*, 1989; Philbin and Gordon, 2005).

5.6   Tool life

Tool life is the time a tool will cut satisfactorily and is expressed in minutes between tool changes. A tool-life criterion is a measurable indication of the end of tool life. It may be defined in terms of critical levels of measurable physical quantities such as cutting power, cutting forces and tool wear. Tool wear is often used as an indication of tool life because it is easy to determine quantitatively. Flank wear land width VB_B is often used as a tool life criterion because of its influence on workpiece surface roughness and accuracy. Under proper cutting conditions, tool life is reached gradually because of progressive wear, as shown in Fig. 5.5. But the tool life could also end prematurely because of excessive chipping or breakage of the cutting edge. This scenario is avoided in all practical machining applications. In cutting metals and MMCs, threshold values of $VB_B = 0.3$ mm for uniform wear or $VB_{Bmax} = 0.6$ mm for irregular flank wear are used for tool life criteria, as described by ISO3685:1993. However, there is no agreement yet on a similar criteria for machining PMCs, but a value of $VB_B = 0.2$ mm has been frequently used. Reliable assessment of tool life is important in machining since considerable time is lost whenever a tool is replaced and reset.

Tool life depends on the tool material and geometry; the cutting parameters (cutting speed, feed, depth of cut, cutting fluids); the workpiece material (chemical composition, hardness, strength); the machining process (turning, drilling, milling); the machine-tool (for example, stiffness and state of maintenance); and other machining parameters. Early work in metal machining by Taylor (1907) has shown that tool life is strongly dependent on cutting speed. An empirical relationship was produced to relate tool life to cutting speed:

$$VT^n = C \qquad \qquad [5.1]$$

where V is the cutting speed, T is tool life, and n and C are empirical constants. This relationship is widely known as the Taylor Tool Life Equation. In the SI system of units, V is measured in m/min, T in minutes and C becomes the cutting speed for a tool life of 1 minute. Values for the emprical constants C and n are determined from tool wear data, as shown schematically in Fig. 5.16. Progressive wear curves (VB_B versus cutting time) for several cutting speeds ($V1$, $V2$ and $V3$) are shown as a function of time in Fig. 5.16a. Using an appropriate tool life criterion (e.g. VB_B limit), the time required by the tool to reach this limit is determined. This time is dependent on cutting speed. The tool life curve (cutting speed versus tool life) is obtained by plotting the data on a log–log scale, as shown in Fig. 5.16b. The coefficients n and C are determined directly from the slope and y-intercept of the straight line in this figure, as explained in an example later. Therefore, each combination of tool material and workpiece and each cutting parameter has its own n and C values, both of which are determined experimentally.

Typical values for n and C are shown in Table 5.3 for machining MMCs and in Tables 5.4 and 5.5 for machining PMCs. The constant C represents

(a)

(b)

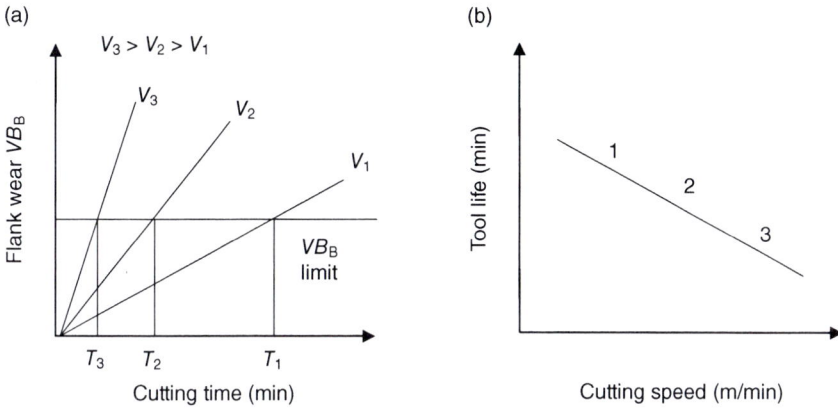

5.16 Schematics of wear curves for (a) several cutting velocities
(1, 2 and 3) and (b) life curve.

the cutting speed for a tool life of 1 minute. Thus, larger values of C indicate better tool wear performance and better machinability at high cutting speeds. The constant n represents the negative reciprocal of the slope of the tool life curve. Smaller values of n correspond to steeper curves and greater dependence of tool life on cutting speed. It is apparent from Table 5.3 that higher values of C are associated with PCD tools (with the exception of K10-Al-20% Al_2O_3 combination). This implies that greater cutting speeds can be achieved with the PCD tools in machining MMCs. This is largely attributed to the high thermal conductivity of diamond as compared to the carbide tool. The values of n are also generally smaller for PCD tools than those for carbide tools, which means that tool life exhibits greater dependence on cutting speed for PCD tools, which is a peculiar behavior. In general, the coefficient n in machining MMCs is generally high, in comparison to machining unreinforced aluminum (where n is in the range of 0.2 to 0.5). This indicates the significant influence of cutting speed on tool wear.

Similar conclusions can be drawn from the data in Tables 5.4 and 5.5 for machining PMCs. Also noted are the smaller values of n and C for machining GFRP in comparison to CFRP, which indicates that the tool wear in machining CFRP is less dependent on cutting speed than when machining GFRP, and thus higher cutting speeds are attainable when machining CFRP. This is attributed to the excellent thermal conductivity of the carbon fibers and their ability to remove heat from the cutting zone. This leads to lower cutting edge temperatures and lower wear rates. The same phenomenon is also present when machining GFRP with PCD tools and diamond coated tools. The excellent thermal conductivity of diamond allows heat to be

Table 5.3 Values of n observed in practice for several cutting tool materials in machining MMCs

MMC	Tool	Process	n	C	V (m/min)	f (mm)	d (mm)	Note
Al-15% SiC – 5% Al$_2$O$_3$	K10	Facing	1.048	9.7	5–50	0.1	0.25	a,b
Al-20% Al$_2$O$_3$	K10	Facing	0.919	1079.5	200–500	0.1	0.25	a,b
Al-20% Si	K10	Facing	0.252	405.9	150–500	0.1	0.25	a,b
Al-15% SiC – 5% Al$_2$O$_3$	PCD	Facing	0.660	1124.8	50–500	0.1	0.25	a,b
Al-20% Al$_2$O$_3$	PCD	Facing	0.854	1864.6	200–500	0.1	0.25	a,b
Al-20% Si	PCD	Facing	0.108	2380.7	1000–2000	0.1	0.25	a,b
Al-20%SiC	K10	Milling	0.832	11.5	5–50	0.1	1.0	a,b,c
Al-20%SiC	PCD	Milling	0.709	3218.9	200–1000	0.1	1.0	a,b,c
Al2024–10% Al$_2$O$_3$	TP30	Turning	0.681	191.7	100–210	0.1	2.0	d,e,f
Al2024–20% Al$_2$O$_3$	TP30	Turning	0.667	145.3	100–210	0.1	2.0	d,e,f
Al2024–10% Al$_2$O$_3$	K10/TiN	Turning	0.780	287.9	100–210	0.1	2.0	d,e
Al2024–20% Al$_2$O$_3$	K10/TiN	Turning	0.696	205.6	100–210	0.1	2.0	d,e

Notes: (a) Data extracted from Heath (2001). (b) $VB_B = 0.2$ mm. (c) Feed per tooth reported in table. (d) Data extracted from Sahin et al. (2002). (e) $VB_B = 0.3$ mm. (f) P30 carbide coated with TiN, Ti(C,N), TiN.

Table 5.4 Taylor's tool life equation coefficients for several tool–workpiece combinations in turning FRPs. Feed = 0.1 mm/rev, *VB* = 0.2 mm (Spur and Lachmund, 1999)

Workpiece	Tool	Cutting speed (m/min)	*d* (mm)	*n*	*C*
UD-GFRP (V_f = 0.7)	K10	30–50	2.0	0.2334	90
UD-GFRP (V_f = 0.7)	PCD	200–250	2.0	0.1684	398
CFRP filament wound	K10	80–300	2.0	0.7813	1640
CFRP filament wound	PCD*	500–1500	2.0	0.4237	2900
GFRP (V_f = 0.5)	K10[†]	100–300	1.0	0.4069	565.6
GFRP (V_f = 0.5)	K10	100–300	1.0	0.2710	152.66

* *VB* = 0.1 mm
[†] Diamond coated K10

Table 5.5 Taylor's tool life equation coefficients for up-milling of multidirectional CFRP laminate with TiN coated four-flute carbide end mill. Tool diameter = 11.11 mm, depth of cut = 1 mm (Ucar and Wang, 2005)

Cutting speed (m/min)	Feed (m/min)	*n*	*C* (m/min)
18–35	0.076	0.798	533.7
18–35	0.127	0.583	141.3
18–35	0.178	0.445	69.7

conducted away from the cutting zone through the cutting tool. The small slope and flatter profile in tool life with cutting speed is related to the low temperature developed during machining GFRP with diamond tools.

The effects of feed and the depth of cut on tool life are expressed in a generalized tool life equation:

$$VT^n f^a d^b = C \qquad [5.2]$$

where *d* is the depth of cut (mm) and *f* is the feed rate (mm/rev). The exponents *a* and *b* must be determined experimentally for each cutting parameter. Figure 5.17 demonstrates the effects of feed rate and depth of cut on tool wear (width of flank wear) in machining Al-SiC MMC. An empirical relation describing this dependence is given below. Even though this is not a formal tool life equation (as given Eq. 5.2), it does adequately describe the influence of process parameters on tool wear. It can be seen that cutting speed has the most significant effect on tool wear, followed by feed rate and, lastly, depth of cut.

$$VB_B = 0.087 V^{0.573} f^{0.252} d^{0.129} \qquad [5.3]$$

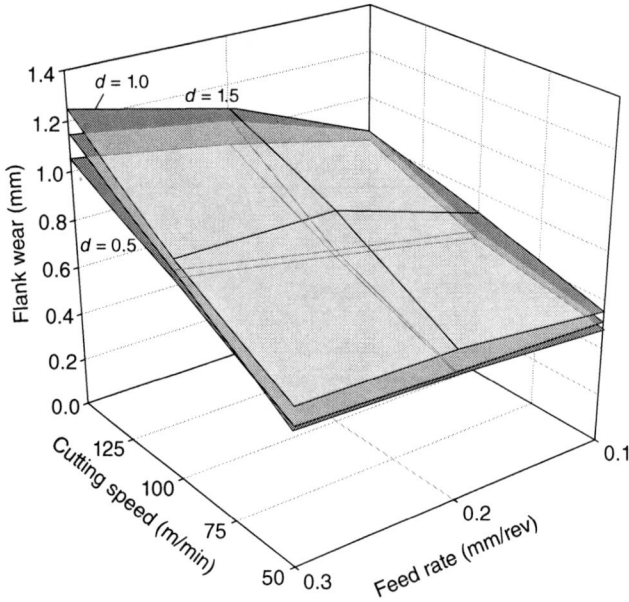

5.17 Effect of cutting parameters on tool wear in machining Al-5%SiC$_p$ MMC with K10 tool under dry conditions. The machining time is 90 seconds (Kilickap *et al.*, 2005); *d*, depth of cut (mm).

5.6.1 Tool life equation example

A typical procedure for determining a tool life equation for a work–tool pair consists of conducting tool wear tests at different cutting speeds, feed rates and depths of cut. The cutting experiment is stopped at pre-designated stops and measurement of wear is made. This measurement could be in the form of uniform flank wear width, crater depth, or maximum notch depth. A wear criterion is determined, based on acceptable machining performance, and the tool life for each cutting condition is determined using this criterion. The tool life data is then used to determine the coefficients of Taylor's tool life equation.

For demonstrating this process, consider the tool wear experimental results presented in Fig. 5.7, and reproduced in Fig. 5.18. Furthermore, consider a tool life criterion of $VB = 0.2$ mm. Using this criterion, the tool life at each cutting speed is determined as shown in the figure. The resulting data is shown in Table 5.6.

The experimental data is then used to determine the coefficients of Taylor tool life equation. Using standard mathematical procedures, the coefficients n and C can be determined using only two points from the table.

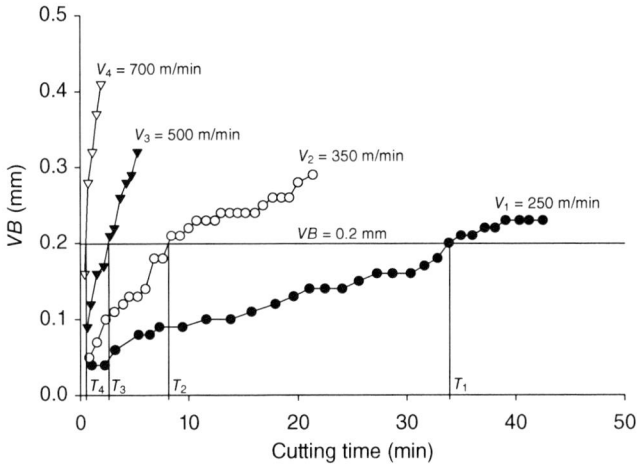

5.18 Tool wear vs. cutting time for the data in Fig. 5.7.

Table 5.6 Tool life data obtained from Fig. 5.18

Data point	Cutting speed (m/min)	Tool life (min)
1	250	34.50
2	350	8.15
3	500	2.64
4	700	0.48

For example, choosing the two extreme points in the table, points 1 and 4, $V1 = 250$ m/min, $T1 = 34.50$ min and $V4 = 700$ m/min and $T4 = 0.48$ min, respectively, we have:

$$250 \times (34.50)^n = C \qquad [5.4]$$

$$700 \times (0.48)^n = C \qquad [5.5]$$

Applying the natural logarithm to each term of these equations and making them equal gives:

$$\ln(250) + n\ln(34.50) = \ln(700) + n\ln(0.48) \qquad [5.6]$$

Solving for n gives:

$$n = 0.241 \qquad [5.7]$$

Substituting this value of n in equation 5.4 or 5.5, we obtain the value of C

$$C = 250 \times (34.50)^{0.241} = 586.6 \text{ m/min} \qquad [5.8]$$

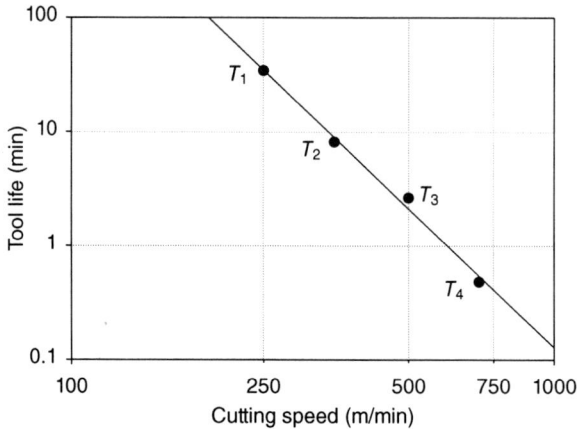

5.19 Tool life equation for data in Table 5.6.

Alternatively, a more representative estimation of the tool life equation coefficients may be obtained by curve fitting of the data in Table 5.6 as shown in Fig. 5.19. Plotting the data on a logarithmic scale displays the equation as a straight line and linear regression can be used to determine the slope and y-intercept of this straight line. For the data shown, the slope is $m = -4.046$ and the y-intercept is $b = 11.244$. The coefficient n is related to the slope of the line in the figure as $m = -1/n$. The coefficient C is related to the y-intercept by $b = (\log C)/n$. Using these relationships, the coefficients of Taylor tool life equation for the data in Fig. 5.19 are $n = 0.247$ and $C = 598.8$, and the tool life equation is

$$VT^{0.247} = 598.8 \qquad\qquad [5.9]$$

5.7   Conclusions

Tool wear in machining MMCs and PMCs occurs primarily by abrasion and microchipping. The latter is greatly assisted by adhesion of the matrix to the tool material in the case of MMCs. Abrasive wear is caused by sliding of the tool surfaces under pressure against the abrasive reinforcement material embedded in the matrix and by three-body abrasion caused by the debris from fibers, whiskers and particles.

Abrasive wear occurs in two modes: soft abrasion and hard abrasion. In hard abrasion, the tool material is removed by microcutting or polishing. The wear surface is smooth. In soft abrasion, the hard particles of the cutting tool are fractured and dislodged from the tool surface by the cutting loads. At the microstructural level, the wear surface is characterized by hard particles standing in relief, and by cracked and broken hard particles. This is brought about by removing the binder phases from between the hard particles by extrusion and erosion. This causes the hard phase particles to break and fall off the tool surface.

These wear lands caused by microchipping are often marked by shallow grooves running in the cutting direction and by large pits or voids on the surface, where larger than grain size material is removed. Microchipping is caused by the oscillating cutting forces that result from repeated impacts of the reinforcement particles against the cutting edge.

In general, the cutting tool material requirements to resist these types of wear are high hardness and high toughness. Unfortunately, cutting tool materials that have high hardness generally suffer from low toughness. An example of this is the family of sintered polycrystalline diamond tools (PCD) commonly used in machining composite materials. These tools commonly fail by chipping due to the lack of toughness. This problem is often avoided by designing the cutting edge with a large included angle, small rake and a chamfer. Cemented carbides, on the other hand, offer higher toughness but lower hardness than PCDs.

Tool life is influenced by cutting tool parameters (such as rake angle, clearance angle) and by process parameters (such as cutting speed, feed speed and depth of cut). The effect of tool geometry on tool wear stems from its effect on cutting forces and cutting edge strength. Tool life decreases with an increase in cutting speed, an increase in feed rate and an increase in depth of cut. The effect of cutting speed on tool life is the most significant. This is perhaps due to the direct relationship between cutting speed and cutting temperature.

5.8   References

Andrewes, C.J.E., Feng, H-Y, Lau, W.M. (2000) Machining of an aluminium/SiC composite using diamond inserts. *Journal of Materials Processing Technology*, **102**, 25–29.

Bergman, F., Jacobson, S. (1994) Tool wear mechanisms in intermittent cutting of metal matrix composites. *Wear*, **179**, 89–93.

Chou, Y.K., Liu, J. (2005) CVD diamond tool performance in metal matrix composite machining. *Surface & Coatings Technology*, **200**, 1872–1878.

Ciftci, I., Turker, M., Seker, U. (2004a) CBN cutting tool wear during machining of particulate reinforced MMCs. *Wear*, **257**, 1041–1046.

Ciftci, I., Turker, M., Seker, U. (2004b) Evaluation of tool wear when machining SiC$_p$-reinforced Al-2014 alloy matrix composites. *Materials and Design*, **25**, 251–255.

Davim, J.P. (2001) Turning particulate metal matrix composites: Experimental study of the evolution of cutting forces, tool wear and workpiece surface roughness with the cutting time. *Journal of Engineering Manufacture, Proceeding of the Institute of Mechanical Engineers*, **215** part B, 371–376.

Davim, J.P. and Baptista, A.P. (2000) Relationship between cutting force and PCD cutting tool wear in machining silicon carbide reinforced aluminium. *Journal of Materials Processing Technology*, **103**, 417–423.

Davim, J.P. (2002) Diamond tool performance in machining metal-matrix composites. *Journal of Materials Processing Technology*, **128**, 100–105.

Ding, X., Liew, W.Y.H., Liu, X.D. (2005) Evaluation of machining performance of MMC with PCBN and PCD tools. *Wear*, **259**, 1225–1234.

El-Gallab, M., Sklad, M. (1998) Machining of Al/SiC particulate metal-matrix composites. Part I: Tool performance. *Journal of Materials Processing Technology*, **83**, 151–158.

El-Gallab, M., Sklad, M. (2000) Machining of Al/SiC particulate metal-matrix composites. Part III: Comprehensive tool wear models. *Journal of Materials Processing Technology*, **100**, 194–199.

Heath, P. (2001) Developments in applications of PCD tooling. *Journal of Materials Processing Technology*, **116**, 31–38.

Hu, J., Chou, Y.K., Thompson, R.G., Burgess, J., Street, S. (2007) Characterization of nano-crystalline diamond coating cutting tools. *Surface & Coating Technology*, **202**, 1113–1117.

Hung, N.P., Boey, K.A., Khor, K.A., Phua, Y.S., Lee, H.F. (1996) Machinability of aluminium alloy reinforced with silicon carbide particulates. *Journal of Materials Processing Technology*, **56**, 966–977.

Kilickap, E., Cakir, O., Aksoy, M., Inan, A. (2005) Study of tool wear and surface roughness in machining of homogenised SiC-p reinforced aluminium metal matrix composite. *Journal of Materials Processing Technology*, **164-164**, 862–867.

Köpf, A., Feistritzer, S., Udier, K. (2006) Diamond coated cutting tools for machining of non-ferrous metals and fibre reinforced polymers. *International Journal of Refractory Metals & Hard Materials*, **24**, 354–359.

Kremer, A., Devillez, A., Dominiak, S., Dudzinski, D., El Mansori, M. (2008) Machinability of Al/SiC particulate metal-matrix composites under dry conditions with CVD Diamond-coated carbide tools, *Machining Science and Technology*, **12**, 214–233.

Lane, G. (1992) The effect of different reinforcement on PCD tool life for aluminium composites, *Proceedings of the Machining of Composite Materials Symposium*, Chicago, 3–15.

Larsen-Basse, J., Devani, N. (1986) Binder extrusion as a controlling mechanism in abrasion of WC-Co cemented carbides. in E. Almond, C. Brooks and R. Warren (eds.), *Science of Hard Metals*, Adam Higler, pp. 883–895.

Li, X., Seah, W.K.H. (2001) Tool wear acceleration in relation to workpiece reinforcement percentage in cutting of metal matrix composites. *Wear*, **247**, 161–171.

Looney, L.A., Monaghan, J.M., O'Reilly, P., Taplin, D.M.R. (1992) The turning of an Al/SiC metal-matrix composite. *Journal of Materials Processing Technology*, **33**, 453–468.

Masuda, M., Kuroshima, Y., Chujo, Y. (1993) Failure of tungsten carbide–cobalt alloy tools in machining of carbon materials. *Wear* **169**, 135–140.

Ozben, T., Kilickap, E., Cakir, O. (2008) Investigation of mechanical and machinability properties of SiC particle reinforced Al-MMC. *Journal of Materials Processing Technology*, **198**, 220–225.

Philbin, P., Gordon, S. (2005) Characteristics of the wear behavior of polycrystalline diamond (PCD) tools when machining wood-based composites. *Journal of Materials Processing Technology*, **162–163**, 665–672.

Qin, F., Hu, J., Chou, Y.K., Thompson, R.G. (2009) Delamination wear of nano-diamond coated cutting tools in composite machining. *Wear*, **267**, 991–995.

Ramulu, M., Faridnia, M., Garbini, J.L., Jorgensen, J.E. (1989) Machining of graphite/epoxy materials with polycrystalline diamond (PCD) tools. In *Machining Characteristics of Advanced Materials*. Ramulu, M., Hashish, M. (Editors), ASME Publication MD-Vol. 16, pp. 33–40.

Sahin, Y. (2005) The effects of various multilayer ceramics coatings on the wear of carbide cutting tools when machining metal matrix composites. *Surface and Coating Technology*, **199**, 112–117.

Sahin, Y., Kok, M., Celik, H. (2002) Tool wear and surface roughness of Al_2O_3 particle-reinforced aluminium alloy composites. *Journal of Materials Processing Technology*, **128**, 280–291.

Sahin, Y., Sur, G. (2004) The effect of Al_2O_3, TiN and Ti (C,N) based CVD coatings on tool wear in machining metal matrix composites. *Surface and Coating Technology*, **179**, 349–355.

Shaw, M.C. (2005) *Metal Cutting Principles*, 2nd edition, Oxford University Press, Oxford.

Sheikh-Ahmad, J., Bailey, J.A. (1999) The wear characteristics of some cemented tungsten carbides in machining particleboard. *Wear*, **225–229**, 256–266.

Sheikh-Ahmad, J., Stewart, J.S., Feld, H. (2003b) Failure characteristics of diamond-coated carbides in machining wood-based products. *Wear*, **255**, 1433–1437.

Sheikh-Ahmad, J.Y., Lewandowski, C.M., Bailey, J.A., Stewart, J.S. (2003a) Experimental and numerical method for determining temperature distribution in a wood cutting tool. *Experimental Heat Transfer*, **16**, 255–271.

Spur, G., Lachmund, U. (1999) Turning of fiber-reinforced plastics. In Jahanmir, S., Ramulu, M., Koshy, P. (editors), *Machining of Ceramics and Composites*. Marcel Dekker, Inc., pp. 209–248.

Spur, G., Wunsch, U.E. (1988) Turning of fiber-reinforced plastics. *Manufacturing Review*, **1**, 124–129.

Taylor, F.W. (1907) *Trans. ASME*, **28**, p. 21.

Tomac, N., Tonnessen, K. (1992) Machinability of particulate aluminium matrix composites. *Annals of the CIRP*, **41**, 55–58.

Usuki, H., Narutaka, N., Yamane, Y. (1991) A study of the cutting performance of diamond coated tools – tool wear of diamond coated tools in machining of CFRP. *International Journal of Japan Society of Precision Engineers*, **25**(1), 35–36.

6

Analyzing surface quality in machined composites

K. PALANIKUMAR, Sri Sairam Institute of Technology, India

Abstract: Composite materials are finding increased applications in many fields due to their excellent properties. Analyzing surface quality in machined composites is very important. This chapter discusses the surface quality in the turning of glass fiber reinforced polymer composites, carbon fiber reinforced polymer composites and silicon carbide reinforced aluminum metal matrix composites. First discussed is the influence of cutting parameters such as cutting speed, feed and depth of cut. Results of experiments are analyzed using response graphs. The chapter then discusses the use of scanning electron micrograph images in determining surface quality of these materials.

Key words: surface roughness, surface quality of composites, fiber and particle reinforced composites, scanning electron microscopy (SEM).

6.1   Introduction

Fiber and particle reinforced composite materials are finding increased applications in various areas of technology due to their mechanical and physical properties and good performance. Composite materials have been widely used in a variety of structures such as aircraft, robots, tennis rackets, bicycles, manufacturing machinery, etc. In fiber and particle reinforced composite materials, fibers/particles act as a load carrying medium and the matrix acts as a load transporting medium. The two or more phases are utilized to take advantage of the best properties of each, and minimize their weaknesses. Composite components are fabricated by various processes such as filament winding, hand lay-up, stir processing, etc. After fabrication, they may require machining to facilitate dimensional control for easy assembly and for functional aspects.

The structures of the composite materials are complex compared to conventional materials. Because of the complexity of the structure of composite materials, their deformation mechanisms under cutting are still far from deeply understood. The behavior of composites is anistropic. The quality of machined products depends upon the reinforcements, matrix materials used, bond strength between reinforcing materials and matrix, type of weave, etc. (Palanikumar, 2008). Surface quality in machined composites is an important design feature in many situations, such as for parts subject to

154

fatigue loads, precision fits, fastener holes and aesthetic requirements. In addition to tolerances, surface roughness also imposes one of the most critical constraints for the selection of machines and cutting parameters in process planning (Wang and Feng, 2002).

This chapter discusses the surface quality observed in machined composites. Three different composites, namely glass fiber reinforced polymer composites, carbon fiber reinforced polymer composites and silicon carbide reinforced aluminum (Al/SiC) metal matrix composites (MMC) are considered. The machining process considered is turning. This chapter first discusses the influence of cutting parameters such as cutting speed, feed and depth of cut. The results of measurements are analyzed using response graphs. The chapter then discusses the use of scanning electron micrograph images in assessing surface quality.

6.2   General concepts of an engineering surface

Modern engineering products require good surface texture with dimensional accuracy. The surface texture of the products highly influences the functioning of machined parts. The aesthetic appearance, load carrying capacity, resistance against corrosion and wear, lubrication, and ability to hold pressure in the produced components depends on the surface texture. In common practice, it is not possible to produce a perfect surface. The surface produced from manufacturing operations always has irregularities and imperfections and thus never attains perfection, having hills and valleys varying in height and spacing. The irregularities found in the manufactured/machined surface are defined by height, spacing and direction, and other random characteristics not of a geometric nature; this is termed surface roughness, surface finish, surface texture or surface quality. The important components of surface topography are (Dagnall, 1996; Whitehouse, 1996; Petropoulos, 2010):

- *Roughness*: these are closely spaced irregularities such as peaks and valleys caused by a combination of tool geometry, feed rate and other sources during cutting, or grit from the grinding wheel.
- *Waviness*: waviness is widely spread irregularities and is caused by the run-out of the workpiece or tool, vibration and chatter in machine tools, wear of the cutting tool and/or inhomogeneity of the work material.
- *Form errors*: this is due to long-period or non-cyclic deviations and is caused by flatness, roundness, and straightness errors of the manufacturing equipment.
- *Flaws*: cracks, pits, scratches and grooves are called flaws. These are infrequent, discrete irregularities due to tool form, process kinematics, method of chip formation, tool nose wear, etc.

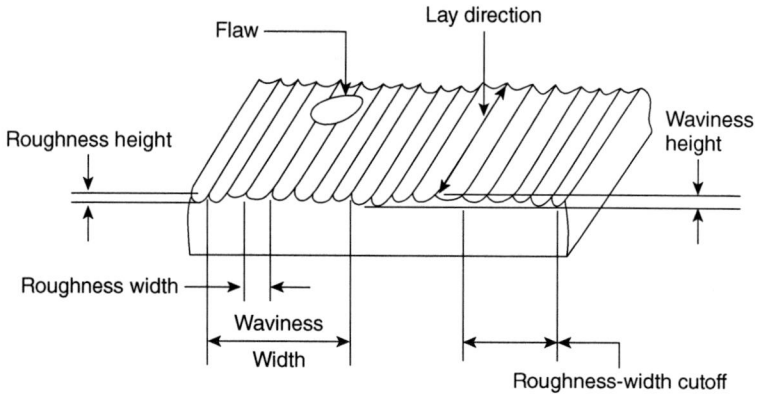

6.1 Surface characteristics in machining (Courtesy of ANSI B46.1-1962).

In general, the surface finish of the manufactured components refers to the texture, flaws, material and coating applied. The requirement of surface texture depends on the specific application of the part. The surface is usually assessed by taking a sampling length at an angle perpendicular to the lay direction. The surface characteristics are presented in Fig. 6.1. In this figure, the surface height indicates the irregularities with respect to a reference line. The roughness width is the measurement of the distance parallel to the nominal surface between successive peaks or valleys in the machined surface that constitute the predominant pattern of the roughness. Roughness width indicates the spacing between the surface irregularities and is to be included in the measurement of the average roughness height. It is always more than the roughness width in order to obtain the total roughness height rating. Lay represents the direction of the predominant surface pattern produced, and it reflects the machining operation used to produce it. Waviness is the surface unevenness. This may be due to workpiece or tool deflection during machining caused by vibration and/or tool run out.

6.2.1   Surface texture parameters

Surface quality can be characterized by a variety of quality indices. Many of these surface quality indices have been developed to characterize particular applications and a good number of surface parameters have been adopted by international standards. To control a manufacturing process and to understand the degree of accuracy required on a component's surface, it is necessary to quantify the surface in both two and three dimensions. Surface texture parameters can be characterized by four different basic types:

- *Roughness parameters*: roughness parameters measure the vertical characteristics of the surface deviations. These are the finer irregularities found on the workpiece surface that are inherent in the production process. Examples of such parameters are center line average roughness, root mean square roughness, kurtosis and peak-to-valley height.
- *Waviness parameters*: these are the surface irregularities that are of greater spacing than the roughness and they occur in the machined surfaces as waves. They are normally caused by malfunctions of the machine tool system, vibration, machine or work deflections, etc.
- *Spacing parameters*: these are the measures of the horizontal or lateral characteristics of the surface deviations and examples of such parameters are mean line, peak spacing, high spot count and peak count.
- *Hybrid parameters*: hybrid parameters are a combination of both the vertical and horizontal characteristics of surface deviations and are combinations of spacing and roughness parameters. Examples of such parameters are root mean square slope of profile, root mean square wavelength, core roughness depth, reduced peak height, valley depth, peak area and valley area.

Out of the above four parameters, surface roughness parameters are of most interest to industries. Some of the important surface roughness parameters are given below.

6.2.2   Roughness parameters

- R_a – arithmetical mean value of deviations about the centerline within the evaluation length. This is also called arithmetic average height. It is the most popular parameter for measuring machined surface and for quality control. It is available for profile and areal data. The definition of arithmetic average height is presented in Fig. 6.2.
- R_q (rms) – R_q is also known as root-mean-square or 'rms' roughness. It is the average of the measured height deviations taken within the evaluation length or area, and measured from the mean linear surface. R_q is the root mean square parameter of R_a and is more sensitive than R_a. Figure 6.3 shows the definition of R_q.
- R_p (*peak*) – this is the value of the highest peak measured above the center line. It is the maximum data point height above the mean line through the entire data set.
- R_v (*valley*) – This is the value of lowest valley measured below the center line. It is the maximum data point depth below the mean line through the entire data. Figure 6.4 shows the definition of R_p and R_v.
- R_t – This is the measure of peak-to-valley height – the absolute value between the highest and lowest peaks. It is available for profile and areal data and has been commonly used with R_a. Normally $R_t = R_p + R_v$.

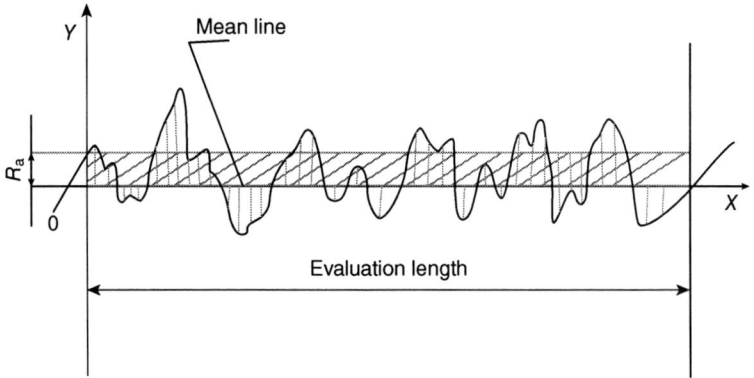

6.2 Definition of average arithmetic height (R_a).

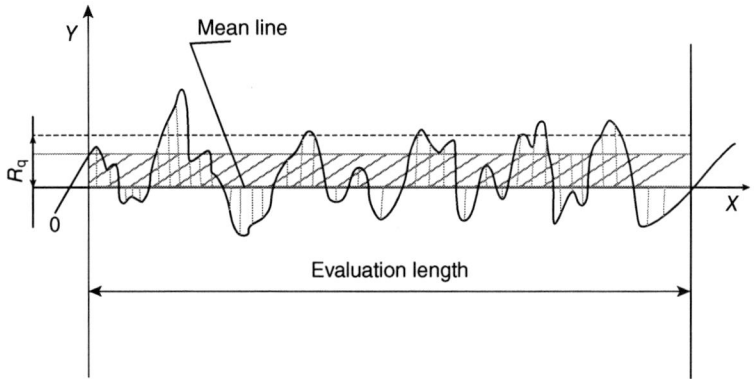

6.3 Definition of R_q.

- R_z – This is the measure of average peak-to-valley height. The average peak-to-valley height has been measured normally for one peak and one valley per sampling length. The single largest deviation is found in five sampling lengths and then averaged. Figure 6.5 shows the definition of R_z.
- R_{max} – This is the value of maximum peak-to-valley profile height. It is the greatest peak-to-valley distance within any one sampling length. Figure 6.6 shows the definition of R_{max}.

6.3   Surface quality in machining

Surface quality is one of the important concerns in machining. The quality of the machined surface is characterized by the accuracy of its manufacture

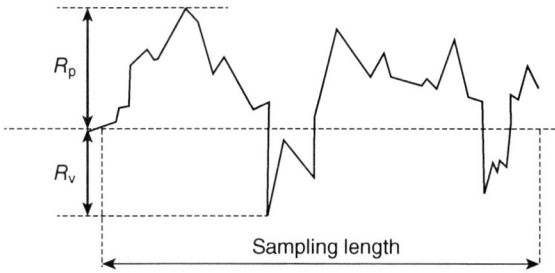

6.4 Definition of R_p and R_v.

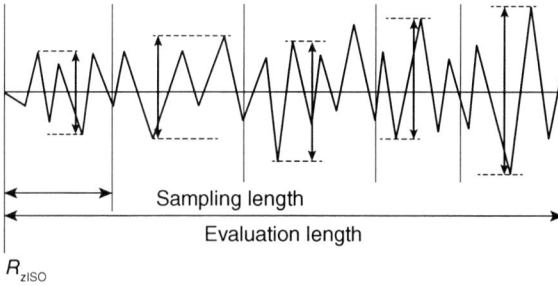

6.5 Definition of R_z.

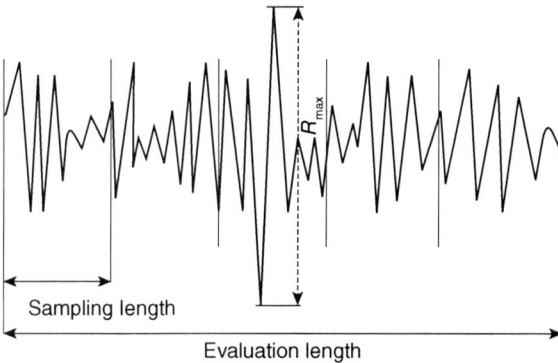

6.6 Definition of R_{max}.

with respect to the dimensions specified by the designer. Every machining operation produces its own characteristic evidence on the machined surface. This evidence is in the form of finely spaced micro irregularities left by the cutting tool on the workpiece. Each type of cutting tool leaves its own individual pattern which therefore can be identified. This pattern is known as surface finish or surface roughness. Surface roughness is an assessment of the material's response to the system of machining. Surface quality of the machined surface is directly proportional to machinability. The term machinability refers to the ease with which a workpiece can be machined to an acceptable surface finish. Work material having good machinability requires a small amount of power to machine, is quickly machined, and gives a good surface finish. For manufacturing the components economically, the engineers and technocrats are challenged to find ways of improving machinability without reducing performance. Machinability is normally analyzed by tool wear, cutting force and surface finish. A good surface finish always indicates good machinability. Achieving a good surface in composite machining processes is very difficult, because the material response is different from plain metal. Under ideal conditions, the surface roughness profile is formed by repetition of the tool profile at intervals of feed per workpiece revolution. Normally the experimental tested values are much different from the theoretical model. The theoretical surface roughness (R_{at}) can be calculated by the following expression (Boothroyd and Knight, 1989):

$$R_{at} = \left[\frac{0.0321 f^2}{R} \right] \qquad [6.1]$$

where f = feed in mm/rev and R = tool nose radius in mm. The theoretical model used above (R_{at}) cannot be used for predicting the surface roughness in the machining of composites. Apart from feed and corner radius, factors such as work piece, tool material combination and their mechanical properties, quality and type of the machine tool used, auxiliary tooling and lubricant used, and vibration between the workpiece, machine tool and cutting tool also affect the quality of the surface (Palanikumar et al., 2008).

6.3.1 Factors affecting surface roughness in machining

The major factors that affect the surface roughness in machining are shown in Fig. 6.7. There are many of these, the most important being:

- Machine tool rigidity and accuracy (vibrations on machine tool, spindle run-out, etc.).
- Work material used.
- Method of chip removal (type of machining).
- Cutting tool geometry and condition of the cutting tool.
- Cutting conditions such as cutting speed, feed, and depth of cut.
- Type of cutting fluid used.
- Finishing required on the workpiece material.

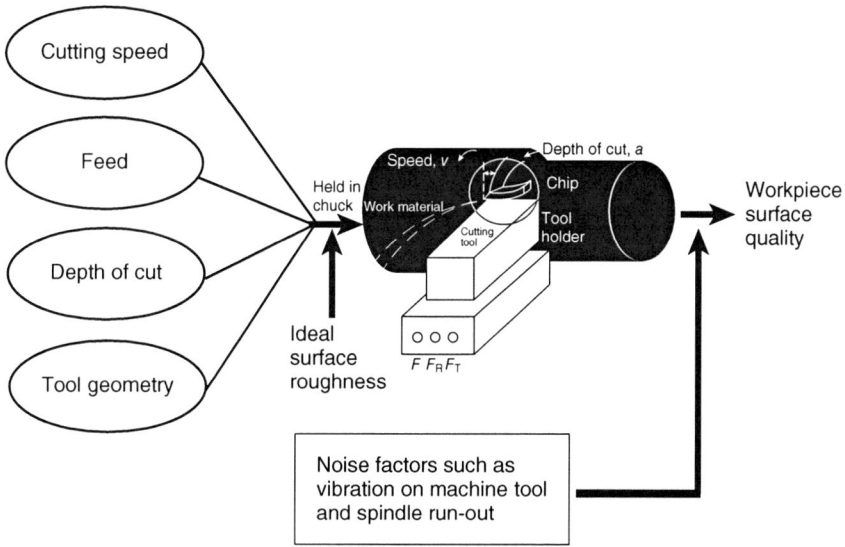

6.7 Mechanism of surface generation in machining.

6.3.2 Surface roughness in machining of composite materials

The properties of composite materials are different from plain metals and are primarily determined by their constituents, such as the nature of the reinforcing fibers or particles used, the matrix materials, and the volume fraction between the matrix and fiber/particles, and fiber orientation or arrangement. In machining, most of the composite materials produce discontinuous, powdery chips. The powdery chips pose health hazards to the operator. Collection and disposal of the chips is also a problem. Proper safety equipment must be used while working with these composite materials.

In the machining of composite materials, the tool wear must be kept at a minimum. During machining, the tool continuously encounters alternate matrix and fiber/particle materials, whose response to machining can vary greatly. The cutting zone experiences both thermal and mechanical stresses. The cutting nose may be subjected to localized dynamic loading, due to the variations in properties of matrix and fiber. This dynamic load may cause tool failure due to possible low cycle fatigue apart from the usual types of wear (Santhanakrishnan *et al.*, 1988). The surface quality is concerned with the geometrical features of the generated surface. For achieving the desired surface finish, it is necessary to understand the mechanisms of the material removal and the kinetics of machining processes that affect the performance of the cutting tool (Sreejith *et al.*, 2000). From the published results, it is known that the cutting mechanism of most of the composite materials is due to a combination of plastic deformation, shearing and bending rupture. The occurrence of the above mechanisms depends on the flexibility, orientation and toughness of the reinforcing fibers/particles. These constitute a surface texture on the work piece. The surface roughness is quantitatively measured by the vertical deviations of a real surface from its ideal form. In measuring surface roughness, if the deviations are large, the surface is rough; if they are small, the surface is smooth. The surface roughness can be evaluated by the different surface roughness parameters such as arithmetic average height (R_a), root mean square roughness (R_q), maximum height of peaks (R_p), maximum height of the profile (R_t), or mean of the third point height (R_{3z}). A typical area graph of surface roughness parameters observed with respect to experiment numbers in the machining of a GFRP composite by a PCD tool is presented in Fig. 6.8. Area graphs are used to evaluate the trends in multiple variable series as well as contribution to the sum. The area graph clearly indicates the magnitude of different surface roughness parameters in the turning of a GFRP composite.

Surface roughness evaluation is very important for many fundamental problems such as friction, contact deformation, heat and electric current conduction, tightness of contact joints and positional accuracy. For this reason, surface roughness has been the subject of experimental and theo-

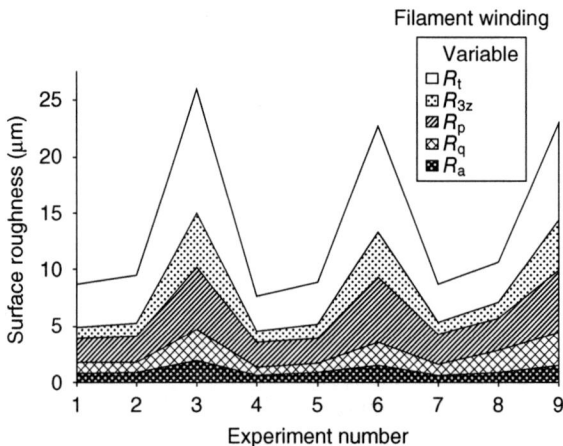

6.8 Area graph for different surface roughness parameters in machining a GFRP composite with respect to experiment number (Palanikumar *et al.*, 2008a).

retical investigations for many decades (Gadelmawla *et al.*, 2002). Also, surface roughness imposes one of the most critical constraints for the selection of machines and cutting parameters in process planning. Even though many surface roughness parameters are used, the average surface roughness (R_a), which is most used in industries, is taken-up for analysis in the present chapter. On a composite machined surface, the result of the roughness test depends mainly on the stylus path with respect to fiber direction because the main direction of fibers may change from layer to layer (Santhanakrishnan, 1990) The roughness is measured a number of times and averaged. The average surface roughness is the integral of the absolute value of the roughness profile height over the evaluation length and is denoted by the following equation (Fig. 6.2):

$$R_a = \frac{1}{l}\int_0^L [Y(x)]\,dx \qquad\qquad [6.2]$$

where l is the length taken for the observation and Y is the ordinate of the profile curve. During machining, tool wear must be maintained within the limit of the ISO recommendation.

Most of the studies on composite machining show that maintaining the surface quality is a serious task. In order to know surface quality and dimensional properties, it is necessary to characterize and quantify the quality of the machined surface and the effect of process parameters on the surface quality; this is discussed in the following section.

6.4   Influence of cutting parameters on surface quality

In the machining of composite materials, many factors influence the surface quality. The product to be produced, its size, shape, machine tool used, machine factors such as precision, slide straightness, temperature stability, vibration on machine tool, tool shape, quality of the tool and tool materials all have influence on the surface quality in machining of composite materials. The physical and chemical properties of the material used and short-range dimensional accuracy of a product also influence the surface finish of the product. Although many factors affect the surface condition of a machined part, parameters such as cutting speed, workpiece condition, feed and depth of cut have more influences on the surface roughness for a given machine tool and workpiece set-up (Palanikumar, 2007). The influence of machining parameters on surface roughness in machining glass fiber reinforced polymer (GFRP) composites, carbon fiber reinforced polymer composites (CFRP) and aluminum silicon reinforced metal matrix composites (Al/SiC–MMC) are discussed in the following sections.

6.4.1   Glass fiber reinforced polymer (GFRP) composites

The use of GFRP composite materials has increased in various areas of technology due to their mechanical and physical properties such as high specific stiffness, creep resistance and corrosion resistance. Fiber reinforced plastics (FRPs) have been widely used in a variety of structures, such as ships, robots and machines. High quality machined surface is essential for such applications and surface microstructure depends on the machining quality (El-Sonbaty *et al.*, 2004). The structures of FRPs are complex. Because of the complexity of their structures, the deformation mechanisms of the materials under cutting are still far from a deep understanding (Wang and Zhang, 2003). When GFRP composites are machined, it is clearly seen that the fibers are cut across and along their lay direction, leaving deformed projecting and partially disclosed fibers on the machined surface (Santhanakrishnan *et al.*, 1988). Conventional machining of fiber-reinforced composites is difficult due to diverse fiber and matrix properties, fiber orientation, the inhomogeneous nature of the material, and the presence of a high volume fraction (volume of fiber over total volume) of hard abrasive fiber in the matrix. Spur and Wunsch (1988) found that, during turning of GFRP composites, surface roughness increases with an increase in the feed but with no dependence on the cutting velocity. In contrast to this, Ramulu *et al.* (1994) achieved better surface roughness at high velocity, so the machining of FRP is an area still full of open questions (Davim and Mata, 2005). In the machining of fiber reinforced composite materials, there are a number of problems associated with it. For example, the varying material properties and degrees of anisotropy cause difficulty in predicting the behavior of the material being machined. This can lead to specific problems of FRP machining (Konig *et al.*, 1985). Most of the studies on GFRP composite machining shows that minimizing the surface roughness was a difficult task.

The relationship between the surface roughness and cutting parameters is caused by the inhomogeneous microstructure of reinforcing fibers/particles in the composite material, which results in surface details including deformations and fractures at micro level, e.g. fiber ends sticking out, peaks of deformed matrix material and holes from debonding between fibers and matrix (Eriksen, 1999).

Figure 6.9 shows the microstructure of the specimen used for experimentation. In the investigation (Palanikumar, 2007), experiments were conducted for a turning operation on a geared lathe. The ISO specification of the tool used for the turning operation is a WIDAX tool holder PC LNR 1616 K12. The insert used is a coated cermet tool having composition: Co/Ni 12.2%; WC 15.0%; (TaNb) C10.0%; TiCN rest. The coating layer system is: CVD-TiN + Ti(C,N) + Ti(C,N) + Al_2O_3 + Ti(C,N,B), and the layer thickness used was 6 μm. The average surface roughness (R_a), which is mostly used in industries,

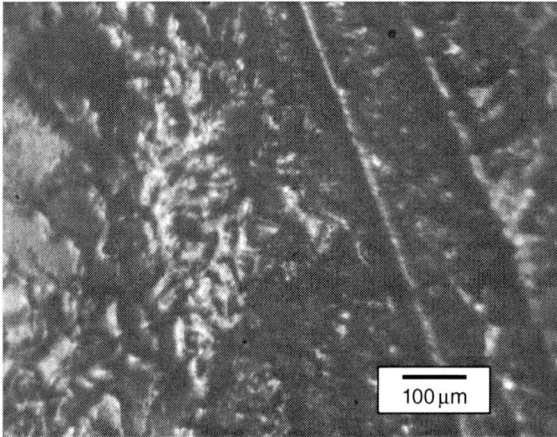

6.9 Microstructure of the GFRP specimen used for the machining experiment.

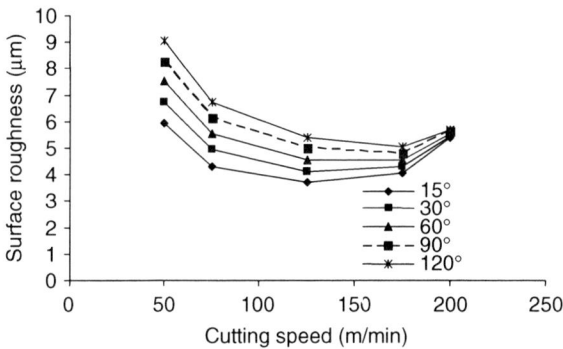

6.10 Variation of arithmetic average surface roughness (R_a) with cutting speed for varying fiber orientation (feed = 0.25 mm/rev, depth of cut = 1 mm) (Palanikumar, 2007).

was taken for this study. This surface roughness was measured using a Surtronic 3+ stylus type instrument manufactured by Taylor Hobson. With composite machined surfaces, the result of the roughness test depends mainly on the stylus path with respect to fiber direction, because the main direction of fibers may change from layer to layer (Konig *et al.*, 1985). For this reason, the roughness was measured several times and averaged.

Figure 6.10 shows the variation of the surface roughness with cutting speed with varying fiber orientation. The figure indicates that the surface roughness decreases nonlinearly with increase of cutting speed. The surface roughness reduces in this proposed manner up to a cutting speed of around

175 m/min. After 175 m/min, the surface roughness increases with increase of cutting speed, and hence 125–175 m/min can be regarded as a critical velocity range for this cutting GFRP composite. Normally, a higher cutting speed is suitable for machining GFRP composites to obtain a good surface finish. But the higher cutting speed should not exceed the critical speed (Palanikumar, 2007). In composite machining at high cutting speed, preferential removal of the workpiece takes place at a faster traverse, which in turn reduce the surface roughness. Also, the heat generation in machining of composites softens the polymer matrix, which in turn reduces the surface roughness.

From the previous results, it is asserted that the surface roughness increases with increase in the fiber orientation angle due to the compressive strain generated in the work material (Takeyama and Lijima, 1988). In the present case, the surface roughness almost always increases with increase of fiber orientation, but in some places it fluctuates with different fiber orientation. This may be due to the imperfection and uneven distribution of glass fibers or uneven layer thickness in the work material used (Sakuma and Seto, 1983).

Figure 6.11 illustrates the variation of arithmetic average surface roughness with respect to cutting speed at different feed rates. The result indicates that the surface roughness shows almost the same trend as discussed before, in Fig. 6.10. But the observed surface roughness is minimum at minimum feed rate. An increase of feed rate increases the load on the tool and increases the surface roughness. But, at low feed rate, fracture is less violent and can be more controllable. For this reason, surface roughness observed is minimum at low feed rates. Also at low feed rates, the strain rate is low. From the experimental data, it is seen that the surface finish at a feed rate of 0.05 mm/rev is better than that of a feed of 0.75 mm/rev.

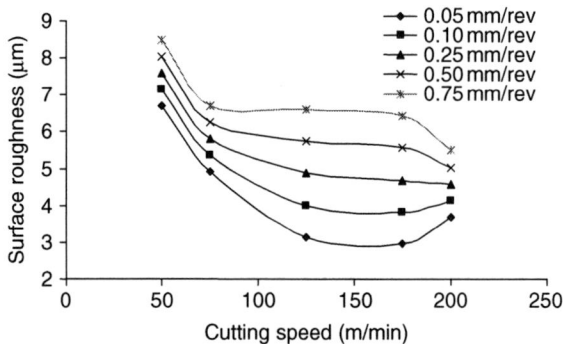

6.11 Variation of the arithmetic average surface roughness (R_a) with cutting speed with varying feed (fiber orientation angle = 60°, depth of cut = 1 mm) (Palanikumar, 2007).

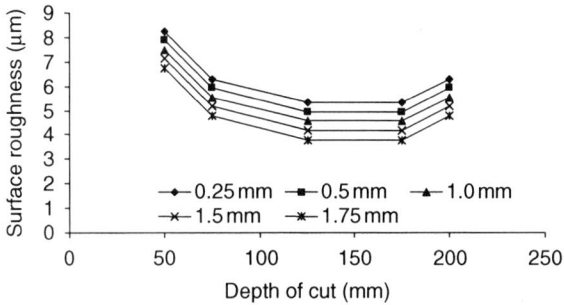

6.12 Variation of the arithmetic average surface roughness (R_a) with cutting speed with varying depth of cut (feed = 0.25 mm/rev, fiber orientation angle = 60°) (Palanikumar, 2007).

Figure 6.12 illustrates the variation of arithmetic average surface roughness at different depths of cut, keeping the feed rate at 0.25 mm/rev and the fiber orientation angle at 60°. Normally, the effect of the depth of cut on composite machining processes is minimum. From Fig. 6.12, it is asserted that a high depth of cut is preferred for machining GFRP composites, the reason being that, at a low depth of cut, the removal of fibers from the matrix is partial and leads to high surface roughness, whereas at a high depth of cut, complete removal of fibers can be possible and leads to a good surface finish. From the analysis, it is asserted that a medium cutting speed, low feed rate and reasonable amount of depth of cut are to be preferred for the machining of GFRP composites.

Figure 6.13 shows SEM micrographs of the workpiece surface profiles observed in the machining of GFRP composites. The micrograph shows that the distribution of fiber in the composite material is not even. There are voids (incomplete penetration) of resin in the composite materials observed. This is due to the uneven distribution of fibers and resin in the manufacturing of the composites and preferential removal of fibers in the composite materials during machining.

Figure 6.14 illustrates the top surface of the specimen observed after the machining operation. Normally, in top layers of composite pipes, the amount of fiber is minimal and resin content is more than that depicted in the figure; here this top layer has been machined off.

Figure 6.15 shows typical machined surfaces obtained in machining of GFRP composites at two different cutting conditions (Palanikumar, 2007). These figures indicate that the surface quality observed is better at high cutting speed, whereas the surface roughness increases with a decrease in the cutting speed. In these figures, tiny fractured fiber particles are stuck to the top surface of the specimen, which in turn produces a coarse structure.

6.13 SEM micrograph of the machined specimen.

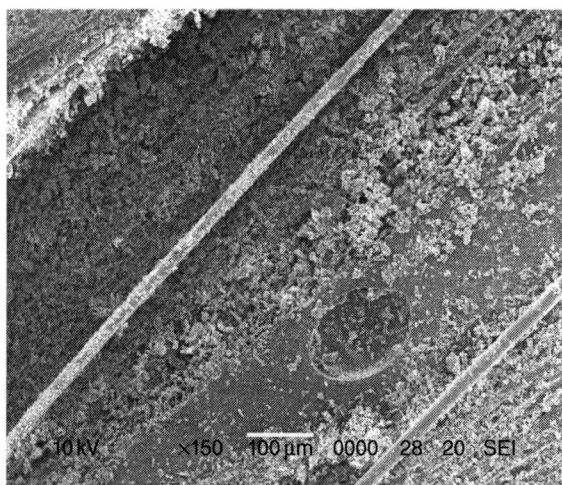

6.14 Surface profile observed at the top layer of the composite (Palanikumar, 2007).

During machining at the lower cutting speed, a large material flow with the cut fibers was noticed which produced high surface roughness, whereas at the higher velocity, the chips mainly consisted of less deformed matrix material and cut fibers. The surface roughness in machining can be improved by adopting proper cutting conditions and choosing appropriate variables.

Figure 6.16 shows an SEM micrograph of the powdery chips along with fractured fibers. In the figure, the black region indicates the lumped masses

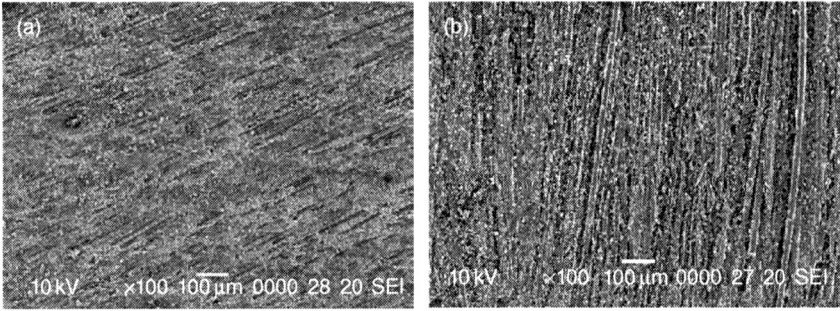

6.15 Surface profile observed at two different cutting speed conditions: (a) 175 m/min; (b) 75 m/min (Palanikumar *et al.*, 2006).

6.16 Micrograph of the chip observed in turning of composites (Palanikumar, 2007).

of matrix material in powdery form, and white spots indicate the fractured fibers. There is no bonding between the matrix and fibers; this may be due to the weak bond strength and to the probable layer lattice structure of the fiber material.

6.4.2 Carbon fiber reinforced polymer (CFRP) composites

Carbon fiber reinforced polymer (CFRP) composite materials are finding increased applications in many areas. They are strong and light, and have usage in aerospace as well as in sail boats, and notably in modern bicycles and motor cycles, where high strength-to-weight ratios are required. They are also used in laptops, tripods, fishing rods, racquet frames, stringed instrument bodies and golf clubs (Palanikumar, 2010). Process variables may be varied to vary stiffness and strength of the carbon fibers and they are available in high modulus and intermediate modulus forms. Since the high material damping of carbon fiber–epoxy composite materials can dissipate any vibration of the composite structure that is induced, they are used for manufacturing high-speed transmission shafts (Reugg and Habermeir, 1980), machine-tool spindles (Lee *et al.*, 1985), and robot arms (Lee *et al.*, 1991). If carbon fiber–epoxy composites are used in robot arms or machine elements, accurate machining of the materials by processes such as turning, milling and drilling are required to provide bearing-mounting and adhesive-joining surfaces. Machining of carbon fiber–epoxy composite materials is not the same as machining conventional plain metals. The wear of sintered-carbide tools and high-speed steel tools is very severe. Hence, the cutting speed and feed rate of the machining operation should be selected carefully in the machining of carbon fiber–epoxy composites (Ki Soo Kim *et al.*, 1992). Also, surface damage such as cracking and delamination of the machined surfaces is severe and obtaining a low surface-roughness is not easy (Lubin, 1982).

For some machining experiments (Palanikumar, 2010), carbon fiber in the form of roving was dipped into a catalyzed polyester resin and filament wound with an orientation angle of ±45° to produce a cylindrical CFRP tube. After curing and removal from the mandrel. The cutting tool used for the machining was a cemented carbide type (TPGN 16 03 04 H13A) and its geometry was with the rake angle of 6°, a clearance angle of 11°, an edge major tool cutting angle of 91°, and a cutting edge inclination angle of 0°. A tool holder having the specification PCLNR 1616 K12 was used. The work material and cutting tool with tool holder used in the investigation is presented in Fig. 6.17.

The influence of surface characteristics in the machining of CFRP composites is assessed by conducting machinability tests. A typical surface profile observed in turning of CFRP composites is presented in Fig. 6.18. The figure represents a schematic of the surface profile obtained when measuring surface roughness on the machined material, and it represents a roughness value of 1.32 μm. The nature of the surface after machining is clearly seen from the picture.

The variation of surface roughness with respect to the cutting speed for a varying feed rate at a constant depth of cut is presented in Fig. 6.19. The figure indicates that an increase of cutting speed slightly reduces the surface roughness in the machining of CFRP composites. At high feed rates, no definite trend is seen. An increase of spindle speed slightly increases the surface roughness and then reduces it. The results indicate that an increase

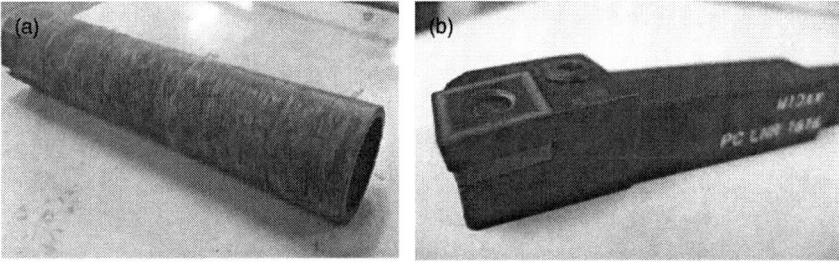

6.17 (a) CFRP tube and (b) cutting tool used for the experiments.

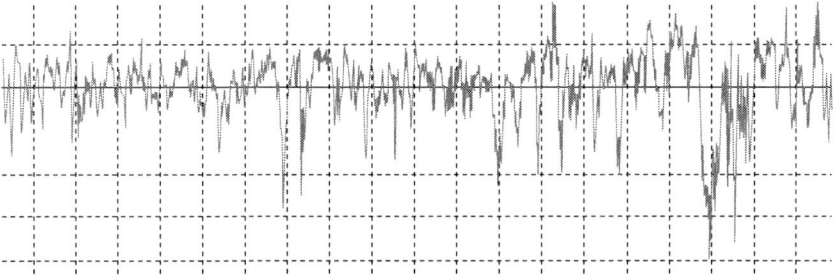

6.18 Typical surface roughness profile observed in machining of CFRP composites.

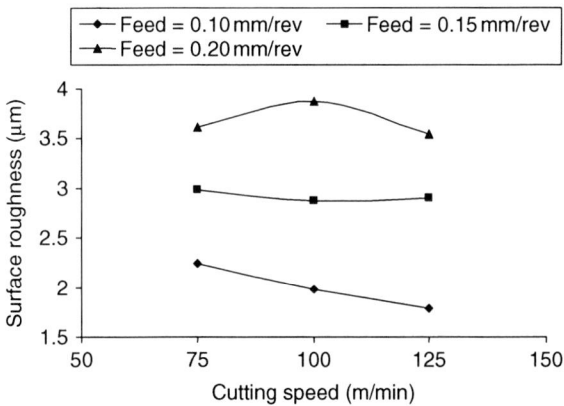

6.19 Variation of the arithmetic average surface roughness (R_a) with cutting speed for varying feed rates (depth of cut = 1.0 mm).

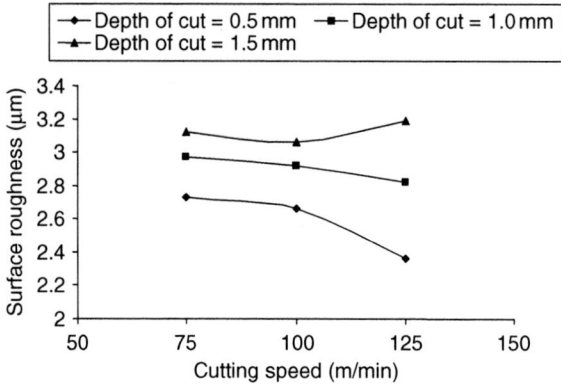

6.20 Variation of the arithmetic average surface roughness (R_a) with cutting speed for varying depth of cut (feed rate = 0.15 mm/rev).

or decrease of cutting speed does not show any variation in the surface roughness. When the material is cut at a low cutting speed, the cutting is carried out by the plowing effect of the tool, which results in a perfect shearing of the fibers, which results in a good surface finish; hence, there is no definite cutting speed recommented for the machining of CFRP composites, but a medium speed is best. An increase of the feed rate increases the surface roughness in machining of CFRP composites. This increase increases the load on the tool and increases the cutting force generated during turning, which in turn increases the surface roughness.

The variation of surface roughness with cutting speed at varying depths of cut is presented in Fig. 6.20. The figure indicates that an increase in depth of cut increases the surface roughness in the turning of CFRP composites. The increase in the depth of cut increases the width of the cut and the forces on the cutting tool, which in turn produces a rough surface. Increase in the depth of cut also increases the contact between the tool and the workpiece, which leads to a rougher surface. The variation of surface roughness with feed rate for varying depths of cut is presented in Fig. 6.21. The figure indicates that an increase of feed and depth of cut increases the surface roughness in machining of CFRP composites. Among the parameters studied, feed rate is the main parameter that influences the surface roughness in turning of CFRP composites, whereas the cutting speed has only little effect. Figure 6.22 shows a micrograph of the turned surfaces at different feed rates by keeping the cutting speed and depth of cut at a constant medium level.

Figure 6.22a shows the surface profile observed at a low feed rate. This figure indicates the comparatively better surface. In Fig. 6.18, the white spot shows the machined resin that is used for the fabrication of composites. Figure 6.22b shows the surface profile observed in the machining of CFRP

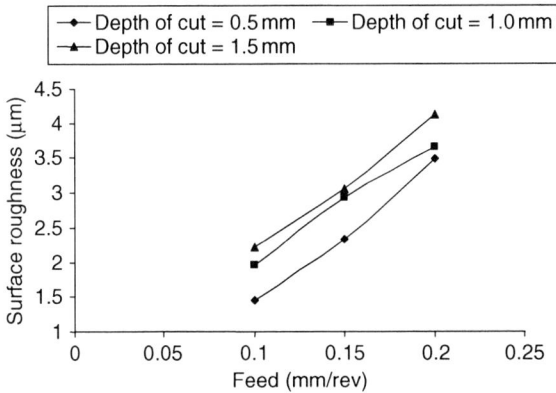

6.21 Variation of the arithmetic average surface roughness (R_a) with feed for varying depth of cut (feed = 0.15 mm/rev).

composites at a medium feed rate. The figure shows the feed marks on the specimen. Figure 6.22c shows the surface profile observed at a high feed rate. The figure indicates the violent fractures of the matrix and fibers in the composites in the longitudinal direction. This may be due to the high tooth load of the tool on the workpiece at high feed rates.

6.4.3 Al/SiC-metal matrix composites

Metal matrix composites (MMC) are attractive materials because of their high specific strength, stiffness and wear resistance. Among modern composite materials, particulate reinforced MMCs are finding increased application due to their favorable mechanical properties. SiC reinforced aluminum is used widely and other compositions for the matrix are available commercially (Antonio and Davim, 2002). With the increasing use of MMCs in applications such as the aerospace industry, the automotive industry and in sports equipment, the machining of such materials have become a very important subject for study (Tosun and Muratoghi, 2004).

The main concern when machining MMC is extremely high tool wear due to the abrasive action of the ceramic reinforcing fibers or particles. Therefore, materials of very high resistance to abrasive wear are generally recommended is tools for the machining of these composites (Cronjager and Meister, 1992; Ping Chen, 1992). HSS tools are inadequate, cemented carbide tools are preferred for rough machining and PCD tools for finish machining operations (Tomac et al., 1992). The cost of PCD tools increases the cost of production so it is necessary to carryout basic machinability studies, in order to find cutting conditions to enable carbide tools to be used, which can result in high productivity at low cost.

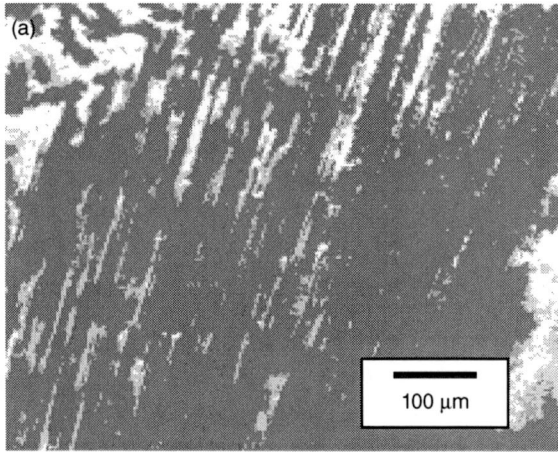

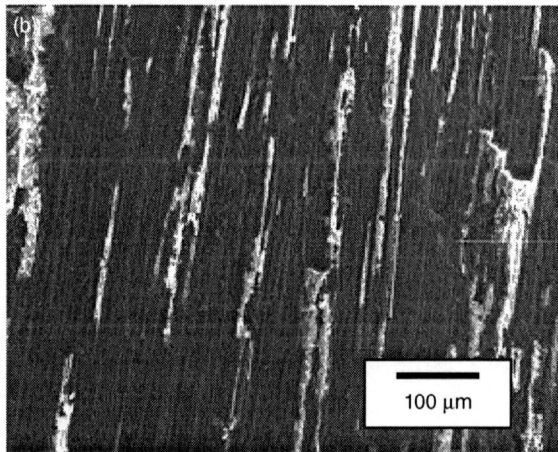

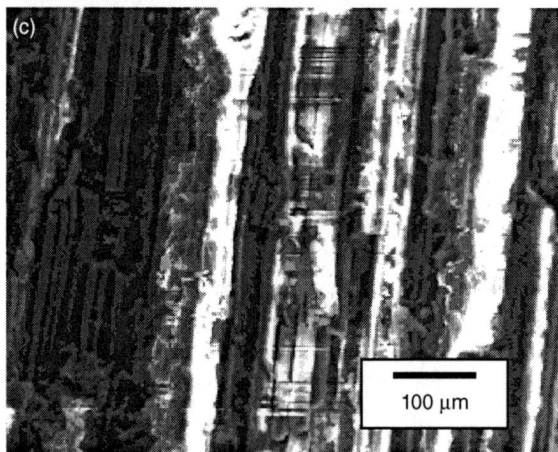

6.22 SEM micrographs of machined surfaces in machining CFRP composites: (a) surface observed at low feed rate; (b) surface observed at medium feed rate; (c) surface observed at high feed rate.

The machining parameters considered are: % volume fraction of SiC, cutting speed, feed and depth of cut – out of which % volume fraction of SiC is specially applied to MMC composites. Previous studies on the effect of machining parameters on Al/SiC composites indicate that higher cutting speeds cause a large deformation rate and severe tool wear (Karthikeyan, 2000). Higher feed rates and depths of cut are also found to result in a deleterious effect on the surface quality in Al/SiC composites. Hence, possible intermediate limits are chosen in order to get the desired surface on the workpiece material.

LM 25 aluminum alloy conforms to BS 1490:1988. LM 25 (7Si; 0.33Mg; 0.3Mn; 0.5Fe; 0.1Cu; 0.1Ni; 0.2Ti) reinforced with green-bonded silicon carbide particles having an average dimension of 2 μm with different volume fractions manufactured through a stir casting route was used for experimentation. The turning tests were performed on a PSG 141 lathe with the following specifications: height of center 177.5 mm, swing in gap 520 mm, spindle speed range 30–1600 rpm, feed range 0.05–3.5 mm/rev, main motor 2.25 kW. The tool used for the turning operation was a WIDAX tool holder ISO specification (PT GNR 2525 M16). The insert used was a carbide tool insert (K10) with the following specifications: TNMG 160404. The tool geometry was rake angle 5°, clearance angle 0°, approach angle 60° and tool nose radius 0.4 mm. The average surface roughness (R_a), which is mostly used in industry, was utilized for this study. The surface roughness was measured by using a MITUTOYO SURF III surface tester.

A study of the machining characteristics of Al/SiC particulate composites is all the more important because of the presence of an abrasive phase in the metal matrix. The presence of SiC in the metal matrix increases the hardness, tensile strength and heat resistance (Karthikeyan, 2000). The machinability of the aluminum alloy used is rated as C grade by ASM (Anon., 1978), which means it can produce continuous chips with a good finish. Nevertheless, when SiC is added, the chip formation entirely differs and the machining characteristics change drastically (Karthikeyan *et al.*, 2000). Figure 6.23 shows the microstructure of 15% volume fraction composites in which the distribution of SiC in the matrix alloy is seen.

During the machining of Al/SiC composites, high tool wear and surface roughness are common phenomena. This is due to the formation of built-up edge (BUE). BUE is a complex phenomenon which is normally formed at low speeds and high feed rates. The shape of the BUE is another important criterion which affects tool life as well as surface roughness. When it is formed as a uniform layer, the surface finish will be better; when it is formed as a heap over the tool surface, it will affect the surface finish. A micrograph of BUE formation in the machining of an Al/SiC composite at high depth of cut and low cutting speed is shown in Fig. 6.24.

6.23 Microstructure of Al-15%SiC particulate composites (Palanikumar and Karthikeyan, 2006).

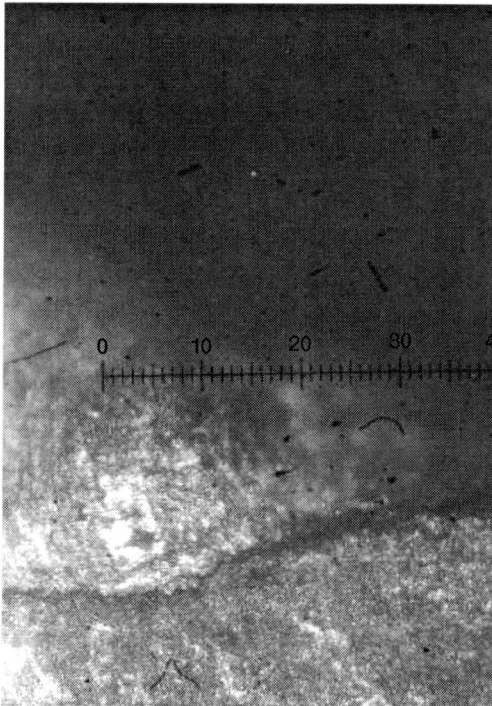

6.24 BUE formation in turning of Al/SiC–MMC (Palanikumar and Karthikeyan, 2006).

The influence of control factors on surface roughness (R_a) can be analyzed by using response graphs. In these graphs, if the difference between the minimum and the maximum values of the average surface roughness for each factor is higher, then the effect on the surface roughness is higher. Figure 6.25 shows the effect of the volume fraction of SiC with respect to different cutting speeds. The results indicate that the roughness of the composite machined surface is highly influenced by the % volume fraction of SiC and the feed rate. Increase in the SiC volume fraction decreases the surface roughness value due to a tendency against the formation of built-up edge. An increase in cutting speed decreases the surface roughness and *vice versa*. At low cutting speeds, chip fracture occurs, which in turn produces imperfections on the surface of the workpiece and leads to higher surface roughness. At high cutting speeds, thermal softening of the tool materials may occur, which in turn removes the BUE formed on the tool and subsequently reduces the surface roughness.

The results also indicate that the surface roughness is at a minimum at minimum feed rate. It is found that, at low feed rate, less, fracture takes place in MMC composites compared with high feed rate. Very small amounts of fracture in the surface always leads to less surface roughness and *vice versa*. An increase in feed rate increases the roughness linearly as a consequence of the increased forces, chatter and incomplete machining at faster traverse; this is presented in Fig. 6.26. An increase in the depth of cut results in high normal pressure and seizure on the rake face, and also promotes BUE formation. Hence, the surface roughness increases along with an increase in the depth of cut. The effect of the volume fraction of SiC with respect to the varying depth of cut is presented in Fig. 6.27.

Typical SEM images of the machined MMC surface topography are shown in Fig. 6.28. The images show the profile of the top surface. In some

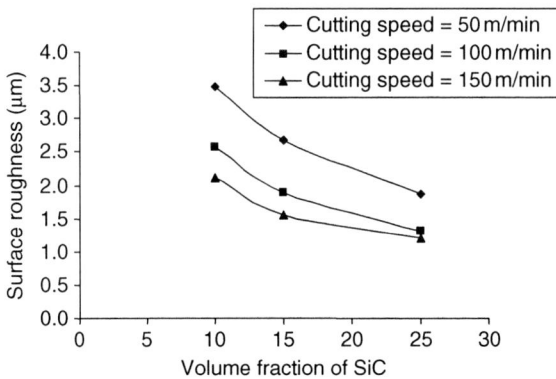

6.25 Effect of cutting velocity with respect to volume fraction of SiC for arithmetic average surface roughness (R_a).

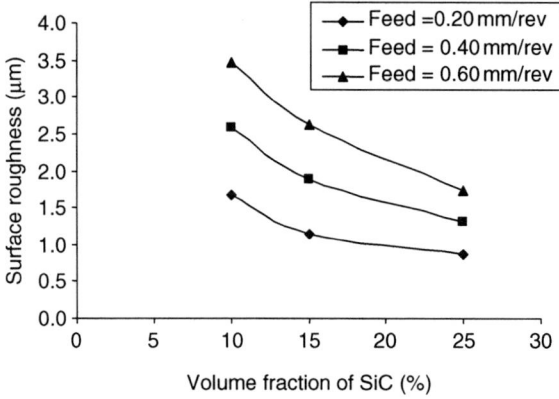

6.26 Effect of volume fraction of SiC at different feed rates for arithmetic average surface roughness (R_a).

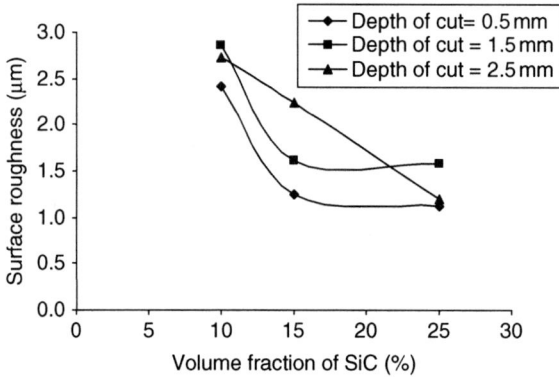

6.27 Effect of volume fraction of SiC for varying depth of cut for arithmetic average surface roughness (R_a).

places on the specimen, tiny surface cracks and very small pit holes are formed. This is due to insufficient penetration of the aluminum matrix in the fabrication of the SiC composite via a stir casting route. In Fig. 6.28, the white layers shows the transformation of aluminium powder during the machining of the Al/SiC particulate composite.

6.5   Conclusions

In this chapter, an analysis of the surface quality in machining of composite materials has been reported. The analysis concerned the machining of three different composite materials, namely a glass fiber reinforced polymer com-

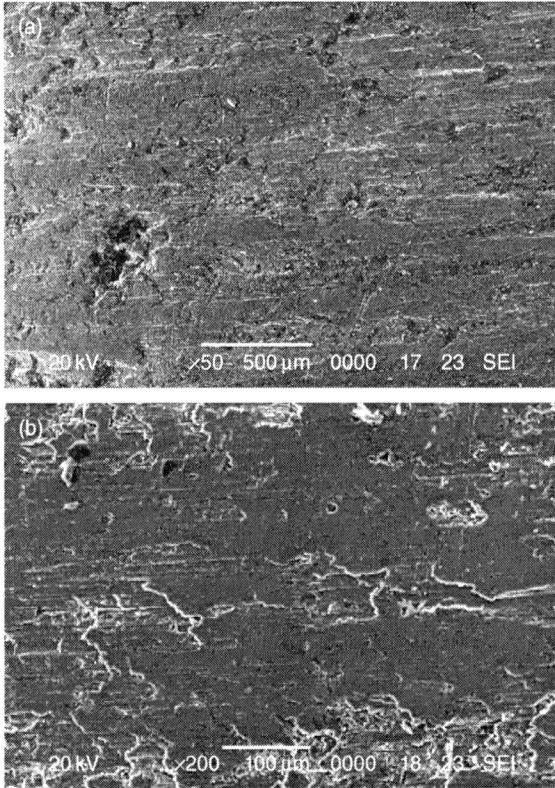

6.28 Typical surface texture observed on the machined workpiece (Palanikumar and Karthikeyan, 2006).

posite, a carbon fiber reinforced polymer composite and an aluminum silicon carbide reinforced metal matrix composite. The effect of machining parameters on the surface quality of the composites was analyzed using the effect graphs and microstructures. The results indicated that the maintenance of the surface quality of a composite is an important concern. Normally, a medium cutting speed is recommended for minimal surface roughness in the machining of GFRP composites, where as for CFRP and Al-SiC metal matrix composites, an increase of cutting speed reduces the surface roughness. An increase of feed rate always increases the surface roughness in the machining of all composite materials. The effect of depth of cut in the machining of composite materials normally has only little effect and can be maintained within acceptable limits. An increase of fiber orientation in the composite materials increases the surface roughness and *vice versa*. For Al-SiC metal matrix composites, an increase in the volume fraction of reinforcement reduces the surface roughness. By using proper cutting conditions and machine tools, the surface quality in composite materials can be maintained.

6.6   References

Anon. (1978), *ASM Handbook on Machining 3*, The Materials Information Society, p. 180.

Antonio CC, Davim JP (2002), Optimal cutting conditions in turning of particulate metal matrix composites based on experiment and a generic search model, *Composites, Part A*, **33**:213–219.

Boothroyd G, Knight W (1989), *Fundamentals of Machining and Machine Tools*, Marcel Dekker, New York, pp. 155–173.

Cronjager L, Meister D (1992), Machining of fibre and particle-reinforced aluminium, *Ann. CIRP*, **41**(1):63–66.

Dagnall H (1996), *Exploring Surface Texture*, Taylor Hobson Ltd, Leicester, UK.

Davim JP, Mata F (2005), Optimization of surface roughness on turning fibre-reinforced plastics (FRPs) with diamond cutting tools. *Int J Adv Manuf Technol*, **26**:319–323.

El-Sonbaty I, Khashaba UA, Machaly T (2004), Factors affecting the machinability of GFR/epoxy composites. *Composite Structures*, **63**:329–338.

Eriksen E (1999), Influence from production parameters on the surface roughness of a machined short fibre reinforced thermoplastic. *Int J Mach Tools Manuf*, **39**:1611–1618.

Gadelmawla ES, Koura MM, Maksoud TMA, Soliman HH (2002), Roughness parameters. *J Mater Process Technol*, **123**:133–145.

Karthikeyan R (2000), *Analysis and Optimization of Machining Characteristics of Al/SiC Particulate Composites'*, Ph.D. Thesis, Annamalai University, Chidambaram, India.

Karthikeyan R, Raghukandan K, Nagarazan RS, Pai BC (2000), Optimizing the milling characteristics of Al-SiC particulate composites, *Metals and Materials*, **6**(6):539–547.

Kim KS, Lee DG, Kwak YK, Namgung S (1992), Machinability of carbon fiber–epoxy composite materials in turning, *J Mat Proc Technol*, **32**:553–570.

Konig W, Wulf Ch, Grab P, Willerscheid H (1985), Machining of fibre reinforced plastics. *Ann CIRP*, **34**:537–548.

Lee DG, Sin HC, Suh NP (1985), Manufacturing of a graphite epoxy composite spindle for a machine tool, *Ann. CIRP*, **34**(1):365–369.

Lee DG, Kim KS, Kwak YK (1991), Manufacture of a SCARA type direct-drive robot with graphite fiber epoxy composite material, *Robotica*, **9**:219–229.

Lubin G (1982), *Handbook of Composites*, Van Nostrand Reinhold, New York, pp. 625–629.

Palanikumar K, Karunamoorthy L, Karthikeyan R (2006), Assessment of factors influencing surface roughness on the machining of glass fiber-reinforced polymer composites, *Materials and Design*, **27**:862–871.

Palanikumar K, Karthikeyan R (2006), Optimal machining conditions for the turning of particulate metal matrix composites using taguchi and response surface methodologies, *Mach Sci Technol*, **10**(4):417–433.

Palanikumar K (2007), Modeling and analysis for surface roughness in machining glass fibre reinforced plastics using response surface methodology, *Materials and Design*, **28**:2611–2618.

Palanikumar K, Sivakumar G, Davim JP (2008), Development of an empirical model for surface roughness in the machining of Al/SiC particulate composites by PCD tool, *Int J Mater Prod Technol*, **32**(2/3):318–332.

Palanikumar K (2008), Application of Taguchi and response surface methodologies for surface roughness in machining glass fiber reinforced plastics by PCD tooling, *Int J Adv Manuf Technol*, **36**:19–27.

Palanikumar K, Mata F, Davim JP (2008a), Analysis of surface roughness parameters in turning of FRP tubes by PCD tool, *J Mater Process Technol*, **204**:469–474.

Palanikumar K (2010), Optimization of surface roughness parameters in turning CFRP composites by carbide tool using response surface and desirability based approach, *Int J Adv Manuf Technol*, submitted for publication.

Petropoulos GP, Pandazaras CN, Davim JP (2010), Surface texture characterization and evaluation related to machining. In: *Surface Integrity in Machining*, Springer, London.

Ping Chen (1992), High performance machining of SiC whisker reinforced aluminium composite by self propelled rotary tools, *Ann CIRP*, **41**(1):59–62.

Ramulu M, Arola D, Colligan K (1994), Preliminary investigation of effects on the surface integrity of fiber reinforced plastics. *Eng Systems Anal ASME*, PD-Vol-**64**(2):93–101.

Reugg C, Habermeir J (1980), Composite propeller shafts design and optimization. In: A. Bunsell *et al.* (Ed), *Advances in Composite Material*, (Proc. ICCM 3), Vol. 2, Paris, Pergamon Press, pp. 1740–1755.

Sakuma K, Seto M (1983), Tool wear in cutting glass fiber reinforced plastics (the relation between fiber orientation and tool wear). *Bull JSME*, **26**(218):1420–1427.

Santhanakrishnan G, Krishnamoorthy R, Malhotra SK (1988), Machinability characteristics of fibre reinforced plastics composites. *J Mech Working Technol*, **17**:95–204.

Santhanakrishnan G (1990), *Investigations on Machining of FRP Composites and their Tribological Behaviour*, PhD thesis, IITMadras, Chennai, India.

Spur G, Wunsch UE (1988), Turning of fiber-reinforced plastics. *Manuf Rev*, **1**(2):124–129.

Sreejith PS, Krishnamoorthy R, Malhotra SK, Narayanasamy K (2000), Evaluation of PCD tool performance during machining of carbon/ phenolic ablative composites. *J Mater Proc Technol*, **104**:53–58.

Takeyama H, Lijima (1988), Machinability of glass fiber reinforced plastics and application of ultrasonic machining. *Ann CIRP*, **37**(1):93–96.

Tomac N, Tonnessen T, Rasch FO (1992), Machinability of particulate aluminium matrix composites, *Ann CIRP*, **41**(1):55–58.

Tosun G, Muratoghi M (2004), The drilling of Al SiCp metal matrix composites. Part II: Workpiece surface integrity, *Composites Sci & Technol*, **64**:1413–1418.

Wang X, Feng CX (2002), Development of empirical models for surface roughness prediction in finish turning, *Int J Adv Man Technol*, **20**(5):348–356.

Wang XM, Zhang LC (2003), An experimental investigation into the orthogonal cutting of unidirectional fiber reinforced plastics. *Int J Machine Tools Manuf*, **43**:1015–1022.

Whitehouse D (1996), *Handbook of Surface Metrology*, Institute of Physics Publishing for Rank Taylor-Hobson Co., Bristol.

Part II
Non-traditional methods for machining composite materials

7

Ultrasonic vibration-assisted (UV-A) machining of composites

Q. FENG and C. Z. REN, Tianjin University, China and
Z. J. PEI, Kansas State University, USA

Abstract: Ultrasonic vibration-assisted (UV-A) machining is a group of machining processes during which ultrasonic vibration is applied on either the cutting tool or the workpiece. This chapter presents UV-A machining processes that have been reported to machine three types of composite materials: metal matrix composites, ceramic matrix composites, and plastic matrix composites. These processes include UV-A turning, UV-A drilling, UV-A grinding, ultrasonic machining, rotary ultrasonic machining, UV-A laser-beam machining, and UV-A electrical discharge machining. The machining principles and major process variables for each process are discussed.

Key words: composite, ultrasonic vibration, turning, drilling, grinding, ultrasonic machining, rotary ultrasonic machining.

7.1   Introduction

Ultrasonic vibration-assisted (UV-A) machining is a group of machining processes during which ultrasonic vibration is applied on either the cutting tool or the workpiece. In these machining processes, the vibration changes the direction and/or speed of relative movement between the cutting tool and workpiece. The tool may periodically lose its direct contact with the workpiece under an appropriate combination of cutting speed, vibration frequency, and vibration amplitude, resulting in material removal mechanisms different from those of conventional machining processes. The cutting force can be reduced and smaller chips obtained when ultrasonic vibration is applied to a machining process. This in turn leads to an improved finished surface and extended tool life (Brehl and Dow, 2008).

   This chapter presents UV-A machining processes that have been reported as suitable for machining composite materials. These processes include UV-A turning, UV-A drilling, UV-A grinding, ultrasonic machining, rotary ultrasonic machining, UV-A laser-beam machining, and UV-A electrical discharge machining. The chapter describes the machining principle and major process variables for each process and discusses published work on UV-A machining of three types of composite materials: metal matrix composites, ceramic matrix composites, and plastic matrix composites. It also discusses possible research needs.

7.2   Ultrasonic vibration-assisted (UV-A) turning

In UV-A turning, a single-point cutting tool is fed toward a rotating cylindrical workpiece at a certain feed rate and depth of cut, to machine the external surface of the workpiece. In addition, ultrasonic vibration is applied to the tip of the cutting tool, as illustrated in Fig. 7.1. There are three independent principal directions in which ultrasonic vibration can be applied during the turning process: cutting direction (or tangential direction), feed direction (or horizontal direction), and depth of cut direction. Vibration in a single direction (1D system) is the most commonly used in UV-A turning (Brehl and Dow, 2008). Ultrasonic vibration can also be applied in both cutting and feed directions (2D system) (Babitsky and Kalashnikov, 2003; Overcash and Cuttino, 2009), causing the tool tip to move in a tiny circle or ellipse. Cutting speed, vibration amplitude, and vibration frequency need to be selected such that the tool tip is separated from the workpiece (and chip) in each cycle of the vibration (Zhong and Lin, 2005). Major process variables in UV-A turning include tool type, cutting speed, feed rate, depth of cut, and vibration amplitude.

UV-A turning was first used in the late 1950s for traditional metal-cutting applications (Brehl and Dow, 2008). UV-A turning is shown to offer distinct advantages over conventional turning when machining ductile, brittle, and 'hard metal' materials over a broad range operating conditions, depths of cut, and tool materials (Brehl and Dow, 2008). These advantages include reduced cutting forces, extended tool life, reduced surface roughness, improved form accuracy, greater depth of cut for ductile regime machining of brittle materials, and suppression of burr formation.

Workpiece materials in most published papers on UV-A turning of composites are metal matrix composites (MMC). Turning of Al-based MMC

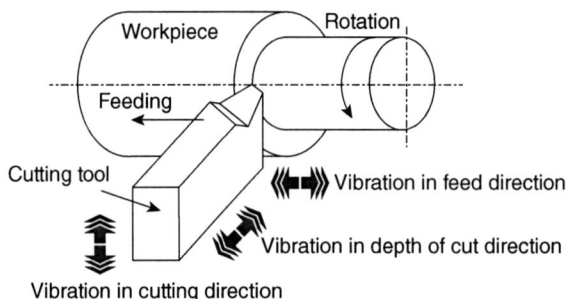

7.1 Illustration of UV-A turning (after Babitsky and Kalashnikov, 2003).

reinforced with SiC particles was studied using diamond tools (Zhong and Lin 2005, 2006; Liu *et al.*, 2002; Zhao *et al.*, 2002). Cutting experiments were performed on conventional lathes. Piezoelectric transducers were attached to the lathes to provide ultrasonic vibration to the tool tip in the cutting direction. Cutting performances of UV-A turning and conventional turning could be compared by switching on and off the ultrasonic transducer. Liu *et al.* (2002) reported that UV-A could reduce cutting force by up to 73%, compared with conventional turning. The cutting force reduction was more significant when adopting a lower cutting speed, feed rate, and depth of cut. Similarly to conventional turning, the cutting force in UV-A turning increased with increasing cutting speed, feedrate, and depth of cut. The roughness of the MMC surface turned with vibration was better than that without vibration. From the results of designed experiments, the influence of cutting speed on cutting force was more obvious than those of feed rate and depth of cut. Surface roughness values R_a, R_q, and R_t for UV-A turning were 21%, 25%, and 33%, respectively; lower than those in conventional turning when other process variables were the same. The residual compression stress on the surface machined by UV-A turning was higher than that on the conventionally turned surface (Zhao *et al.*, 2002).

Plastic matrix composites (PMC) demonstrated different machinability by UV-A turning compared with MMC. Kim and Lee (1994) carried out experiments on the turning of carbon fiber-reinforced plastics (CFRP) with ultrasonic vibration applied in the cutting direction. Three tool materials (tungsten carbide, poly-crystal diamond, and single-crystal diamond) were used. The single-crystal diamond tool yielded lower cutting force and lower surface roughness in both UV-A turning and conventional turning when compared with the other tool materials. In UV-A turning, the machined surface roughness was not related to depth of cut (different from conventional turning) but was related strongly to cutting speed and feed rate. UV-A turning resulted in a better machined surface than conventional turning when the cutting speed was lower than a critical value.

7.3   UV-A drilling

UV-A drilling uses a drill to cut or enlarge a hole in a workpiece with ultrasonic vibration applied on either the drill or the workpiece in the feed direction, as illustrated in Fig. 7.2. Ultrasonic vibration is usually applied on the workpiece because it is more difficult to apply it on a rotating drill. However, when the workpiece is too large or too heavy, a vibrating tool will be preferred. Major process variables in UV-A drilling include drill type, rotation speed, feed rate, and vibration amplitude.

There are many documented advantages of UV-A drilling (Thomas and Babitsky, 2007). These include reduced cutting force and torque, increased

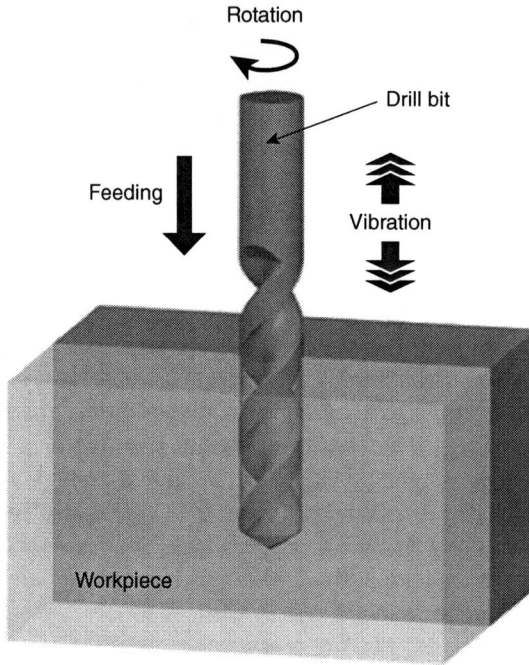

7.2 Illustration of UV-A drilling (after Xu *et al.*, 2009).

positional accuracy, reduced or eliminated burr formation, improved chip expulsion, improvement in both hole roundness and size, extended drill bit life, increased material removal rate, improved hole surface finish, and reduced or eliminated built up edge.

Experiments to drill an SiC particle-reinforced Al matrix composite were carried out by UV-A drilling with ultrasonic vibration applied in the feed direction using carbide twist drills (Liu *et al.*, 2005; Xu *et al.*, 2009). Experimental results indicated that the cutting force could be reduced and the tool life prolonged when ultrasonic vibration was applied. Spiral chips were found with a larger curl radius and looser spire by UV-A drilling compared with conventional drilling (Liu *et al.*, 2005). Surface roughness of the machined surface by UV-A drilling was lower than that by conventional drilling. The twist drill wore severely when the SiC particle content in the composite was high. The main wear form was grinding abrasion, since the hardness of the matrix is lower than that of the SiC particles. The main edges and the chisel edge suffered more severe abrasion in conventional drilling than in UV-A drilling. Thrust force and torque increased with the rise of feed rate and drilling depth. The drilling torque in UV-A drilling could be about 20% to 30% lower than that in conventional drilling.

Carbon fiber-reinforced polymer (CFRP) composites are hard, brittle, and have low shear strength between layers. Zhang *et al.* (1994) found that the thrust force was reduced and delamination-free holes could be obtained in drilling CFRP with ultrasonic vibration assistance.

Glass fiber-reinforced polymer (epoxy) (GFRP) was also drilled by UV-A drilling with vibration of the tool in the feed direction (Ramkumar *et al.*, 2004; Aoki *et al.*, 2005, 2006; Arul *et al.*, 2006). Surface roughness of the surfaces machined by UV-A drilling was lower than that by conventional drilling under many conditions. The reduction of surface roughness (R_a) was more than 50% when ultrasonic vibration was applied. Arul *et al.* (2006) found that the trend of thrust, flank wear, and delamination with the number of holes in conventional and vibration drilling are similar. There was a good correlation between thrust and delamination.

7.4   UV-A grinding

In UV-A grinding, vibration can be applied to the grinding system in three independent principal directions, as illustrated in Fig. 7.3: longitudinal direction (along the workpiece feeding direction), transverse direction (along the axial direction of the grinding wheel), and normal direction (along the normal direction of the workpiece surface). Ultrasonic vibration can also be applied in a combination direction of two directions (Zhao *et al.*, 2005, 2008). Important process variables include wheel depth of cut, wheel rotation speed, feed rate, and vibration amplitude.

Ceramic matrix composite (Al_2O_3/ZrO_2) is the only composite material machined by UV-A grinding reported in the literature (Zhao *et al.*, 2005, 2006, 2008; Wu *et al.*, 2007, 2009; Tong *et al.*, 2009; Xiang *et al.*, 2006). Ultrasonic vibration was applied to the workpiece in a combination direction of the longitudinal direction and the transverse direction, as illustrated in Fig. 7.3.

The critical depth of cut (when depth of cut is lower than this value, the material removal mechanism will change from brittle mode to ductile mode) of the Al_2O_3/ZrO_2 composite in UV-A grinding (25 μm) was larger than that in conventional grinding (15 μm). The grinding force (normal force and tangential force) in UV-A grinding was about 20–30% lower than that in conventional grinding under the same conditions. The material removal rate (MRR) in UV-A grinding of Al_2O_3/ZrO_2 composite, when the wheel depth of cut was 10 μm, was twice as high as that in conventional grinding when other process variables were the same. UV-A grinding also yielded a lower surface roughness than conventional grinding. The tool wear in UV-A grinding was about 10% of that in conventional grinding. As in conventional grinding, the grinding force and MRR in UV-A grinding increased with increasing wheel depth of cut. Surface roughness increased

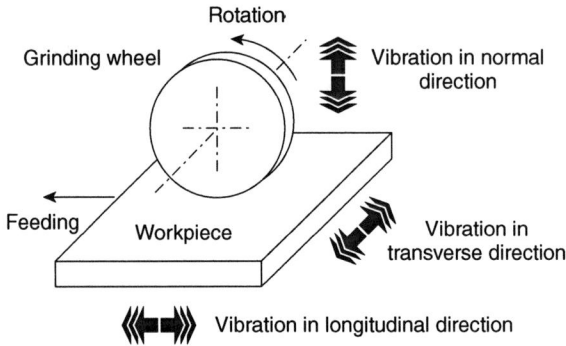

7.3 Illustration of UV-A grinding (after Zhao *et al.*, 2008).

with increasing depth of cut and feed rate, and decreased with increasing vibration frequency, abrasive grain size, and wheel speed.

7.5   Ultrasonic machining (USM)

In USM, abrasive slurry is delivered to the interface between the workpiece and the tool, as illustrated in Fig. 7.4. The tool is fed toward the workpiece under a certain static load. Either the tool or the workpiece vibrates at an ultrasonic frequency. The abrasive slurry is a mixture of abrasive material (e.g. silicon carbide and boron carbide) suspended in water or oil. The vibration causes the abrasive particles in the slurry between the tool and the workpiece to impact the workpiece surface, causing material removal by micro-chipping. Common tool materials used in USM include soft steel and stainless steel. Important process variables include tool material, abrasive type, abrasive concentration, abrasive grain size, static load, and vibration amplitude.

The first patent concerning USM was granted in 1945 (Thoe *et al.*, 1998). It has several advantages (Thoe *et al.*, 1998). It can be used to produce either a through or blind hole with a complex shape by die sinking, using complex form tools. It is a non-thermal process and does not require the workpiece material to be conductive. It is believed to be a stress- and damage-free process, preferable for machining workpiece materials with low ductility and high hardness. The drawback is the very slow machining speed for a single hole; its production rate can be justified only in multi-hole process (Hocheng and Tsao, 2005).

USM has been used to drill several ceramic matrix composites (CMC). Experiments on USM of carbon fiber-reinforced SiC were carried out by Hocheng *et al.* (2000). The effects of machining parameters (abrasive type, abrasive concentration, electric current, and static load), on material

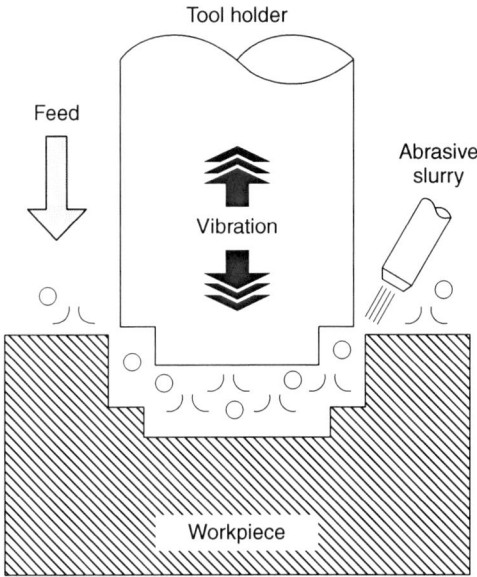

7.4 Illustration of USM (after Hocheng *et al.*, 2000).

removal rate (MRR), hole clearance, edge quality, and tool wear were studied. MRR was calculated by measuring the machining time and the volume of removed material. The hole clearance was measured by a profile projector. The edge quality was evaluated by hole-edge photography. The tool wear was measured by a vernier caliper. Results showed that MRR is proportional to the static load applied to the tool. MRR was at a maximum using an optimal abrasive grain size. The optimal value depended upon the amplitude of the tool oscillation. Increasing abrasive grain size when the abrasive concentration was kept unchanged would mean a reduction in the number of abrasive grains, causing MRR to decrease. As vibration amplitude increased up to a certain value, MRR increased. A further increase in vibration amplitude beyond this value resulted in a reduction in MRR. The hole accuracy can be studied by hole clearance (diameter clearance between the drilled hole and the tool). Hole clearance decreased with an increase in the static load and a decrease in abrasive grain size. When the static load increased, the machining time decreased and the transverse vibration of the tool depressed. As the vibration amplitude increased, the hole clearance increased as a result of the enhanced vibration of the tool and the abrasive in the gap. As the vibration amplitude increased, the hole clearance increased. In addition, delamination and splintering at the hole exit were more severe with an increase in static load. When machining is progressing toward the hole exit, the uncut thickness of the workpiece becomes insufficient to resist the push-out of material, causing delamination and splintering.

Hocheng conducted experiments on the USM of epoxy/C and PEEK/C to explore the cutting nature and cutting mechanism (Hocheng and Hsu, 1995). No delamination occurred at the hole edge. The major cutting mechanism was abrasive particle hammering or impacting on the workpiece to remove material in micro craters. Brittle fracture of the fibers and plastic deformation of the matrix were observed. The machined surface roughness increased with grain size, vibration amplitude, and abrasive concentration, and was independent of feed rate or fiber direction. The grain size and vibration amplitude significantly affected the hole clearance.

Experiments on USM of SiC whisker-reinforced Al_2O_3 were also conducted (Lee and Chan, 1997; Lee and Deng, 2001). MRR varied when machining with different direction angles θ (the angle between machined surface and the surface normal to the hot pressing direction, as illustrated in Fig. 7.5). The maximum MRR was observed when $\theta = 0°$, and the minimum MRR was observed when $\theta = 90°$. The effects of whisker orientation on the machined surface roughness were similar to those on MRR. Higher SiC whisker content tended to produce lower surface roughness. Furthermore, surface roughness decreased with increasing workpiece fracture toughness.

Similar results on MRR and surface roughness were obtained in USM of $Al_2O_3/(Ti,W)C$, Al_2O_3/TiC, Al_2O_3/TiB_2, and Al_2O_3/SiC whisker composites (Deng and Lee, 2002; Ramulu, 2005). When machining of SiC whisker-

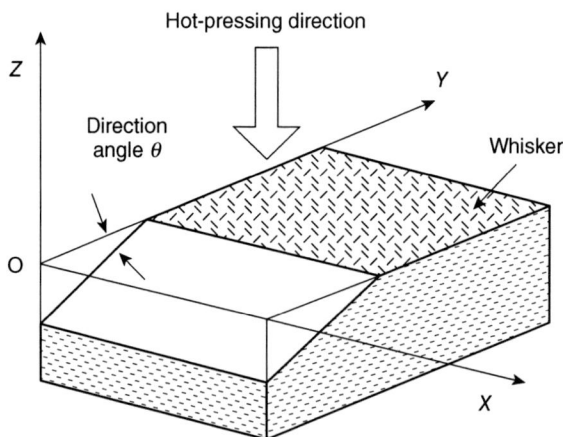

7.5 Illustration of the anisotropy of whisker distribution with direction angle θ in hot-pressing whisker reinforced ceramic composite (after Lee and Deng, 2001).

reinforced Al_2O_3, the minimum MRR value and lowest surface roughness value were observed when $\theta = 90°$. For a $\theta = 0°$ surface, little or no whisker toughening takes place; since the whiskers share no load, lateral cracks can readily propagate through the matrix. For a $\theta = 90°$ surface, the bridging whiskers impose closure stress at the crack tip and resist the crack propagation. MRR and surface roughness increased with abrasive grain size. Harder abrasives result in higher MRR and higher surface roughness. Furthermore, MRR was lower when the workpiece material had higher fracture toughness. MRR was lower for whisker-reinforced ceramic composites than for particle-reinforced ceramic composites. The flexural strength of USM machined specimens varied narrowly from the mean value, and the composites with higher fracture toughness had higher Weibull modulus. Weibull modulus is a parameter characterizing the surface integrity of the machined surface. A high value of Weibull modulus implies a narrow distribution of strength, and denotes a good material with a high degree of homogeneity of properties. Low Weibull modulus indicates that the distribution of the strength-controlling crack size is broader (Deng and Lee, 2000).

Experiments on the USM of hydroxyapatite/SiC bioceramic composites were conducted by Niu *et al.* (2007). MRR and machined surface roughness increased with increase of the direction angle θ. Under the same machining conditions, the higher the fracture toughness of the material, the lower the MRR.

Experiments on the USM of Al_2O_3/$LaPO_4$ composites were carried out by Abdul Majeed *et al.* (2008a, 2008b, 2009). A critical content (30% by weight) of $LaPO_4$ yielded the best machinability. Hollow tools resulted in higher MRR than solid tools.

7.6   Rotary ultrasonic machining (RUM)

Rotary ultrasonic machining (RUM) is a hybrid machining process that combines the material removal mechanisms of diamond grinding and of ultrasonic machining. Figure 7.6 illustrates the RUM process. The drilling tool is a core drill made of metal-bonded diamond abrasives. The rotating tool is ultrasonically vibrated and fed toward the workpiece. Meanwhile, coolant is pumped through the hole in the middle of the drill to flush away the debris. Important process variables are tool rotation speed, tool feed rate or pressure, vibration amplitude, abrasive grain size, and abrasive concentration.

Experiments on RUM of SiC fiber-reinforced C and SiC fiber-reinforced SiC were conducted (Li *et al.*, 2004, 2005a). Compared with drilling holes without ultrasonic vibration, the thrust force was reduced by about 50% and MRR could be improved by about 10% with RUM. Furthermore, thrust force curves with RUM of CMC had much significant undulation

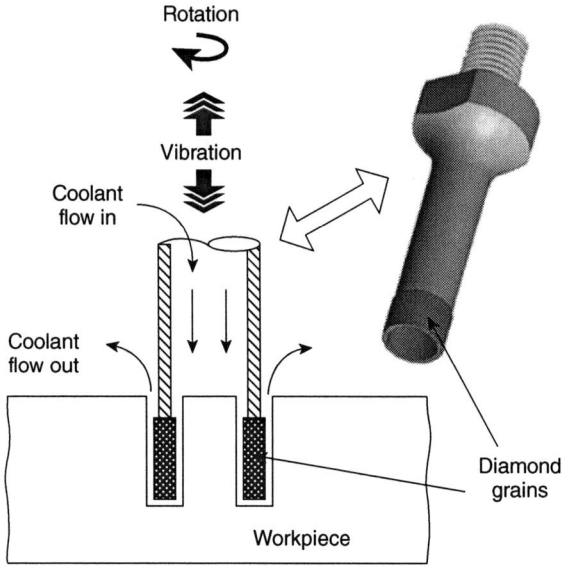

7.6 Illustration of rotary ultrasonic machining (after Li *et al.*, 2005a).

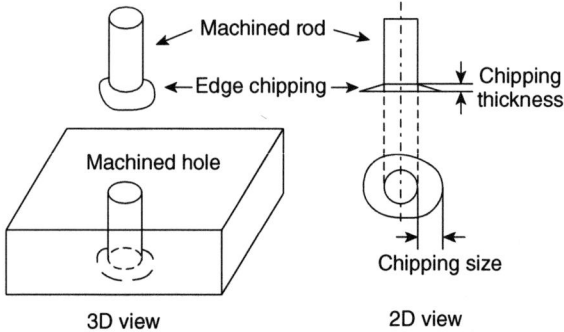

7.7 Illustration of chipping size and chipping thickness (after Li *et al.*, 2005a).

than RUM of pure ceramics. MRR increased as tool rotation speed, feed rate, and vibration amplitude increased. The hole quality was quantified by the thickness and size of edge chipping, as shown in Fig. 7.7. Lower values of chipping size and thickness suggest better hole quality. Higher tool rotation speed would be effective to reduce chipping thickness. Chipping thickness increased as feed rate increased. The effect of tool rotation speed was stronger at the high level of feed rate. Vibration amplitude had less significant effects on chipping thickness.

Experiments on RUM of a ZrO_2/Al_2O_3 composite with five different mixture ratios (0%, 25%, 50%, 75%, and 100% of Al_2O_3) were conducted (Li *et al.*, 2005b). The maximum thrust force during the drilling process was the highest when the samples were 50% Al_2O_3 and 50% ZrO_2. It was about 4 times that of pure Al_2O_3 or ZrO_2.

7.7   UV-A laser-beam machining (LBM)

In laser-beam machining (LBM), the source of energy is a laser, which focuses optical energy on the surface of the workpiece. The highly focused, high-density energy melts and evaporates portions of the workpiece in a controlled manner. It is widely used in cutting sheet and drilling holes. UV-A LBM is a combination of LBM and ultrasonic vibration. The vibration is applied to the laser generator or workpiece in the feed direction.

One of the major advantages of laser drilling over conventional drilling is that there is no contact between the tool and the workpiece, eliminating delamination caused by the thrust force and vibration of the drill bit. Drilling small diameter holes is possible, down to 0.025 mm. For large holes, above 0.5 mm diameter, the laser beam is controlled to cut the outline of the hole. Laser cutting for composite materials is complicated, since the constituents of the composite material usually have very different thermal conductivities, heat capacities, and vaporization temperature (Hocheng and Tsao, 2005).

Experiments on drilling Al-Cu based composites reinforced by SiC particles were carried out by UV-A LBM, with the ultrasonic vibration applied in the feed direction (Liu *et al.*, 2005; Xu *et al.*, 2009). Experimental results indicated that with ultrasonic vibration, the quality of the machined holes was improved, the efficiency of machining was improved, the dimensions of the heat affected area were reduced, and the depth of the machined hole was increased.

7.8   UV-A electrical discharge machining (EDM)

The EDM process is based on thermoelectric energy travelling between an electrode and the workpiece. A pulse discharge occurs in a small gap between the workpiece and the electrode and removes the unwanted material from the workpiece through melting and vaporizing (Mohd Abbas *et al.*, 2007). However, the dielectric liquid flow in the narrow discharge gap is slow and the removal of discharged debris is difficult due to lower pulse energy and lack of electrode rotation, resulting in lower machining efficiency and accuracy (Tong *et al.*, 2008).

Introduction of ultrasonic vibration to the electrode is one of the methods used to improve dielectric circulation, which facilitates debris removal and the creation of a large pressure change between the electrode and the workpiece, as an enhancement of molten material ejection from the surface of the workpiece (Guo *et al.*, 1997).

EDM was developed in the late 1940s and a study of UV-A EDM has been undertaken since the mid 1980s by Mohd Abbas *et al.* (2007). UV-A EDM has been applied to steel and alloys (Murti and Philip, 1986; Hocheng and Pa, 2003). Composites machined by EDM include PMC, CMC, and MMC, but most of the experiments on composites have been limited to EDM without vibration (Lau *et al.*, 1990; Guu *et al.*, 2001; Hocheng *et al.*, 1997, 1998).

7.9   Conclusions

This chapter has introduced UV-A machining processes for composite materials. Conclusions can be drawn as follows.

(i)   The use of ultrasonic vibration in different machining processes has been documented for more than 50 years, but no reports on UV-A machining of composite materials existed until the 1990s. Ninety-seven percent of the reported work on UV-A machining of composite materials was published in the 2000s. Application of ultrasonic vibration in conventional machining processes starts from single direction (linear vibratory tool path) to dual-direction (circular/elliptical tool path) and then multi-direction (three dimensional tool path) and from traditional machining (e.g. cutting and abrasive machining) to nontraditional machining (e.g. LBM and EDM).

(ii)   Reported composite materials machined by UV-A machining are very limited in type. Table 7.1 shows the published work on UV-A machining of composite materials. In the literature, ceramic matrix composites (CMC) have been studied the most, followed by polymer matrix composites (PMC) and metal matrix composites (MMC). MMCs and PMCs are mostly studied by UV-A turning and drilling, while CMCs are mostly studied by UV-A grinding, USM, and RUM.

(iii)   UV-A machining demonstrates many advantages in the machining of composite materials over conventional methods. Benefits of UV-A machining include decrease in cutting force, improvement in machined surface, increase in machining efficiency, and increase in tool life.

(iv)   Most reported effects of process variables in UV-A machining are similar to those in conventional methods. But there are some exceptions. For example, in the UV-A turning of carbon fiber-reinforced plastics, machined surface roughness was related strongly to cutting speed and feed rate, but not related to depth of cut; while in conventional turning, surface roughness was related to cutting speed, feed rate, and depth of cut.

Table 7.1 Number of published experimental works on UV-A machining of composites

Material		UV-A turning	UV-A drilling	UV-A grinding	USM	RUM	UV-A LBM	UV-A EDM
MMC (matrix/reinforcement)	Al/SiC	4						
	Al-Cu/SiC		2				2	
CMC (matrix/reinforcement)	$Al_2O_3/LaPO_4$				3			
	Al_2O_3/SiC				3			
	Al_2O_3/TiC				1			
	Al_2O_3/TiB_2				1			
	$Al_2O_3/(Ti,W)C$				1			
	Al_2O_3/ZrO_2			7				
	C/SiC					1		
	SiC/TiB_2				1			
	SiC/C				1			
	SiC/SiC					1		
	Hydroxyapatite/SiC				1	1		
	ZrO_2/Al_2O_3				1	1		
PMC (matrix/reinforcement)	Epoxy/C				1			
	Epoxy/glass		4					
	PEEK/C				1			
	Plastic/C	1	1					

(v)   The reported investigations have been limited to experimental studies. Mechanisms in UV-A machining of composite materials have not been thoroughly studied.

Future research needs include extending UV-A machining to a much wider range of composite materials. It is also desirable to conduct theoretical investigations into machining mechanisms of UV-A machining of composite materials.

7.10   References

Abdul Majeed, M.; Vijayaraghavan, L.; Malhotra, S.K.; Krishnamoorthy, R. (2008a), 'A.E. monitoring of ultrasonic machining of $Al_2O_3/LaPO_4$ composites', *Journal of Materials Processing Technology*, **207**, 321–329.

Abdul Majeed, M.; Vijayaraghavan, L.; Malhotra, S.K.; Krishnamurthy, R. (2008b), 'Ultrasonic machining of $Al_2O_3/LaPO_4$ composites', *Journal of Machine Tools and Manufacture*, **48**, 40–46.

Abdul Majeed, M.; Vijayaraghavan, L.; Malhotra, S.K.; Krishnamoorthy, R. (2009), 'Characterization and machining of alumina ceramic reinforced with lanthanum phosphate', *Journal of Materials Processing Technology*, **209**, 2499–2507.

Aoki, S. Hirai, S.; Nishimura, T. (2005), 'Prevention from delamination of composite material during drilling using ultrasonic vibration', *Key Engineering Materials*, **291**, 465–470.

Aoki, S.; Nishimura, T.; Hirai, S. (2006), 'Improvement of machined surface of composite material during drilling using ultrasonic vibration', *Transactions of the Japan Society of Mechanical Engineers, Part C*, **72**, 2629–2633.

Arul, S.; Vijayaraghavan, L.; Malhotra, S.K.; Krishnamurthy, R. (2006), 'The effect of vibratory drilling on hole quality in polymeric composites', *International Journal of Machine Tools and Manufacture*, **46**, 252–259.

Babitsky, V.I.; Kalashnikov, A.N. (2003), 'Ultrasonically assisted turning of aviation materials', *Journal of Materials Processing Technology*, **132**, 3157–3167.

Brehl, D.E.; Dow, T.A. (2008), 'Review of vibration-assisted machining', *Precision Engineering*, **32**, 153–172.

Deng, J.; Lee, T. (2002), 'Ultrasonic machining of alumina-based ceramic composites', *Journal of the European Ceramic Society*, **22**, 1235–1241.

Deng, J.; Lee, T. (2000), 'Surface integrity in electro-discharge machining, ultrasonic machining, and diamond saw cutting of ceramic composites', *Ceramics International*, **26**, 825–830.

Guo, Z.N.; Lee, T.C.; Yue, T.M.; Lau, W.S. (1997), 'Study of ultrasonic-aided wire electrical discharge machining', *Journal of Materials Processing Technology*, **63**, 823–828.

Guu, Y.H.; Hocheng, H.; Tai, N.H.; Liu, S.Y. (2001), 'Effect of electrical discharge machining on the characteristics of carbon fiber reinforced carbon composites' *Journal of Materials Science*, **36**, 2037–2043.

Hocheng, H.; Hsu, C.C. (1995), 'Preliminary study of ultrasonic drilling of fiber reinforced Plastics', *Journal of Materials Processing Technology*, **48**, 255–266.

Hocheng, H.; Lei, W.T.; Hsu, H.S. (1997), 'Preliminary study of material removal in electrical-discharge machining of SiC/Al', *Journal of Materials Processing Technology*, **63**, 813–818.

Hocheng, H.; Guu, Y.H.; Tai, N.H. (1998), 'Feasibility analysis of electrical-discharge machining of carbon-carbon composites', *Materials and Manufacturing Processes*, **13**, 117–132.

Hocheng, H.; Tai, N.H.; Liu, C.S. (2000), 'Assessment of ultrasonic drilling of C/SiC composite material', *Composites Part A: Applied Science and Manufacturing*, **31**, 133–142.

Hocheng, H.; Pa, P.S. (2003), 'Continuous secondary ultrasonic electropolishing of an SKD61 cylindrical part', *International Journal of Advanced Manufacturing Technology*, **21**, 238–242.

Hocheng, H.; Tsao, C.C. (2005), 'The path towards delamination-free drilling of composite materials' *Journal of Materials Processing Technology*, **167**, 251–264.

Kim, J.D.; Lee, E.S. (1994), 'Study of the ultrasonic-vibration cutting of carbon-fiber reinforced plastics', *Journal of Materials Processing Technology*, **43**, 259–277.

Lau, W.S.; Wang, M.; Lee, W.B. (1990), 'Electrical discharge machining of carbon fiber composite materials', *International Journal of Machine Tools and Manufacture*, **30**, 297–308.

Lee, T.C.; Chan, C.W. (1997), 'Mechanism of the ultrasonic machining of ceramic composites', *Journal of Materials Processing Technology*, **71**, 195–201.

Lee, T.C.; Deng, J. (2001), 'Ultrasonic erosion of whisker-reinforced ceramic composites', *Ceramics International*, **27**, 755–760.

Li, Z.; Jiao, Y.; Deines, T.W.; Pei, Z.J.; Treadwell, C. (2004), 'Experimental study on rotary ultrasonic machining of poly-crystalline diamond compact', *IIE Annual Conference and Exhibition*, 769–774.

Li, Z.C.; Jiao, Y.; Deines, T.W.; Pei, Z.J.; Treadwell, C. (2005a), 'Rotary ultrasonic machining of ceramic matrix composites: Feasibility study and designed experiments', *International Journal of Machine Tools and Manufacture*, **45**, 1402–1411.

Li, Z.C.; Pei, Z.J.; Zeng, W.M.; Kwon, P.; Treadwell, C. (2005b), 'Preliminary experimental study of rotary ultrasonic machining on zirconia toughened alumina,' *Transactions of the North American Manufacturing Research Institution of SME*, **33**, 89–96.

Liu, C.S.; Zhao, B.; Gao, G.F.; Jiao, F. (2002), 'Research on the characteristics of the cutting force in the vibration cutting of a particle-reinforced metal matrix composites SiCp/Al', *Journal of Materials Processing Technology*, **129**, 196–199.

Liu, C.S.; Zhao, B.; Gao, G.F.; Zhang, X.H. (2005), 'Study on ultrasonic vibration drilling of particulate reinforced aluminum matrix composites', *Key Engineering Materials*, **291**, 447–452.

Mohd Abbas, N.; Solomon, D.G.; Fuad Bahari, M. (2007), 'A review on current research trends in electrical discharge machining (EDM)', *International Journal of Machine Tools and Manufacture*, **47**, 1214–1228.

Murti, V.S.R.; Philip, P.K. (1987), 'Analysis of the debris in ultrasonic-assisted electrical discharge machining', *Wear*, **117**, 241–250.

Niu, Z.W.; Zhao, Q.Z.; Wei, X.T. (2007), 'Study on the ultrasonic machining of hydroxyapatite/SiCw composite bioceramics material', *Precision Surface Finishing and Deburring Technology – 9th International Symposium on Precision*

Surface Finishing and Deburring Technology, Suzhou, China, ICSD2007, **24**, 61–64.

Overcash, J.L.; Cuttino, J.F. (2009), 'Design and experimental results of a tunable vibration turning device operating at ultrasonic frequencies', *Precision Engineering*, **33**, 127–134.

Ramkumar, J.; Malhotra, S.K.; Krishnamurthy, R. (2004), 'Effect of workpiece vibration on drilling of GFRP laminates', *Journal of Materials Processing Technology*, **152**, 329–332.

Ramulu, M. (2005), 'Ultrasonic machining effects on the surface finish and strength of silicon carbide ceramics', *International Journal of Manufacturing Technology and Management*, **7**, 107–126.

Thoe, T.B.; Aspinwall, D.K.; Wise, M.L.H. (1998), 'Review on ultrasonic machining', *International Journal of Machine Tools and Manufacture*, **38**, 239–255.

Thomas, P.N.H.; Babitsky, V.I. (2007), 'Experiments and simulations on ultrasonically assisted drilling', *Journal of Sound and Vibration*, **308**, 815–830.

Tong, H.; Li, Y.; Wang, Y. (2008), 'Experimental research on vibration assisted EDM of micro-structures with non-circular cross-section', *Journal of Materials Processing Technology*, **208**, 289–298.

Tong, J.L.; Zhao, B.; Yan, Y.Y. (2009), 'Research on chip formation mechanisms of nano-composite ceramics in two-dimensional ultrasonic grinding', *Key Engineering Materials*, **416**, 614–618.

Wu, Y.; Sun, A.G.; Zhao, B.; Zhu, X.S. (2007), 'Study on surface integrity of ultrasonic vibration grinding for $Al_2O_3/ZrO_{(2(n))}$ micro-nanocomposites', *Acta Aeronautica et Astronautica Sinica*, **28**, 1009–1013.

Wu, Y.; Zhao, B.; Zhu, X.S. (2009), 'Brittle–ductile transition in the two-dimensional ultrasonic vibration grinding of nanocomposite ceramics', *Key Engineering Materials*, **416**, 477–481.

Xiang, D.H.; Ma, Y.P.; Zhao, B.; Chen, M. (2006), 'Study on critical ductile grinding depth of nano ZrO_2 ceramics by the aid of ultrasonic vibration', *Key Engineering Materials*, **304**, 232–235.

Xu, X.X.; Mo, Y.l.; Liu, C.S.; Zhao, B. (2009), 'Drilling force of SiC particle reinforced aluminum-matrix composites with ultrasonic vibration', *Key Engineering Materials*, **416**, 243–247.

Zhang, Q.X.; Sun, S.; Luo, J.; Feng, Y.; Ma, C.; Tu, X. (1994), 'Study on ultrasonic vibration drilling in carbon fiber reinforced polymers', *Chinese Journal of Mechanical Engineering*, **7**, 72–77.

Zhao, B.; Liu, C.S.; Zhu, X.S.; Xu, K.W. (2002), 'Research on the vibration cutting performance of particle reinforced metallic matrix composites SiCp/Al', *Journal of Materials Processing Technology*, **129**, 380–384.

Zhao, B.; Zhang, X.H.; Liu, C.S.; Jiao, F.; Zhu, X.S. (2005), 'Study on ultrasonic vibration grinding character of nano ZrO_2 ceramics', *Key Engineering Materials*, **291**, 45–50.

Zhao, B.; Wu, Y.; Liu, C.S.; Gao, A.M.; Zhu, X.S. (2006), 'The study on ductile removal mechanisms of ultrasonic vibration grinding nano-ZrO_2 ceramics', *Key Engineering Materials*, **304**, 171–175.

Zhao, B.; Wu, Y.G.; Jiao, G.F. (2008), 'Research on micro-mechanism of nano-composite ceramic in two-dimensional ultrasound grinding', *Key Engineering Materials*, **359**, 344–348.

Zhong, Z.W.; Lin, G. (2005), 'Diamond turning of a metal matrix composite with ultrasonic vibrations', *Materials and Manufacturing Processes*, **20**, 727–735.

Zhong, Z.W.; Lin, G. (2006), 'Ultrasonic assisted turning of an aluminium-based metal matrix composite reinforced with SiC particles', *International Journal of Advanced Manufacturing Technology*, **27**, 1077–1081.

8

Electrical discharge machining of composites

B. LAUWERS, J. VLEUGELS, O. MALEK, K. BRANS
and K. LIU, Katholieke Universiteit Leuven, Belgium

Abstract: In this chapter the electrical discharge machining (EDM) of ceramic composites is discussed. First, an introduction to the machining methods and the ceramic materials used is given. Then a detailed study of the electrical discharge behavior of three ceramic systems is given: ZrO_2, Si_3N_4 and boride-based composites. It is shown that material properties such as secondary phase type, composition and grain size have a substantial influence on the material removal mechanisms, surface roughness, material removal rate and flexural strength. Finally, the application of EDM on high-end ceramics is demonstrated by two case studies, namely a Si_3N_4-TiN turbine impeller and a B_4C spray nozzle.

Key words: electrical discharge machining, ceramic composites, electrical conductivity, surface properties.

8.1   Introduction

Electrical discharge machining (EDM) has become the most popular, non-traditional, material removal process in today's manufacturing practice. This is due to a number of reasons.[1,2] EDM enables one to machine extremely hard materials, and complex shapes can be produced with high precision. Its inherent capability for automation is another feature fulfilling the expectations of modern manufacturing. EDM is therefore mostly applied in the die and mold-making industry and in the construction of prototypes. The advent of numerically controlled equipment enabled various new EDM technologies such as deep sinking EDM along several axes, contouring EDM, wire EDM and milling EDM. This, together with a higher performance and better accuracy, yielded a functional expansion that is partially responsible for the growing interest in EDM. Within the above mentioned industries, EDM is mostly applied to machine metals, such as high alloyed steels. One of the conditions required to make EDM feasible is electrical conductivity of the material.

Advanced engineering ceramics are more and more employed in modern industries because of their excellent mechanical properties such as high hardness, high compressive strength, and chemical and abrasive resistance. An important requirement for these ceramics is that they can be shaped in

202

an economical way. For simple geometries, conventional sawing and grinding under optimized conditions will be suitable. Most ceramics are prepared by the conventional powder metallurgical (PM) process. Although PM is a near-net shape process, investigation of the PM-market reveals that about 60% of all components need some kind of post-machining operation. One explanation for this is the inability to produce geometries with transverse holes, undercuts, bevels, slots and threads during the powder pressing operation. Another explanation is the general trend to design more complex parts with tighter tolerances that call for machining. Machining operations can account for more than 20% of the total production costs of a ceramic component.

EDM is a process that can machine these hard materials, provided that the ceramics have a sufficiently small electrical resistivity (<100 Ωcm).[3] The electrical resistivity of sintered SiC (0.05 Ωm) and B_4C (0.01 Ωm) is just sufficient to allow EDM machining. But in practice, for some EDM technologies such as sinking EDM, the process is very difficult due to the difficult flushing and cooling conditions.[4] For most ceramic materials destined for EDM, other electro-conductive phases are added (e.g. TiN), which result in so called ceramic composites (e.g. ZrO_2-TiN, Si_3N_4-TiN, B_4C-TiB_2).[5]

The most studied electro-erodable ceramic composite is Si_3N_4-TiN. Although this is available on the market, literature reports are very fragmented and difficult to compare, especially with respect to the settings of the EDM machine used. Electro-erodable ZrO_2-TiN and Al_2O_3-SiC_w-TiC have also recently been introduced to the market, but their method of use (how to set generator settings, machining strategies to apply, etc.) has not been reported so far. So, even though electro-erodable ceramic composites are commercially available, suitable EDM technologies are often missing. Although state-of-the-art EDM generators can machine electro conductive ceramic materials, they are not developed to do so.

In order to allow efficient EDM of ceramic composites, a good understanding of the process material interaction is required. Within several research projects, such as the EU-project MONCERAT, the K.U.Leuven has worked together (mainly as coordinator) with other institutes and industries to broaden the application field of ceramic components by joint and interactive research on EDM technology and novel ceramic materials.

The next section briefly describes the principles of EDM. This knowledge is necessary in order to fully understand the process material interaction in the machining of ceramic composites. It is then followed by Section 8.3, dealing with electrical conductive ceramic materials and composites suited for EDM. Section 8.4 gives a detailed description of EDM of ceramic composites. The influence of the micro-structure of the material on the EDM performance (material removal rate, surface quality, etc.) is explained. This explanation is given for the most common classes of ceramic composites based on Si_3N_4, ZrO_2, B_4C and several transition metal borides. The need for advanced machine generators is briefly described in Section 8.5. Some applications that have been worked out in the MONCERAT EU-project are described in Section 8.6.

8.2   Principles of electrical discharge machining (EDM)

In EDM, material removal is achieved by controlled electrical discharges between two electrodes, one being the tool, the other the workpiece (Fig. 8.1). In general, both electrodes are submerged in a non-conductive fluid (dielectric).

The EDM process is known in different variants, of which 'wire EDM' and 'die sinking EDM' have been used for many years (Fig. 8.2). Another, more recent, technique is 'milling EDM', where a standard rotating electrode (diameters varying from 50 μm to 10 mm) follows a programmed path similar to a classical milling operation. This technology has also found important applications in micro machining. By dressing the electrode tip,

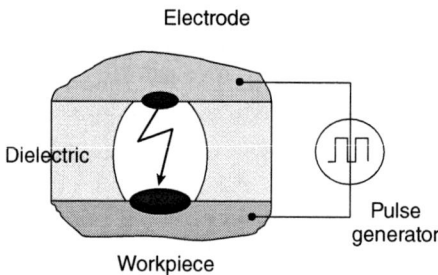

8.1 Principle of the EDM process.

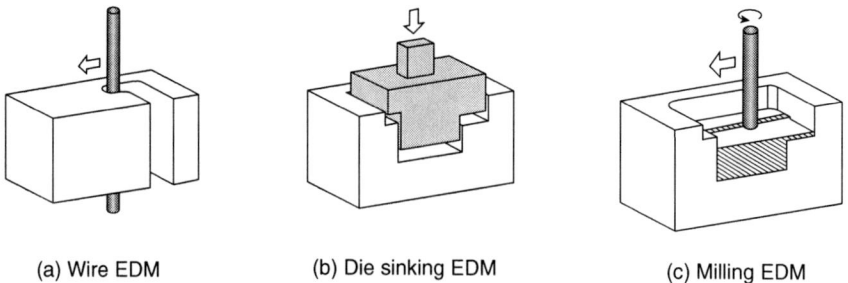

(a) Wire EDM   (b) Die sinking EDM   (c) Milling EDM

8.2 EDM technologies.

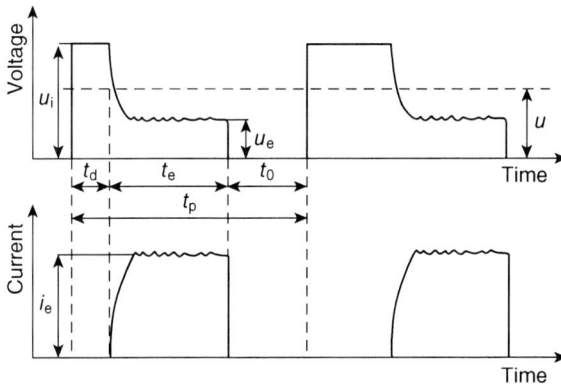

8.3 Typical pulse shape (iso-energetic). Notation: u_i, open generator voltage; u_e, discharge voltage; i_e, discharge current; t_d, ignition delay time; t_e, discharge duration; t_0, pulse interval time; t_p, pulse cycle time.

diameters less than 10 μm can be obtained, so small features can be machined.

Figure 8.3 shows a typical pulse shape (voltage and current behavior measured over the gap) for a die sinking EDM operation. The shown pulse shape is a so called 'iso-energetic pulse', where various parameters, such as discharge current and pulse duration, are controlled to obtain a constant energy per pulse. Depending on the EDM technology, the workpiece material and strategy, the pulse frequency ($1/t_p$) varies from 100 kHz up to 1 MHz. Pulse currents can vary from a few nano amps up to 150 A, and higher.

The machining is realized without contact between the electrodes. Movement of the tool is controlled by the servo feed control system of the machine. The servo control software can use several parameters to keep a constant sparking gap. The parameters that are most frequently used are the discharge delay time (t_d) and/or the average voltage over the gap.

In normal processing conditions, each discharge locally melts and evaporates material from the workpiece as well as the electrode (Fig. 8.4). Multiple discharges create a typical surface topography like a moon landscape. The surface topography shown is typical for EDM machining of metallic materials such as iron and steel. Besides removal of material, some of the molten material can re-solidify on the surface, giving a typical white layer (see Fig. 8.4c). This zone can also contain micro-cracks (due to the melting–re-solidification process of often high alloyed steels) which degrade the mechanical properties of the final part. Therefore, an initial 'roughing regime' has always to be followed by a number of finishing regimes in which the energy applied is gradually decreased (smaller discharge current, smaller

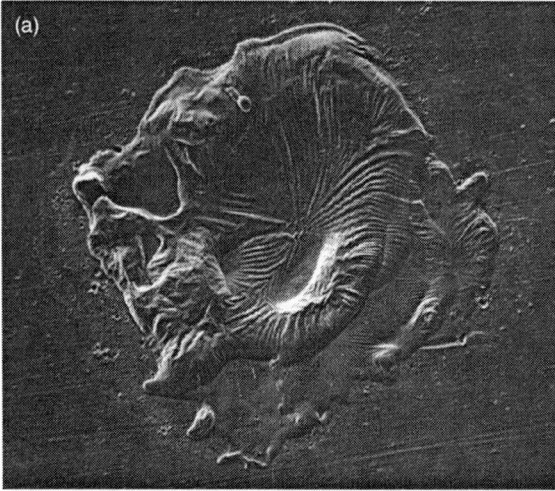

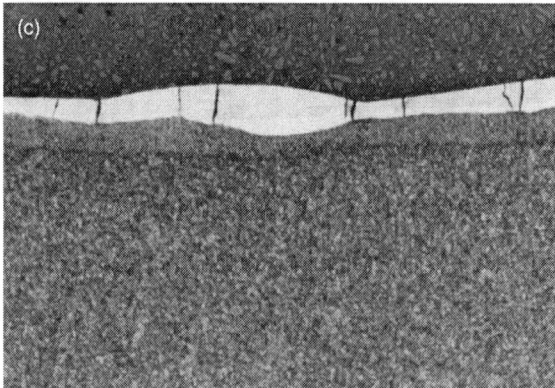

8.4 Typical surface topography/quality of EDM machined surface (metallic material, high energy regime). (a) Surface topography of one discharge, (b) surface topography of multiple discharges, (c) white layer.

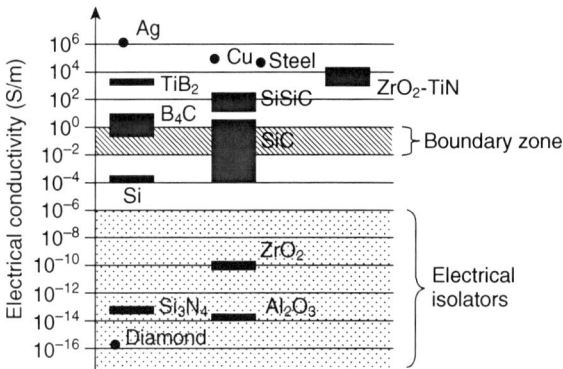

8.5 Electrical conductivity (resistivity) of various materials.

pulse durations). Shorter pulses with low current give, in general, smaller craters and hence a better surface quality and precision.

In micro machining (e.g. micro milling EDM), the applied energy is always lower (also in the case of roughing) due to the small features and to reduce electrode wear. Therefore, generators used for micro-machining are often of the relaxation type where energy is released from a charged capacitor.[6]

As described earlier, the material should be electrically conductive to allow EDM machining. A generally accepted rule is an electrical resistivity smaller than 100 Ωcm. Figure 8.5 shows the electrical conductivity of various materials. Metals such as steel, and copper can be machined very well, while ceramic materials such as ZrO_2, and Al_2O_3 are isolators and cannot be machined. Other ceramic materials such as TiB_2 have sufficient conductivity, while materials such as B_4C and SiC are situated in a border region between machineability and non-machineability. This means that the rule of 100 Ωcm is certainly not exact, but indicates a transition region.

8.3   Electrically conductive ceramic materials and composites

Advanced technical ceramics such as ZrO_2, Al_2O_3, SiC and Si_3N_4 are becoming increasingly important materials of choice in industrial applications such as metal forming, part machining and as construction materials in mechanical components. In addition, due to their high wear resistance and chemical inertness, ceramics are also present in medical applications such as hip, knee and dental implants. Most ceramic components are prepared by conventional powder metallurgy (PM), which, in most cases, has to be followed by some kind of post-sintering machining operation. However,

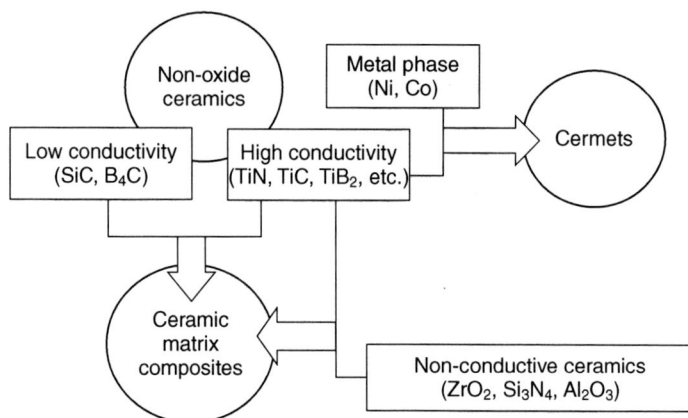

8.6 Classification of electrically conductive ceramics.[7]

these materials are extremely difficult to machine using conventional techniques, and grinding is often only possible with cubic boron nitride and diamond tooling. These tools limit the geometry of the component considerably due to the inflexible nature of the grinding process. Therefore, EDM is a potential technique, but, as described above, the material should be electrically conductive.

Figure 8.6 shows a classification of electrical conductive ceramic materials suited to be machined by EDM. They can be classified in 'non-oxide ceramics', 'cermets' and 'ceramic composites'. Non-oxide ceramics can be conductive (e.g. TiN, TiC, TiB$_2$, etc.) or less conductive (e.g. SiC, B$_4$C). Cermets are created by the addition of metal phases (Ni, Co) to non-oxide ceramics. Ceramic matrix composites consist of (at least) two different main phases (with small amounts of sinter additives or stabilizers). Generally speaking, a composite is created out of a low conductive (SiC, B$_4$C) or non-conductive phase (Si$_3$N$_4$, ZrO$_2$, Al$_2$O$_3$), and a very electrically conductive (non-oxide) ceramic such as TiN, TiCN, WC, NbC. Adding a certain percentage of these conductive phases drastically lowers the electrical resistivity of the composite to a level that easily allows EDM machining. Moreover, it can also increase the hardness and strength of the composites.[8] Tuning the amounts and distribution of both phases allows one to obtain a ceramic composite with specific required mechanical properties; but, as shown in the following sections, it will also have an important influence on the EDM process behavior.

Silicon nitride (Si$_3$N$_4$) ceramic composites are regarded as some of the most important high-temperature structural materials.[9] Recently, high-strength and high-toughness Si$_3$N$_4$ matrix composites have been developed to improve the mechanical reliability of Si$_3$N$_4$ ceramics. It has been reported

8.7 Micro turbine in Si_3N_4-TiN.

that the introduction of an electro conductive second phase can improve the mechanical properties and electro conductivity of silicon nitride ceramics.[10] TiN has some attractive properties as the second phase material, e.g. high hardness, good chemical durability, high electrical conductivity, and good compatibility with Si_3N_4 when sintered in nitrogen. Si_3N_4 based materials have a wide range of possible applications and are very promising for cutting tools or components that operate at very high temperatures, e.g. micro turbines. The part shown in Fig. 8.7 is an example of a micro turbine, EDM'ed in Si_3N_4-TiN (Kersit). Details of the applied machining strategy will be explained in Section 8.6.

Zirconia (ZrO_2) has an excellent strength and high fracture toughness, but a modest hardness. Incorporation of a secondary hard electroconductive phase can increase the hardness as well as the electrical conductivity of the insulating ZrO_2 matrix, without reduction of the strength and fracture toughness.[8] *Zirconia-based ceramic composites* offer a very good compromise between hardness and toughness, which makes them very useful for a wide range of applications. Moreover, the electrical conductivity allows machining by EDM so that very complex parts can be manufactured accurately. Many different types of EDM machineable ZrO_2-based

Table 8.1 Mechanical properties of ZrO_2-based composites[7]

	Ratio (%)	Density (g/cm³)	E-modulus (GPa)	Hardness (HV10) (kg/mm²)	Toughness (MPa/m^0.5)	3 pt Bending strength (GPa)
ZrO_2-WC	60–40	9.80	328	1653 ± 20	7.5 ± 0.3	1964 ± 88
ZrO_2-NbC	50–50	6.26	330	1510 ± 12	5.7 ± 0.2	1574 ± 157
ZrO_2-TiCN	60–40	5.76	284	1474 ± 24	6.2 ± 0.3	1521 ± 61
ZrO_2-TiN	60–40	5.81	274	1377 ± 20	5.4 ± 0.3	1666 ± 201
ZrO_2-nano TiCN	60–40	5.53	256	1304 ± 8	9.3 ± 0.4	1314 ± 60
ZrO_2-TiN*	66–34	5.80	273	1304 ± 8	6.5 ± 0.4	1298 ± 101

*Standard commercial material (KGS20)

composites can be made using high conductive non-oxide second phase ceramics such as TiC, TiCN, NbC, TiC and WC, covering a wide range of properties (see Table 8.1).

Zirconia-based ceramic composites show a unique combination of high strength, hardness and toughness, and therefore fill the gap between cermets and other ceramics. This toughness is the result of the transformation from one crystal structure to another when loaded (in stress) and is called 'transformation toughness'. Below 1240°C, pure ZrO_2 has a monoclinic state (m), between 1240°C and 2380°C it is tetragonal (t), and beyond this it is cubic. When bulk zirconia is cooled down during the last stage of the sintering cycle the t → m transformation will cause structural failure because of the volume increase of about 4.5%. A solution to this problem is adding stabilizers, such as Y_2O_3, CeO_2 or CaO, that insure that the tetragonal crystal structure is retained at lower temperatures. This metastable t-ZrO_2 phase transforms to m-ZrO_2 at room temperature under the influence of an applied stress. In this way, the compressive stresses introduced by the transformation act as a crack shielding mechanism and tend to close the cracks, increasing the fracture toughness, as can be seen in Fig. 8.8. Crack propagation is considerably slowed down due to this toughening mechanism.

Another typical phenomenon for ZrO_2 which can play an important role in EDM is 'low temperature degradation'.[11] This problem can be compared to stress corrosion in metals. High residual stresses increase the susceptibility of the metals to corrosion. For zirconia-based alloys, the solute can react with the material lattice in the presence of an external agent and cause bond rupture, which induces premature t to m transformation. This increases localized stress and leads to micro- and macrocracking. There are some methods to reduce the low temperature degradation problem. The driving force of the transformation reaction can be reduced by a controlled disper-

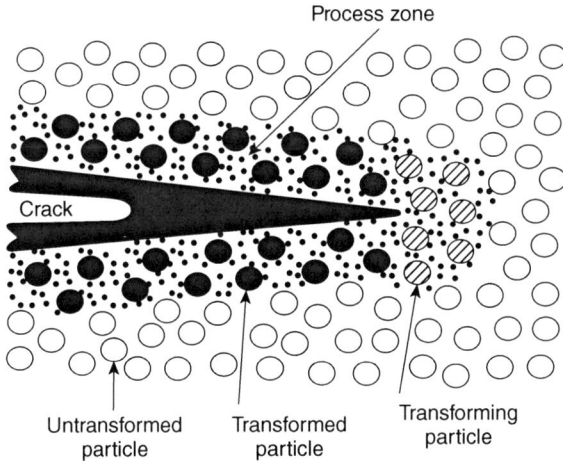

8.8 Stress-induced transformation toughening.

sion of the yttria and addition of different solutes such as alumina. Therefore 0.75 mol% Al_2O_3 is added to most ZrO_2-based composites.

Zirconia-based ceramic composites are very suitable for applications that require high strength and toughness, in combination with sufficient wear resistance. An increasingly large market is the biomedical industry. Pure zirconia is already used for implants such as hips and dental bridges. These implants, or more complex shaped custom-made implants, could be made by EDM out of electrically conductive ZrO_2-based composites. Other applications are in the die making industry and for mechanical components in general. Some applications where EDM is involved will be discussed later in this chapter.

The extremely high hardness of B_4C and TiB_2 (3500 kg/mm^2 and 2200 kg/mm^2) makes them very suitable for applications where abrasive wear resistance is the main concern.[12] However, the strength and toughness of both materials is not very high. By combining them in a *B_4C-TiB_2 composite*, a considerable increase in toughness and strength can achieved.[12] In Fig. 8.9 it is seen that adding TiB_2 to the B_4C matrix greatly improves the fracture toughness. The strength also benefits from the formation of a composite, being substantially higher than the strength of both monoliths. The reason for this is that the presence of a secondary phase in a composite limits grain growth of the matrix by means of grain boundary pinning. Smaller grains are beneficial for the overall strength of a ceramic.

An upcoming material class are composites based upon a *boride matrix*, such as ZrB_2, NbB_2, HfB_2 and TiB_2. These ceramics are particularly interesting in high temperature applications due to their extremely high melting

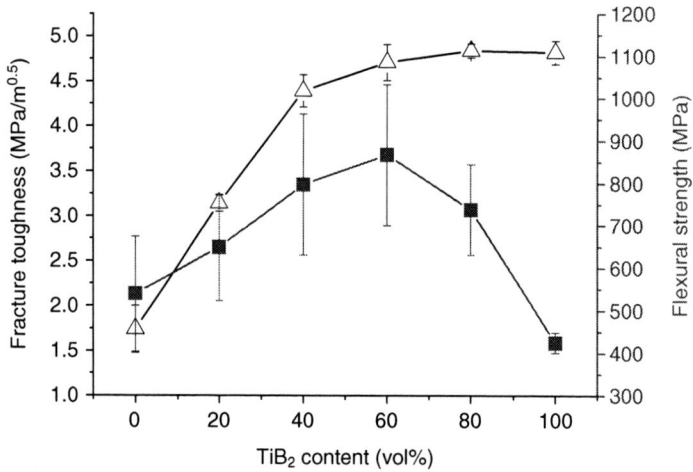

8.9 Fracture toughness and 3-pt bending strength of TiB_2-B_4C composites.[12]

points (>3000°C), high oxidation resistance, and high thermal and electrical conductivity;[13–15] they are therefore the focus of research for a variety of applications, such as refractory linings, electrodes, microelectronic devices and cutting tools. More recently, there is an interest in these materials for application in the leading edge of re-entry vehicles. However, these mono-liths are notoriously difficult to densify during sintering due to excessive grain coarsening. For this reason, a composite is formed with silicon carbide, which acts as a grain growth inhibitor and offers increased oxidation resist-ance due to the formation of a SiO_2 outer layer on the component, which is beneficial for high temperature applications in oxidizing environments such as the earth's atmosphere. These boride-SiC composites are nearly impossible to machine using classical machining methods due to their high hardness (>2000 kg/mm^2) and the tendency to form oxides that stick onto diamond cutting tools. Because of the excellent electrical conductivity of transition metal borides, electrical discharge machining (EDM) is a suitable machining technology for this class of ceramics.

It can be concluded that by creating composites, the mechanical proper-ties and machining behavior by means of EDM can be greatly improved. Compared to some other ceramic composites, the strength and toughness of these materials is still relatively low but tuning the material properties by adding the right ratio of second phase material significantly increases their application range. The main type of application, however, is still com-ponents that suffer low stresses but high abrasive wear. In the following section, Si_3N_4, ZrO_2, B_4C and transition metal boride-based composites are further investigated with respect to EDM machining and in Section 8.6 several cases are presented.

8.4   EDM of ceramic composites: understanding the process–material interaction

As explained above, the material removal mechanism in EDM of metals (steel, copper) is melting/evaporation. Melting as a removal mechanism is also common in the EDM of ceramic materials (Fig. 8.10a). Applying higher energies (higher discharge current, higher pulse duration) yields higher material removal rates but a decreased surface quality. Also material properties such as the thermal conductivity largely influence the material removal. A low thermal conductivity will keep the generated heat concentrated at the surface, resulting in more melting.

However, when machining ceramic materials other mechanisms can occur, such as spalling (Fig. 8.10b); and chemical reactions such as oxidation and decomposition (Fig. 8.10c) can also occur. In spalling, small volumes of material will be separated from the base material. This spalling effect is most often related to the generation (often due to thermal shock) of larger micro-cracks (perpendicular and parallel to the top surface) generated during EDM. These larger micro cracks make the separation of a volume during successive discharges much easier. This means that ceramic materials with low resistance against crack propagation gives more chance for spalling. Dependent on the size of the particles that break out, the surface quality can be very bad.

As the EDM process occurs at high temperature conditions (above 2000°C), chemical reactions can occur. The surface topography of an EDM'ed sample of Si_3N_4-TiN shows a foamy and porous layer. This foamy structure is caused by an oxidation reaction of Si_3N_4 and TiN, generating gas bubbles ($N_2(g)$). Besides oxidation, the ceramic materials can also decompose at high temperatures.

When machining a ceramic composite (or in general a ceramic material), different material removal mechanisms can even occur at the same time. But most often, one of the mechanisms is dominant. The dominant mechanism is not only related to the type of material, but is largely influenced by the micro-structure of the material, the composition and the EDM process settings. This means that for other process conditions, ZrO_2-TiN can have melting as the dominant material removal mechanism (compared to the spalling shown in Fig. 8.10b). In order to optimize an EDM technology for a given material, it is important to understand which mechanisms occur. So the interaction between process and material has to be understood very well. This will now be further explained for different materials classes in

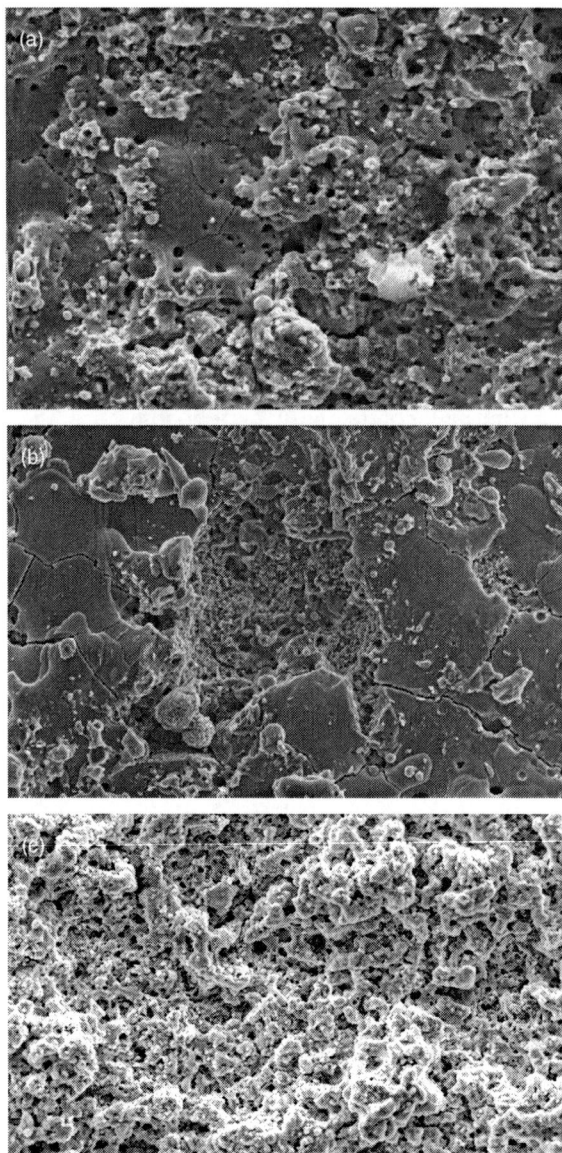

8.10 Examples of EDM material removal mechanisms when machining ceramic composites: (a) melting, (b) spalling, (c) chemical reactions.

the following subsections. The described results are based on detailed experimental investigations performed at the K.U. Leuven.

8.4.1   Silicon nitride (Si$_3$N$_4$)-based composites

Figure 8.11 shows two cross-sectional views of Si$_3$N$_4$-TiN (commercially available Kersit 601, composition 64/36), machined by sinking EDM (i_e: 64 A, t_e: 100 μs, t_0: 50 μs, u_i: 200 V) where various material removal mechanisms can clearly be seen. The dominant mechanism here is melting and, as a consequence of this, a typical recast layer visible. The recast layer contains many micro cracks in a vertical direction. These cracks can connect with horizontal cracks close to the base material, allowing large parts of the recast layer to break off from the surface (spalling). Some vertical cracks run very deep into the base material. Cracks of more than 100 μm can be seen. These deep micro cracks are caused by the difference in thermal expansion coefficient ($\Delta\alpha$) between both phases ($\alpha = 3.3°$K^{-1} for Si$_3$N$_4$ and 9.4°K^{-1} for TiN).

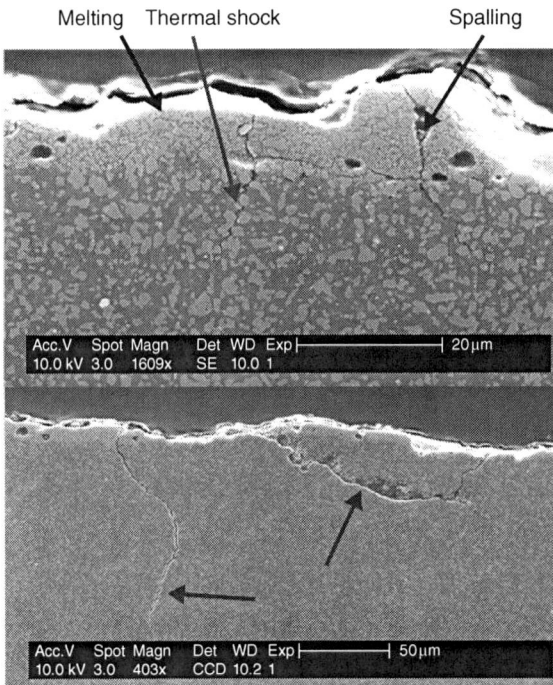

8.11 Melting, thermal shock and spalling during EDM machining of Si$_3$N$_4$-TiN.

The surface shown in Fig. 8.10c (foamy structure) was obtained by wire EDM (rough regime, water dielectric). Under these conditions, Si_3N_4-TiN decomposes at high temperatures (>1700°):

$$Si_3N_4 \rightarrow 3Si(I) + 2N_2(g) \tag{8.1}$$

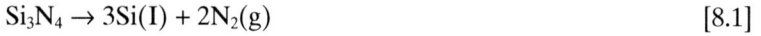

Because nitrogen gas escapes from the surface, a foamy layer is created.[5] If water as the dielectric is used, additional oxidation reactions can occur that produce more nitrogen gas:

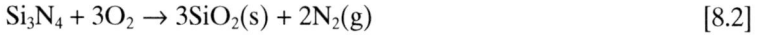

$$Si_3N_4 + 3O_2 \rightarrow 3SiO_2(s) + 2N_2(g) \tag{8.2}$$

$$2TiN + 2O_2 \rightarrow 2TiO_2(s) + N_2(g) \tag{8.3}$$

The surface roughness of such a surface as in Fig. 8.10c is very high and it is difficult to improve the surface with additional EDM finishing steps. The decomposition reaction limits the EDM performance of all Si_3N_4-based ceramics. Because the impact of chemical reactions on the EDM perform-ance of Si_3N_4-based materials is large, many possible options for improving the surface quality can be considered. Improvements can be achieved by optimizing generator parameters or other process parameters (e.g. flushing) or by modifying the material composition or preparation method. It is not possible to completely avoid the chemical reactions but the intensity of the reactions, and thus their effect on the EDM surface quality can change (as explained above, the surfaces in Fig. 8.10c and Fig. 8.11 are different).

Research has explained that the intensity of the chemical reactions depends heavily on the generator parameters. Figure 8.12 shows the change in material removal mechanism dependent on the pulse duration. In this sinking EDM experiment, the discharge duration is varied between 6.4 μs and 100 μs, with a constant discharge current (24 A) and pulse interval time (100 μs). If the dominant material removal mechanism is melting, the expected trend would be a continuous rise in surface roughness for an increasing discharge duration. However, a minimal surface roughness is found for a discharge duration of 24 μs. For longer discharges, a continu-ously increasing roughness is obtained, but for shorter discharges, a strong increase of the surface roughness (up to 10 μm R_a) is noticed. This trend is found for normal and negative polarity.

SEM images reveal that the dominant mechanisms change between short and long discharges and the changing point is situated around the discharge duration that gives the minimum roughness. The intensity of the chemical reactions increases for decreasing discharge durations, and for longer dis-charges the dominating mechanism changes from chemical reactions (decomposition) to melting. The fact that melting is dominating in this region also explains why the normal expected trend for metals (where melting is always dominant) is observed. The steep rise in roughness for

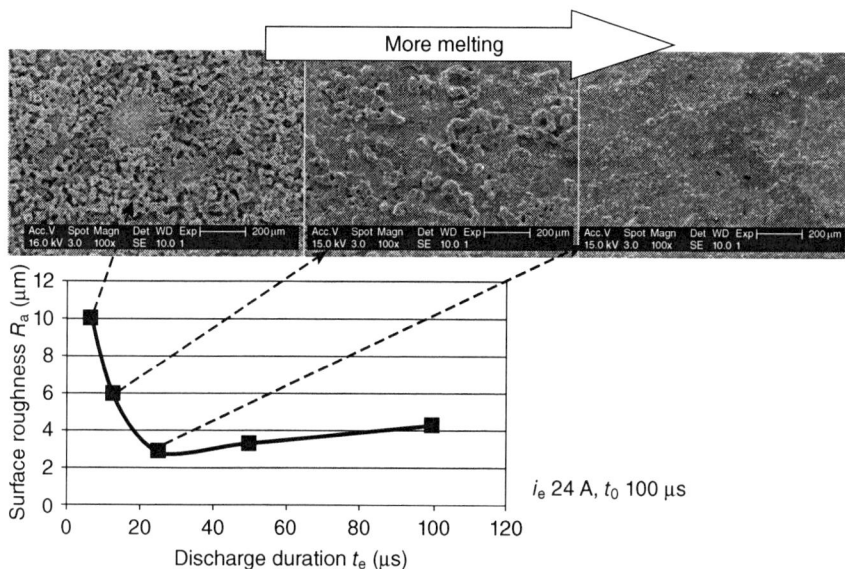

8.12 Influence of discharge duration on the surface roughness of sinking EDM machined Si_3N_4-TiN.

very short discharges is a proof of the changing mechanisms, with SEM images confirming that the shorter the discharge the less melting and the more foamy structures are seen.

Besides the pulse duration (generator parameters), the flushing conditions also largely influence the EDM behavior. A comparison test between milling and sinking EDM proves this. In milling EDM, tubular rotating electrodes ($\emptyset_{outer}$: 5 mm; $\emptyset_{inner}$: 3 mm; with flushing through the electrode) give excellent flushing conditions. The sinking EDM tests were performed with a cylindrical electrode (copper) of 8 mm diameter. Table 8.2 shows the generator settings used for the comparison test. For this material, the selected milling EDM parameters (short discharge durations) could not be applied for sinking EDM, because there was no material removal.

Figure 8.13 shows the results in terms of material removal rate and surface roughness. With these longer discharge durations (100 μs), the removal rate in sinking EDM is good but the disadvantage of the long pulses is that the (sub)surface quality deteriorates and becomes much worse than in the case of milling EDM (Fig. 8.14).

Wire EDM is mostly characterized by short pulse durations. This is because the energy input should be kept low to avoid wire rupture, which means that chemical reactions (decomposition of Si_3N_4) will be dominant, making it very difficult to apply efficient finishing machining. The effect of

Table 8.2 Generator settings for comparison tests (M-EDM and S-EDM) for
Si_3N_4-TiN

	i_e (A)	t_e (μs)	t_0 (μs)	u_i (V)
M-EDM	64	12.8	50	200
S-EDM/32	32	100	50	200
S-EDM/64	64	100	50	200

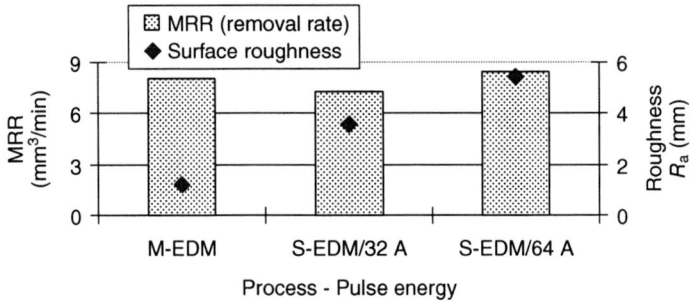

8.13 Comparison: M-EDM and S-EDM for Si_3N_4-TiN.

these chemical reactions can be lessened by reducing the amount of Si_3N_4
(and accordingly increasing the TiN amount). In this research, three different compositions have been compared. Figure 8.15 shows the results of wire
EDM rough cuts in a water dielectric (roughing regime). The surface roughness clearly decreases with a decreasing amount of Si_3N_4 but no 'normal'
value (that would be obtained in case of pure melting/evaporation) is
reached. Even with only 54 vol% of Si_3N_4 the roughness is very high. Further
reducing the Si_3N_4 does not make much sense as the mechanical properties
would not be sufficient any longer.

8.4.2 Zirconia-based materials

As for the previously mentioned Si_3N_4 materials, in the machining of ZrO_2-
based composites several material removal mechanisms such as 'melting/
evaporation' and 'spalling' can be present (Fig. 8.16). Also as before, whether
one or other mechanism is dominant depends on the composition and the
micro-structure of the material.

To demonstrate the influence of the amount of secondary phase in a
ZrO_2-based composite, ZrO_2-TiN is chosen as a case material to determine
the interrelationships between EDM behavior and the microstructure. Five
different compositions are tested with TiN contents of 30, 40, 50, 75 and

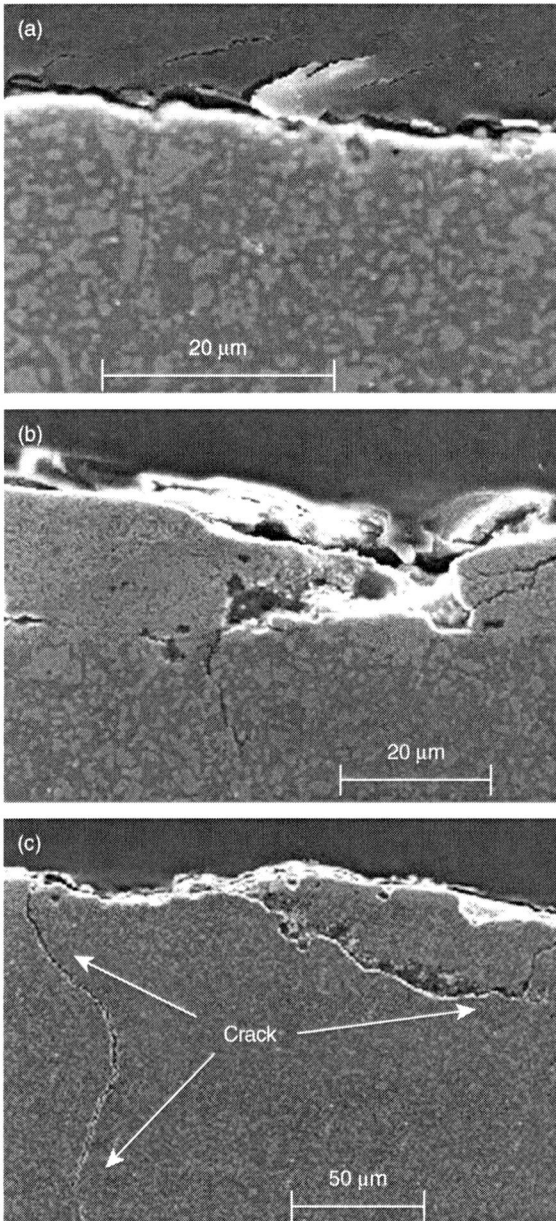

8.14 Surface quality of Si_3N_4-TiN samples which were (a) M-EDM machined and (b) S-EDM machined at $i_e = 32$ and (c) $i_e = 64$.

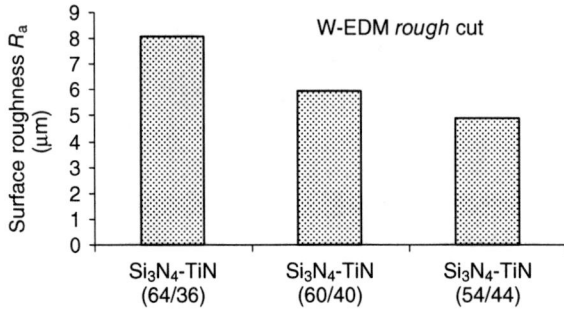

8.15 Comparison of surface roughness between composites with different amounts of second phase material.

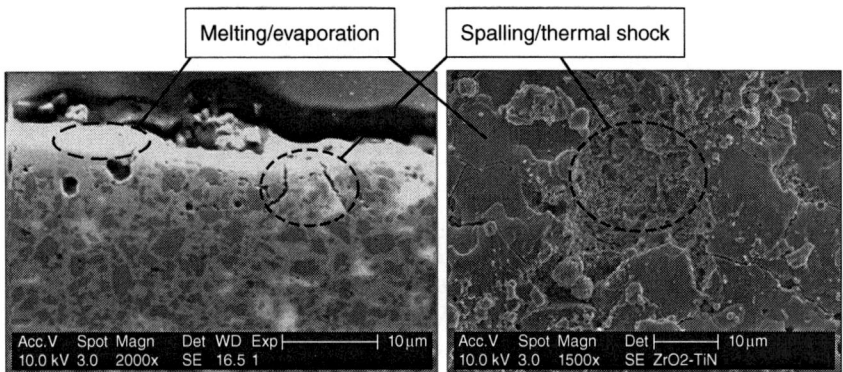

8.16 Melting/evaporation and spalling/thermal shock during W-EDM of ZrO_2-TiN (60/40, t_e: 0.9 μs, t_0: 20 μs).

90 vol%. The electrical resistivity drops steeply between 30 and 50 vol% TiN content. In this region, percolation in the TiN phase is established. This means that a conducting network is formed by means of TiN particles, allowing the current to follow a low resistive path through the TiN phase. This is shown in Fig. 8.17.

Wire EDM tests have been performed to investigate the machineability (material removal rate (MRR) and surface roughness). Among the influence of other process parameters, the MRR decreases with increasing TiN content. This is due to the considerable increase in thermal conductivity associated with an increased amount of TiN (18 W/mK). The ZrO_2 phase acts as an isolator (1–2 W/mK), keeping all the heat at the top surface, facilitating melting. With an increased TiN content, the heat is more readily dissipated into the bulk and melting is inhibited. From the topographies presented in Fig. 8.18 it can be seen that, with an increasing TiN content,

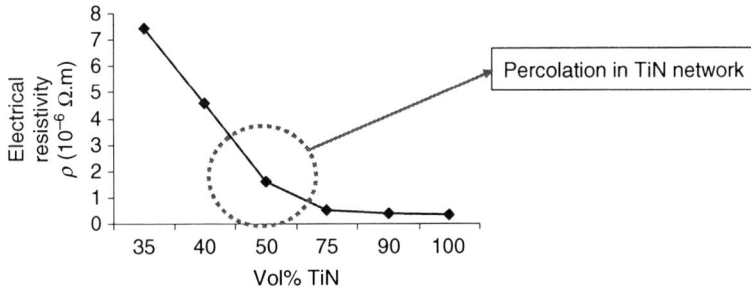

8.17 Electrical resistivity as a function of amount of TiN.

there are more large areas where resolidified material has broken off the surface. The topographies for the 50 vol% TiN composition show mainly melting as the removal mechanism; certainly at low discharge energy. The 90 vol% TiN composition, however, shows only small surface areas with molten and resolidified material, and has spalling as the dominant removal mechanism. In the case of excessive spalling (high TiN content and high discharge energy), the surface quality becomes very bad and difficult to finish to a very low roughness by W-EDM.

The SEM images shown in Fig. 8.19 indicate that superfinishing strategies with eleven cuts leave no remaining surface damage. After the first six cuts, however, cracks can still be observed in the compositions with 75 and 90 vol% TiN. The surface of the 50/50 vol% composite looks fine after only six cuts. It is therefore clear that using higher TiN contents in ZrO_2-TiN composites leads to the requirement of more finishing passes to achieve comparable surface roughnesses with lower TiN content grades (Fig. 8.19e and f).

Not only the amount of secondary phase, but also the type and grain size of it, largely influences the micro structure of the material and hence the removal rate and the obtained surface quality. To show this, various ZrO_2 ceramic composites based on various second phase (WC, TiC and TiCN) in different grain sizes were developed and tested. (To improve the mechanical properties, there is a trend to develop ZrO_2 composites with finer microstructures.) All ZrO_2-based composites with the different second phases (WC, TiC and TiCN) are made of a 60/40 vol% composition. Table 8.3 lists the average particle size of the different grades of the starting powders used.

For WC and TiCN, three compositions were produced: nano, micro and a mixture of 50% nano/50% micro. For TiC, only two compositions (micro and nano) were produced. Inspection of the microstructures of the sintered materials shows that they corresponded to the selected grain sizes of the starting powders (Fig. 8.20).

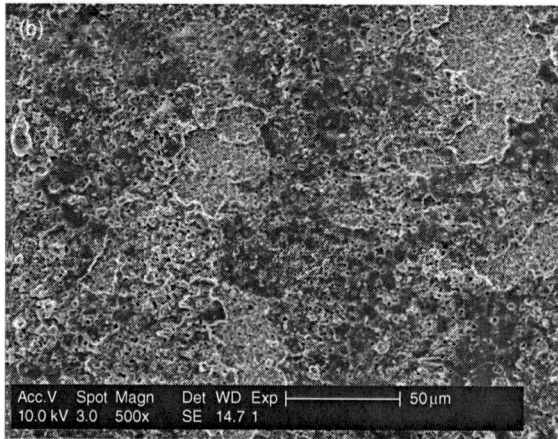

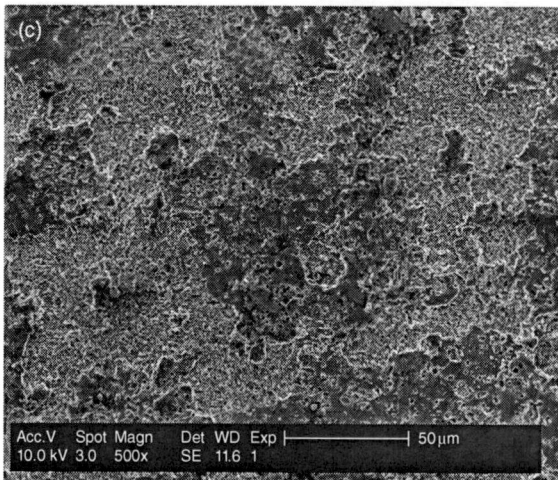

8.18 Topography of W-EDM machined ZrO$_2$-TiN: (a) 50/50 ZrO$_2$-TiN, (b) 25/75 ZrO$_2$-TiN, and (c) 10/90 ZrO$_2$-TiN.

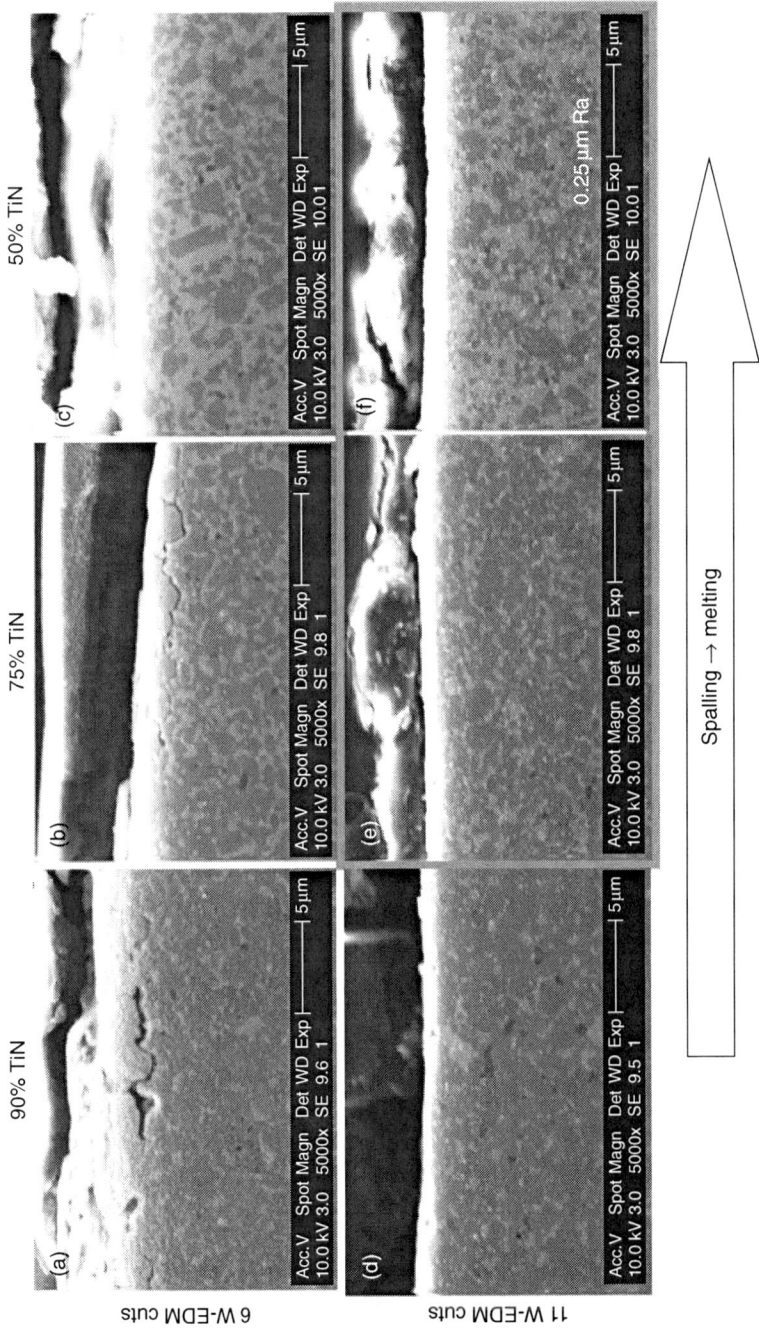

8.19 Surface quality shown in cross-sections as a function of amount of TiN in the composite and the number of W-EDM finishing cuts.

Table 8.3 Average grain size (nm) of the starting powders

ZrO$_2$	TiCN		WC		TiC	
	Nano	Micro	Nano	Micro	Nano	Micro
55	65	770	150	900	62	1300

To test the EDM behavior, various wire EDM experiments were performed. Based on the experimental results obtained, the relative effects of the second phase (for micro and nano) on the cutting speed and the surface roughness were plotted in Fig. 8.21.

As described above, the effect of WC is totally different compared with that of TiC and TiCN. A micro based ZrO$_2$-WC cuts faster than a nano-based one, while the contrary is the case for TiC and TiCN. The effects on the cutting speed and the surface quality is related to the materials properties and the type of material removal mechanism such as melting, spalling and chemical reactions. Figure 8.22 shows the topographies and cross-sections of the machined samples, which are used to identify possible material removal mechanisms. Differences in cutting speed between different materials can also be caused by the variation in electrical resistivity. However, measurements of the peak current during discharge showed maximal variations of only about 15%, which does not explain the measured differences in cutting speed. If the EDM material removal mechanism is mainly based on melting, a cutting speed increase should result in a finer microstructure, because of the lower thermal conductivity. This effect has been identified for wire-EDM of WC-Co.[17] The higher cutting speed for the nano based ZrO$_2$-TiCN and ZrO$_2$-TiC (which behave similar) can be explained by the lower thermal conductivity. But based on the topographies and the cross-sections, spalling occurs as well (Fig. 8.22).

The spread of the surface roughness (Fig. 8.21) is also due to the irregular texture. Especially for ZrO$_2$-TiCN, the topography shows areas of melting and areas where the material has been broken off.

Besides the traditional reasons for material removal through crack formation and consequentially grain fallout, there are two new material removal mechanisms which are typical for ZrO$_2$ composites. Firstly, TiCN can dissolve in ZrO$_2$ when the temperature during sintering becomes too high. These samples were made using the pulsed electric current sintering (PECS) technique which has a risk of overheating when the current follows a preferred path through the conductive particles in the powder mix. If TiCN dissolves in ZrO$_2$, the toughness of the sintered material should be lowered considerably, facilitating crack propagation and spalling. However,

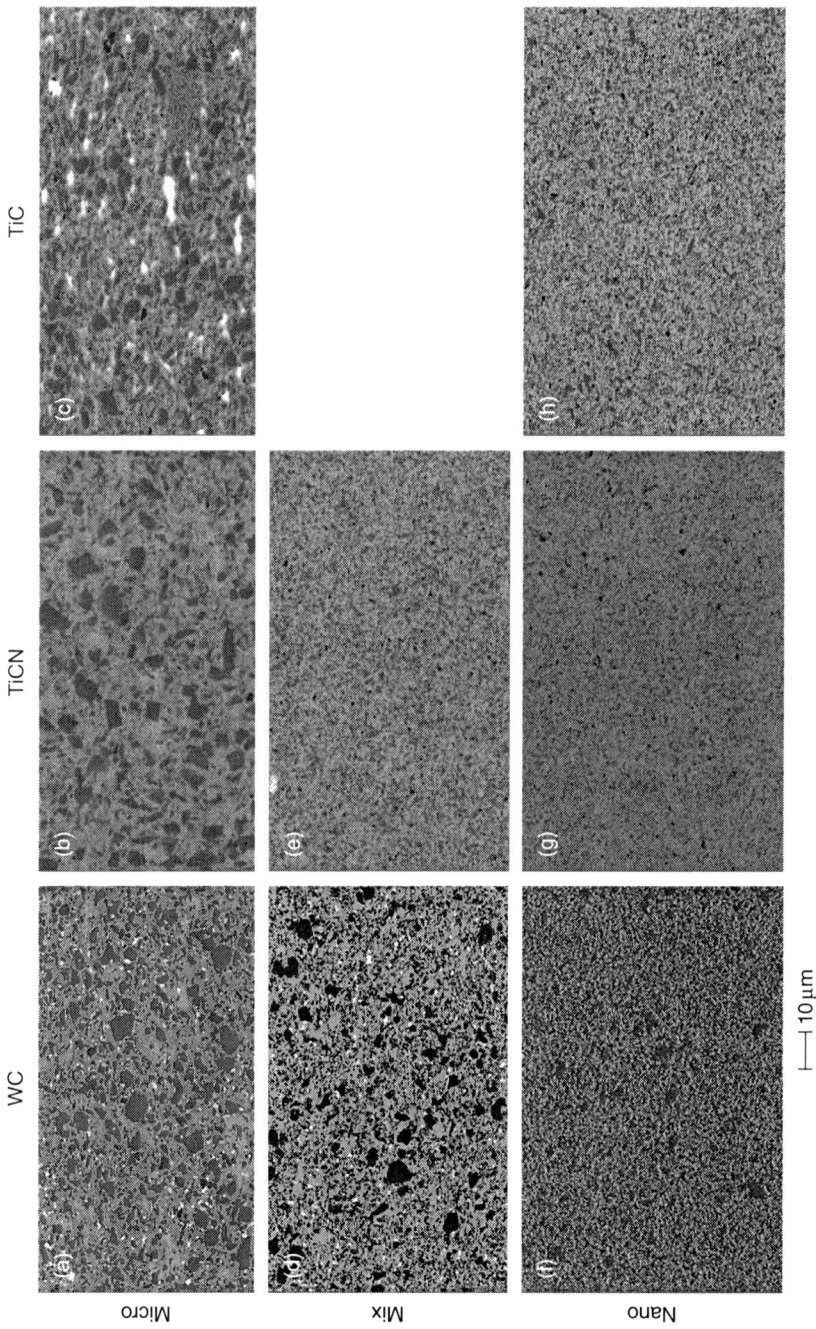

8.20 Microstructure after sintering for ZrO$_2$-WC, ZrO$_2$-TiCN and ZrO$_2$-TiC composites with different secondary phase grain sizes.

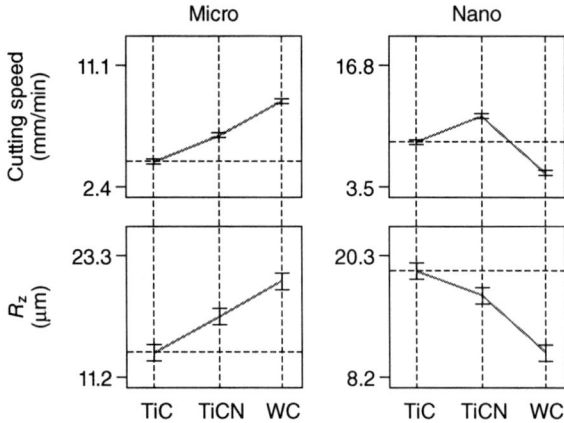

8.21 Effect of the conductive phase.

the measured toughness of this composite is high so the reason must be found elsewhere. A second explanation is a premature ZrO_2 phase transformation that results in micro crack generation. This means that the ZrO_2 grains are not stable enough to stay in the metastable t phase due to high residual stress levels. Unwanted ZrO_2 transformation can also be caused by destabilization of the material lattice in the water dielectric of the W-EDM machine (low temperature degradation).

The nanocomposites show more material removal by means of spalling than the microcomposites. There are three possible reasons for this. Firstly, the thermal conductivity of nano composites is low in comparison to micro composites. Because of the lower thermal conductivity, the temperature gradient is higher and cracks are formed more easily. Secondly, if some agglomerates are left in the microstructure, as is the case for microcomposites, the resulting stress level on the ZrO_2 are lower than in a completely uniform microstructure, as is the case for nano composites. For these yttria-stabilized composites, it is reported that the sensitivity to low temperature degradation increases with an increasing toughness (= higher transformability) which makes the nanocomposites more sensitive to this problem. The nano composites appear to be very close to the point of spontaneous transformation that generates microcracks throughout the part. Any additional heating or induced stress can set off this spontaneous transformation.

Debris analysis confirmed the heavy spalling effect for ZrO_2-TiCN (Fig. 8.23). Larger particles could be collected for the nano material. An Energy-dispersive X-ray (EDX) analysis of the debris showed the presence of Zr, Ti, O, C, N, Cu and Zn, the final two originating from the wire electrode. The presence of wire electrode material within the debris means that some area of the debris particle had been molten during previous discharges.

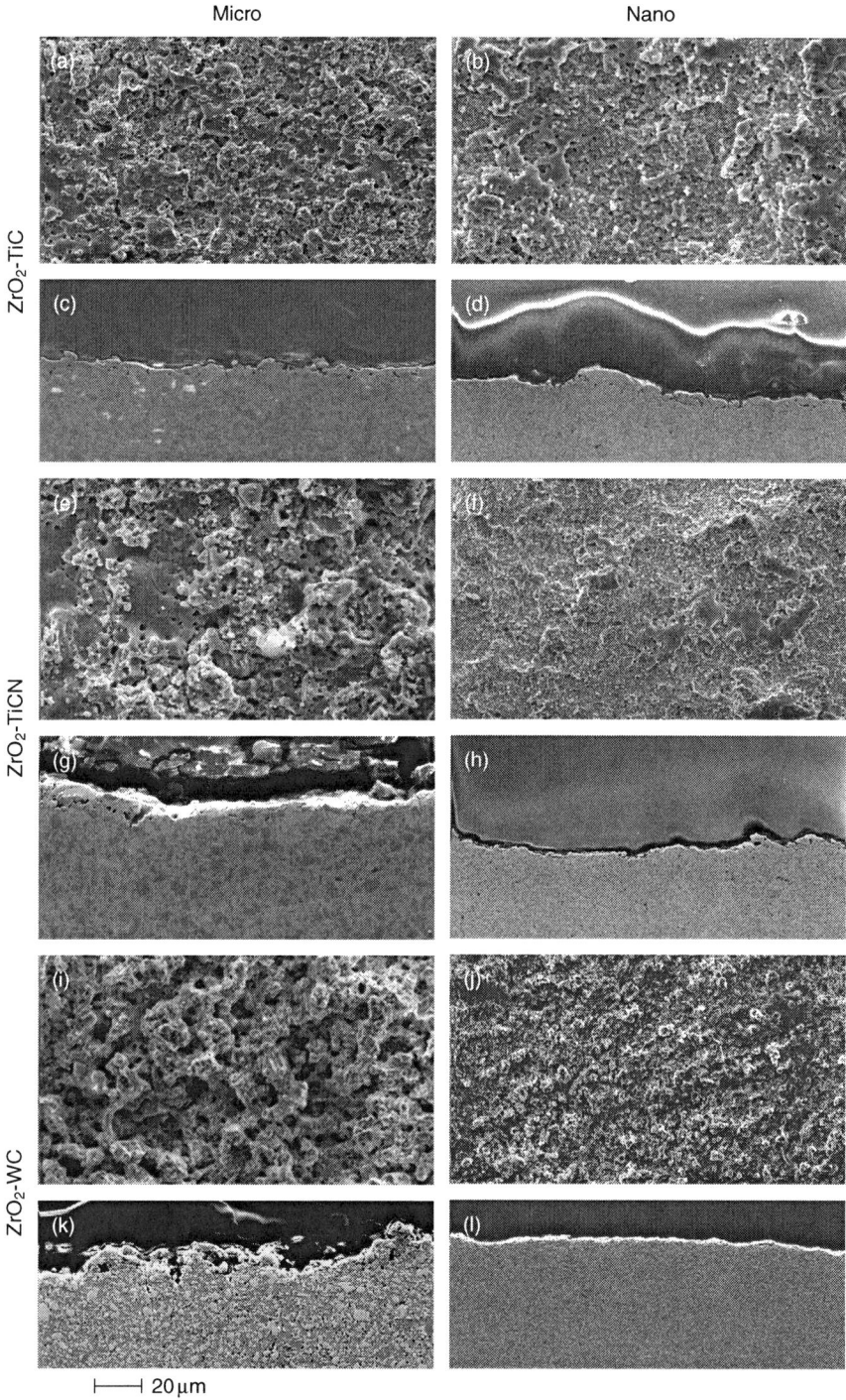

8.22 Topographies and section views of machined samples (EDM parameters: $\hat{u}_i$, 100 V; t_e, 1.5 μs; t_0, 25 μs).

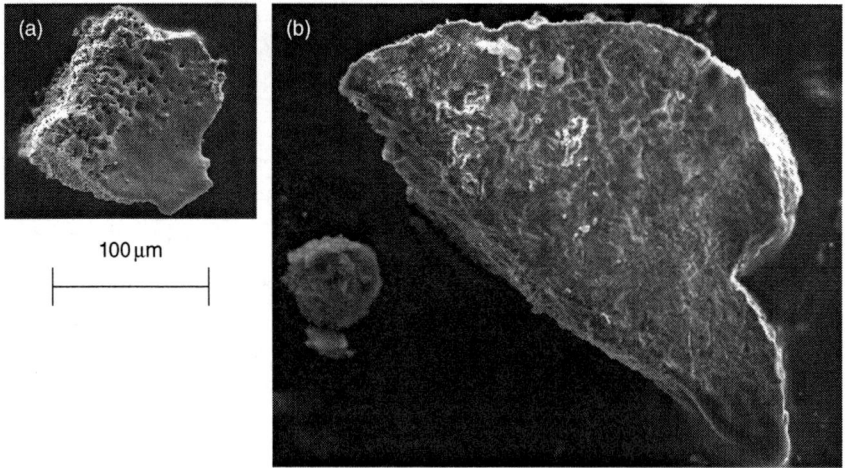

8.23 Debris analysis for ZrO_2-TiCN: (a) micro and (b) nano.

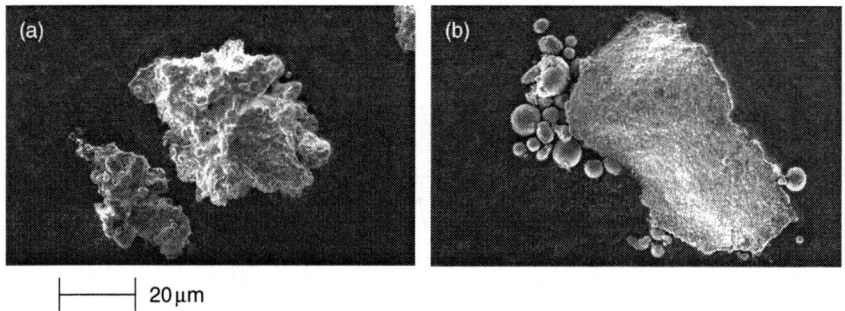

20 μm

8.24 Debris analysis for ZrO_2-WC: (a) micro and (b) nano.

The texture of the nano-based ZrO_2-WC ceramic composite was typical for melting. Some much smaller irregularly shaped particles were identified, which suggested that the occurrence of crack formation was limited (Fig. 8.24).

ZrO_2-WC is less sensitive to crack formation and propagation because of the high strength and toughness of these grades. The lower cutting speed for the nano-based material is, however, not in agreement with the lower thermal conductivity. The reason for the higher cutting speed for the micro-based ZrO_2-WC is due to the existence of chemical oxidation reactions.[18]

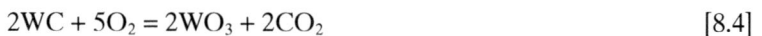

$$2WC + 5O_2 = 2WO_3 + 2CO_2 \qquad [8.4]$$

The machined micro ZrO_2-WC samples (Fig. 8.25) have a different color which might indicate the presence of yellow colored WO_3. In addition, WO_3

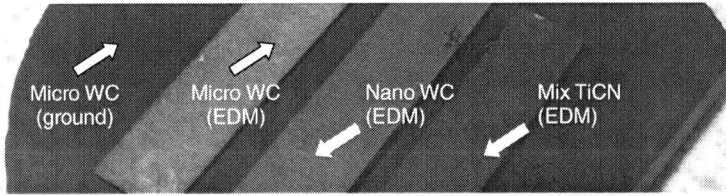

8.25 Wire EDM machined samples with identical generator parameters.

readily evaporates during EDM. The gases evolved generate the foamy structure, which is clearly visible for the micro-based ZrO_2-WC on Fig. 8.22. This oxidation–evaporation is an additional material removal mechanism which contributes to the high cutting speed but also increases the surface roughness. The micro ZrO_2-WC is much more sensitive than the nano material, due to the larger grains, and the reaction occurs at larger depths because of the higher thermal conductivity.

The occurrence of chemical reactions and the crack formation (especially for the nano composites) is caused mainly by a premature ZrO_2 phase transformation. This means that the ZrO_2 grains are not stable enough and the result is a poor surface quality during EDM. No wire EDM finishing technology could be found to achieve a surface roughness that was comparable to that of the micro composites. Increasing the stabilizer content can solve this problem. A doubly stabilized (1 mol% Y_2O_3 + 1 mol% Nd_2O_3) improved the ZrO_2 stability and reduced the sensitivity to low temperature degradation.[19] Wire EDM experiments gave excellent results on this more stabilized material. The effect was that there was no more large scale ZrO_2 transformation during the process, making melting again the dominant MRM. Because of the lower thermal conductivity of the composite in comparison with the micro composite, the cutting rate was even higher. Moreover, this finer grained composite now allowed finishing EDM as well, and the surface quality improved compared to the micro-based composite. Thus, by modifying the grain size and keeping the ZrO_2 stable, the cutting rate and the surface quality can be improved.

8.4.3 Boron carbide (B_4C)-based materials

Investigating the EDM behavior of B_4C-TiB_2 ceramic composite reveals that melting is the most dominant material removal mechanism. The change is most striking for TiB_2 (Fig. 8.26) where a small addition of only 10 wt% of B_4C changes the EDM behavior and causes a shift of the dominant removal mechanism from crack propagation to melting. Figure 8.26 shows a rough wire EDM cut surface and no real signs of spalling can be found.

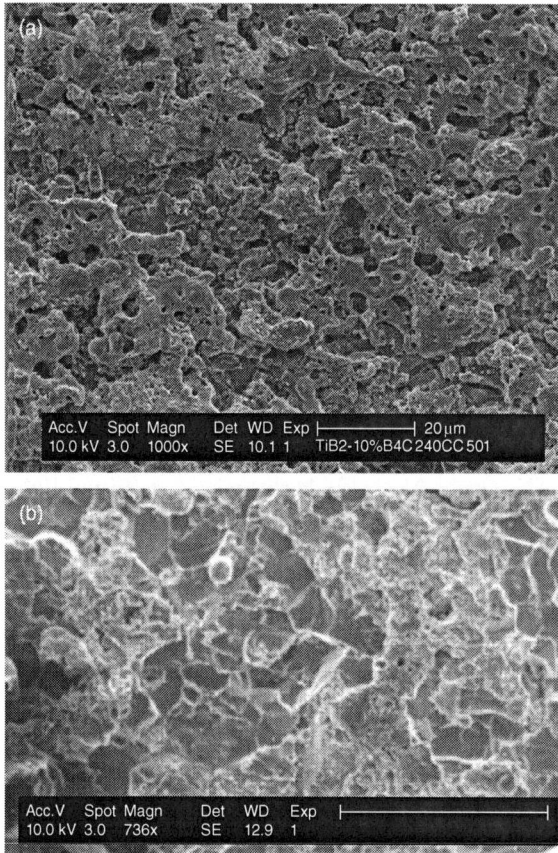

8.26 Topography after rough wire EDM (roughing regime) cut on
(a) B_4C-TiB_2 (10/90) and (b) pure TiB_2.

Another indication of the changed EDM behavior is the relatively low roughness obtained already after the rough cut of 1.29 μm R_a, compared to 2.55 μm R_a for TiB_2. The main reason for the change in EDM behavior is the huge increase in strength. Materials with B_4C as the main phase already have a higher strength and both show significant melting (Fig. 8.27).

The fact that both composite materials have melting as a dominant mechanism creates the potential for a much improved surface finish compared with the pure phases, which are more sensitive to spalling.

Due to the change in material removal mechanism to melting, the rough-cutting rate decreases when compared with pure TiB_2 (about four times). A lot of energy is needed to melt all the material whereas for pure TiB_2 large pieces can break off at once. However, the final achievable roughness

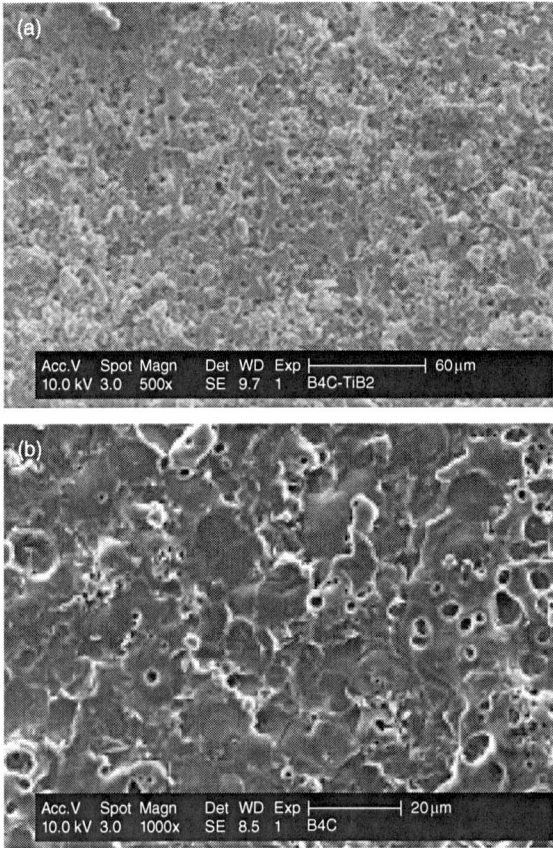

8.27 Topography after rough wire EDM (roughing regime) cut on (a) B$_4$C-TiB$_2$ (60/40) and (b) pure B$_4$C.

by EDM is probably a more important parameter for the EDM performance as it can limit the applicability for this material in EDM'ed components with high surface demands. Therefore, a technology has been developed for wire EDM finishing. After five finishing steps, a very low surface roughness of 0.25 μm R_a can be obtained. Compared to the best roughness for TiB$_2$ (0.66 μm R_a) this is a very significant improvement.

8.4.4 Other boride-based composites

Additional research on the EDM behavior of boride-based composites at the K.U.Leuven is still in an early phase. To investigate the EDM behavior of such composites, four grades have been prepared with 20 vol% of SiC

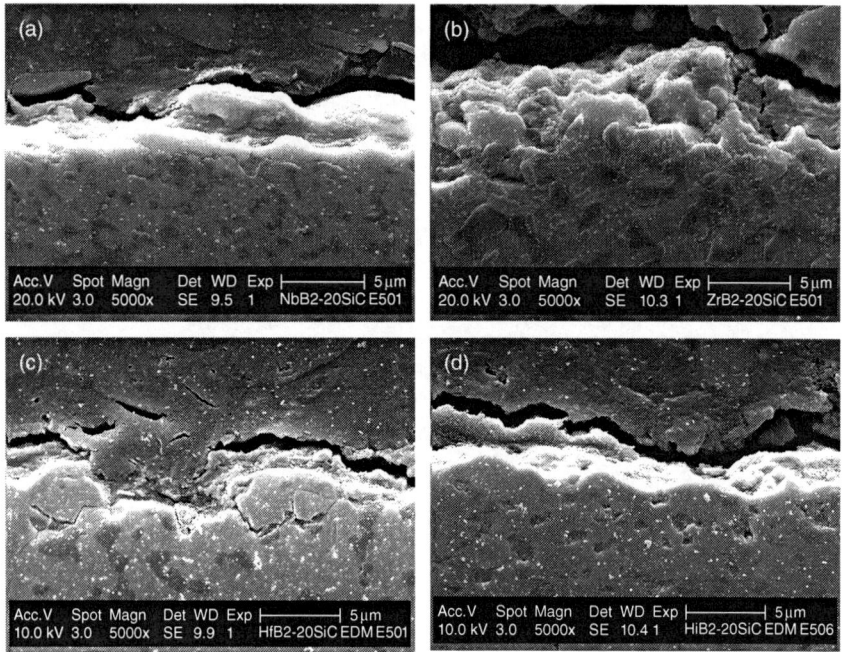

8.28 SEM micrographs of EDM surfaces of (a) NbB$_2$-SiC, (b) ZrB$_2$-SiC, (c) HfB$_2$-SiC and (d) TiB$_2$-SiC.

using pulsed electric current sintering: ZrB$_2$-, NbB$_2$-, TiB2- and HfB$_2$-SiC. The EDM surfaces (machined by a roughing regime) are presented in Fig. 8.28. It can be seen that spalling is the dominant mechanism in the ZrB$_2$-SiC composite (indicated as the fracture surface in Fig. 8.28b), while the others show a mix between thermoshock damage (indicated by the subsurface cracks visible in Fig. 8.28a and b) and melting/evaporation.

The main reason for the spalling behavior of the ZrB$_2$-SiC grade is the oxidation of ZrB$_2$ during sintering and the transformation of this ZrO$_2$ from t-ZrO$_2$ to m-ZrO$_2$ during cool down after EDM in the resolidified outer layer. The volumetric expansion which accompanies this transformation causes the resolidified layer to come off and reveal the spalling surface, as is presented in Fig. 8.28b.

Surface roughness values (R_a) after EDM are presented in Fig. 8.29. It can be seen that the NbB$_2$-SiC grade shows considerably lower surface roughness after five finishing regimes when compared with the other grades. The reason for this is the significantly lower thermal conductivity of the NbB$_2$ monolithic material (24 W/mK) compared with 51, 57 and 66 W/mK for HfB$_2$, ZrB$_2$ and TiB$_2$ respectively. This lower thermal conductivity, as explained in previous sections, concentrates the heat at the top surface and

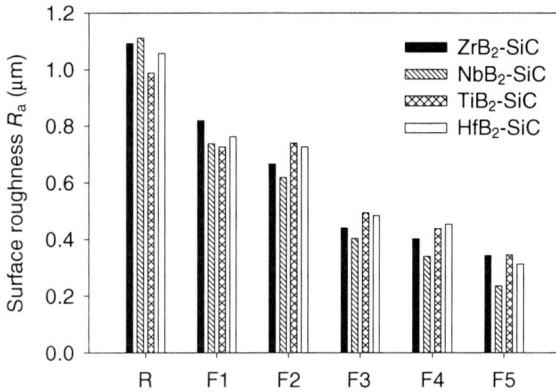

8.29 Surface roughness after EDM.

facilitates melting. It also decreases the depth of the heat-affected zone, which is beneficial for the surface roughness. The material removal rates are also related to the thermal conductivity, with the highest material removal rate in the NbB_2-SiC grade (17 mm^2/min) – the others featured lower MRRs (15, 14 and 15 for HfB_2, ZrB_2 and TiB_2 respectively). It can therefore be concluded that for borides, the thermal conductivity is of great importance to determine the EDM behavior.

8.5   New generator technology for EDM

Good electrically conductive ceramic materials/composites can be machined with standard EDM machines having generators and related technologies developed for metals. However, if the electrical conductivity becomes less (e.g. low electrically conductive ceramics such as SiC, B_4C or ZrO_2-based composites with a lower content of second phase) the efficiency of material removal reduces strongly.

Although the discharge current can be set (in the case of iso-energetic pulse generators), the real discharge current over the gap can drop because of the voltage drop into the ceramic material (most generators have a maximum voltage, e.g. 200 V). Figure 8.30 shows the measured current profiles when machining (wire EDM, standard generator) steel or sintered SiC. Due to the resistivity of the material, the current rise is smaller and therefore smaller peak currents can be obtained.

Within the EU-project MONCERAT, new generators have been developed. Higher voltage sources have been integrated for compensation of voltage drops into the material. In addition, smaller pulses (smaller energy, short pulse durations) can be generated in order to increase the surface quality during finishing.

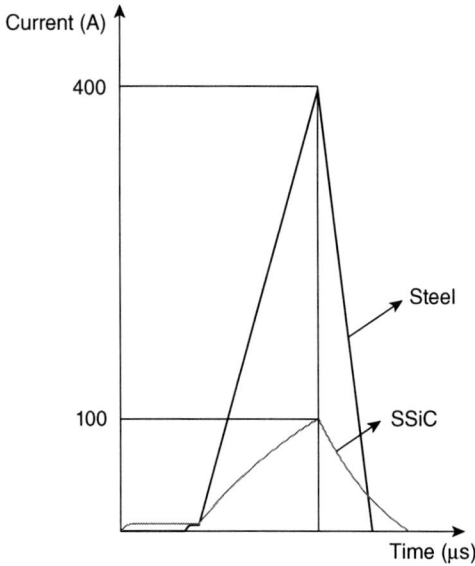

8.30 Reduced discharge current for wire EDM of high electrically resistive ceramics.

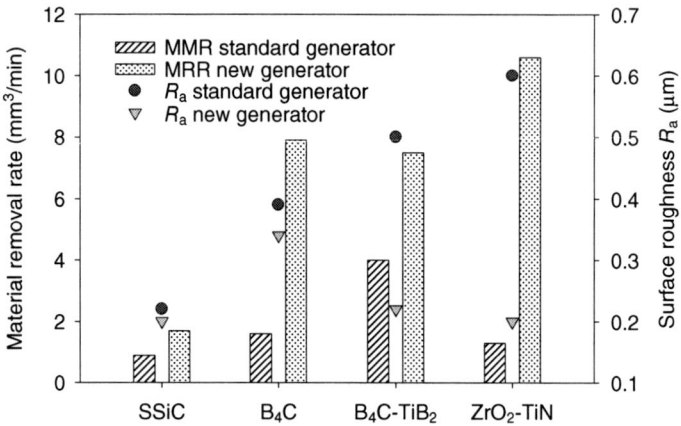

8.31 Impact of new generator technology on EDM performance (material removal, surface quality).

Figure 8.31 shows the impact of classical generators (anno 2007) and the newly developed generators on material removal rate and surface quality for some ceramic materials. This also explains why the development of ceramic materials suited for EDM and EDM generator/technology development should occur in an iterative way.

8.6   EDM strategies and applications

This section describes the EDM machining of some components. It is explained that the knowledge of the 'process–material' interaction is important for developing a proper and efficient machining strategy.

8.6.1   Application 1: turbine impeller – Si_3N_4-TiN

Based on the knowledge of material removal mechanisms, it was decided to make the impeller (Fig. 8.7) by a combined strategy. Using short pulse durations enhances chemical reactions (decomposition, oxidation), giving a much higher material removal (and also low electrode wear). These short pulse durations may, however, not be applied during finishing, because the chemical reactions result in a porous and foamy structure.

On first sight, the use of a micro milling EDM machine to produce the turbine impeller would be the best option. However, the available micro milling EDM machine had only a relaxation type generator, characterized by small energy pulses (small current, short pulse durations) which makes finishing quite difficult. Therefore, the part was made on a sinking EDM machine (AgieCharmilles Roboform 350), having a generator that could produce a wide range of pulse shapes (relaxation type as well as iso-energetic).

Roughing and semi-finishing steps were applied using relaxation pulses (because of the short pulses). For final finishing, a strategy using an iso-energetic pulse (low current, higher pulse duration) was used to give a smooth surface and furthermore avoid the foamy structure. Graphite electrodes having the negative shape of the cavity between two blades were milled on a 3-axis micro-milling machine (machining time ~60 min/electrode). The roughing electrodes were used for two cavities because of the low tool wear. For the semi-finishing and finishing step, all cavities were finished with a fresh electrode. Table 8.4 gives an overview of the machining steps and the obtained machining time (145 min/cavity). Figure 8.32 shows the machining set-up and the produced impeller.

The final produced turbine impeller has a surface roughness as expected: 0.82 μm R_a (7.10 μm R_z). Dimension and geometry control is performed on a coordinate measurement machine (Mitutoyo FN 905). For each EDM machined cavity, 1200 points on three surfaces (pressure, suction and hub surfaces) are collected, processed and meshed to compare with the original CAD model. Based on the deviation map of a top and a side view (Fig. 8.33), all the cavities are fully symmetrical. A small error is noticed at the tip of shroud with a maximum deviation of +0.041 mm at the suction surface and –0.045 mm at the pressure surface, which might be caused by the large curvature at this region.

Table 8.4 Strategies for the EDM machining (sinking EDM) of the micro turbine impeller

Machining step	Pulse type	Regime	Electrode		Machining time/cavity (min)
			Undersize (μm)	No. of electrodes	
1	Relax.	Roughing I	150	4	45
2	Relax.	Semi-finishing II	100	8	60
3	Iso.	Finishing	25	8	40
				Total	145

8.32 Machining of gas turbine impeller by S-EDM.

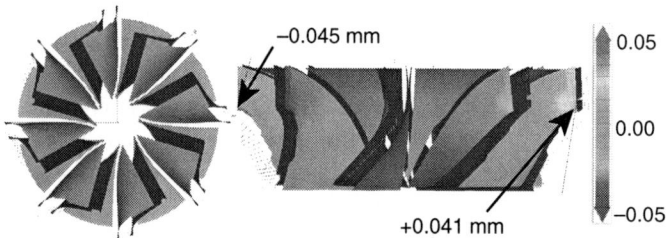

8.33 Dimensional quality check of the turbine impeller produced by sinking EDM.

8.6.2 Application 2: spray nozzle – B_4C-TiB_2

Figure 8.34 shows a spray nozzle (used for the spraying of powder materials) that has been EDM machined in B_4C-TiB_2 (60/40). The EDM machinability of the composite material is much higher than of pure B_4C, mainly because of the much higher electrical conductivity. Compared to the machin-

(a)

(b)

Ø13.4 ± 0.2
6°
Ø12 ± 0.2
45°
0.8 ± 0.05
6.5 ± 0.1
Ø0.7 ± 0.05
Ø12 ± 0.1

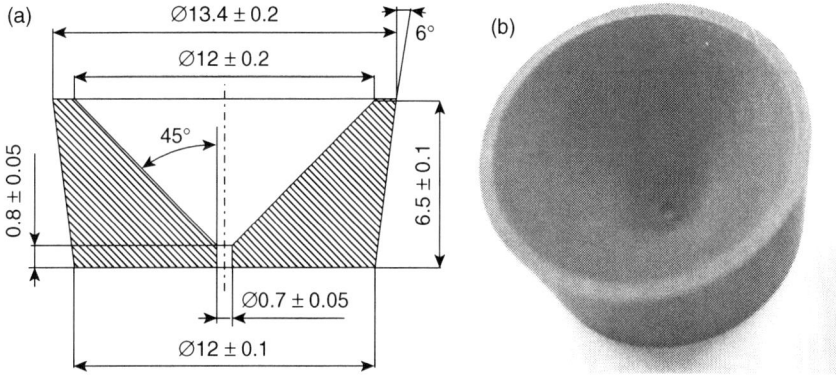

8.34 Spray nozzle in B_4C-TiB_2; (a) dimensions (in mm) and (b) final product.

Table 8.5 Sinking EDM strategy for the B_4C-TiB_2 nozzle

Machining step	Electrodes	Regimes	Programmed depth
1. Rough sinking EDM	1: Cu	Regime A	5 mm
		Regime B	5.75 mm
2. Finishing sinking EDM	2: Cu	Regime B	5.75 mm
		Regime C	6 mm
3. Through hole sinking EDM	3: WCu	Regime D	Through hole

ing of pure B_4C, the composite material gives more melting instead of spalling.[7]

The good machineability of the composite allows a pure sinking EDM-based strategy (while for the machining of the pure B_4C, a combined strategy with milling EDM and sinking EDM[16] is proposed). The applied strategy is shown in Table 8.5. The parameters from the regimes and the machining times per regime/electrode are given in Table 8.6.

The total machining time was around 3.5 hours. If the nozzle were to be made out of pure B_4C (lower electrical conductivity), the machining time would be about 5.5 hours (with a combined milling EDM/sinking EDM strategy as explained in reference 16).

Table 8.6 Machining times for each step (regime–electrode combination) for the sinking EDM operation of the B_4C-TiB_2 nozzle

Regime	Generator parameters				Machining time with electrode (min)		
	U_0 (V)	i_e (A)	t_e (μs)	t_0 (μs)	1	2	3
A	−200	6	12.8	25	17	—	—
B	−200	8	12.8	25	55	30	—
C	−200	6	12.8	25	—	98	—
D	−80	Relaxation: C: 27 nF			—	—	4

The nozzle was afterwards tested in industrial practice. The life-time was threefold compared to a traditionally used carbide nozzle and twofold for one of the nozzles made out of pure B_4C. This, again, shows the potential of ceramic composites. The size of the outgoing hole diameters determines the life time. The shape of all these nozzles after their life-times expired can be compared in Fig. 8.35.

8.7   Conclusions

This chapter has given a brief overview of the challenges related to the EDM machining of ceramic composite materials. By tuning the composition of these materials, the mechanical properties as well as the EDM behavior can be improved.

It has been shown that several material removal mechanisms, such as melting, spalling and chemical oxidation/decomposition, exist. Whether one or the other is dominant depends on the material composition, the materials micro-structure and the EDM settings. This means that for technology development, the process–material interaction has to be well understood. But if this is the case, then various ceramic components can be machined by EDM in an efficient way.

8.8   Acknowledgments

The content of this chapter is based on several research projects – EU FP6 research project MONCERAT (NMP2-CT-2003-505541), EU-FP6-SME project EUROTOOLING21 (NMP-CT-2004-505901), EU FP7 project Integ-μ (CP-IP 214013-2), the Fund for Scientific Research Flanders (FWO) project No. G.0539.08 and industrial collaboration with AgieCharmilles (Switzerland).

Example 1   Example 2

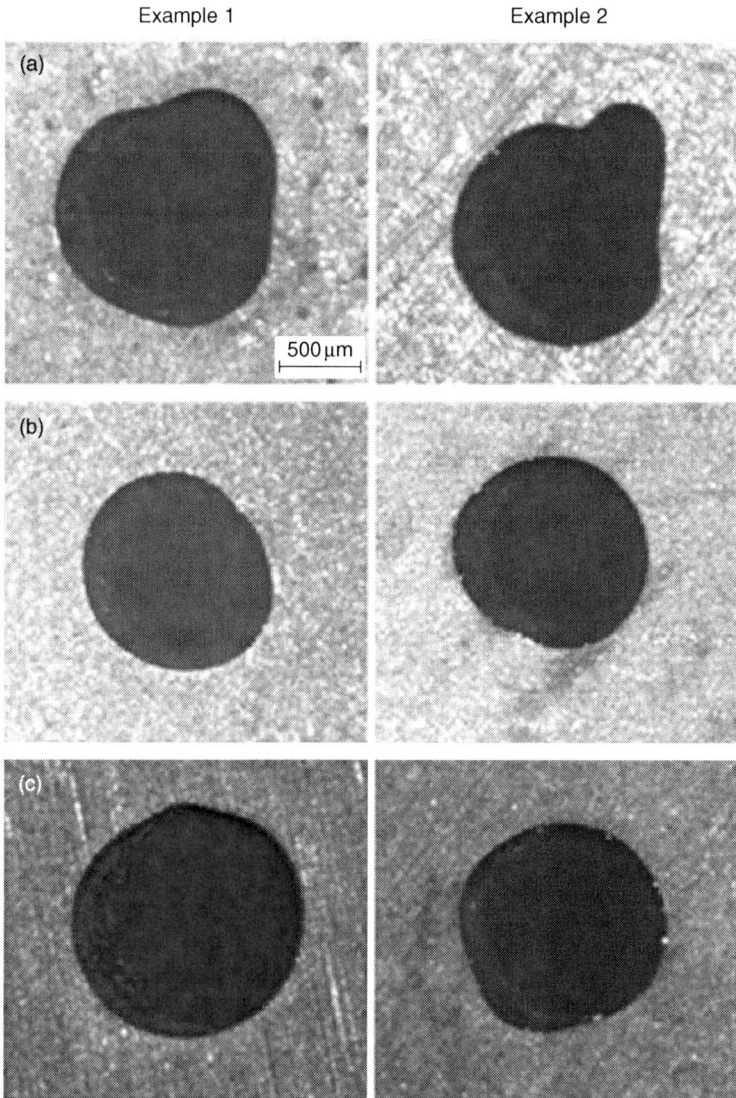

8.35 Comparison of central hole in spray nozzle made from (a) cemented carbide, (b) B$_4$C (Tetra-Bor®) and (c) B$_4$C-TiB$_2$ (VR50) after lifetimes of, respectively, 10, 20 and 30 hours.

8.9   References

1 Snoeys, R., Staelens, F., Dekeyser, W., 1986, Current trends in non-conventional metal removal processes, *Annals of the CIRP*, **35/2**: 467–480.
2 Kunieda, M., Lauwers, B., Rajurkar, K.P., Schumacher, B.M., 2005, Advancing EDM through fundamental insight into the process, *Annals of the CIRP*, **54/2**: 599–622.
3 König, W., Dauw, D.F., Levy, G., Panten, U., 1988, EDM – Future steps towards the machining of ceramics, *Annals of the CIRP*, **37/2**: 623–631.
4 Luis, C.J., Puertas, I., Villa, G., 2005, Material removal rate and electrode wear study on the EDM of silicon carbide, *Journal of Materials Processing Technology*, **164–165**: 889–896.
5 Lauwers, B., Kruth, J.P., Liu, W., Eeraerts, W., Schacht, B., Bleys, P., 2004, Investigation of the material removal mechanisms in EDM of composite ceramic materials, *Journal of Materials Processing Technology*, **49**(1–3): 347–352.
6 Liu, K., Lauwers, B., Reynaerts, D., 2010, Process capabilities of micro-EDM and its applications, *International Journal of Advanced Manufacturing Technology*, **47**: 11–19.
7 Brans, K., 2010, *Electrical Discharge Machining of Advanced Ceramics*, PhD Thesis, K.U.Leuven, ISBN 978-94-6018-178-8.
8 Anne, G., Put, S., Vanmeensel, K., Jiang, D., Vleugels, J., Van Der Biest, O., 2005, Hard, tough and strong ZrO$_2$-WC composites, *Journal of the European Ceramic Society*, **25**: 55–63.
9 Gao, L., Li, J., Kusunose, T., Niihara, K., 2004, Preparation and properties of TiN–Si$_3$N$_4$ composites, *Journal of the European Ceramic Society*, **24**: 381–386.
10 Herrmann, M., Balzer, B., Schubert, C., Hermel, W., 1993, Densification, microstructure and properties of Si$_3$N$_4$-Ti(C,N) composites, *Journal of the European Ceramic Society*, **12**: 287–296.
11 Salehi, A., Yuksel, B., Vanmeensel, K., Van der Biest, O., Vleugels, J., 2009, Y$_2$O$_3$–Nd$_2$O$_3$ double stabilized ZrO$_2$–TiCN nanocomposites, *Materials Chemistry and Physics*, **113**: 596–601.
12 Huang, S., Vanmeensel, K., Malek, O., Van der Biest, O., Vleugels, J., 2010, Microstructure and mechanical properties of pulsed electric current sintered B$_4$C–TiB$_2$ composites, *Materials, Science and Engineering*, **528**: 1302–1309.
13 Shu-Qi, G., 2009, Densification of ZrB$_2$-based composites and their mechanical and physical properties: A review, *Journal of the European Ceramic Society*, **29**: 995–1011.
14 Fahrenholtz, W., Hilmas, G., 2007, Refractory diborides of zirconium and hafnium, *Journal of the American Ceramic Society*, **90**: 1347–1364.
15 Monteverde, F., Guicciardi, S., Bellosi, A., 2003, Advances in microstructure and mechanical properties of zirconium diboride based ceramics, *Materials Science and Engineering*, **A346**: 310–319.
16 Lauwers, B., Kruth, J.P., Brans, K., 2007, Development of technology and strategies for the machining of ceramic components by sinking and milling EDM, *Annals of the CIRP*, **56**: 225–228.
17 Lauwers, B., Liu, W., Eeraerts, W., 2006, Influence of the composition of WC-based cermets on manufacturability by wire-EDM, *Journal of Manufacturing Processes*, **8**: 83–89.

18 Malek, O., Vleugels, J., Perez, Y., De Baets, P., Liu, J., Van den Berghe, S., Lauwers, B., 2010, Electrical discharge machining of ZrO_2 toughened WC composites, *Materials Chemistry and Physics*, **123**: 114–120.

19 Salehi S., Yuksel B., Vanmeensel K., Van Der Biest O., Vleugels J., 2009, Y_2O_3-Nd_2O_3 double stabilized ZrO_2-TiCN nanocomposites, *Materials Chemistry and Physics*, **113**: 596–601.

9
Electrochemical discharge machining of particulate reinforced metal matrix composites

J. W. LIU, South China University of Technology, China and
T. M. YUE, The Hong Kong Polytechnic University, Hong Kong

Abstract: This chapter begins with a brief description of the principle and the characteristics of the electrochemical discharge machining (ECDM) process, and its advantages in shaping particulate reinforced metal matrix composites (MMCs) are highlighted. The design and operation scheme of the ECDM equipment are then given. The chapter goes on to discuss single-factor experiments and the orthogonal analyses that were performed to determine how significant the effects of the various machining parameters on material removal rate (MRR) are. Recognizing that the final machined surface quality is also a critical measure that needs to be evaluated, a study on the effects of machining parameters on surface roughness is included.

Key words: electrochemical discharge machining, composites, material removal, voltage waveform, surface roughness.

9.1   Introduction

It is accepted that metal matrix composites (MMCs) are, in general, much more difficult to machine than their monolithic counterparts, irrespective of whether or not conventional or unconventional techniques are used. This does not cause any surprise, since most of the reinforcement phases are hard ceramic materials, and because of this, cubic boron nitride (CBN) and polycrystalline diamond (PCD) tools are often required.[1-4] Apart from the extreme hardness of most of the reinforcement phases, the vast differences in physical, chemical and mechanical properties between the metal matrix and the reinforcement phase have positioned MMCs as a group of difficult-to-machine materials.

Using non-traditional machining techniques, such as laser and water jet machining, one is able to achieve a fairly high material removal rate, but this is often accompanied by some serious surface and subsurface defects.[5-8] In many cases, these are unacceptable in the finished product and can undermine its fatigue strength and other properties. Moreover, these two machining methods are not really ideal for 3-D shaping purposes. Among

242

the many non-conventional machining methods available, electrical discharge machining (EDM), wire-EDM, and electrochemical machining (ECM), are perhaps the most promising processes for shaping MMCs[9–16] when flexibility of the shaping geometry is considered. Notwithstanding the merits of these two kinds of machining methods for shaping MMCs, there are still problems that need to be solved and improvements to be made before they can be effectively utilized. One of the main pitfalls encountered in EDM and ECM of MMCs is the low machining rate. In addition, wire-EDM also suffers from the frequent encountering of unstable machining conditions and the high risk of tool breakage.[9,17] Good surface finish and accurate dimensional control are other main concerns of ECM when applied to MMCs.

It is envisaged that hybrid processes, which combine the actions of electrical discharge machining (EDM) and electrochemical machining (ECM), can increase the material removal rate to above that of other individual process in the machining of MMCs. Indeed, one of the hybrid processes, electro-chemical arc machining (ECAM), has been explored for machining electrically conductive materials. The process incorporates material removal by electro-chemical action as well as by the electric arc.[18–20] The material removal rate of ECAM can be five to fifty times greater than that using the individual processes of ECM and EDM. A similar method to ECAM is electro-chemical spark machining (ECSM), which again is an electrically-based hybrid process primarily used for shaping electrically non-conductive materials, such as composites, glass and ceramics.[21–25] In ECSM, material from the workpiece is removed by the heat produced by sparking in the vicinity of the workpiece.

With regard to the application of ECDM for shaping MMCs, very little information can be found in the open literature. Nonetheless, preliminary results have shown that it is an ideal method for shaping particulate MMCs.[26] The main advantage of ECDM over EDM, in addition to its higher material removal rate, is that it is a more stable machining process in nature than EDM, because ECDM uses a conducting electrolyte as the working medium, rather than a dielectric as is the case with EDM. This gives rise to a relatively wide machining gap for electrical discharging and such a condition facilitates the removal of machined debris. When compared to ECM, ECDM again provides a higher material removal rate, but more importantly, due to the effect of the electrical discharge, a better surface finish can be obtained. The findings on this aspect are presented in this chapter.

Recently, a research study has been launched to explore the potential of ECDM in shaping MMCs and, in particular, to study the process of ECDM of particulate MMCs. Some major findings on the relationship between machining parameters and material removal rate, as well as surface roughness, are presented here.

9.2   The principles of electrochemical discharge machining (ECDM)

The basic principles of the ECDM process are illustrated in the schematic diagram of Fig. 9.1. The process is a chain of complex physical–chemical actions, where workpiece material, in simple terms, is removed by electrochemical dissolution (ECD) as well as by electrodischarge erosion (EDE). The electrolyte bath provides limited conduction between the tool and the workpiece; sodium nitrite ($NaNO_3$) and sodium hydroxide ($NaOH$) are two commonly used electrolytes. During machining, the tool-electrode is normally made the cathode, while the workpiece acts as the anode. The tool can adopt various forms, such as a travelling wire (similar to W-EDM), a non-rotational solid figure or a rotational solid figure. In the machining of conductive materials, such as MMCs, the employment of a pulsed DC voltage between the two electrodes will cause a current to pass across the machining gap. This results in anodic dissolution of the workpiece according to Faraday's law of electrolysis, while at the same time, due to the electrochemical reactions, the electrical resistance between the cathode and anode increases because the number of hydrogen bubbles at the cathode surface is increased. This causes the voltage between the anode and cathode to increase rapidly, and once it reaches the breakdown voltage of the bubbles, sparks are initiated. In the case of a conductive workpiece, the discharge takes place between the tool and the workpiece in the form of an arc. This phenomenon perhaps can be better understood using the illustration of an ECDM waveform shown in Fig. 9.2, in which φ_a is the applied voltage, and

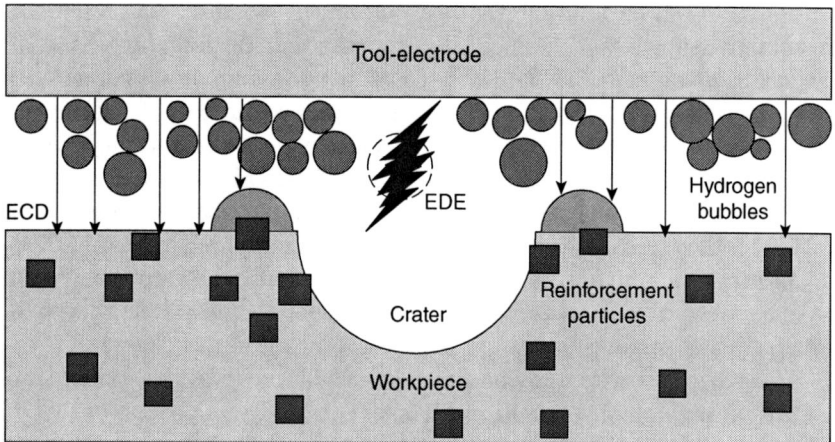

9.1 Schematic diagram showing the combined action of the electrochemical dissolution (ECD) and electrodischarge erosion (EDE).

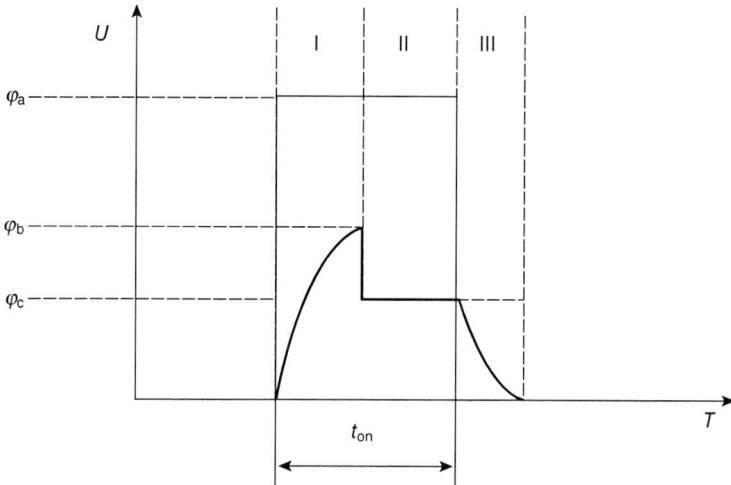

9.2 An illustration showing the characteristics of an ECDM waveform.

φ_b is the breakdown voltage, φ_c is the arc maintaining voltage, and t_{on} is the pulse-on-time. During Stage I, the voltage between the anode and cathode will increase. Once it reaches the breakdown voltage, a spark will be initiated and the voltage will immediately drop to the arc maintaining voltage. At this point in time, Stage II sets in, and the arc discharge will persist untill the end of the pulse-on-time if the energy of the power supply can sustain the current at the arc maintaining voltage. At the end of the pulse-on-time, due to the electric capacity effect, the voltage between the anode and cathode will gradually drop to zero (Stage III). Material removal by EDE is mainly confined to Stage II, while the action of ECD could cover the whole period in which there is a flow of current. But it is understood that the ECD action will be influenced by the condition of the machined surface, such as the form of the anodic surface film and the amount of gas evolution. Most of the previous work on ECDM of non-conductive materials primarily concentrated on the phenomenon of the electrochemical-aided discharge mechanism and the material removal by EDE, and much less attention was made to the material removal through the ECD effect. This is probably due to the insignificant effect of ECD on material removal when dealing with non-conductors, such as ceramics and polymeric composites.

9.2.1 Material removal of MMCs

In ECDM of MMCs, even though the amount of material removed by ECD was found to be much less than that removed by EDE, it still contributes in some measure to the total amount. A previous study on ECDM of

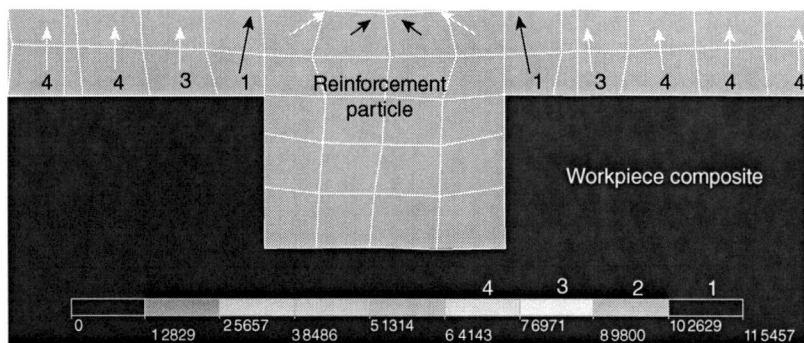

9.3 ANSYS simulation results of the current density distribution for ECDM of a particulate MMC. (The alumina particle is assumed to have a square shape of size 21 μm and its electrical resistance is assumed to be infinitely large. The machining gap is 50 μm and the voltage between the electrodes is 20 V.)

particulate MMCs showed that a relatively high current density is present around the interfaces between the matrix and the particle during the course of the machining[26] (Fig. 9.3). This high current density situation was found to be pronounced for the conditions where the particle protrusion was less than 40%, beyond which the degree of intensification diminished. It is considered that a relatively high current density at the interface enhances the chemical dissolution of the matrix material, thus facilitating the removal of the ceramic particles. Although the ECM action has the effect of dissolving the metal phase around the reinforcement phase, its major function is to enable EDM sparking to be operable under a relatively large spark gap size condition, and thus facilitating the removal of machined debris, including the reinforcing phase(s) of the MMC from the inter-electrode gap. Recognizing this, it is important to employ a discharge gap that is larger than the size of the reinforcement particle.

Similarly to the conventional EDM process, in ECDM, the discharge causes the workpiece material to melt, vaporize and form chips. An analysis of the discharge mechanism in ECDM of a particulate reinforced MMC, together with a model to reveal the electric field acting on a hydrogen bubble in the ECDM process, has been presented in a recent publication.[27] The model is capable of predicting the critical breakdown voltage for spark initiation for a given processing condition. The model shows that the breakdown voltage is independent of the presence of the reinforcement phase. Although, many studies have proposed that gas bubbles at the cathode surface will coalesce into a gas film through which the discharge occurs,[23,28] it is considered that the formation of a gas film or layer will depend on the machining conditions, such as electrolyte concentration, temperature and the geometry of the cathode electrode. Moreover, most of the previous work on ECDM of non-conductors used a small cathode with a large anode-to-cathode surface ratio (typically over 100) and this undoubtedly would promote the formation of a gas film. With regard to material removal, although melting and vaporization is still a major removal mechanism in ECDM of MMCs, an examination of the machining debris showed that spalling is another major material removal mechanism;[27] this is caused by thermal stresses.

9.3   ECDM equipment

Figure 9.4 presents a schematic diagram of a typical ECDM set-up, with essential elements, while Fig. 9.5 shows in-house built ECDM equipment. This has a spindle (Fig. 9.6) that is fixed to the Z-axis, onto which a cylindrical steel tool is held. The motorized spindle is capable of rotating over a wide range of speeds. During the course of processing, the electric current flows to the tool-electrode through a conduction bush that is placed inside an electrically insulating bearing housing. During operation, a spring

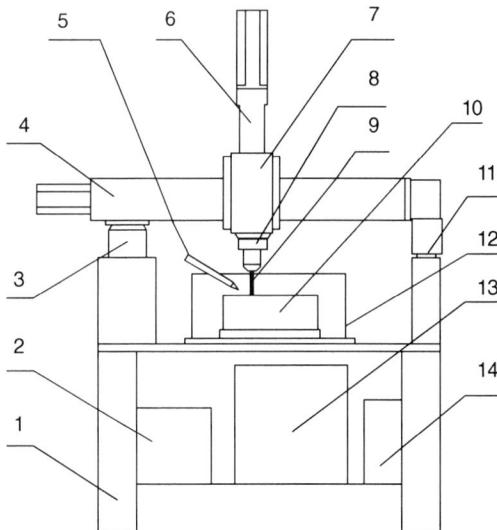

9.4 ECDM set up (working table (1), electrolyte system (2), Y-axis (3), X-axis (4), nozzle (5), Z-axis (6), tool holder (7), electric transmission system (8), tool (9), workpiece (10), slide rail (11), electrolyte bath (12), control system (13), power source (14)).

9.5 In-house built ECDM equipment.

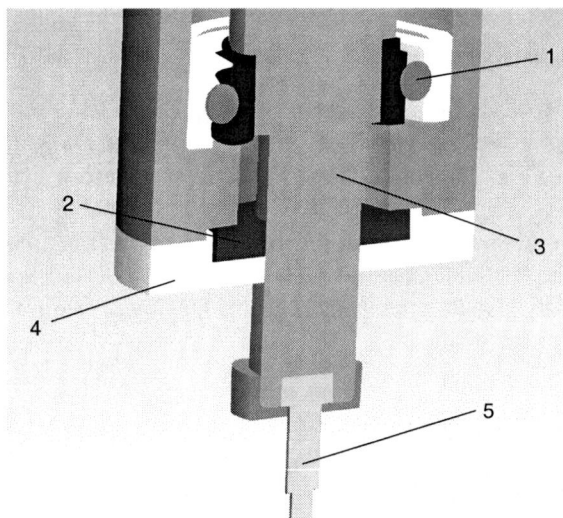

9.6 Design of the spindle where an electric insulating bearing system is employed to insulate the shaft from other components of the spindle. (Ceramic ball bearing (1), insulation cover (2), shaft (3), insulating bearing housing (4), tool (5).)

pushes the bush into tight contact with the shaft. With such a design, unwanted sparking between the bush and the shaft can be avoided. Moreover, with this spindle design, the working current will be confined to the processing area.

During the course of machining, the electric current flows from an electric transmission system, placed on a plastic insulated seat, to the tool. The resistance of the electric transmission system is negligible. This particular equipment has a pulse number controllable electrical source, with which the current can be adjusted without changing the applied voltage. The average current output of this electrical source can be adjusted from 0.5 A

Table 9.1 Major specifications of the ECDM equipment

	X-axis	Y-axis	Z-axis	Spindle	Electrical source	Electrolyte bath
Travel (mm)	250	250	100			
Repeatability accuracy (μm)	5	5	1			
Position accuracy (μm)	15	15	2			
Speed (rpm)				0–20 000		
Power (kW)				1.5		
Peak current (A)					0.5–100	
Voltage (V)					20–120	
Pulse duration (μs)					4–400	
Duty cycle					1:1–1:10	
Maximum liquid level (mm)						200
Electrolyte medium						NaNO$_3$
Circulator flow (L/min)						0–20

to 100 A. The pulse duration ranges between 4 μs and 400 μs, while the duty cycle can be operated between 1:1 and 1:10. Some major specifications of this ECDM equipment are given in Table 9.1.

The electrolyte (NaNO$_3$) is pumped into the processing area through the nozzle with the aid of an electrolyte circulation system. After the electrolyte has been properly filtered, it returns to the electrolyte circulation system.

Associated with the ECDM setup is a tailor-made power source. Its pulse duration can be adjusted from 4 μs to 400 μs and it operates with a 100 A peak current, whereas the working voltage ranges between 20 V–110 V, with the duty cycle capable of working between 1:1 and 1:10. A typical square waveform of this power source is shown in Fig. 9.7.

During ECDM, if the machining debris cannot be discharged from the spark gap in time, then short-circuiting would occur. Figure 9.8 indicates the voltage characteristics during processing under various machining conditions. If it operates under a stable machining Condition (I), the processing voltage fluctuates between φ_1 and φ_2, while for an unstable Condition (II), it would normally be maintained for a short period of time. If the unfavourable processing situation persists, i.e. Condition (II), the voltage will drop further and finally short-circuiting would occur, i.e. Condition (III), in which the voltage drops sharply from φ_3 to φ_4. With this problem in mind, the electrode control system of the ECDM is so designed that it can monitor the change of voltage and provide appropriate remedial actions to avoid short circuiting from happening. Should an unstable processing condition be encountered, the servo control system will command the tool to retract instantly so as to improve the debris discharge through a wider machining

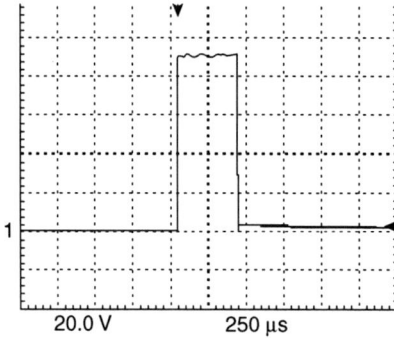

9.7 Square waveform of the power source.

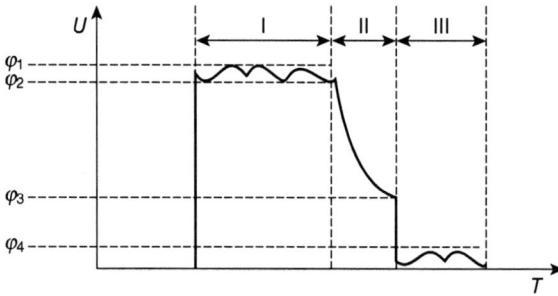

9.8 Voltage behaviour for different processing conditions. (Stable processing condition (I), unstable processing condition (II), short-circuiting (III).)

gap. The retraction speed and the distance can be controlled according to the unstable processing condition encountered.

The following describes the working principle of the process control system. A voltage collection card is employed to monitor the processing voltage. Initially, the processing voltage data are collected and those voltages that are too high or too low are filtered out. The average voltage value is then calculated using the following equation:

$$U = \frac{\sum_{j=1}^{i} U_j}{i}$$
[9.1]

The average voltage obtained is used to determine the status of the processing condition. If the value found matches that of an unstable condition, the control system will signal the tool to retract. The monitoring flowchart of Fig. 9.9 summarizes the working sequences of the process control system.

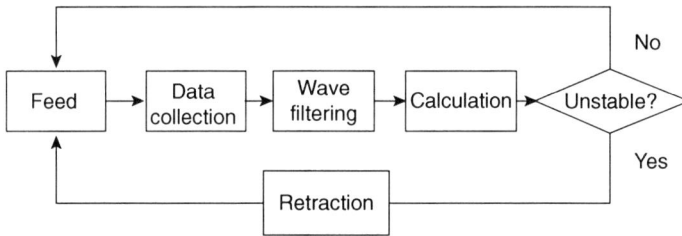

9.9 Monitoring process flowchart.

In fact, the data collection card and the movement control card are integrated into a computerized control system; the processing voltage between the tool and the workpiece is monitored by the data collection card. The collected data are then calculated and analysed by a computer program. If the voltage determined implies a stable processing state, the tool will continue feeding. On the other hand, if an unstable condition is detected, the computer will give a signal to the driver through the movement control card such that the servo-actuator activated and the tool retracts.

9.4   Parameters affecting material removal rate (MRR)

The findings presented at this time are based on the research study of a particulate reinforced aluminium 6061, with 10 vol% Al_2O_3 (10ALO) or 20 vol% Al_2O_3 (20ALO). The materials were supplied by Alcan in the form of rolled plates, having a thickness of 36 mm, with reinforcement particles of nominal size 21 µm. The results cover single-factor experiments on the effects of the various machining parameters on MRR. In addition, the significance of the effects of the various machining parameters on MRR is also presented and is based on the results of an orthogonal analysis.

9.4.1 Effect of pulse duration on MRR

The relationship between pulse duration and MRR of the Al_2O_3 reinforced Al composites under a continuous pulsing condition is summarized in Fig. 9.10. As expected, a higher percentage of ceramic phase resulted in a drop in MRR; however, somewhat unexpectedly, the MRR decreased as the pulse duration increased from 32 µs to 64 µs. This is rather different from the normal behaviour in machining monolithic metallic materials. A drop in MRR when the volume of reinforcement phase was increased can be explained by examining the temperature of the processing area. Figure 9.11 gives an example of the simulation results of the temperature field of a processing area of 10ALO. It clearly shows that the ceramic particles have

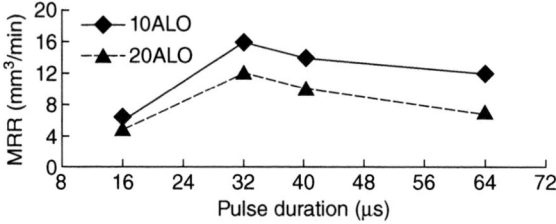

9.10 Effects of pulse duration on MRR (pulse current 4 A, electrolyte concentration 0.5 wt%, applied voltage 80 V).

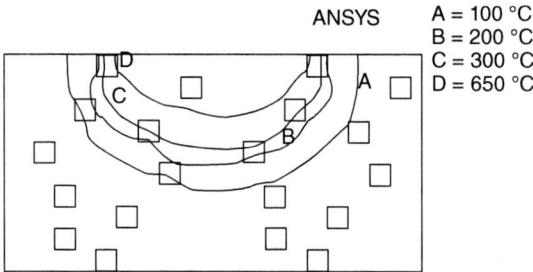

9.11 Simulation results of the temperature field of the composite when subject to an EDM spark.

the effect of restraining heat from passing through. As a result, the volume of the molten pool is smaller than that of the material without the ceramic reinforcement phase. The results are thus obvious if the volume of the reinforcement phase is increased – a smaller molten pool will be produced and this will reduce the amount of material removed. On the other hand, the decrease in MRR when the pulse duration was increased was unanticipated because a longer pulse duration means a higher energy input and should result in a higher erosion rate. The reason for a lower MRR at a larger pulse duration could be, despite a longer pulse duration initially removing a large volume of material from the workpiece, that the MRR drops when the debris (which comprises the reinforcing ceramic phase), is trapped within the spark gap and cannot be carried away quickly enough by the flushing fluid. It is likely that this phenomenon occurred in the experiment, since the nominal size of the ceramic reinforcement phase is 21 µm, which is large when compared to the size of the discharge gap in the experiment. In a situation where a significant number of ceramic particles are trapped in the gap, the cutting process becomes unstable and material removal is reduced. Thus, it appears that in order to benefit from the high cutting rate under a relatively long pulse duration condition, the size of the reinforcement phase should be smaller than the discharge gap.

9.4.2 Effect of current, machining gap and applied voltage on MRR

The relationship between machining current and MRR is presented in Fig. 9.12, where it is apparent that the MRR for both composites increased greatly when the current increased from 3 A to 6 A. One would expect, similarly to the pulse duration effect, that an increase in current would result in a drop in MRR, as a higher energy input is provided. However, the results show that the MRR increased when the current was increased. This means that the machining condition was stable, and effective material removal was achieved at high machining currents.

It is believed that a high current condition will facilitate the action of ECD. The main effect of ECD is to increase the machining gap, and once the machining gap is enlarged, the ceramic particles can be discharged more readily, hence a stable machining condition for electrical discharging is maintained. However, it normally happens that a further increase in spark gap beyond its optimal value leads to a decrease in MRR. To further elaborate this gap-size effect, single-pulse experiments were conducted. A single-pulse condition was employed because it is easier to control the machining gap precisely. Figure 9.13 shows the relationship between the machining gap

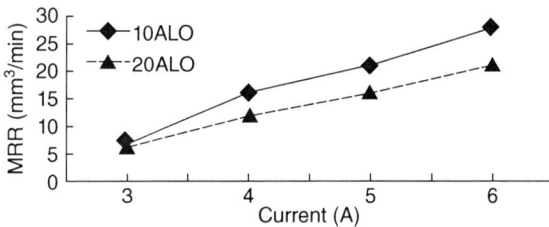

9.12 Effect of current on MRR (pulse duration, electrolyte concentration and the applied voltage were 32 µs, 0.5 wt% and 80 V, respectively).

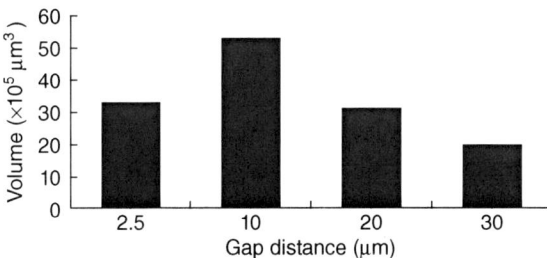

9.13 Crater volume measured under single pulse condition.

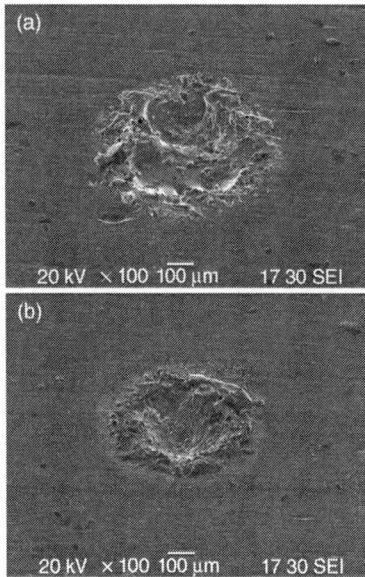

9.14 SEM images of the craters produced using machining gaps of (a) 20 μm and (b) 30 μm.

and crater volume obtained under a single pulse condition, while Fig. 9.14 reveals the SEM images of the craters obtained using machining gaps of 20 μm and 30 μm. The results confirm that beyond a certain gap dimension, in this case 10 μm, MRR was decreased. The reason for a smaller material removal volume when processing at large gaps is elucidated by analysing the voltage waveform with respect to spark ignition-delay. Figures. 9.15a–d show the voltage waveforms obtained for the processing conditions with machining gaps of 2.5 μm, 10 μm, 20 μm and 30 μm, respectively. These figures clearly show that ignition-delay increases with increase of the machining gap. When the ignition-delay time increases, this has the effect of reducing the arc maintaining period. It also means that the discharge energy is reduced. As a result, the crater volume becomes smaller when the processing gap is relatively large.

Similarly to the current effect, Fig. 9.16 shows that the MRR is greatly increased when the applied voltage was increased from 60 V to 110 V. Within this voltage range, machining took place under stable machining conditions, i.e. with effective removal of machining debris. At high applied voltages, a stable machining condition is attained and a relatively large discharge gap is maintained, which again facilitates the removal of debris.

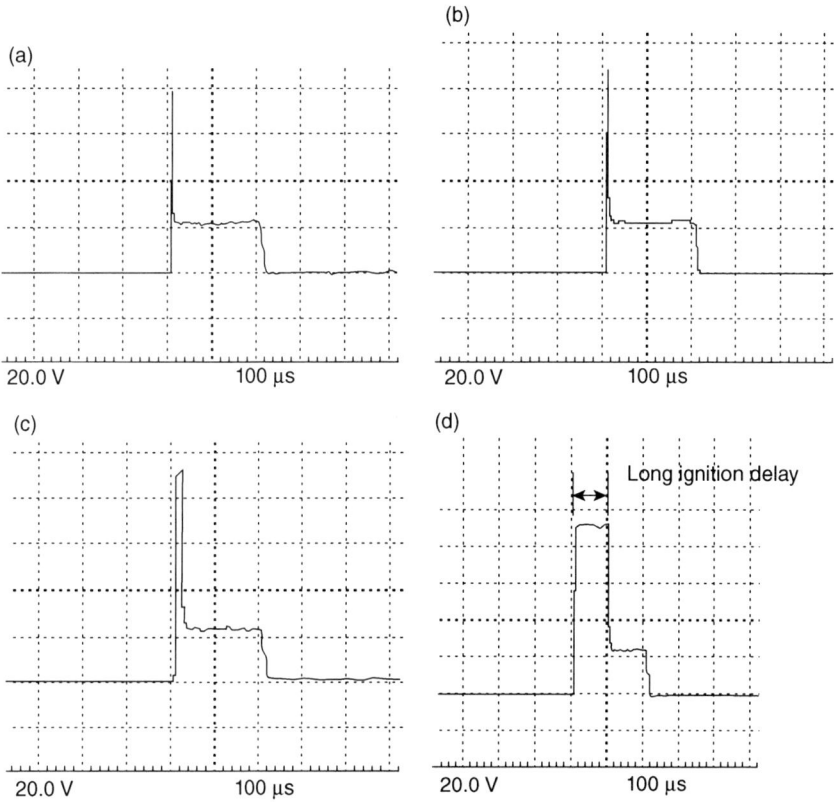

9.15 Voltage waveforms obtained for the conditions of gap dimension of (a) 2.5 µm, (b) 10 µm, (c) 20 µm, (d) 30 µm (applied voltage 90 V).

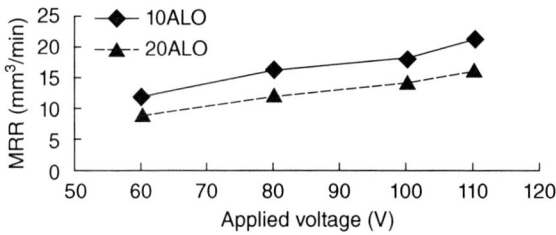

9.16 Effect of applied voltage on MRR (pulse duration, electrolyte concentration and current were fixed at 32 µs, 0.5 wt% and 4 A, respectively).

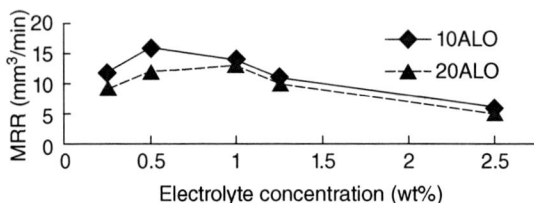

9.17 Effect of electrolyte concentration on MRR.

9.4.3 Effect of electrolyte concentration on MRR

Figure 9.17 shows that for 10ALO, the maximum MRR occurred at an electrolyte concentration of around 0.5%, and it appears that a concentration higher or lower than this level would result in a lower MRR. A similar trend was also obtained for 20ALO, where the maximum MRR occurred at an electrolyte concentration of around 1%. At low concentrations, the hydrogen formation rate at the cathode is low, and as a consequence, a longer time is required during the pulse-on-period for generating enough bubbles between the two electrodes and in reaching the critical breakdown voltage. That is to say, the discharge time is correspondingly reduced, and hence the amount of molten material produced is less than in the high concentration conditions in which a longer discharge time is experienced. On the other hand, under high concentration conditions, the ECD effect is enhanced, and the EDE action is weakened. This behaviour was confirmed when the ECDM waveforms were studied, and the results are presented in the next section. From this, it is clear that EDE plays a more important role in material removal than ECD, and a weakening in EDE would result in a drop in MRR.

9.4.4 Orthogonal analysis

More results on the effects of the various machining parameters on MRR were obtained through orthogonal analysis. The analysis examined the electrolyte concentration, pulse duration and working current. Each factor is presented with three levels (Table 9.2). An L9(34) orthogonal design table was established with the aim of optimizing MRR. Based on this objective, an orthogonal analysis was performed for both the 10ALO and 20ALO materials and the results are given in Table 9.3 and Table 9.4, respectively.

With the 10ALO composite, the impact of the different factors follows the sequence of current > pulse duration > electrolyte concentration. In the conditions of 6 A current, pulse duration 32 μs and 0.5 wt% electrolyte, the highest MRR is obtained. As for the 20ALO composite, the ranking of

Table 9.2 Parameter levels for the numerical analysis

Factors/levels	1	2	3
A (Medium)	0.25	0.5	1
B (Pulse duration)	16 µs	32 µs	64 µs
C (Current)	4 A	5 A	6 A

Table 9.3 Results of the orthogonal analysis on MRR for the 10ALO material

Factors Series no.	A 1	B 2	C 3	D 4	MRR (mm³/min)
1	1	1	1	1	5
2	1	2	2	2	19
3	1	3	3	3	22
4	2	1	2	3	18
5	2	2	3	1	28
6	2	3	1	2	12
7	3	1	3	2	21
8	3	2	1	3	14
9	3	3	2	1	15
Ij	46	44	31		
IIj	58	61	52		
IIIj	50	49	71		
Ij = Ij/3	15	15	10		
IIj = IIj/3	19	20	17		
IIIj = IIIj/3	17	16	24		
R	12	17	40		

the significance of the three factors follows the sequence of current > electrolyte concentration > pulse duration. Within the limits of the study, the optimum conditions for achieving the highest MRR are: current of 6 A, pulse duration of 32 µs, and 1 wt% concentration.

The results of the analysis suggest that in order to achieve a high MRR for Al_2O_3 particulate reinforced MMCs using ECDM, the applied current is the most influential factor among current, pulse duration and electrolyte concentration. This result is considered to be reasonable due to the fact that both the EDE and ECD actions are directly governed by the magnitude of the applied current. Therefore, no matter which one of these two actions is dominant in the machining process, current will still play an important role. Further, with regard to the effect of the concentration of the electrolyte, the results show that the effect becomes more important as the amount of the reinforcing phase is increased.

Table 9.4 Results of the orthogonal analysis on MRR for the 20ALO material

Factors Series no.	A 1	B 2	C 3	D 4	MRR (mm³/min)
1	1	1	1	1	6
2	1	2	2	2	14
3	1	3	3	3	16
4	2	1	2	3	13.5
5	2	2	3	1	21
6	2	3	1	2	7
7	3	1	3	2	22
8	3	2	1	3	13
9	3	3	2	1	17
Ij	36	41.5	26		
IIj	41.5	48	44.5		
IIIj	52	40	59		
Ij = Ij/3	12	14	9		
IIj = IIj/3	14	16	15		
IIIj = IIIj/3	17	13	20		
R	16	8	33		

9.5   Parameters affecting surface roughness

In assessing the potential of using ECDM to shape MMCs, not only is the MRR an important factor that needs to be considered, but the final machined surface roughness is also a critical measure. It was reported that EDM reduced the fatigue strength of a SiC MMC by as much as 20% if a coarse machined surface was produced.[29] With this in mind, a study of the effects of machining parameters on surface roughness was conducted. Further studies of the effects of ECDM surface quality on mechanical properties, in particular fatigue strength need to carried out.

The relationship between pulse duration and surface roughness (R_a) of Al_2O_3 reinforced Al composites is shown in Fig. 9.18. Disregarding the percentage of the reinforcement phase, the trend shows that once a minimum R_a has been reached, only a small increase in roughness is measured, even if the pulse duration is significantly increased. On the other hand, at short pulse durations, in this case 16 μs, a high R_a value was obtained. This can be explained by analysing the ECDM pulse waveform. The waveforms for different pulse durations (Fig. 9.19) in processing the 20% Al_2O_3 composite show that an increase in pulse duration promotes ECDM activity, i.e. both EDE and ECD activities, particularly the EDE activity that was strong at a duration of 32 μs. This can be explained on the basis that an increase in pulse duration leads to an increase in the number

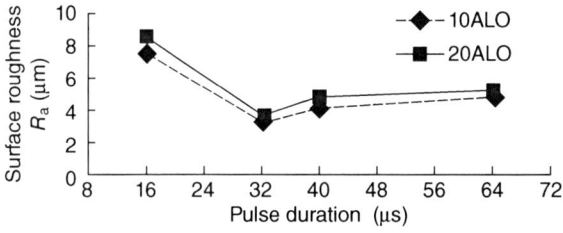

9.18 Effect of pulse duration on surface roughness (pulse current 4 A, electrolyte concentration 0.5 wt% and applied voltage 80 V).

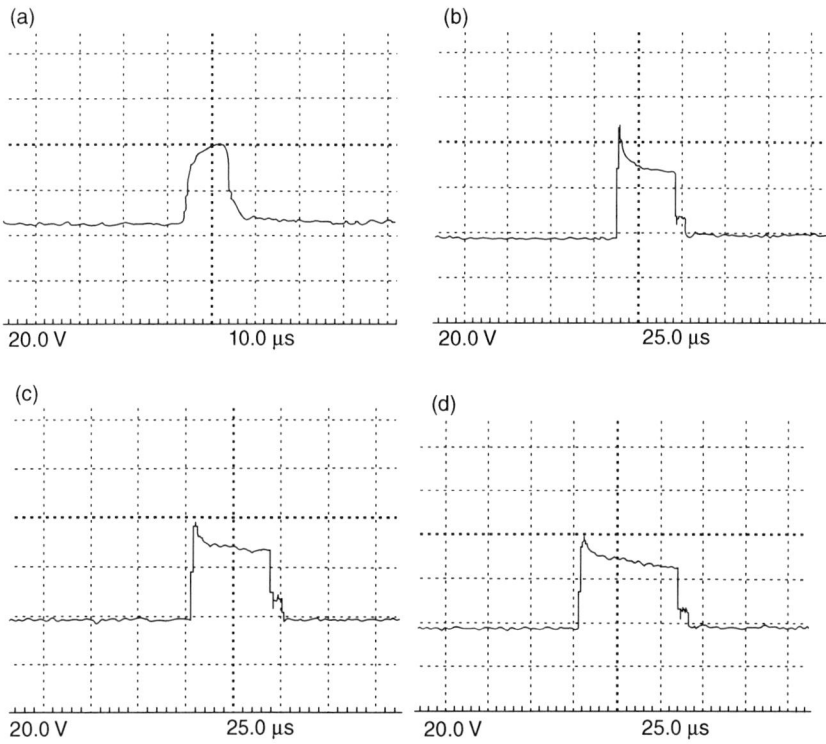

9.19 Effect of pulse duration on ECDM waveforms: (a) 16 µs, (b) 32 µs, (c) 40 µs, (d) 64 µs.

9.20 Machined surfaces produced using (a) 16 μs pulse duration and (b) 64 μs pulse duration.

of hydrogen bubbles generated, and this causes the voltage between the two electrodes to be increased, thus creating a favourable condition for the bubbles to reach the breakdown voltage and to initiate sparks. On the other hand, in examining the waveform of the 16 μs pulse duration condition, only ECM activity was realized (Fig. 9.19a). In the case of a pure ECD condition, i.e. without the action of EDE, not only was the MRR found to be on the low side, but the roughness value was also high. It is considered that without the aid of the EDE action, the ceramic particles exposed at the workpiece surface are not readily removed (Fig. 9.20a) by ECD alone. The consequence is twofold; first, a low MRR is obtained, and second, due to the accumulation of particles at the surface, the surface roughness is increased significantly. Conversely, for the conditions of long pulse duration, most of the loosely attached ceramic particles at the workpiece surface would have been removed with the aid of the EDE action. An example of such a machined surface processed at a pulse duration of 64 μs is given in Fig. 9.20b.

Other effects, such as current and electrolyte concentration, on surface roughness are shown in Fig. 9.21 and Fig. 9.22, respectively. The surface roughness of the Al_2O_3 composites increases slightly with increasing processing current. An increase in current will result in a higher pulse energy,

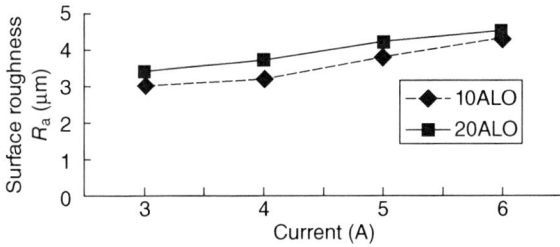

9.21 Effect of current on surface roughness (pulse duration 32 μs, electrolyte concentration 0.5 wt%, applied voltage 80 V).

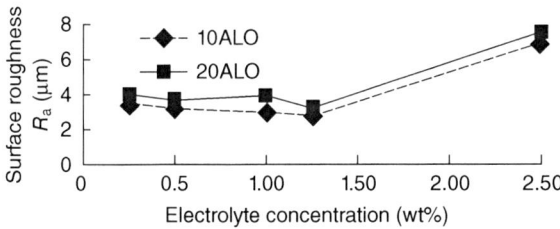

9.22 Effect of electrolyte concentration on surface roughness (pulse current 4 A, pulse duration 32 μs and applied voltage 80 V).

which means larger discharge craters will form and thus increase the surface roughness, while for the electrolyte concentration effect, the results show that above a certain concentration, in this case 1.25 wt%, the roughness increases sharply. Again, this is considered to be due to the lack of EDE activity and the domination of ECD action at high electrolyte concentrations, a similar reason to that given for the pulse duration effect. Figure 9.23 shows the different ECDM waveforms obtained for various electrolyte concentrations, which clearly show that as the concentration is increased, the ECD action is enhanced, and the EDE activity is weakened. At a concentration of 2.5 wt%, only ECD activity is observed. Examples of the machined surface obtained by using low and high electrolyte concentrations are given in Fig. 9.24.

9.6   Conclusions

Although only a limited number of research studies have been conducted on ECDM of MMCs, the results obtained so far are very promising. Not only can the material removal rate be higher than that of ECM and EDM alone, but more importantly, due to having a wider discharging gap than

(a)

(b)

20.0 V   25.0 µs   20.0 V   25.0 µs

(c)

(d)

20.0 V   25.0 µs   20.0 V   25.0 µs

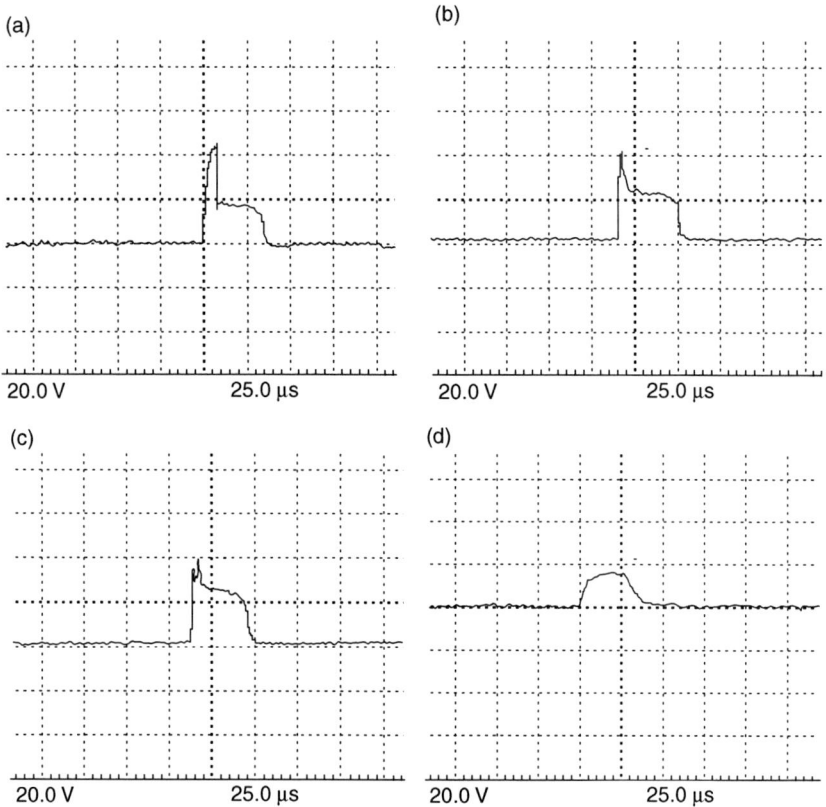

9.23 Effect of electrolyte concentration on ECDM waveforms:
(a) 0.25 wt%, (b) 1 wt%, (c) 1.25 wt%, (d) 2.5 wt%.

EDM, it is a more stable machining process than EDM; a better surface finish can also be obtained than by ECM. The authors are confident that a surface finish in the sub-micron range can be achieved with proper equipment development and electrode design. The process is particularly suitable for shaping particulate types of composites. It is considered that the process has not been widely explored for shaping MMCs, not because of any major inherent problems of the process itself, but because of difficulty in identifying some suitable engineering products or components for ECDM, in order to demonstrate its strengths. It is hoped that the information provided here will serve as the initiation point for many more research and practical studies. Finally, it is the authors' belief that ECDM should be seriously considered by industry as a promising means for shaping MMCs.

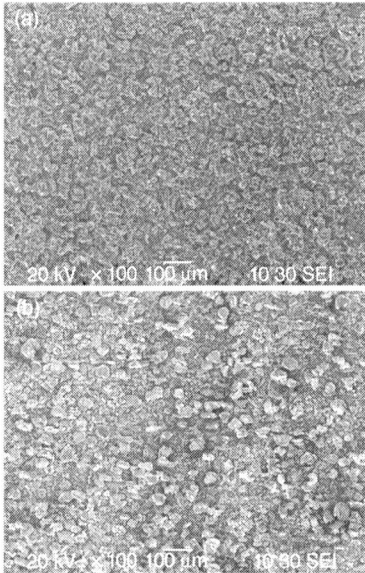

9.24 Machined surfaces produced using (a) 0.25 wt% electrolyte concentration and (b) 2.5 wt% electrolyte concentration.

9.7   Acknowledgement

The research work was supported by the Hong Kong Polytechnic University.

9.8   References

1 L.A. Looney, J.M. Monaghan, P. O'Rielly and D.M.R. Taplin, The turning of an Al/SiC metal-matrix composite, *J. Materials Processing Technology*, **33**(4) (1992) 453–468.
2 N.P. Hung, F.Y.C. Boey, K.A. Khor, Y.S. Phua and H.F. Lee, Machinability of aluminum alloys reinforced with silicon carbide particulates, *J. Materials Processing Technology*, **56**(1–4) (1996) 966–977.
3 N. Tomac and K. Tønnessen, Machinability of particulate aluminium matrix composites, *Annals CIRP*, **41** (1992) 55–58.
4 G.A. Chadwick and P.J. Heath, Machining metal matrix composites, *Metals and Materials*, **Feb** (1990) 73–76.
5 T.M. Yue, and W.S. Lau, Pulsed Nd: YAG laser cutting of SiC/Al-Li metal matrix composite, *Materials and Manufacturing Processes*, **11** (1996) 17–29.
6 W.S. Lau, T.M. Yue and M. Wang, Ultrasonic-aided laser drilling of aluminium-based metal matrix composites, *Annals CIRP*, **43** (1994) 177–180.
7 F. Müller and J. Monaghan, Non-conventional machining of particle reinforced metal matrix composite, *International J. of Machine Tools & Manufacture*, **40** (2000) 1351–1366.

8 E. Savrun and M. Taya, Surface characterization of SiC wisker/2124 aluminium and Al$_2$O$_3$ composites machined by abrasive water jet, *J. Materials Science*, **23** (1988) 1453–1458.

9 M. Ramulu and M. Taya, EDM machinability of SiCw/Al composites, *J. Materials Science*, **24** (1989) 1103–1108.

10 T.M. Yue and Y. Dai. Wire electrical discharge machining of Al$_2$O$_3$ particle and short fibre reinforced Al-based composites, *Materials Science and Technology*, **12** (1996) 831–835.

11 M. Rozenek, J. Kozak, L. Dabrowski and K. Lubkowski, Electrical discharge machining characteristics of metal matrix composites, *J. Materials Processing Technology*, **109**(3) (2001) 367–370.

12 B.H. Yan, C.C. Wang, H.M. Chow and YC Lin, Feasibility study of rotary electrical discharge machining with ball burnishing for Al$_2$O$_3$ 6061Al composite, *International J. of Machine Tools & Manufacture*, **40** (2000) 1403–1421.

13 B. Mohan, A. Rajadurai and K.G. Satyanarayana, Electric discharge machining of Al–SiC metal matrix composites using rotary tube electrode, *J. Materials Processing Technology*, **153–154** (2004) 978–985.

14 L.H. Hihara and P. Panquites IV, Method of electrochemical machining (ECM) of particulate metal-matrix composites (MMCs), *United States Patent* 6110351 (2000).

15 C. Senthilkumar, G. Ganesan, and R. Karthikeyan, Bi-performance optimization of electrochemical machining characteristics of Al/20%SiCp composites using NSGA-II, *Proc. IMechE, Part B: J. Engineering Manufacture*, **242** (2010) 1399–1407.

16 C. Senthilkumar, G. Ganesan, R. Karthikeyan and S. Srikanth, Modelling and analysis of electrochemical machining of cast Al/20%SiCp composites, *Materials Science and Technology*, **26**(3) (2010) 289–296.

17 B.H. Yan, H.C. Tsai, F.Y. Huang and L.C. Lee, Examination of wire electrical discharge machining of Al$_2$O$_3$p/6061Al composites, *International J. of Machine Tools & Manufacture*, **45** (2005) 251–269.

18 I.M. Crichton and J.A. McGeough, Studies of discharge mechanisms in electrochemical arc machining, *J. Applied Electrochemistry*, **15** (1985) 113–119.

19 A.B. Khayry and J.A. McGeough, Analysis of electrochemical arc machining by stochastic and experimental methods, *Proceedings of the Royal Society of London – Series A, Mathematical and Physical Sciences*, **412** (1987) 403–429.

20 H. El-Hofy, *Advanced Machining Processes: Nontraditional and Hybrid Machining Processes*, New York: McGraw-Hill (2005).

21 S. Tandon, V.K. Jain, P. Kumar, and K.P. Rajurkar, Investigations into machining of composites, *Precision Engineering*, **12**(4) (1990) 227–238.

22 V.K. Jain, S. Tandon, and P. Kumar, Experimental investigations into electro chemical spark machining of composites, *Trans. ASME, J. of Engineering for Industry*, **112**(2) (1990) 194–197.

23 R. Wüthrich and V. Fascio, Machining of non-conducting materials using electrochemical discharge phenomenon – an overview, *International J. of Machine Tools & Manufacture*, **45** (2005) 1095–1108.

24 N. Gautam and V.K. Jain, Experimental investigations into ECSD process using various tool kinematics, *International J. of Machine Tools & Manufacture*, **38**(1–2) (1998) 15–27.

25 Y.P. Singh, V.K. Jain, P. Kumar and D.C. Agrawal, Machining piezoelectric (PZT) ceramics using an electrochemical spark machining (ECSM) process, *J. Materials Processing Technology*, **58**(1) (1996) 24–31.

26 J.W. Liu, T.M. Yue and Z.N Guo, Wire electrochemical discharge machining of Al_2O_3 particle reinforced aluminum Alloy 6061, *Materials and Manufacturing Processes*, **24**(4) (2009) 446–453.

27 J.W. Liu, T.M. Yue and Z.N Guo, An analysis of the discharge mechanism in electrochemical discharge machining of particulate reinforced metal matrix composites, *International J. of Machine Tools & Manufacture*, **50** (2010) 86–96.

28 M.S. Han, B.K. Min and S.J. Lee, Modeling gas film formation in electrochemical discharge machining processes using a side-insulated electrode, *J. Micromechanics and Microengineering* **18**(4) (2008) Article Number: 045019.

29 M. Ramulu, G. Paul and J. Patel, EDM surface effects on the fatigue strength of a 15 vol% SiCp/Al metal matrix composite material, *Composite Structures*, **54** (2001) 79–86.

10
Fundamentals of laser machining of composites

G. CHRYSSOLOURIS and K. SALONITIS,
University of Patras, Greece

Abstract: Composite materials, in most cases, are cured in a mold to a 'near final shape'; however, machining is still required at both the preparation and the finishing stages. The advantages of laser machining are its high rates, no tool wear, no contact forces, and a relatively high precision. The effectiveness of lasers depends on the thermal nature of the machining process. Nevertheless, some difficulties arise because of the difference in the thermal properties of the various components of the composite. In this chapter, the state-of-the-art of laser machining of composite materials is presented.

Key words: composite materials, laser machining.

10.1 Introduction

Composites are made up from a combination of materials, each of them providing a unique characteristic to the final product; as a result, a new material, whose properties would not be attainable by conventional means, is produced. The composites constitute individual phases; those of the matrix and the reinforcement. The matrix phase surrounds that of the reinforcement, thus providing a supporting structure for maintaining its position. The reinforcement phase contributes special properties which enhance those of the matrix. As an example, in the case of carbon fiber reinforced composites, the strength and rigidity of the material can be controlled by varying the amount of carbon fiber incorporated into the polymer matrix. This ability to tailor properties, combined with the inherent low density of the composite and its (relative) ease of fabrication, makes this material an extremely attractive alternative for many applications.

For the structuring of the present chapter, the composites are categorized based on the material type of the composite's matrix. Thus, they can be classified into non-metallic, and metallic matrix composites.

A typical example of a non-metallic matrix composite is the fiber-reinforced carbon fiber–epoxy composite, which consists of carbon fibers embedded in an epoxy resin matrix. The carbon fibers themselves exhibit high strength and rigidity but lack ductility. Due to their brittleness, they

266

cannot be used individually and thus an epoxy resin matrix is used. The epoxy itself is not very strong, but it plays two important roles. It acts as a medium to transfer load to the fibers, and the fiber–matrix interface deflects and stops small cracks, thus making the composite more crack-resistant to cracks than either of its constituent components. Fiber reinforced composites are used mainly as structural materials because they present increased rigidity, strength, and low density.

Metal matrix composites are structured by incorporating strong ceramic fibers in a metal matrix in order to produce a strong, rigid material, e.g. SiC fibers embedded in an aluminum matrix. This type of composite finds applications from highly stressed airframe materials to components with moderate loads, such as fuselage skins. Composites with metal fibers embedded in a ceramic matrix (ceramic-matrix composites) are produced to take advantage of the ceramic's strength while obtaining an increase in toughness from the metal fibers, which can deform and deflect cracks.

Due to their light weight, high strength, and directional properties, composites have a great manufacturing potential. Although in most applications, composites are cured in a mold to a 'near final shape,' machining is still required both at the preparing and the finishing stages. A finishing process is usually necessary for the final surface quality and dimensional accuracy to be achieved, if the required tolerances are high and cannot be achieved through the initial processing of the composite part. The finishing or secondary processes usually include deburring, trimming, drilling and boring. For deburring, a typical accuracy to be achieved is 1 mm; while accuracies of less than 0.1 mm are required for the drilling and boring operations. Conventional machining operations are difficult to perform on composite materials since their behavior is not only non-homogeneous and anisotropic, but also depends on diverse reinforcement and matrix properties, and the volume fraction of matrix and reinforcement (Teti, 2002). Delamination, debonding, and fiber breakage can lead to strength reduction of the finished part. The cutting tool alternately encounters matrix and reinforcement materials, whose response to machining can be entirely different. Thus, the machining of composite materials imposes special demands on the geometry and wear resistance of the cutting tools.

Laser machining offers advantages over conventional methods, such as high machining rates, no tool wear, no contact forces, and relatively high precision. The effectiveness of lasers depends on the thermal nature of the machining process, which does not involve any mechanical force applied to the material. Nevertheless, some difficulties arise because of the difference in the thermal properties of the various components of the composite. The effects of the laser beam on the material are generally connected with the following characteristics of the beam and material properties: power density, emission wavelength, interaction time, beam polarization, absorption coefficient at the given wavelength, melting and vaporization temperature, thermal conductivity and heat capacity. Fiber reinforced plastics (FRP) generally exhibit a high absorption of infrared rays typical of those produced by CO_2 lasers. Moreover, their thermal properties are such that the vaporization process occurs at much lower specific powers (10^3–10^5 W/cm²) than it does in metals. The thermal degradation mechanism, leading to material removal, is strongly influenced by the nature of the composite's components. The thermal properties of various polymeric resins that constitute 40–60% by volume of an FRP, are similar to each other. They are characterized by low values of thermal conductivity, thermal diffusivity, and decomposition heat. The differences in the thermal properties of the fibers and the matrix are extremely high for both graphite (carbon) and glass fibers, while such differences are low for the aramid ones. The energy required for the vaporization of the fibers is higher than that for the matrix; hence the laser power required for cutting FRP will be strongly dependent upon the kind of fibers and their volume fraction. With high laser beam densities concentrated on a very small focal point, the time to vaporize the FRP constituents is very short but, due to their different thermal properties, the fibers and the matrix can exhibit very different values of vaporization times. It is possible to observe two limit conditions under a constant specific power: both the fibers and the matrix exhibit only slightly different vaporization times (as in the case of polyester resin and aramid fibers) and therefore the composite's behavior can be considered homogeneous; or the fibers and the matrix may show very different vaporization times (graphite–resin and glass–resin) so that the resin reaches its vaporization temperature while the fibers are still unaffected. It has been shown that the heat penetration into the laminae and the extension of the heat-affected zone are larger by almost an order of magnitude for CFRPs due to the high conductivity of their fibers (Tagliaferri *et al.*, 1985).

10.2   Fundamentals of laser machining

10.2.1 Laser technology basics

Laser light differs from ordinary light in that it consists of photons that are all at the same frequency and phase (coherence). A laser's ability to produce coherent light is based upon the principle that the photons of light can stimulate the electrons of atoms; consequently, they emit photons of the exact same frequency. The possibility of this stimulation was postulated by Einstein in 1917, but the first working laser was not built until almost half a century later. Other significant characteristics of the laser beam include monochromaticity, diffraction and radiance. Monochromaticity implies that the range of frequencies emitted by the light source is small, and in most

Table 10.1 Types of lasers

Laser type		Wavelength (nm)	Typical performance
Solid	Ruby	694	PM, 5 W
	Nd:YAG	1064	PM, CW, 1–800 W
	Nd:glass	1064	PM, CW, 2 mW
Semiconductor	GaAs	800–900	PM, CW, 2–10 mW
Molecular	CO_2	10.6 μm	PM, CW, <15 kW
Ion	Ar	330–520	PM, CW, 1 W–5 kW
	Excimer	200–500	PM
Neutral gas	He-Ne	633	CW, 20 mW

cases it is only one spectral line. The laser light presents very low diffraction, and as a result, it can be projected over a distance with little divergence of the beam or loss in beam intensity. Finally, lasers have very high radiance, i.e. the amount of power per unit area emitted is high.

A typical laser machining system is complex, with the laser source being the most critical component. The basic mechanisms required to produce laser light include stimulation, amplification and population inversion; these requirements have been thoroughly described by Chryssolouris (1991). Based on the lasing medium used, the laser sources can be classified into solid, liquid or gas, either in a continuous wave (CW) or in pulsed mode (PM) of operation. Nowadays, although there is a variety of different laser sources (Table 10.1), the most common ones used for machining are CO_2, Nd:YAG and the excimer lasers. The efficiency of the laser machining depends on the ability of the material to absorb the energy emitted by the laser beam. The absorption of a material depends on the wavelength of the laser beam (Fig. 10.1), so for the laser machining of composites, the correct type of a laser source has to be selected.

Besides the laser source type, other factors that affect the machining process are the beam's power, the wavelength, the temporal and spatial mode and the focal spot size. A typical laser beam machining system is shown in Fig. 10.2.

10.2.2 Basics of laser machining

As indicated in the introduction, laser machining can replace mechanical removal methods in many industrial applications. The fact that laser machining is a non-contact, thermal process allows the machining of difficult-to-machine materials; however, the power densities introduced on the surface of the workpiece are high and can lead to elevated temperatures and thus to material decomposition when composites are being machined.

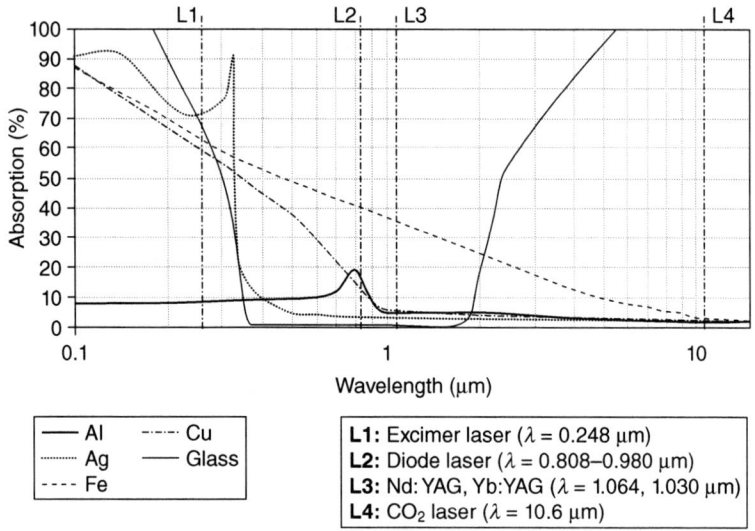

10.1 Absorptivity as a function of material and laser beam wavelength.

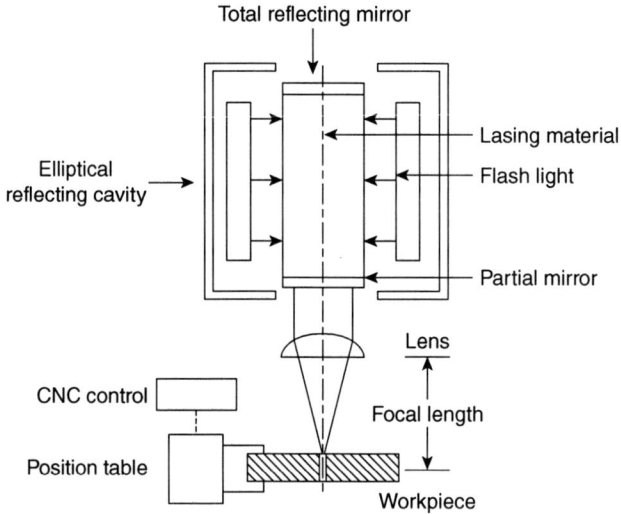

10.2 Laser beam machining system (El-Hofy, 2005).

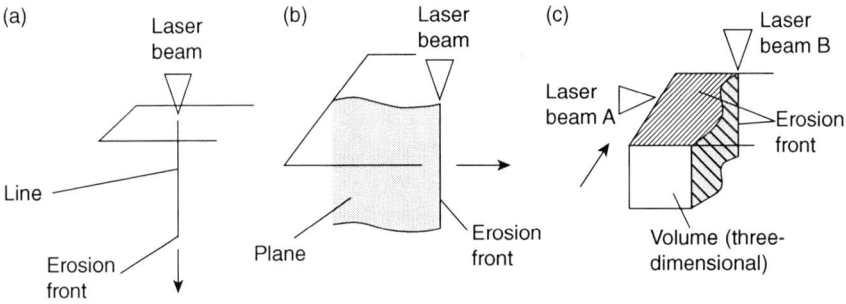

10.3 Schematic of laser machining: (a) one-dimensional (drilling), (b) two-dimensional (cutting) and (c) three-dimensional (milling) (Chryssolouris, 1991).

In general, laser machining can be divided into one-, two- and three-dimensional processes, as can be seen in Fig. 10.3. These categories can be identified by examining the shape and the kinematics of the erosion front, namely in the region of the workpiece where material removal takes place.

In the case of a one-dimensional process (drilling), the laser beam is stationary relative to the workpiece. The erosion front, located at the bottom of the drilled hole, propagates in the direction of the line source in order to remove the material. In the case of a two-dimensional process (cutting), the laser beam is in relative motion with respect to the workpiece. Material removal occurs by moving the line source in a perpendicular direction to the line direction, thereby forming a two-dimensional surface. The erosion front is located at the leading edge of the line source. For three-dimensional machining, two or more laser beams are used, and each beam forms a surface through relative motion with the workpiece. The erosion front of each surface is found at the leading edge of each laser beam. When the surfaces intersect, the three-dimensional volume bounded by the surfaces is removed.

Three major issues in any laser machining process are:

- material removal rate,
- dimensional accuracy,
- surface quality.

The material removal rate is governed, in each case, by the propagation speed of the erosion front. In laser drilling (a one-dimensional process), the material removal rate is determined by the speed that the erosion front moves towards the beam. In laser cutting (a two-dimensional process), the scanning velocity determines the rate at which the two-dimensional surface increases in the workpiece. In three-dimensional laser machining, the two-dimensional surfaces produced by two laser beams define a three-dimensional volume of the material to be removed. The speed at which these two surfaces propagate determines the time required for the removal of a material's given volume.

Dimensional accuracy is particularly determined by the taper of the hole for laser drilling, the kerf geometry for laser cutting, and the groove shape for three-dimensional machining. The surface quality of all laser machining processes is related to factors such as surface roughness, dross formation, and the heat-affected zone.

10.2.3 Laser machining operations

Drilling

Laser drilling involves a stationary laser beam that uses its high power density to melt or vaporize material from the workpiece. In principle, laser drilling is governed by an energy balance (Fig. 10.4) between the irradiating energy from the laser beam and the conduction heat into the workpiece, the energy losses to the environment, and the energy required for a phase change in the workpiece. Energy losses occur for a number of different reasons, some of which are (i) when the material is being heated above the required temperature for melting, (ii) plasma formation, (iii) the low absorptivity of the material, (iv) the convection of heat due to the use of gas jet, and so on. However, the advantages ensuing from the use of laser drilling instead of mechanical drilling have to do with (i) its thermal nature (which does not depend on the mechanical properties of the workpiece), (ii) the higher accuracies achieved and (iii) the higher machining rates. The

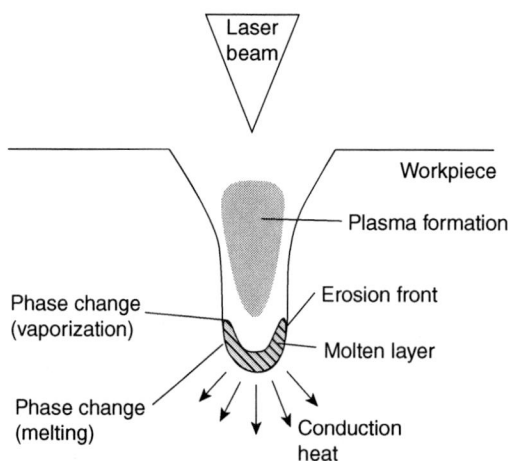

10.4 Laser drilling (Chryssolouris, 1991).

main limitations of the process comprise its inability to produce stepped diameter holes and the lack of accurate depth control.

Cutting

In the laser through-cutting process, a kerf is created by the relative motion between the laser beam and the workpiece surface. The physical mechanisms for material removal and energy losses (Fig. 10.5) are similar to those for drilling, where the incoming laser beam energy is balanced by the conduction heat, the energy for the melting or vaporization of material, and the heat losses to the environment.

The basic advantages of laser cutting are the higher material removal rates and the narrower kerf widths compared with those of mechanical cutting or shearing. On the other hand, the effectiveness of the process is reduced as the workpiece thickness is increased.

Similar to the laser cutting process is the laser grooving process, where a groove is produced by scanning a laser beam over the workpiece surface. In this case though, the laser beam does not penetrate the entire workpiece thickness. The physical mechanisms are similar to those of drilling and cutting. The major challenge of such a process has to do with controlling the groove depth.

3-D laser machining

Various concepts have been presented for laser machining bulk workpieces and for producing three-dimensional geometries. Early attempts were

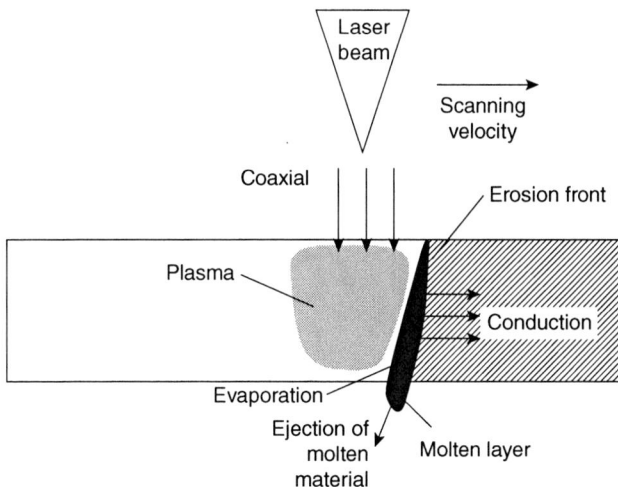

10.5 Laser cutting (Chryssolouris, 1991).

reported in the 1990s, when two intersecting beams were used for removing a volume of material. This could be achieved by the so called laser milling process, during which two laser beams were positioned at oblique angles from the workpiece to create converging grooves in it (Chryssolouris *et al.*, 1990; Chryssolouris, 1991; Tsoukantas *et al.*, 2002). In a similar way, laser turning can be achieved by the removal of the ring or helix volume of the material. More recently, fiber lasers mounted on robot arms have been used commercially.

10.3   Laser machining of metal matrix composites (MMCs)

Metal matrix composites (MMCs) are composed of a metal matrix phase and a reinforcing one. Based on the type of the reinforcing phase they can be divided into three classes: particle reinforced composites, long-fiber rein-forced composites and short-fiber reinforced composites. In comparison with their base metals, they offer a number of advantages, such as higher specific strengths and moduli, higher elevated temperature resistance, lower coefficients of thermal expansion, and, in some cases, better wear resistance. MMCs also have some advantages compared with polymer matrix compos-ites, including higher matrix dependent strength and moduli, higher ele-vated temperature resistance, no moisture absorption, higher electrical and thermal conductivities, and non-flammability. The machining of such ma-terials though, is a difficult task. Machining with the use of conventional processes such as turning and drilling, generally results in excessive tool wear, due to the presence of hard particles resulting in a very abrasive nature for these materials (Teti, 2002). Consequently, laser cutting is being increasingly used for the machining of MMCs.

Quality characteristics of interest in laser machining include material removal rate (MRR), machined geometry (kerf width, hole diameter, taper), surface quality (surface roughness, surface morphology), metallurgical characteristics (recast layer, heat-affected zone, dross inclusion) and mechanical properties (hardness, strength, etc.) (Fig. 10.6).

Kerf width

One of the advantages of laser cutting is the narrow kerf width. Mueller and Monaghan (2000) have experimentally shown that CO_2 laser cutting of particle-reinforced metal matrix composites can produce cuts with a narrow kerf width (less than 0.4 mm). With high laser powers and lower feed rates, the cuts have a wider underside, probably due to side burning.

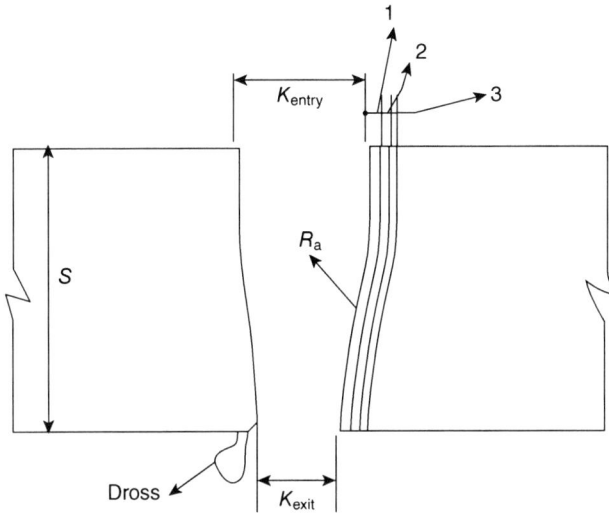

10.6 Cut quality attributes of interest for laser cutting. K_{entry}: kerf width at entry side; K_{exit}: kerf width at exit side; R_a: surface roughness; *S*: thickness of material; 1: oxidized layer; 2: recast layer; 3: heat affected zone (HAZ) (Li *et al.*, 2004).

Surface quality

The surface quality of a laser-made cut is usually relatively poor, and it is characterized by its surface roughness. It is related to the appearance of typical surface striations. These striations occur as a result of the intermittent flow of the molten material during cutting. The surface quality of the laser cut depends on a number of different parameters. Goeke and Emmelmann (2010) identified the most significant ones and depicted them on a cause–effect diagram for the case of a non-metallic composite (Fig. 10.7), however the applicability of such a diagram is more general and can be used for all kind of composites. A similar cause–effect diagram was also produced by Mathew *et al.* (1999) for the case of heat affected zone and taper. Mueller and Monaghan (2000) found that by decreasing the cutting feed rate, a smoother surface could be obtained. In the case of two different particle-reinforced metal matrix composites, it was experimentally proven that with a higher feed rate, laser cutting tends to be intermittent, thus resulting in a rougher surface. A smoother surface can also be obtained by machining at a higher output power. The poor surface quality is also due to the damage produced by excessive heating of the material's sub-surface layer.

Slocombe and Li (2000) investigated the effect of the laser power, the scanning speed, and the assist gas on the surface quality. They used a

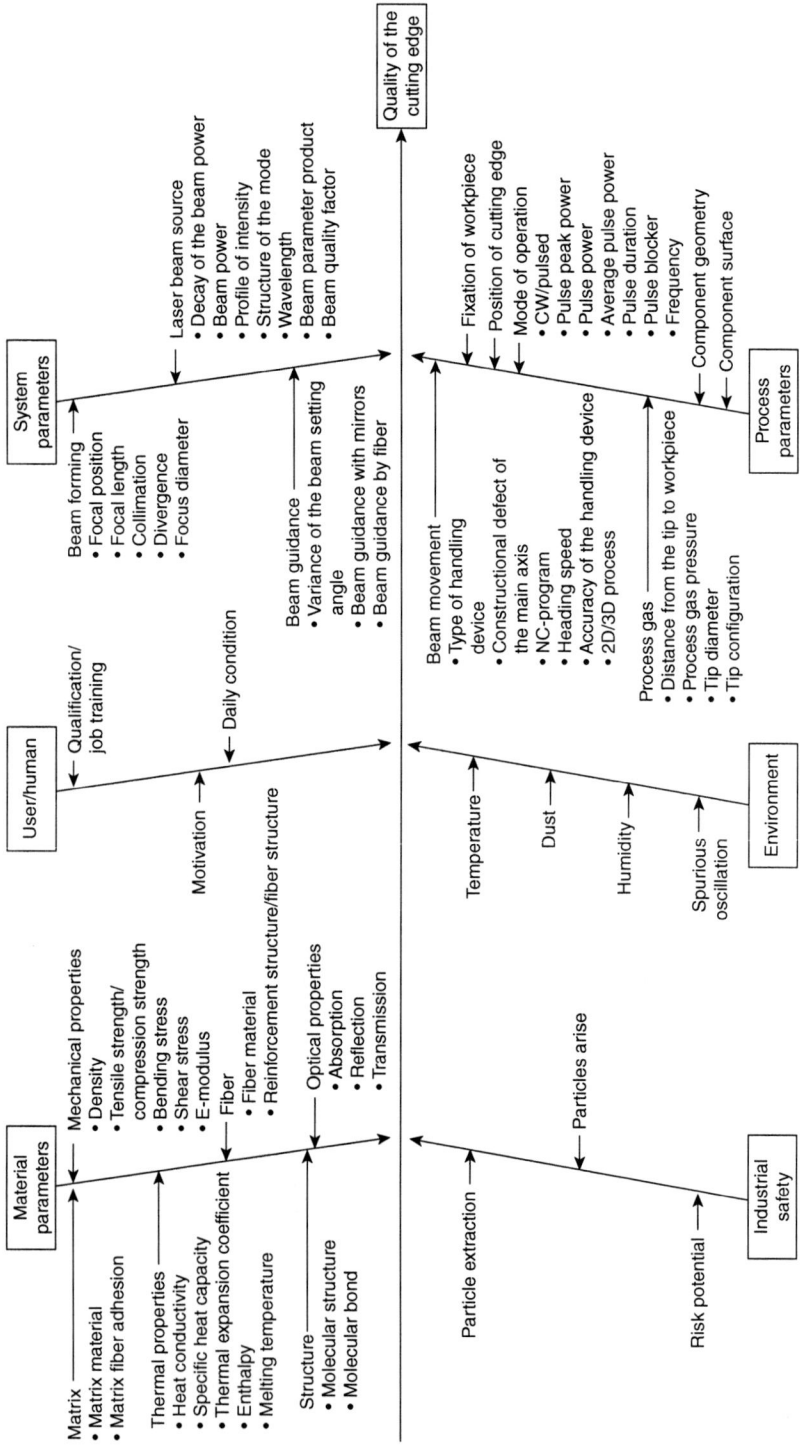

10.7 Cause–effect diagram of laser beam cutting of a composite (Goeke and Emmelmann, 2010).

Q-switched Nd:YAG laser for cutting metallic–polymer blends. Slocombe *et al.* (2000), focused on the effect of the powder geometry. Regular metal powders in a polymer matrix could not be laser machined without the formation of dross due to the mechanical locking of irregular powders when compacted. In machining metal/polymer composites, it becomes evident that the use of a laser beam is more beneficial when using spherical metal powders rather than irregular metal ones.

Productivity issues

Laser machining offers significant productivity advantages for rough cut-off applications. Mueller and Monaghan (2000) have experimentally shown that laser cutting is suitable for high feed rates (up to 3000 mm/min), when compared with those obtainable by electrical discharge machining and abrasive water jet cutting. They concluded that by reinforcing an aluminum matrix with SiC ceramic particles, the machinability of the composite is improved over that of raw aluminum, due to the reduction in the optical reflectively of the material. On the other hand, the material removal rate can be further increased if higher energy levels are selected during the pulsed Nd:YAG laser cutting (Lau *et al.*, 1995).

Slocombe and Li (2000), experimentally investigated the effect of the scanning speed on the groove depth of different composites. As can be seen in Fig. 10.8, for a fixed laser power, as the grove depth is increased, the

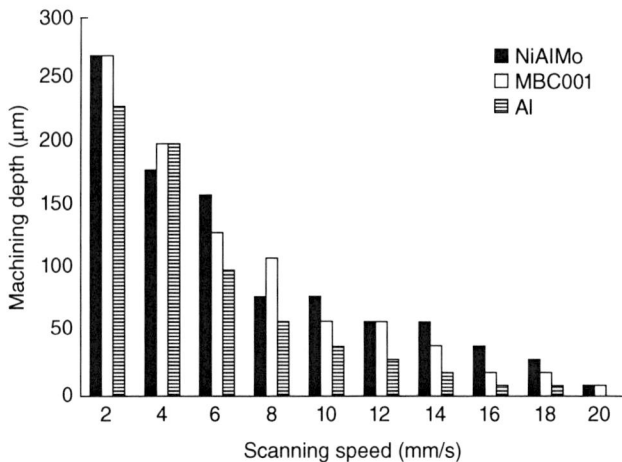

10.8 Relationship between scanning speed and laser machining depth for three different composite samples, for a fixed laser power of 70 W and frequency of 20 kHz (Slocombe and Li, 2000).

scanning speed is lowered. It has also been shown that the material's composition affects the groove depth. This is due to the difference in the laser absorption and thermal properties of the materials, with Al being highly reflective and conductive. The laser energy is reflected back and the heat is dissipated faster, thus reducing the penetration depth.

10.3.1 Laser assisted machining (LAM) of MMCs

As indicated, the machining of MMCs results in excessive tool wear due to the presence of hard particles, resulting in the very abrasive nature of this material (Teti, 2002). On the other hand, the laser machining of MMCs cannot as yet provide clean cuts with the quality characteristics required. An innovative solution is the so called laser-assisted machining (LAM) process, which uses a laser beam as a heat source, focused on the unmachined section of the workpiece, directly in front of the cutting tool (Fig. 10.9). The heat softens the material's surface layer without the composite melting or subliming, so that ductile deformation rather than brittle deformation occurs during processing. As a result, the cutting forces and the cutting tool wear are effectively reduced, and hence the tool life is increased and the surface quality improved. Wang *et al.* (2002) investigated the laser-assisted cutting of the Al_2O_3 particle reinforced aluminum matrix composite and achieved a 30 to 50% reduction in the cutting forces and a 20 to 30% reduction in the tool wear (Fig. 10.10).

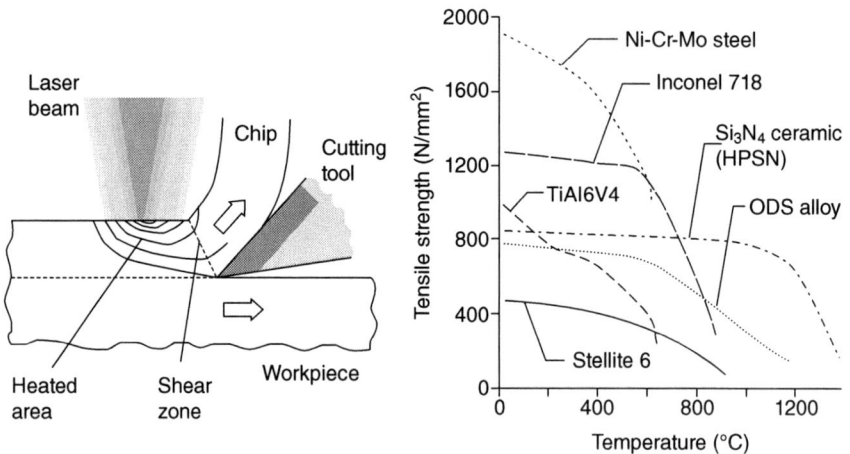

10.9 Principle of laser assisted machining (Pollmann and Becker, 2001).

(a)

(b)

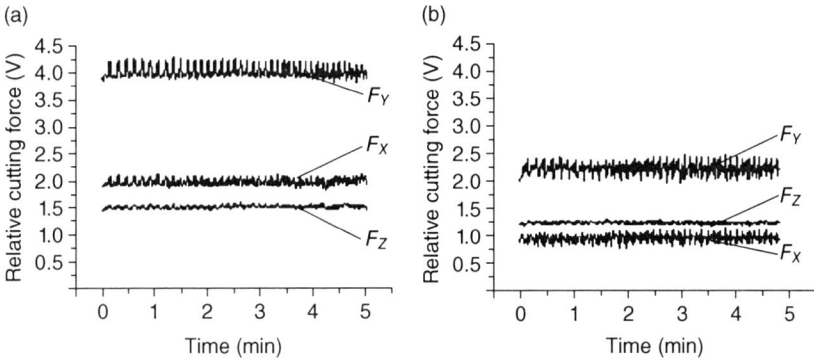

10.10 Comparison cutting force of (a) conventional cutting and (b) laser-assisted hot cutting (Wang et al., 2002). $V = 25$ m/min, $a_p = 0.3$ mm, $f = 0.1$ mm/rev. F_X = axial force, F_Y = radial force, F_Z = main cutting force.

10.4   Laser machining of non-metallic composites

Non-metallic composites, and especially fiber-reinforced polymer composites such as aramid (AFRP), glass (GFRP), carbon (CFRP), and boron fiber composites can be found in various high tech applications. One of the first experimental studies concerning the use of lasers for processing non-metallic composites was presented by Koenig et al. (1985), using GFRP, AFRP and CFRP. Since 1980, laser processing of non-metallic composites, and especially CFRP composites, has been the topic of several research projects (Herzog et al., 2008; Pan and Hocheng, 1996; Cenna and Mathew, 2001).

Laser source type significance

One of the most important factors, as indicated also in the introduction of this chapter, is the laser source type. Goeke and Emmelmann (2010), used both CO_2 and fiber laser beam sources for cutting CFRP specimens. The significantly higher absorption of 10.6 µm-wavelength of a CO_2 laser was proved to pose advantages compared with those of a fiber laser with a wavelength of 1.07 mm (Fig. 10.11), resulting in higher cutting depts. This was accredited to the two different mechanisms of pyrolysis of the polymer and the fiber. When using fiber lasers, the radiation is absorbed by the material surface and heat is transmitted to the material, whilst on the other hand, the radiation of CO_2 lasers is absorbed by the material's volume. The emitted laser radiation is therefore absorbed by polymer chains and is converted into heat energy by vibration excitation.

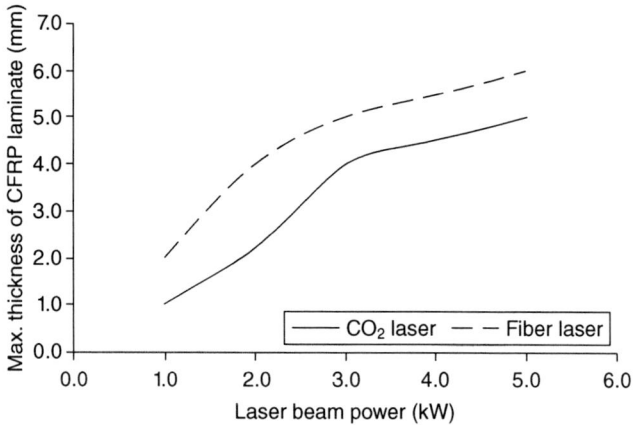

10.11 Comparison of maximum CFRP laminate thickness that could be processed reliably by means of fiber or CO_2 laser with a constant feeding rate of 5.0 m/min (Goeke and Emmelmann, 2010).

Lau *et al.* (1995) compared the effectiveness of Nd:YAG and excimer laser sources for processing different composites. Although the excimer laser can produce a cleaner cut surface with less material damage, its cutting rate needs to be improved, due to the unavailability of high power machines.

Herzog *et al.* (2008) also dealt with the issue of the effect of the laser source on the cutting of CRFPs. Three different types of laser source (CO_2, disk and pulsed Nd:YAG lasers) were used for the experimental investigation of the cutting quality. In Fig. 10.12, cross-sections of the laser-cut samples, as well as two reference specimens from abrasive water-jet cutting and milling are shown. It is evident that the abrasive water-jet cutting produces high aesthetic quality; no delamination or free fiber ends are observed. The milled specimen also shows an overall high optical quality of the cut. In all laser-cut specimens, the heat affected zone (HAZ) is shown. When comparing the laser-cut samples with each other, the pulsed Nd:YAG laser obviously produces the smallest HAZ. The kerf widths of the laser cut samples are all considerably lower than those of the water-jet and milling samples.

Surface quality

The cutting quality, in terms of the presence of charred material, the extension of delamination, and the slope of the kerf surface, depends on the scanning velocity, the laser power, and the assist gas. A recent study has

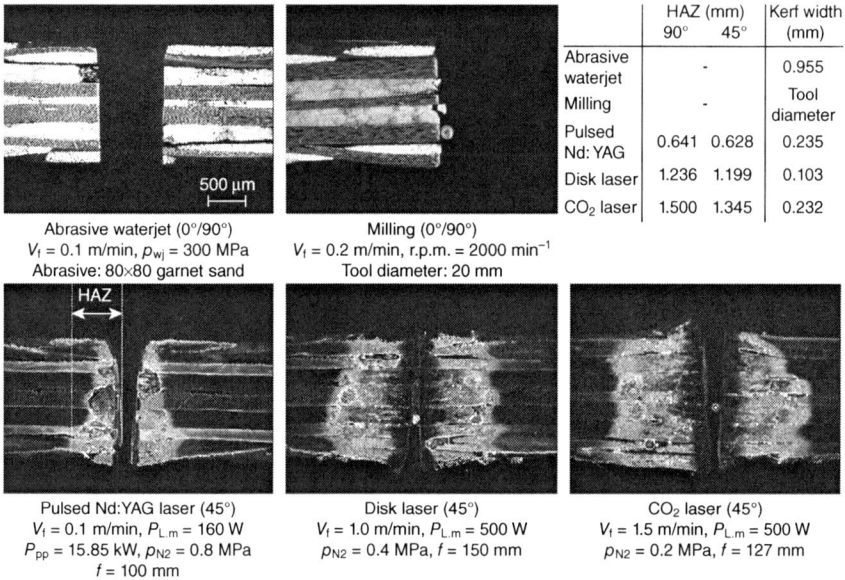

	HAZ (mm) 90°	HAZ (mm) 45°	Kerf width (mm)
Abrasive waterjet	-		0.955
Milling	-		Tool diameter
Pulsed Nd:YAG	0.641	0.628	0.235
Disk laser	1.236	1.199	0.103
CO_2 laser	1.500	1.345	0.232

Abrasive waterjet (0°/90°)
$V_f = 0.1$ m/min, $p_{wj} = 300$ MPa
Abrasive: 80×80 garnet sand

Milling (0°/90°)
$V_f = 0.2$ m/min, r.p.m. = 2000 min^{-1}
Tool diameter: 20 mm

Pulsed Nd:YAG laser (45°)
$V_f = 0.1$ m/min, $P_{L.m} = 160$ W
$P_{pp} = 15.85$ kW, $p_{N2} = 0.8$ MPa
$f = 100$ mm

Disk laser (45°)
$V_f = 1.0$ m/min, $P_{L.m} = 500$ W
$p_{N2} = 0.4$ MPa, $f = 150$ mm

CO_2 laser (45°)
$V_f = 1.5$ m/min, $P_{L.m} = 500$ W
$p_{N2} = 0.2$ MPa, $f = 127$ mm

10.12 Cross-sections of samples processed using different laser beam types (Herzog *et al.*, 2008).

identified the major parameters that affect the quality of the cutting edge in the case of laser processing of CFRP (Goeke and Emmelmann, 2010). In Fig. 10.7, these parameters are presented in a cause–effect diagram.

The most significant factors affecting the surface quality are the material properties, and especially the thermal anisotropy of the composite. The difference in thermal properties of the carbon fiber and the resin matrix is responsible for the poor quality of the laser-cut surfaces in CFRP composites, whereas the best results are achieved with AFRP composites due to the similar polymeric nature of the fiber and matrix (Caprino and Tagliafferri, 1988).

The composite material anisotropy is also the major issue in laser grooving. Chryssolouris *et al.* (1988) investigated the laser grooving of carbon/ Teflon, glass/Teflon and glass/polyester materials. The groove depth, groove shape, and heat-affected zone depend on the laser beam direction relative to the fibers' orientation. In most grooving applications, the laser beam is directed perpendicularly to the fibers' orientation since the fibers are usually aligned in the plane perpendicularly to the workpiece thickness; however, in some cases, such as in laser turning, the laser beam can be directed in an arbitrary angle relatively to the fibers' orientation. Heat losses to the environment and matrix decomposition result in groove depth

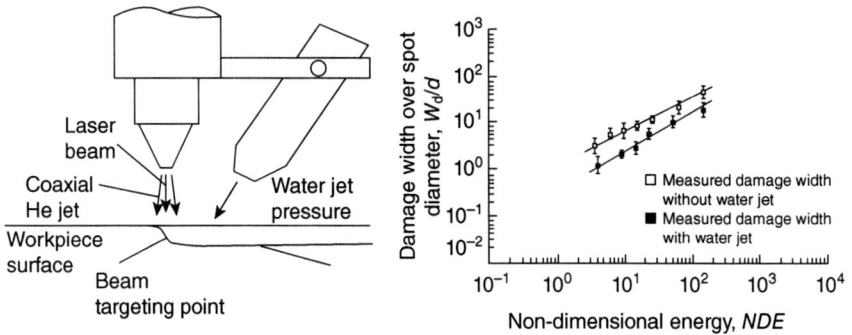

10.13 Water jet set-up and comparison of damage width for laser grooving with and without water jet (Chryssolouris *et al.*, 1993).

reduction, especially at high energy densities. The thermal damage can be reduced by using a cooling medium in tandem with the laser beam. The use of a water jet has been investigated by Chryssolouris *et al.* (1993), and resulted in a reduction of up to 70% in the heat affected zone dimensions (Fig. 10.13).

A coaxial inert gas jet (He and Ar) can be used in the cutting process to minimize charring. However, experimental studies by Lau and Lee (1992) showed that compressed air removed more material in comparison with argon inert gas during the laser cutting of carbon fiber composites.

Kerf width

The kerf width is strongly dependent on the laser processing parameters. It has been shown that the kerf width, at the inlet and exit, decreases with increased cutting speed, and the inlet kerf width is less sensitive to cutting speed (Di Ilio *et al.*, 1990; Tagliaferri *et al.*, 1985). On the other hand, the kerf width is less sensitive to power changes (Di Ilio *et al.*, 1990). It has been proven experimentally though, that increase of the laser power increases the kerf width (Al-Sulaiman *et al.*, 2006; Lum *et al.*, 2000) for the CO_2 laser cutting of different fiber composites (Fig. 10.14).

In the case of laser drilling, it has been experimentally proven that the setting of the focal plane position, with regard to the workpiece surface, affects the hole taper. Tuersley *et al.* (1998) experimentally investigated the Nd:YAG laser drilling of fiber-glass composites for minimizing the hole taper. They showed that the most significant factor was the material thickness. The CO_2 and Nd:YAG laser drilling of polyester foils and glass fiber reinforced epoxy laminates gave a larger hole diameter at an increased laser power (Vitez, 2000).

(a)

Power = 1200 W

Power = 1800 W

(b)

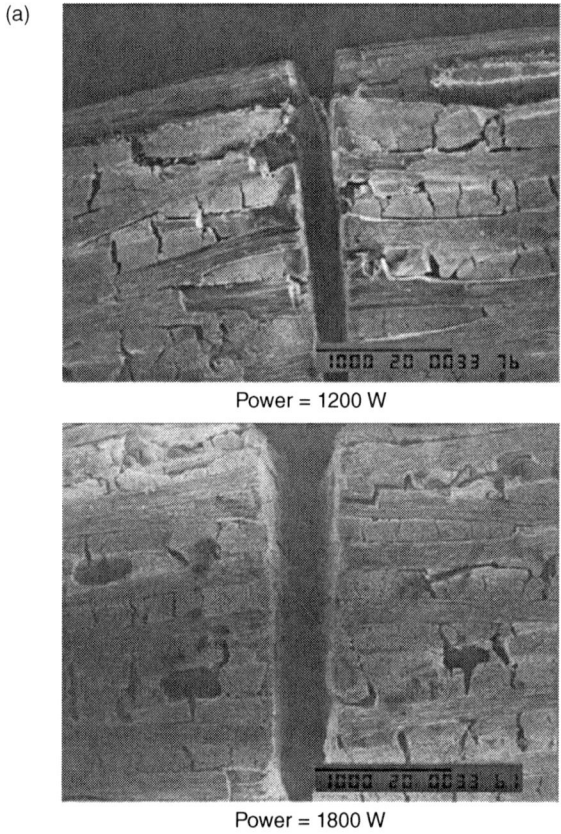

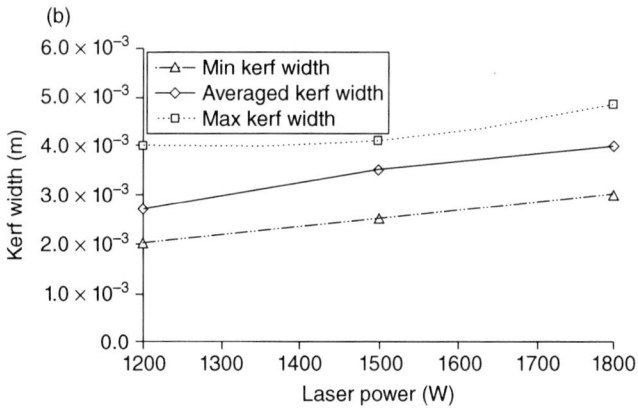

10.14 (a) Micrographs of laser cut cross-sections for two laser power settings. (b) Variation of maximum, averaged and minimum kerf width sizes with laser power intensity (Al-Sulaiman et al., 2006).

Theoretical simulation of laser processing of composites

A characteristic of laser cutting is the relation between cutting efficiency and beam scanning direction relative to fiber orientation. Due to material anisotropy, the thermal response of a workpiece to the laser beam depends on the cutting direction. This effect is most apparent in graphite-fiber composites, where the thermal properties between the constituent materials are very different. Furthermore, nonuniform material loss is a function of the matrix and fiber materials selected (aramid versus glass and graphite materials) and depends strongly on the thermal conductivity of the constituent materials. Aramid fibers have thermal characteristics similar to those of the polymer matrix materials, so the material behavior during the beam/material interaction is similar to that of homogeneous materials.

A number of theoretical attempts have been reported in the literature for predicting the characteristics of the cut. Chryssolouris (1991) has presented a number of theoretical simulation models for the estimation of the temperature field and the geometry of the drill, cut and groove during laser processing. The laser grooving process for composite materials has been simulated theoretically for the prediction of the maximum groove depth and heat affected zone (HAZ) by Sheng and Chryssolouris (1993 and 1995). The model they presented took into consideration the anisotropy of the materials' properties and was verified experimentally on graphite/vinylester composites. In Fig. 10.15, it is obvious that the radial grooving results in larger groove depths than does the axial grooving, for a corresponding energy density.

Caprino and Tagliaferri (1988) have developed a thermal model for the prediction of the maximum laser cutting speed during the laser cutting of fiber reinforced polymer composites. They assumed that no melting takes place in the HAZ, and that material is removed by an instantaneous evapo-

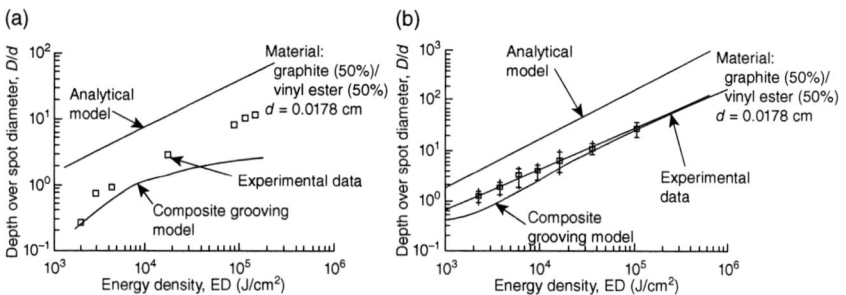

10.15 Depth vs. energy density for (a) axial and (b) radial laser grooving from model and experiments (Sheng and Chryssolouris, 1995).

rative process when the vaporization temperature is reached. Cheng *et al.* (1998) have developed a 3-D finite-difference-based heat-flow model for the prediction of a hole's shape and size in carbon fiber composites during laser drilling. Cena and Mathew (2001) also presented a theoretical model for the prediction of the kerf width at the inlet, at the exit, and the angle of the cut surfaces in the case of cutting of GFRP and AFRP. In the case of AFRP, the model presented good agreement with the experimental results, whereas in the case of GFRP, the model had the tendency to underestimate the results.

10.5   Conclusions

In this chapter, the state of the art of laser machining of both metal and non-metallic matrix composite materials, has been presented. Laser machining of composites presents a number of advantages, compared with other competing technologies such as abrasive jet cutting and conventional cutting. However, there is much space for further improving the surface quality of the laser cuts. For this reason, a number of ongoing projects exist both in the US and in Europe for improving the performance of the laser machining of composites. Furthermore, a theoretical simulation of the process allows profound understanding of the physics of the process.

10.6   References

Al-Sulaiman FA, Yilbas BS and Ahsan M (2006), 'CO_2 laser cutting of a carbon/carbon multi-lamelled plain-weave structure', *Journal of Material Processing Technology*, **173**, 345–351.

Caprino G and Tagliaferri V (1988), 'Maximum cutting speed in laser cutting of fibre reinforced plastics', *International Journal of Machine Tools and Manufacturing*, **28**, 389–398.

Cenna A and Mathew P (2001), 'Analysis and prediction of laser cutting parameters of fibre reinforced (FRP) composite materials', *Journal of Machine Tools and Manufacturing*, **42**, 105–113.

Cheng CF, Tsui YC and Clyne TW (1998), 'Application of a 3-D heat flow model to treat laser drilling of carbon fibre composites', *Acta Metallurgica et Materialia*, **46**, 4273–4285.

Chryssolouris G, Sheng P and Choi WC (1988), 'Investigation of laser grooving for composite materials', *CIRP Annals – Manufacturing Technology*, **37**, 61–164.

Chryssolouris G, Sheng P and Choi WC (1990), 'Three dimensional laser machining of composite materials", *Journal of Engineering Materials and Technology ASME*, **112**, 387–392.

Chryssolouris G (1991), *Laser Machining: Theory and Practice*, Springer-Verlag, New York.

Chryssolouris G, Sheng P and Anastasia N (1993), 'Laser grooving of composite materials with the aid of a water jet', *Journal of Engineering for Industry ASME*, **115**, 62–72.

Di Ilio A, Tagliaferri A and Veniali F (1990), 'Machining parameters and cut quality in laser cutting of aramid fibre reinforced plastics', *Materials and Manufacturing Processes*, **5**, 591–608.

El-Hofy H. (2005), *Advanced Machining Processes*, McGraw Hill.

Goeke A and Emmelmann C (2010), 'Influence of laser cutting parameters on CFRP part quality', *Physics Procedia*, **5**, 253–258.

Herzog D, Jaeschke P, Meier O and Haferkamp H (2008), 'Investigation on the thermal effect caused by laser cutting with respect to static strength of CRFP', *Journal of Machine Tools and Manufacturing*, **48**, 1464–1473.

Koenig W, Wulf Ch, Grass P and Willerscheid H (1985), 'Machining of fibre rein-forced plastics', *CIRP Annals – Manufacturing Technology*, **34**, 537–548.

Lau WS and Lee WB (1992), 'Pulsed Nd:YAG laser cutting of carbon fibre compos-ite materials', *CIRP Annals – Manufacturing Technology*, **39**, 179–182.

Lau WS, Yue TM, Lee TC and Lee WB (1995), 'Unconventional machining of com-posite materials', *Journal of Materials Processing Technology*, **48**, 199–205.

Li L and Achara A (2004), 'Chemical assisted laser machining for the minimization of recast and heat-affected zone', *CIRP Annals – Manufacturing Technology*, **53**, 175–178.

Lum KCP, Ng SL and Black I (2000), 'CO_2 laser cutting of MDF 1: Determination of process parameter settings', *Optics and Laser Technology*, **32**, 67–76.

Mathew J, Goswami GL, Ramakrishnan N and Naik NK (1999), 'Parametric studies on pulsed Nd:YAG laser cutting of carbon fibre reinforced plastic composites', *Journal of Materials Processing Technology*, **89–90**, 198–203.

Mueller F and Monaghan J (2000), 'Non-conventional machining of particle rein-forced metal matrix composite', *International Journal of Machine Tools and Man-ufacture*, **40**, 1351–1366.

Pan C and Hocheng H (1996), 'The anisotropic heat affected zone in the laser groov-ing of fiber-reinforced composite material', *Journal of Material Processing Tech-nology*, **62**, 54–60.

Pollmann W and Becker W (2001), 'Laser-assisted manufacturing pacemaking tech-nology in the 3rd millennium', *Proc. of the LANE 2001*, 39–54.

Sheng P and Chryssolouris G (1993), 'Comparison of surface quality improvement techniques for laser grooving of composite materials', *Journal of Manufacturing Science and Engineering ASME*, **64**, 795–801.

Sheng P and Chryssolouris G (1995), 'Theoretical model of laser grooving for com-posite materials', *Journal of Composite Materials*, **29**, 96–112.

Slocombe A and Li (2000), 'Laser ablation machining of metal/polymer composite materials', *Applied Surface Science*, **154–155**, 617–621.

Slocombe A, Taufik A and Li L (2000), 'Diode laser ablation machining of 316L stainless steel powder/polymer composite material: Effect of powder geometry', *Applied Surface Science*, **168**, 17–20.

Tagliaferri V, Di Ilio A and Crivelli Visconti I (1985), 'Laser cutting of fibre-reinforced polyesters', *Composites*, **16**, 317–325.

Teti R (2002), 'Machining of Composite Materials', *CIRP Annals – Manufacturing Technology*, **51**, 611–634.

Tsoukantas G, Salonitis K, Stavropoulos P and Chryssolouris G (2002), 'An over-view of 3D laser materials processing concepts', *Proceedings of SPIE, 3rd GR-I International Conference on New Laser Technologies and Applications*, **51715**, 224–228.

Tuersley IP, Hoult TP, Tony P and Pashby IR (1998), 'Nd-YAG laser machining of SiC fibre/borosilicate glass composites. Part II. The effect of process variables', *Composites Part A*, **29A**, 955–964.

Vitez ZI (2000), 'Laser processing of adhesives and polymeric materials for micro-electronics packaging applications', *Proceedings of the 4th IEEE International Conference on Adhesive Joining and Coating Technology in Electronics Manufacturing*, Espoo, Finland, 289–295.

Wang Y, Yang LJ and Wang NJ (2002), 'An investigation of laser-assisted machining of Al_2O_3 particle reinforced aluminum matrix composite', *Journal of Materials Processing Technology*, **129**, 268–272.

11

Laser machining of fibre-reinforced polymeric composite materials

R. NEGARESTANI and L. LI,
The University of Manchester, UK

Abstract: Fibre-reinforced polymeric composites (FRPs) are heterogeneous and anisotropic structures. These properties pose significant challenges in the selection of their machining processes, tools and process parameters to minimise defects and maintain high processing rates. This chapter introduces the fundamentals of laser-based machining science and technologies for processing FRP, particularly carbon fibre-reinforced plastics (CFRPs). The effects of laser pulse length, wavelength, power density, scanning speed and process gas on the material behaviour are included and a comparison with other machining techniques such as mechanical and water jet machining is given.

Key words: laser, machining, cutting, carbon fibre-reinforced plastics (CFRPs), fibre-reinforced polymeric composites (FRPs), composite, carbon, fibre.

11.1 Introduction

Machining of composite materials is dominated by mechanical processes. Usually, the reinforcing fibres such as carbon, glass, boron, alumina and silicon carbide are highly abrasive and hard. Therefore, special cutting tool materials and geometry must be used to minimise tool wear. Around 60% of part rejections are due to machining errors.[1]

Machining challenges are particularly experienced in the case of CFRPs. This is due to the large difference between mechanical and thermal properties of the constituents. Machining defects, including inter- and intra-laminar delamination, fibre pullout, and poor surface quality, often occur. Additionally, an inability to meet dimensional tolerances may require rework, or even part rejections. Proper selection of tool material, grade and geometry, as well as particular process data are required for high quality mechanical machining of CFRPs. These sometimes require specific equipment for a particular application which can be a costly practice.[2,3] The cost for machining CFRP can hence be as high as half of the total final cost.[4] Therefore, alternative machining processes are being studied, including abrasive water jet (AWJ) machining, laser machining, electrical discharge

288

machining (EDM) (applicable to CFRPs only, since carbon fibres are electrically conductive) and ultrasonic machining (USM). Lasers are found to be of more use in the high-speed cutting of thin laminates and AWJ in the cutting of thicker polymer laminates.[5]

Abrasive waterjet has been shown to be an effective method for composite machining. The low thermal and mechanical forces of AWJ machining are ideal for FRPs. Process parameters, including supply pressure, standoff distance, abrasive size, water flow rate, and cutting speed, may be adjusted to achieve the desired cut surface quality and kerf taper.[6] Since AWJ involves low thermal and mechanical forces, it is ideal for composite materials. In the case of CFRPs, generally, high jet pressures, low standoff distances, low to medium transverse speeds, small abrasive particle size and small nozzle diameter are used to reduce taper angle and surface roughness and irregularities including delamination and waviness.[6–11] New techniques including cutting with forward angling the jet in the cutting plane,[12] multiple-pass cutting,[9] and controlled nozzle oscillation[13–16] have also been used in AWJ to enhance the cutting performance, such as the depth of cut and surface finish. However, delamination and trapping of abrasives in the composite laminates are issues of concern. Other issues related to AWJ machining are the noise level and abrasive slurry generated during the process, which are potential health hazards to the operators and the environment.

Lasers as non-contacting and non-abrasive machining tools exhibit unique advantages in materials processing, eliminating tool wear, vibrations and cutting forces. Laser cutting can be easily automated and can be performed at high cutting speed. Therefore, lasers have often been proposed as a promising tool for the machining of composites over the past 30 years. The challenges to laser processing are to minimise or eliminate thermal damage and maintain high processing speed. Defects, such as the heat affected zone (HAZ), charring, resin recession and delamination due to intense thermal effects, are major obstacles for industrial applications of laser machining of CFRP composites.[17] Quality improvements achieved in laser cutting of CFRPs using techniques such as an additional coolant (water)[18] or cryogenic assist gas,[19,20] pulsed and/or UV beam processing[21,22] show the potential in laser machining of CFRPs.

Recent developments in laser materials processing technology have opened new opportunities. These include availability of high power, high beam quality, and short and ultra short pulsed systems, as well as modern high precision CNC stages and galvanometer mirror scanner systems that allow rapid laser–material interaction to improve process productivity, quality and accuracy. The visible light and near infrared wavelength laser beams can also be transmitted through fibre optics and manipulated by industrial robots (to distances over 200 m from the laser unit). A summary of the desirability of different machining processes for CFRPs is given in Table 11.1.[23] As can be seen, laser machining offers some desirability of characteristics. Compared with the abrasive waterjet process, lasers can achieve narrower kerf widths and higher cutting speeds while offering capabilities of cutting near the edges for fibre-reinforced polymer composites (FRPs).[24–26] Laser cutting can also be utilised to trepan holes in CFRPs (where the required hole diameter is larger than the beam spot diameter). However, as a thermal process and owing to the heterogeneous properties of FRPs, laser machining can introduce thermal damage such as delamination, matrix recession and tapered cut kerf.[27]

Since FRPs are heterogeneous, their constituents have different thermal properties, as shown in Table 11.2. Material removal depends on the properties of the constituents. Most thermoplastic matrix materials cutting by laser beam is based on the shearing of a localised melt using a gas jet. Thermoset resins are removed by chemical degradation which requires higher temperature and energy, when compared to thermoplastics. Reinforcing fibres generally require higher temperatures and energy to vaporize, when compared to the resin. In addition, the anisotropy of FRPs generates non-uniform thermal gradients inside the laminate.[28] These are the main factors responsible for the formation of a large HAZ, cavities, matrix recession and delamination in nanosecond, pulsed to continuous wave, visible to infrared wavelength, laser cutting of FRPs. These defects deteriorate the performance of composites in both static and fatigue conditions. Laser beam and assist gas characteristics, operating conditions and material properties that influence these defects are presented as a cause–effect diagram in Fig. 11.1.

11.2   Effect of laser and process gas

11.2.1 Power density and interaction time

Laser power density and interaction time show major effects on the extent of thermal damages in laser processing.[30] The relationship between the vaporisation of the common constituents of FRPs and the beam power density versus interaction time is illustrated in Fig. 11.2.[27] When fibres and matrix exhibit only slightly different vaporisation times (e.g. polyester resin and aramid fibre) the composite thermal behaviour can be microscopically homogeneous. Therefore, aramid fibre-reinforced polymer composites (AFRPs) behave better under laser cutting.[28] Up to 9.5 mm thick AFRP laminates are cut by a laser[5] and laser machining rates of 2.5 times the mechanical cutting speeds can be achieved.[25] Glass and graphite fibres show vaporisation times much higher than the matrix. Therefore, fibres remain unchanged while the matrix reaches its vaporisation temperature. This, together with the high thermal conductivity of carbon fibres, leads to poorer

Table 11.1 Desirability characteristics of machining processes for CFRP composite material[23]

	Machining processes					
	Mechanical	EDM	Wire EDM	USM	Abrasive waterjet	Laser
Capital cost	5	2	3	1	6	4
Running cost	4	2	3	1	6	5
Tool/consumables cost	3	1	2	4	5	6
Machining cost	4	2	3	1	6	5
Process time	5	1	2	3	6	4
Compactness and mobility	5	1	2	3	4	6
Total	26	9	15	13	33	30
Advantages	Good surface finish	Cutting of complex parts	Cutting of curved surfaces	Very good surface, minimal damage, reliable products	No thermal damage, thick section cutting	Narrow kerf, high processing speed, high flexibility and automation capability
Drawbacks	Tool wear, delamination and fibre pullout	Expensive tooling and equipment, low MRR, some thermal damage	Expensive tooling, low MRR, some thermal damage	Tool wear, very low MRR	Roundness of the cut edge, noise level, waste water, large workshop	Thermal damage, fumes and dust generation

NB. Scale: 1, least desirable; 6, most desirable (i.e. desirability increases as number increases).

Table 11.2 Typical thermal properties of selected constituents of FRP composites. L: longitudinal, i.e. along fibre length; T: transverse, i.e. in radial direction[5,19,21,27-29]

	Density (g/cm³)	Decomposition temperature (K)	Coefficient of thermal expansion (m/m.K)	Thermal conductivity (W/m.K)	Specific heat capacity (J/kg.K)	Thermal diffusivity (cm²/s) × 10^{-3}	Heat of vaporization (J/g) × 10^3
Aramid fibre	1.44	820	−2 L 59 T	0.05	1420	0.24	4
Glass fibre	2.55	2570	5	1	850	4.61	31
Carbon fibre	1.85	4000	−0.5 L 5 T	50	710	380	45
Polyester	1.25	670	80	0.2	1200	1.33	1
Epoxy	1.20	700	65	0.1	1100	0.76	1.1
Vinyl ester	1.25	650	75.4	0.2	1200	1.33	1

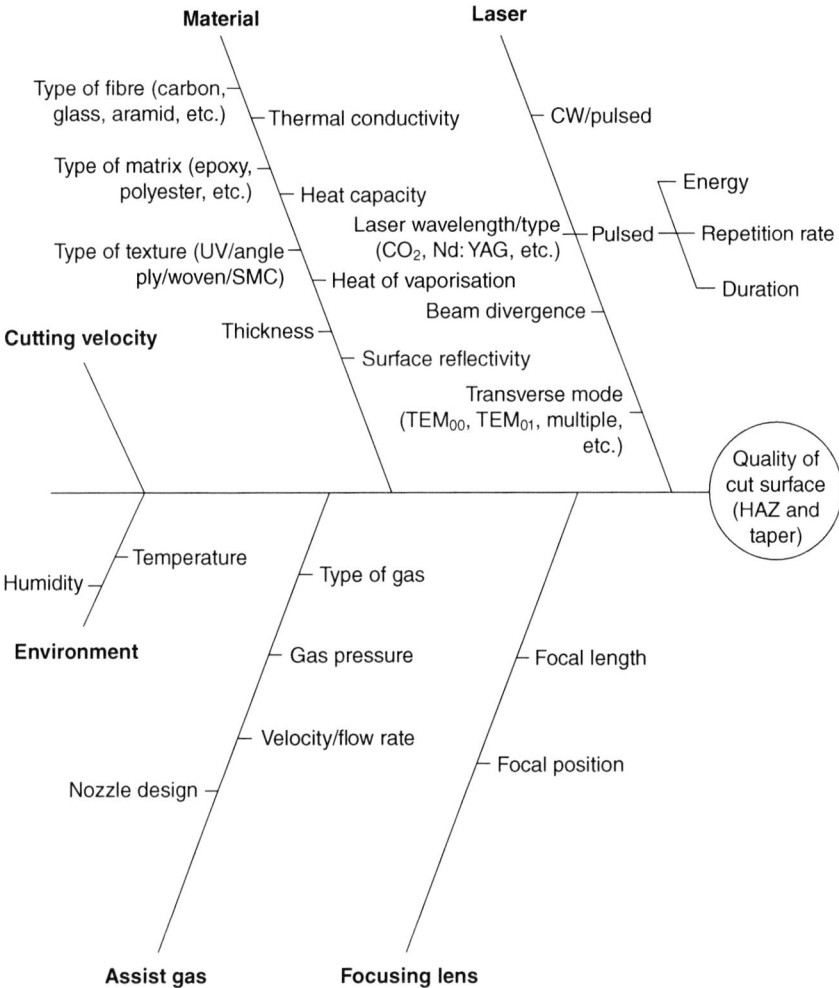

11.1 Cause–effect diagram on the quality in laser machining of FRPs (adapted from Reference 21).

cut quality of glass and carbon fibre composites as compared to aramid fibre composites.[28]

Laser power (controlling the power density) and cutting speed (controlling the interaction time) are the dominant factors influencing the quality in laser cutting of FRPs.[28,31] The extent of HAZ, kerf widths and depth reduce with increasing cutting speed and decreasing beam power. Increasing the speed also reduces charring in machining burnt fibre and matrix.[27] The phenomena can be explained by the energy per unit length of laser cut, P_0/V_B ratio, where P_0 is the laser power and V_B is the scanning speed.[27] At

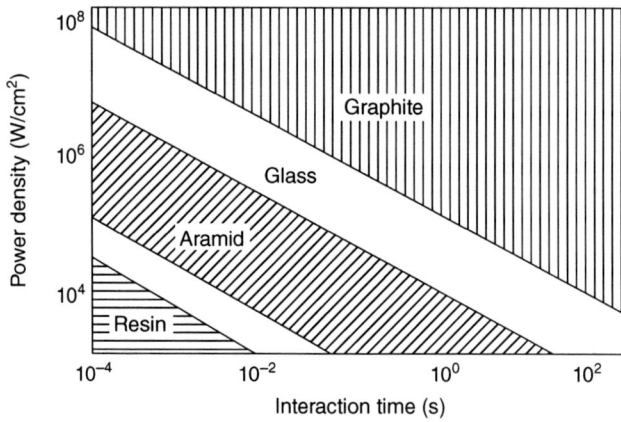

11.2 Limit conditions for vaporisation of common constituents of FRPs; power intensity versus interaction time (adapted from Reference 27).

Table 11.3 Representative data for laser cutting of FRP composites[27,28,32,33]

Material	Power (W)	Scanning speed (mm/s)	Depth of cut (mm)
Carbon fibre/epoxy	1000–2000	15–120	1–4
Carbon fibre/epoxy	300	5	1
Carbon fibre/polyester	800	8	2
Glass fibre/epoxy	1000	30	5
Glass fibre/polyester	800	8	2
Kevlar/epoxy	150–950	30	3.2–9
Aramid/polyester	800	8	2

higher power levels, the range of speed for a better quality cut is wider than that at lower power levels. The minimum required P_0/V_B ratio is particularly influenced by the fibres as they have higher reaction temperatures compared to the polymer resin. AFRPs require the lowest and CFRPs require the highest ratio for the same thickness and fibre volume fraction. Table 11.3 summarises the process conditions required to cut different FRPs.[27,28,32,33]

In laser cutting, the main process conditions can be incorporated into a single parameter (energy density, E_d) as:[20,25,34,35]

$$E_d = \frac{n\eta P_0}{V_B d_B}$$ [11.1]

where n is the number of beam passes, η is absorptivity, P_0 (W) is the beam power, V_B (mm/s) is scanning speed and d_B (mm) is the beam spot diameter. For the same material and beam diameter, this can be summarised as the ratio of power to scanning speed (energy per unit length) multiplied by the number of passes. Thus, a reduction in energy per unit length can be achieved by decreasing power and/or increasing scanning speed, which would conse-quently increase the number of passes required for a through-cut. As the speed governs the power input per unit length, a minimum critical energy input exists at which a through cut can be achieved with minimal thermal damage. This maximum speed limit above which no through-cut occurs depends on the material thickness and power density. Caprino and Tagliaf-feri[36] developed and experimentally evaluated a simple one-parameter model for the maximum cutting speed of FRPs as:

$$V_{Bmax} = \frac{P_0}{\delta \cdot D \cdot d_B} \qquad [11.2]$$

where

$$\delta = \frac{\pi \rho [L_v + C_p (T_v - T_0)]}{4\eta} \qquad [11.3]$$

V_{Bmax} (mm/s) is the maximum cutting speed, P_0 (W) is the beam power, d_B (mm) is the focal spot diameter, D (mm) is the material thickness, L_v (J/kg) is the latent heat of vaporisation, C_p (J/(kg.K)) is the specific heat capacity, T_v (K) is the vaporisation temperature, T_0 (K) is the initial temperature and η is the beam absorption coefficient. δ (J/mm³) is a constant for a given material and given laser.

It is normally recommended to use high beam power at maximum cutting speed to reduce the thermal defects in continuous wave (CW) laser beam cutting of FRPs.[31,36] For example, a multiple-pass cutting using a 1 kW CW beam ytterbium-doped fibre laser (1070 nm) shows a notable reduction of delamination as scanning speed increases.[23] This was determined by a sys-tematic sensitivity analysis on the effect of variation of power and scanning speed on the laser cut quality, keeping the energy per unit length constant at the optimum level (17 J/mm) for a 2 mm thick multiple lamina angular ply CFRP composite. The optimum ratio was confirmed by the power and scanning speed level that resulted in minimum thermal damage and maximum processing rate. Simultaneous variation of power and scanning speed at the constant energy per unit length ratio (i.e. 17 J/mm), showed that the scanning speed (interaction time) is the most influential factor in laser cutting quality of CFRPs. Figure 11.3 illustrates cross-sectional SEM images of laser cut kerfs, showing considerable delamination reduction with an increase in scanning speed in multiple-pass cutting.[23]

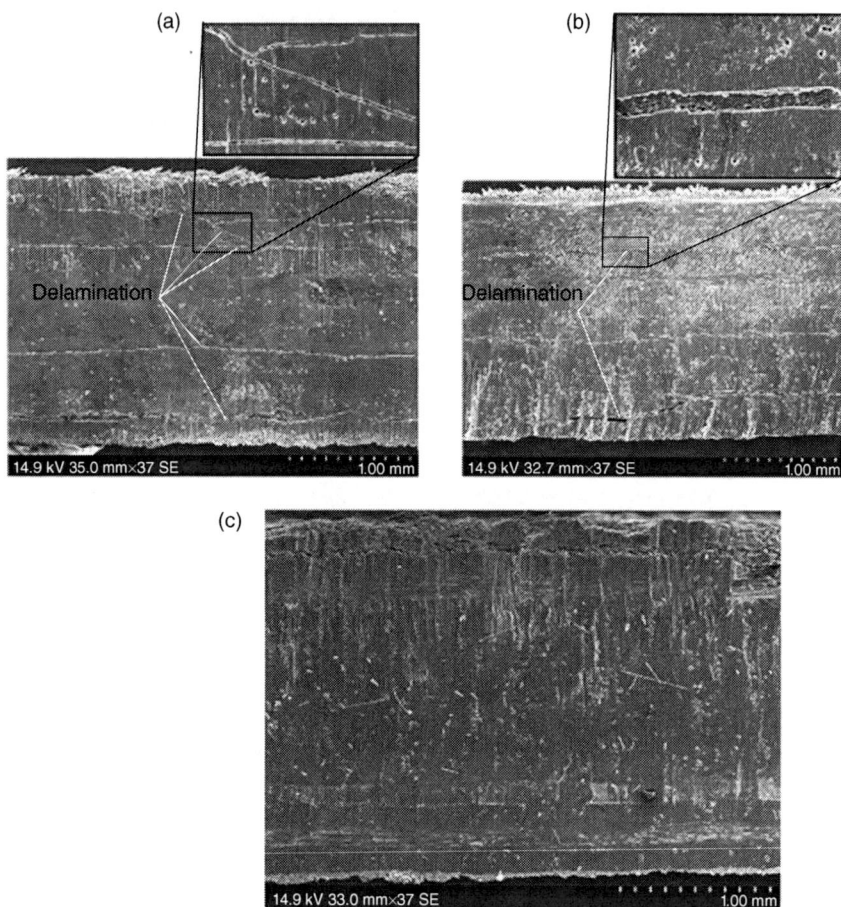

11.3 Influence of increasing speed in multiple-pass cutting on delamination at (a) 20 mm/s (1 pass), (b) 40 mm/s (2 passes) and (c) 80 mm/s (12 passes) at 340 W power, −2.38 mm focal point position (FPP) and 8 bar nitrogen assist gas using 1 kW CW fibre laser cutting of 2 mm thick CFRP laminates.[23]

Another preferred method over CW laser cutting of FRPs, particularly for machining carbon fibre composites, is using short pulsed lasers.[5] Pulsed lasers deliver high values of irradiance in a short period of time. A few-nano-second-duration, pulsed Nd:YAG laser, for instance, can produce irradiances exceeding 10^9 W/cm^2.[37] It is only electron beam machining that can compete with laser machining in this respect.[37] Besides an accelerated removal mechanism, the other advantage of pulsed beams is the cooling rate involved, which is higher compared with a CW laser beam, thus improv-

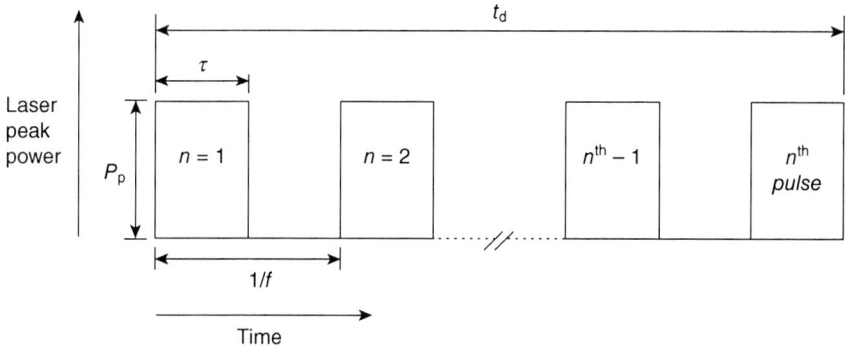

11.4 Relationship of laser pulse parameters for standard rectangular pulses.[23] Notation: P_p, peak power; τ, pulse width; f, repetition frequency; t_d, cutting time; n, number of pulses; $1/f$, period between pulses.

ing the cutting quality of FRPs.[5,21] Figure 11.4 illustrates the relationship of the laser pulse parameters for standard rectangular pulses.

The laser pulse variables as presented in Fig. 11.4 are interrelated to each other. The peak power delivered in a laser pulse is dependent on the pulse energy and the pulse duration according to the following equation:

$$P_p = \frac{E_p}{\tau} \qquad [11.4]$$

Therefore, the irradiance in a pulsed laser beam can be expressed as:

$$I_0 = \frac{E_p}{\tau A_0} \qquad [11.5]$$

The maximum pulse energy is limited by the mean power delivered by the laser, defined as:

$$P_0 = E_p \cdot f \qquad [11.6]$$

where f is the repetition rate (number of pulses per second of the laser). Equation 11.5 shows the average value of the laser irradiance in time and space within each pulse. In fact, a more accurate laser power description considering the variation of laser energy in both temporal and spatial dimensions involves the beam transverse electromagnetic mode (intensity distribution) and coupling with the target material.[30] Consequently, the higher beam intensity, shorter interaction time and better focusing behaviour of pulsed Nd:YAG lasers for instance, leads to less thermal load and hence less thermal damage to CFRPs as compared to a continuous wave CO_2 laser.[5] Mathew *et al.*[21] conducted a systematic study on pulsed Nd:YAG

laser cutting of CFRP composites. It was observed that the HAZ was proportional to the pulse energy – the higher the pulse energy, the larger the
HAZ. The effects of laser–material interaction time are more complicated.
They are represented by few parameters: laser energy delivery mode (continuous wave or pulsed laser beam), repetition rate, pulse duration and
cutting speed. High repetition rate, long pulse duration and slow cutting
speed generally increase interaction time and produce larger HAZ.

11.2.2 Assist gas

Thermal degradation characteristics in the laser cutting of FRPs are influenced by the type and pressure of the assist gas.[38] In CFRPs, thermosetting
epoxies (being heavily cross-linked and amorphous) do not show true
melting or viscous flow upon heating and, once exposed to excess heating
(e.g. during laser machining), they will decompose.[39] Carbon fibres, on the
other hand, are highly crystalline structures. Because of the high bonding
energies (of various carbon atom-to-atom bonds) at atmospheric pressure
(as is common during laser studies), the carbon elements undergo direct
vaporisation (degrade directly from solid to gas phase) at temperatures
around 4000 K.[40] The decomposition of the matrix occurs at a relatively
much lower temperature (around 700 K). Control of thermal degradation,
and hence damage of CFRPs in thermal processing, is best observed in an
inert atmosphere.[41] Considering the higher thermal diffusivity of the carbon
fibre and the rapid heating–cooling rates involved in pulsed laser cutting,
quicker but controlled decomposition of fibres, in particular, can result in
less thermal damage to the cut surface through conduction. In an oxidative
medium, the decomposition of the fibres,[41] as well as the epoxy matrix,[42] is
enhanced with the heat released from the exothermic reactions. Oxidative
decomposition of CFRPs is mainly influenced by carbon fibre oxidation in
the form of the following two exothermal reactions.[41]

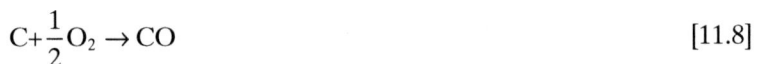

$$C + O_2 \rightarrow CO_2 \qquad\qquad [11.7]$$

$$C + \frac{1}{2}O_2 \rightarrow CO \qquad\qquad [11.8]$$

Figure 11.5 illustrates thermal gravimetric analysis (TGA) for a 60% carbon
fibre/40% epoxy polymer CFRP in oxidative and inert media at two different heating rates. As can be observed from Fig. 11.5a, the material decomposes more quickly in air (through oxidation of fibres) as compared to a
nitrogen medium. From Fig. 11.5b, it is clear that in nitrogen the decomposition shows only one weight loss peak representing devolatilisation (at
around 673K). In air, on the other hand, a different decomposition mechanism is evident through more stages of weight loss (i.e. devolatilisation, char

(a)   (b)

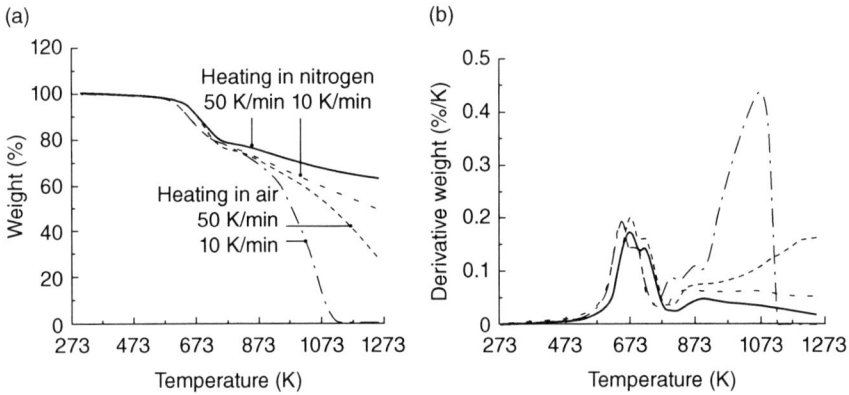

11.5 (a) Weight loss and (b) derivative weight loss TGA of 1.2 mm thick cross-ply CFRPs in nitrogen (inert) and air (oxidative) at 10 K/min and 50 K/min heating rates.[38]

oxidation and then fibre oxidation at around 1073K). It can also be seen from the figure that, although increasing the heating rate decreases the weight loss, the difference between oxidative and non-oxidative environment is still valid. Therefore, the observed difference can still be expected to be valid at even higher heating rates.

Although the presence of a reactive gas can deteriorate the quality (through excessive degradation) in laser processing, nevertheless, once controlled, it can be useful for effective material removal and reduced thermal damage.[38] Thereby, a low oxygen content in an inert assist gas can considerably reduce thermal damage in laser cutting CFRPs.[38,43]

11.2.3 Wavelength

When a high power laser beam interacts with a material, light absorption, transmission and reflection occur. Depending on the laser wavelength, power density and interaction time, as well as material properties, heat generation, conduction and material removal take place with different mechanisms. In general, longer wavelength (i.e. 532 nm to 10.64 μm) laser–material interaction is based on a thermal process (except in the case of a femtosecond laser) due to the fact that only molecular vibrations are excited by the long, low-energy wavelength photons. However, when an electronic excitation is induced by a laser with a higher photon energy UV wavelength, photo-chemical reactions can take place that allow the photons to break the molecular bond (if the photon energy is higher than the bond energy) without introducing heat. Laser processing of FRP composites has been mostly studied by means of industrial lasers, such as CO_2 (10.6 μm wavelength), fundamental YAG (1.6 μm wavelength), diode pumped solid state (DPSS) lasers (UV, visible to IR) and excimer (UV range) lasers. FRPs generally show high absorption of infrared spectrum light so that deep penetration occurs at significantly lower power intensities (10^{2-3} W/cm^2) than for metals (10^6 W/cm^2).[5,28] Deep penetration is a removal mechanism by which the power intensity of the beam is high enough to vaporise material and a vapour column is formed.[37] FRPs do not generally undergo a fusion reaction and hence the vapour column is not surrounded by melted material, unlike the case with metals.[28]

CO_2 lasers (10.6 μm) have been used to investigate laser cutting of CFRPs both in the CW[28,36] and pulsed mode.[44] Nevertheless, Nd:YAG lasers (1.06 μm) have been reported to give less thermal damage due to more rapid pulse-off cooling.[5] Lau et al.[45] studied the quality factors in response to different process parameters using a pulsed Nd:YAG system. They demonstrated the effectiveness of pulse width and the cooling gas on the quality. Mathew et al.[21] also studied Nd:YAG laser cutting of CFRPs on the basis of optimising the process factors. None of the studies though resulted in good quality according to the quality classes by Caprino and Tagliaferri,[36] who suggested the acceptable extent of fibre pull out as less than 150 μm and the kerf width close to the beam spot diameter with no fibre swelling.

Ablative photodecomposition (i.e. photo-ablation) is an alternative mechanism to reduce the thermal damage of thermal decomposition. It can be defined as a mechanism of UV (high photon energy beam) laser–material interaction in which the atomic and/or molecular bonds are broken down.[46] Each material exhibits an ablation threshold (depending on its molecular bond energy). A beam with a photon energy equal or above this threshold causes a photo-ablation mechanism.[46] For instance, deep-UV (e.g. 193 nm) wavelength laser pulses cause ablative photo-decomposition (APD) of organic polymers while longer wavelengths (e.g. 532 nm) also ablate but via melting.[47] A comparison of cut surface quality in CO_2 (i.e. IR beam) and excimer (i.e. UV beam at 248 nm wavelength) cutting for Kevlar (i.e. AFRP) and CFRP laminates shows that AFRPs exhibit a homogeneous surface in both IR and UV beam processing owing to the closer thermal properties of the fibres and the resin as compared to CFRPs.[46] Charring in IR beam cutting of AFRPs is observed while IR cutting of CFRPs experiences more defects, such as large cavities and matrix recession. UV beam cutting of AFRPs shows a smoother surface finish, which could be attributed to closer molecular/atomic bond strengths of fibres and polymer resin to the beam photon energy[46] and hence fewer thermal reactions involved as compared to CFRPs. Decreasing the laser energy and frequency, increases the homogeneity of the cut surface.[46]

Ablation and photochemical reactions in the laser processing of CFRPs using UV beam systems have been reported[22,48] to considerably reduce

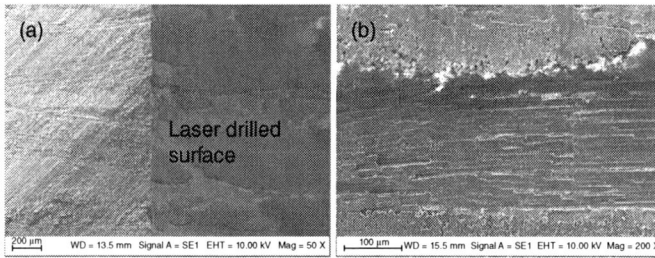

11.6 Cross-sectional SEM images of laser drilling: (a) edge and (b) surface of 3.1 mm thick CFRP laminate using a 355 nm Nd:YVO$_4$ DPSS laser system.[51]

thermal damage. Third harmonic Nd:YVO$_4$ (355 nm) DPSS lasers in particular have shown exceptional laser processing quality in machining, trepan drilling and laser-based repair of CFRPs.[22,49–52] Figure 11.6 shows the cross-sectional SEM images of 6 mm diameter holes drilled in 3.1 mm thick CFRP cross-ply laminate using a 10 W third harmonic Nd:YVO$_4$ DPSS laser system.[50] The right side of Fig. 11.6a is the laser drilled surface. Different contrast shows the different fibre orientations. No obvious fibre extruding (matrix recession) or delamination was observed at the edge of the drilled hole. Figure 11.6b is a higher magnification image of the laser drilled surface, which also shows minimal delamination, cavities and matrix recession.

11.3   Effect of materials

Constituent materials, fibre orientation, fibre volume and stacking sequence of laminates influence laser cutting of FRPs.[5,18,36] Aramid and carbon fibres directly vaporise/decompose at the elevated power densities of a laser beam. GFRPs, however, shows a different mechanism.[36,53] Fenoughty *et al.*[5] suggested two distinct modes for glass fibre degradation. In the first case, a central cavity is formed at the end of the fibre, whilst in the second case the fibres partially melt at their ends, forming little glass beads. As glass fibres (unlike aramid and carbon fibres) are transparent to Nd:YAG laser beams (1.06 μm), they are not ideal for machining by these systems.[5] The fibre orientation and stacking sequence of laminates affect heat transfer behaviour of FRPs and hence the laser machining quality.[18,36] FRPs, as anisotropic structures, show three different thermal conductivities in x, y and z directions (see Fig. 11.7).

Thermal penetration (δ) depends on the thermal diffusivity (α) and time (t) as:[30]

$$\delta = 2\sqrt{\alpha t}$$ [11.9]

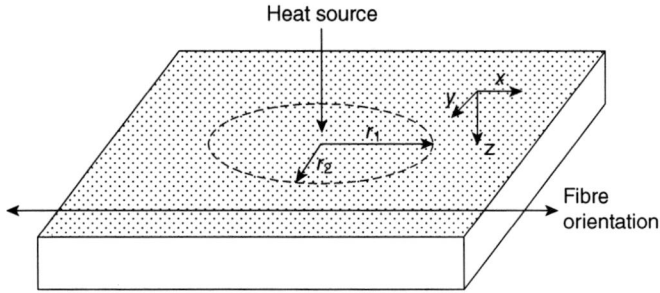

11.7 Schematic of HAZ ellipse formation in FRPs.

Thermal diffusivity follows:[20]

$$\alpha = \frac{k}{\rho C_p} \qquad [11.10]$$

where k (W/(m.K)) is the thermal conductivity, ρ (kg/m^3) is the density and C_p (J/(kg.K)) is the specific heat capacity. Thereby, Pan and Hocheng[20] applied an isotherm method using the rule of mixtures, and an estimate of the ratio between depth of penetration in the parallel (x) and cross (y) directions as:

$$\frac{r_2}{r_1} = \sqrt{\frac{k_y}{k_x}} \qquad [11.11]$$

Some geometric configurations of the beam–fibre interaction in the laser cutting of FRPs are discussed by Chryssolouris et al.[18] When cutting in a direction across the fibre orientation, the heat is conducted away to the kerf sides by the fibres (and is lost); hence the efficiency of the process is lower and matrix recession is higher as compared with cutting along the fibre orientation, where the heat conducted by the fibres serves to preheat the material.

Different thermal expansion coefficients of carbon fibres in the radial and longitudinal directions[5] as well as considerable differences between thermal properties of the fibres and the polymer matrix and high thermal conductivity of carbon fibres[28] cause severe thermal damage to CFRPs during laser processing. Beside a large heat-affected zone and matrix recession, these complexities induce particular surface morphologies such as fibre swelling, matrix squeezing and cavities.[54] These are similar to the EDM surface morphology features[55] as laser and EDM are both thermal processes.

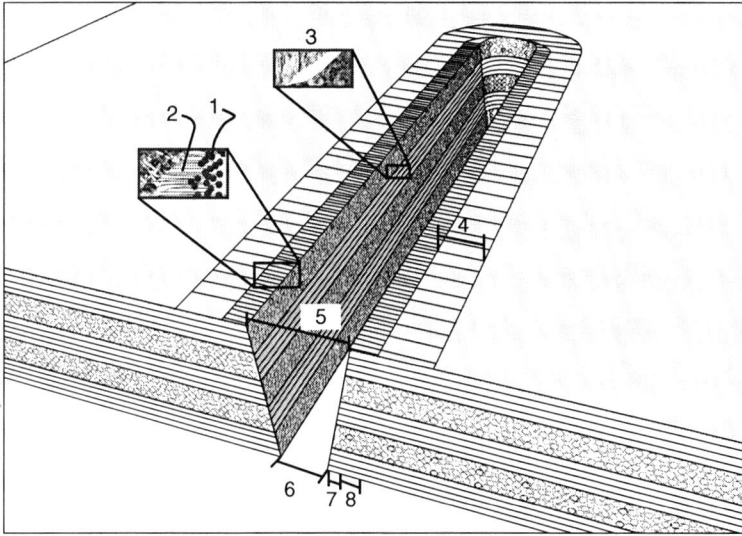

11.8 Typical cut surface quality characteristics in laser cutting of FRPs. 1, Fibre end charring/swelling; 2, matrix recession at the beam entrance; 3, delamination; 4, heat affected matrix at the beam entrance; 5, kerf width at the beam entrance; 6, kerf width at the beam exit; 7, matrix recession at the beam exit; 8, heat affected matrix at the beam exit.

11.4 Quality criteria

Assessment of quality in the laser cutting of composites is a challenging task. Damage of the material in laser machining involves thermal alteration of the matrix and the fibres, as well as interface failure (causing delamination), that are difficult to detect.[56] Therefore, the quality is usually assessed by visual methods or measurements performed using optical microscopy.[26,28] Figure 11.8 is a schematic of the cut surface quality characteristics in the laser cutting of FRPs. As can be seen, charring/fibre swelling is followed by a zone where fibres are protruding from the matrix (i.e. matrix recession). Matrix recession is the region in which temperatures exceed the vaporisation temperature of the matrix.[57] Heat-affected matrix (partially degraded by the heat conducted within fibres) and kerf width are larger at the beam entrance as compared with the exit side, leading to tapered-cut kerf. Figure 11.9 shows scanning electron microscope (SEM) images of major quality defects in the laser cutting of CFRPs.[38]

Based on the above discussion, quality assessment of laser cutting of CFRPs can be categorised within three criteria which include the thermal damage and the geometry defects as well as the processing duration, as presented in Fig. 11.10.[23]

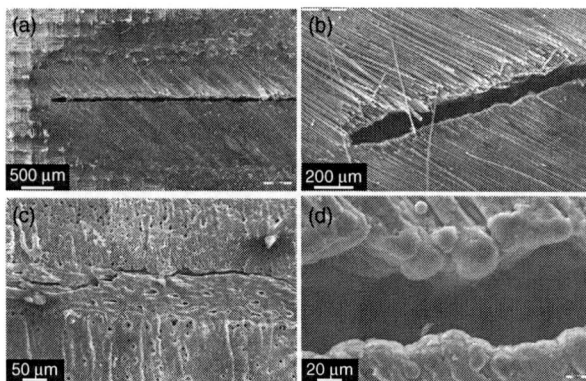

11.9 SEM images of typical quality defects in continuous wave fibre laser cutting of CFRP composites. (a) Large heat-affected zone, (b) matrix recession, (c) delamination between two lamina, (d) fibre end swelling.[38]

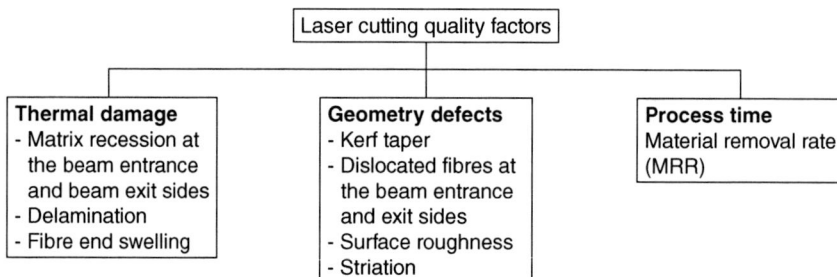

11.10 Quality factors in laser cutting of CFRP composites.[23]

One aspect which has not been completely solved in the laser cutting of FRPs is defining a standard quality criteria, owing to their complex properties and cut quality.[26] Caprino and Tagliafferi[36] suggested three cut quality classes as:

- *Class A (good quality):* top kerf width W_{ka} ~$\leq$ beam spot diameter; length of protruding fibres $W_f \leq 50$ µm; absence of visible charring;
- *Class B (acceptable quality):* W_{ka} ~$\geq d$; 50 µm $\geq W_f \leq 150$ µm; presence of visible charring;
- *Class C (unacceptable quality):* $W_{ka} > d$, $W_f > 150$ µm, high charring.

Generally, matrix recession is the critical factor in assessing the quality.[27] These concepts are commonly used as the benchmark for quality assessment and evaluation in the laser machining of FRPs.

11.5   Conclusions

Laser machining of composite materials requires short beam–material interaction times (short pulse and high scanning speed) and/or short wavelengths (e.g. UV), to minimise heat effects. A multiple pass approach is normally used for material removal. Compared with mechanical and water-jet cutting techniques, laser machining is slower, but some problems such as tool wear and water penetration can be avoided. With the availability of high power picosecond/femtosecond lasers, and high power third or fourth harmonic diode pumped solid state lasers, laser machining of composite materials may become commercially viable and highly competitive.

11.6   References

1 Abrate, S. and D.A. Walton, Machining of composites. Part I: Traditional methods. *Composite Manufacturing*, 1992, **3**(2): p. 75–83.
2 Hempstead, B., B. Thayer, and S. Williams, Composite automatic wing drilling Equipment (CAWDE), in *SAE World Congress*, 2006: Detroit, USA.
3 Kishore, R.A., R. Tiwari, P.K. Rakesh, I. Singh, and N. Bhatnagar, Investigation of drilling in fibre-reinforced plastics using response surface methodology. *Proceedings of the Institution of Mechanical Engineers, Part B: Journal of Engineering Manufacture*, 2011, **225**(3): p. 453–457.
4 Koren, D. *Machining Composites by Conventional Means*, 2005 [viewed 20.07.2010]: Available from: http://www.mmsonline.com/articles/machining-composites-by-conventional-means.
5 Fenoughty, K.A., A. Jawaid, and I.R. Pashby, Machining of advanced engineering materials using traditional and laser techniques. *Journal of Material Processing Technology*, 1994, **42**: p. 391–400.
6 Arola, D. and M. Ramulu, A study of kerf characteristics in abrasive waterjet machining of graphite/epoxy composite. *Journal of Engineering Materials and Technology*, 1996, **118**(2): p. 256–265.
7 Ramulu, M. and D. Arola, The influence of abrasive waterjet cutting conditions on the surface quality of graphite/epoxy laminates. *International Journal of Machine Tools and Manufacture*, 1994, **34**(3): p. 295–313.
8 Shanmugam, D.K. and S.H. Masood, An investigation on kerf characteristics in abrasive waterjet cutting of layered composites. *Journal of Materials Processing Technology*, 2009, **209**(8): p. 3887–3893.
9 Wang, J., T. Kuriyagawa, and C.Z. Huang, An experimental study to enhance the cutting performance in abrasive waterjet machining. *Machining Science and Technology: An International Journal*, 2003, **7**(2): p. 191–207.
10 Ho-Cheng, H., A failure analysis of water jet drilling in composite laminates. *International Journal of Machine Tools and Manufacture*, 1990, **30**(3): p. 423–429.
11 Wang, J. and D.M. Guo, A predictive depth of penetration model for abrasive waterjet cutting of polymer matrix composites. *Journal of Materials Processing Technology*, 2002, **121**: p. 390–394.

12 Wang, J., Abrasive waterjet machining of polymer matrix composites – cutting performance, erosive and predictive models. *International Journal of Advanced Manufacturing Technology*, 1999, **15**: p. 757–768.

13 Siores, E., W.C.K. Wong, L. Chen, and J.G. Wager, Enhancing abrasive waterjet cutting of ceramics by head oscillation techniques. *CIRP Annals – Manufacturing Technology*, 1996, **45**(1): p. 327–330.

14 Wang, J., Predictive depth of jet penetration models for abrasive waterjet cutting of alumina ceramics. *International Journal of Mechanical Sciences*, 2007, **49**(3): p. 306–316.

15 Lemma, E., L. Chen, E. Siores, and J. Wang, Optimising the AWJ cutting process of ductile materials using nozzle oscillation technique. *International Journal of Machine Tools and Manufacture*, 2002, **42**(7): p. 781–789.

16 Wang, J., Depth of cut models for multipass abrasive waterjet cutting of alumina ceramics with nozzle oscillation. *Frontiers of Mechanical Engineering in China*, **5**(1): p. 19–32.

17 Campbell, F.C., *Manufacturing Processes for Advanced Composites*. 2004, Oxford: Elsevier.

18 Chryssolouris, G., P. Sheng, and N. Anastasia, Laser grooving of composite materials with aid of a water jet. *Transactions of ASME*, 1993, **115**: p. 62–72.

19 Pan, C.T. and H. Hocheng, The anisotropic heat-affect zone in the laser grooving of fibre-reinforced composite material. *Journal of Material Processing Technology*, 1996, **62**: p. 54–60.

20 Pan, C.T. and H. Hocheng, Evaluation of anisotropic thermal conductivity for unidirectional FRP in laser machining. *Composites Part A: Applied Science and Manufacturing*, 2001, **32**(11): p. 1657–1667.

21 Mathew, J., G.L. Goswami, N. Ramakrishnan, and N.K. Naik, Parametric studies on pulsed Nd:YAG laser cutting of carbon fibre reinforced plastic composites. *Journal of Material Processing Technology*, 1999, **89–90**: p. 198–203.

22 Denkena, B., F. Völkermeyer, R. Kling, and J. Hermsdorf. Novel UV-laser applications for carbon fibre reinforced plastics. In *Applied Production Technology APT'07*, 2007, Bremen.

23 Negarestani, R., *Laser Cutting Of Carbon Fibre-reinforced Polymer Composite Materials, School of Mechanical, Aerospace and Civil Engineering*, 2010, The University of Manchester: Manchester. p. 239.

24 DeGarmo, E.P., J.T. Black, and R.A. Kohser, *Materials and Processes in Manufacturing*, 8th ed. 1999, New York: Wiley and Sons.

25 Chryssolouris, G., Laser machining: Theory and Practice. In *Mechanical Engineering* (ed. F.F. Ling), 1991, Springer-Verlag, Berlin.

26 Abrate, S. and D.A. Walton, Machining of composite materials. Part II: Non-traditional methods. *Composite Manufacturing*, 1992, **3**(2): p. 85–94.

27 Cenna, A.A. and P. Mathew, Evaluation of cut quality of fibre reinforced plastics – A review. *International Journal of Machine Tools and Manufacture*, 1997, **37**(6): p. 723–736.

28 Tagliaferri, V., A. Di Ilio, and I.C. Visconti, Laser cutting of fibre-reinforced polyesters. *Composites*, 1985, **16**(4): p. 317–325.

29 Voisey, K.T., S. Fouquet, D. Roy, and T.W. Clyne, Fibre swelling during laser drilling of carbon fibre composites. *Optics and Lasers in Engineering*, 2006, **44**: p. 1185–1197.

30 Steen, W.M., *Laser Material Processing*, 3rd ed. 2005, London: Springer-Verlag.

31 Tagliaferri, V., I. Crivelli, and A. Di Illio. Machining of fibre reinforced materials with laser beam: Cut quality evaluation. In *International Conference on Composite Materials, ICCM*, 1987. London.

32 Schucher, D. and G. Vees, *Laser material processing of composite materials. Machining of composite materials II.* In *Proceedings of the ASM Materials Congress*l, 1993, ASM: Pittsburg, 17–23 Oct., p. 153–158.

33 König, W., C. Wulf, P. Grass, and H. Willerscheid, Machining of fibre reinforced plastics. *CIRP Annals – Manufacturing Technology*, 1985, **34**(2): p. 537–548.

34 Chryssolouris, G., P. Sheng, and W.C. Choi, Three-dimensional laser machining of composite materials. *Journal of Engineering Materials and Technology*, 1990, **112**(4): p. 387–392.

35 Chryssolouris, G., P. Sheng, and W.C. Choi, Investigation of laser grooving for composite materials. *Annals of the CIRP*, 1988, **37**(1): p. 161–164.

36 Caprino, G. and V. Tagliaferri, Maximum cutting speed in laser cutting of fibre reinforced plastics. *International Journal of Machine Tools & Manufacture*, 1988, **28**(4): p. 389–398.

37 Ready, J.F., *Industrial Applications of Lasers*, 2nd ed., 1998, Academic Press: New York.

38 Negarestani, R., L. Li, H. Sezer, D. Whitehead, and J. Methven, Nano-second pulsed DPSS Nd:YAG laser cutting of CFRP composites with mixed reactive and inert gases. *International Journal of Advanced Manufacturing Technology*, 2010, **49**(5): p. 553–566.

39 Ashby, M.F. and D.R.H. Jones, *Engineering Materials 2: An Introduction to Microstructures, Processing and Design*, 2nd ed., 1998, Butterworth-Heinemann: Oxford.

40 Bundy, F.P., Pressure–temperature phase diagram of elemental carbon. *Physica A: Statistical Mechanics and its Applications*, 1989, **156**(1): p. 169–178.

41 Yin, Y., J.G.P. Binner, T.E. Cross, and S.J. Marshall, The oxidation behaviour of carbon fibres. *Journal of Materials Science*, 1994, **29**(8): p. 2250–2254.

42 Chen, K.S., R.Z. Yeh, and C.H. Wu, Kinetics of thermal decomposition of epoxy resin in nitrogen–oxygen atmosphere. *Journal of Environmental Engineering*, 1997, **123**(10): p. 1041–1046.

43 Tetsuo, I., *Laser cutting and working device for composite material*: Japan patent 58020390A.

44 De Iorio, I., V. Tagliaferri, and A.M. De Ilio. Cut edge quality of GFRP by pulsed laser: Laser–material interaction analysis. in *LAMP'87*, 1987. Osaka.

45 Lau, W.S., W.B. Lee, and S.Q. Pang, Pulsed Nd:YAG laser cutting of carbon fibre composite materials. *Annals of the CIRP*, 1990, **39**: p. 179–182.

46 Dell'Erba, M., L.M. Galantucci, and S. Miglietta, An experimental study on laser drilling and cutting of composite materials for the aerospace industry using excimer and CO_2 sources. *Composites Manufacturing*, 1992, **3**(1): p. 14–19.

47 Garrison, B.J. and R. Srinivasan, Laser ablation of organic polymers: Microscopic models for photochemical and thermal processes. *Journal of Applied Physics*, 1985, **57**(8): p. 2909–2914.

48 Lau, W.S., T.M. Yue, T.C. Lee, and W.B. Lee, Unconventional machining of composite materials. *Journal of Materials Processing Technology*, 1995, **48**: p. 199–205.

49 Fischer, F., L. Romoli, and R. Kling, Laser-based repair of carbon fiber reinforced plastics. *CIRP Annals – Manufacturing Technology*, 2010, **59**(1): p. 203–206.

50 Li, Z.L., P.L. Chu, H.Y. Zheng, G.C. Lim, L. Li, S. Marimuthu, R. Negarestani, M.A. Sheikh, and P. Mativenga, Laser machining of carbon fibre-reinforced plastic composites. In *Advances in Laser Materials Processing Technology: Technology, Research and Application* (J. Lawrence *et al.*, editors.) 2010, Woodhead Publishing Limited: Cambridge.

51 Li, Z.L., H.Y. Zheng, G.C. Lim, P.L. Chu, and L. Li, Study on UV laser machining quality of carbon fibre reinforced composites. *Composites Part A: Applied Science and Manufacturing*, 2010, **41**(10): p. 1403–1408.

52 Li, Z.L., H.Y. Chu, G.C. Lim, L. Li, S. Marimuthu, R. Negarestani, M. Sheikh, and P. Mativenga. Process development of laser machining of carbon fibre reinforced plastic composites. In *International Congress on Applications of Lasers and Elctro-optics, ICALEO*, 2008, Temecula, CA, USA.

53 Müller, R., R. Nuss, and M. Geiger. CO_2 laser cutting fibre reinforced polymers. In *High Power Lasers and Laser Machining Technology*, 1989, SPIE, Vol. 1132, p. 222–229.

54 Cheng, C.F., Y.C. Tsui, and T.W. Clyne, Application of a three-dimensional heat flow model to treat laser drilling of carbon fibre composites. *Acta Materialia*, 1998, **46**(12): p. 4273–4285.

55 Lau, W.S., M. Wang, and W.B. Lee, Electrical discharge machining of carbon fibre composite materials. *International Journal of Machine Tools and Manufacture*, 1990, **30**(2): p. 297–308.

56 Di Illio, A., V. Tagliaferri, and F. Veniali, Machining parameters and cut quality in laser cutting of aramid fibre reinforced plastics. *Materials and Manufacturing Processes*, 1990, **5**(4): p. 591–608.

57 Negarestani, R., M. Sundar, M.A. Sheikh, P. Mativenga, L. Li, Z.L. Li, P.L. Chu, C.C. Khin, H.Y. Zheng, and G.C. Lim, Numerical simulation of laser machining of carbon-fibre-reinforced composites. *Proceedings of the Institution of Mechanical Engineers, Part B: Journal of Engineering Manufacture*, 2010, **224**(B7): p. 1017–1027.

12

Laser-based repair for carbon fiber reinforced composites

F. FISCHER, Laser Zentrum Hannover e. V., Germany,
L. ROMOLI, University of Pisa, Italy and R. KLING and
D. KRACHT, Laser Zentrum Hannover e.V., Germany

Abstract: Recent progress in laser system technology enables innovative techniques for the machining of CFRPs. A representative application is the layer-by-layer removal of damaged composite material to provide a cavity for refilling with repair plies. Results show that it is possible to achieve a reliable and automatable removal rate to perform arbitrary repair cavity geometries, obtaining a relevant time reduction with respect to the conventional manual grinding process. The combination of modern UV-laser sources with a scanning technology enables scanning speeds of up to 4.0 m s^{-1}, and suppresses heat affected zones and detachment of fibers from the polymer matrix. The interlaminar shear strength of repaired laminates and reference specimens have been measured and evaluated according to DIN 65148, and the results are reported here.

Key words: laser, repair, heat affected zone, interlaminar shear strength.

12.1 Introduction

Carbon fiber reinforced plastics (CFRPs) are gaining wide acceptance in the aeronautic sector, for the excellent strength-to-weight ratio offered by the laminates. In spite of this, the brittle nature of the fibers combined with the very low toughness of the thermoset resins conventionally used as matrix, makes CFRP sensitive to dynamic loads and surface wear.[1] An improvement of the ductility and of the resistance to abrasive wear is represented by the recent use of polyetheretherketone (PEEK) as a matrix for unidirectional laminae. This thermoplastic material offers excellent mechanical and chemical resistance and retains these properties up to remarkably high temperatures (150°C).[2] Nevertheless, thermosets such as epoxies still form a major portion of the matrix materials used.

The maintenance of CFRP parts during service life represents a critical issue since it happens often that structures undergo damage from accidental impact or hydraulic fluid absorption. These phenomena lead to partial or total delamination between adjacent layers with different oriented plies and highly reduce the mechanical strength of the structures. Thus, maintenance problems associated with composites cannot be underestimated and may

309

well be regarded as the weak link in the new technology chain. In this regard, repairing composite parts is the main concern of end-users as well as manufacturers since recycling difficulties and high replacement costs make repairing very advantageous. On-site scarf repair is, until now, the best alternative for aeronautic parts where one side of the assembled structure is not reachable. This is done by manual grinding of the damaged part and by refilling the obtained cavity with a patch of new plies. The material removal needs expert manpower and, as a main drawback, cannot readily be generalized for different repair applications. It would be, then, advantageous to develop a faster removal technique to be employed in a wide range of applications, allowing a high accuracy in the machining of the cavity. For this purpose, a promising technique is represented by laser machining, since the contactless removal allowed by the use of laser light completely avoids two critical drawbacks that are typical in the use of conventional grinding tools: rapid tool wear due to the abrasive action of the carbon fibers and mechanical stresses generated in the workpiece which may cause delamination.

12.2   Carbon fiber reinforced polymer (CFRP) repair principles

The main focus of an aircraft repair is to recover the mechanical integrity and the aeronautical shape. Furthermore, the repair process must ensure that the repaired parts will not lose their function of resisting the design loads, whereby the force allocation will not change. The repair technique for CFRP parts has to be applicable, reliable and cost effective.

12.2.1 Types of repair techniques

Numerous types of repair techniques exist and they can be roughly classified into two categories: the superposition of external overlap plies and the substitution of damaged plies.

The first category includes all lap techniques in which the repair material is applied either on one or on both sides of the laminate over the damaged area, as is shown in Fig. 12.1. The external superposition of pre-preg plies is the fastest and cheapest repair strategy, but crack propagation is not prevented by eliminating the damaged area and strongly limits the reliability of the residual mechanical properties of the structure. In spite of this, the design and analysis of adhesive lap joints is now a relatively mature discipline, especially for isotropic adherends.[3] Similarly, for simple lap joint configurations, the adhesive stresses when bonding composite adherends are also well understood,[4,5] to the extent that analytic software tools are available for the design of such joints that also account for the nonlinear adhesive behaviour.[6]

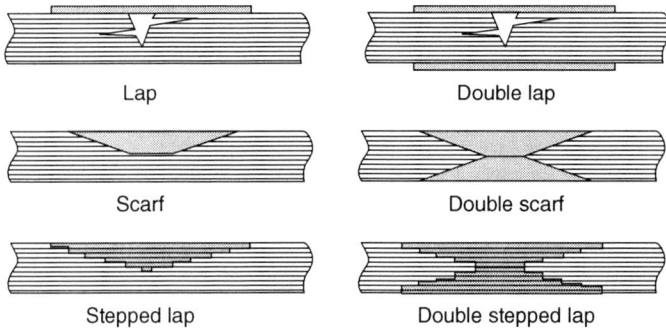

Lap	Double lap
Scarf	Double scarf
Stepped lap	Double stepped lap

12.1 Common bonded joints.

The second group includes scarf and stepped lap repairs, each of which can be constructed in single or double shear arrangements, as shown in Fig. 12.1. In most cases, access limitations or structural design (e.g. sandwich configurations) limit repairs to a single side instead of superior double-scarf or double stepped lap repairs. The repair material is inserted into the laminate in place of the material removed due to the damage. It is evident that in this second group, machining the cavity inside the laminate is the main concern and this starts from a detailed analysis of the damaged area in order to precisely identify its extension. The cavity can be formed either with a uniform surface at a fixed angle (scarf repair) or with a step structure (stepped lap repair).

Adhesively bonded scarf repair and stepped lap repairs are normally used in very important structures requiring a full or significant strength recovery, or when a flush surface is imposed by aerodynamic or stealth reasons.[7] The higher efficiency of these repair methods, compared with the easy-execution lap repairs, is justified by the larger bond length and the reduction of stress concentrations at the bond length edges caused by the adherend thickness grading at the bond region, characteristic of these geometries. An optimum-designed scarf or stepped lap joint is, in fact, significantly stronger than an optimum lap joint, with failure occurring in the adherend outside of the joint instead of adhesive peel or shear failures.[8] On the other hand, the realization of this repair strategy offers several drawbacks from a manufacturing point of view. First of all, the use of small scarf angles, necessary to obtain higher efficiencies, may not be obtainable, since they require an excessive repair area. Secondly, unlike lap or stepped-lap joints, the stiffness of the bonded surface varies along the bondline. The complexity in designing an optimum scarf repair for anisotropic composite structures is then related to a large number of parameters (different fiber orientation, dimensions of the repair plies, thickness of the adhesive layer, etc.) that influence the joint performance.

Simple methods for the design of such joints have been proposed: however, they typically assume that the adhesive stresses can be approximated as uniform along the bondline.[9] More recent analyses have reported that the peel and shear stresses can vary significantly along the bondline in response to changes in ply orientation.[10,11] In order to predict the mechanical properties of the repaired laminate, the refilling plies should be disposed following the laminate stacking sequence.[12] As a direct consequence, this procedure implies that every cavity step corresponds to the thickness of a lamina, which is normally in the range of 125 to 375 μm. It is then possible to understand how the machining strategy and its resolution are important to guarantee a correct removal rate and a regular step geometry.

12.2.2 Repair procedure

As previously stated, the main focus of the repair is to recover the mechanical integrity and the aeronautical shape, while ensuring that the repaired parts will not lose their function to resist the design loads, whereby the force allocation will not change.

The repair process includes several steps. Firstly, the localization of the failure has to be accomplished by an experienced staff member. In addition, a non-destructive testing method (NDT) has to assess the degree of the damage. Regarding the degree of the damage, the geometrical shape of the repair structure has to be defined. This shape may have a complex 3-dimensional geometry. To ensure that the repair is able to bear the loads, a stepped-lap joining between the patch and the structure is a suitable method. Investigations have shown that high stresses can be achieved with this kind of joint.[13]

The next step of the repair process is to remove the defected material via applicable procedures. Until today, a mechanical milling process or a manual grinding procedure have been the conventional methods to perform a complex repair structure. But the removal by mechanical milling generates appreciable thermal stresses, so that the risk of a delamination failure becomes more relevant. Therefore, manual grinding removal is the more suitable application for this process and is carried out by using a hand-held pneumatic router or grinder, as shown in Fig. 12.2.[14]

The manual grinding procedure during a CFRP repair process is a difficult and time-consuming task. The quality and the resolution of the grinding strongly depend on the expertise of the trained personnel and are limited by the restriction of the grinding tool. This implies a large potential for 'human error', which can have significant implications for the high level of safety required for industries such as the aerospace sector.

The repair process is continued with the refilling of the generated step structure cavity with all the above mentioned requirements such as

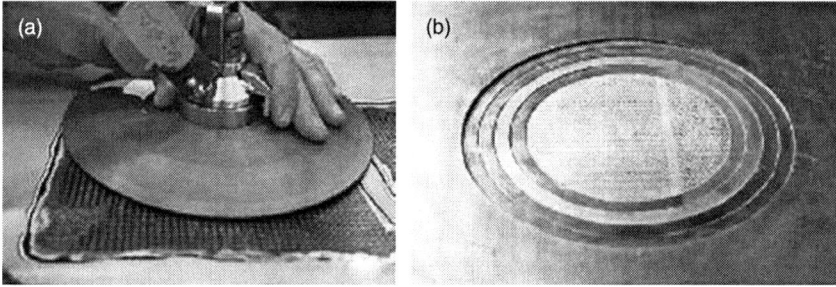

12.2 (a) Diamond angle grinder (by GMI AERO, France), (b) circled repair step structure performed by the diamond angle grinder.

correspondence with the laminate thickness and the fiber orientation. The repair itself will be realized using the vacuum consolidation technique (VCT). (This technique is also the initial point for manufacturing large CFRP structures. Especially in case of prototype structures, this technology offers a high potential for building complex structures without high equipment investment.) The principle of this technique is shown in Fig. 12.3.

The vacuum consolidation technology is based on a simple heated mat and a foil on top of the stacked lay-up; these provide the upper mold in combination with their function as a vacuum bag. The consolidation process itself is characterized by heating up the laminate under vacuum conditions. Obviously, complex 3-D shapes are feasible as well. For the repair, each single ply is cut with its fiber orientation according to the ply-book. This can easily be done by an automatic cutting system as well as by the laser-system used for generating the repair zone. All plies, which are similar to a repair patch, are fixed by an ultrasonic pistol onto the specimen. The consolidation of the repair itself is again realized using VCT, which offers the potential for doing so outside the repair shop.

Each repaired part requires specific tests to assess and certify the residual strength after repairing. It is, then, advantageous to develop a material removal technique to be employed in a wide range of applications, allowing a high accuracy in the machining of the cavity and repeatability of the obtained results. The possibility of establishing an automatable and precise machining process that enables one to standardize the refilling strategy, is a main concern for manufacturers since it represents the bases for the certification of the whole repair procedure.

12.3   UV laser–CFRP interaction

The difficulties in machining with conventional tools inspire the use of alternative machining technologies with the twofold objectives of achieving

(a)

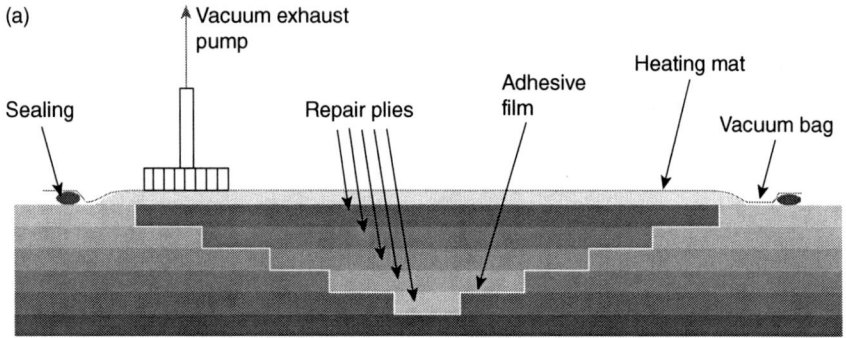

(b)

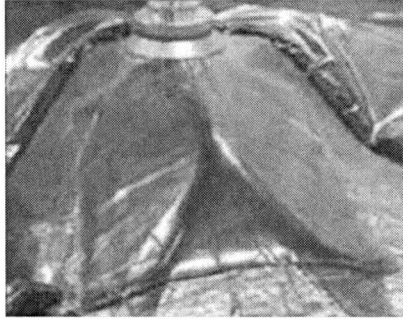

12.3 Typical build up for a vacuum consolidation.

high cutting speed and product reliability by preventing damage of the material. Lasers, as non-contact and wear-less machining tools, exhibit unique advantages in processing anisotropic and inhomogeneous materials such as CFRPs. Nevertheless, the fact that polymers used as matrixes are characterized by vaporization temperatures and thermal conductivities that are one or two orders of magnitude lower than carbon fibers, typically leads to extended thermal damage. This aspect are clearly evidenced in Table 12.1, which lists the thermal properties of four common polymer matrixes with respect to those of carbon fibers.

The key challenge in laser processing is, then, to minimize the heat affected zone (HAZ) in which the matrix is melted or even burned, and fibers that are thermally degraded, while retaining high processing speed.

12.3.1 Theoretical aspects

UV lasers are nowadays adopted to machine CFRPs because the high energy (about 3.49 eV) of photons combined with the short pulse duration generates a vigorous expulsion of material on the surface due to breaking of the chemical bonds that constitute the structure of the plastic matrix and

Table 12.1 Thermal properties of fibers and matrix materials[15,16]

Material	Conductivity (W m^{-1} K^{-1})	Density (g cm^{-3})	Specific heat (J kg^{-1} K^{-1})	Diffusivity (cm^2 s^{-1})	Vaporization temperature (°C)
Epoxy resin[1]	0.1	1.21	1884	0.0004	400–450
Polyphenylene-sulfide PPS[1]	0.29	1.66	795	0.0022	350–500
Polyester PS[2]	0.2	1.25	1200	0.0013	350–500
Polyamide PA6[1]	0.13	0.84	2500	0.0006	350–450
Polyetheretherketone PEEK[1]	0.25	1.32	320	0.0059	350–500
Carbon fiber T300[2]	50*	1.85	710	0.66*	3000–3300

* Along fiber axis

fibers. This phenomenon, known as ablation and already treated in Chapter 11, is mainly due to the photochemical action of the laser radiation: the dynamic impact of laser pulses hinders heat transfer to the workpiece and the ns-pulse duration transforms a relatively low average beam power to a peak-pulse intensity in the GW cm^{-2} regime able to remove both fiber and matrix. As a result, ablation is confined to a narrow area of impact, ensuring excellent resolution and high edge quality without a large development of HAZ.

Aspects relative to the material removal mechanism in the ablation process have been subject to intense theoretical and experimental investigation, as reported by Prasad et al. (2007).[17] These authors reported that the photochemical/photothermal nature of the bond scission depended on the identity of the workpiece material, wavelength, and pulse duration. Energy balance was generally observed up to a threshold value, above which the absorbed energy generated chemical–physical changes in the irradiated material. In the case of a 'pure' photochemical excitation scenario, the photon energy is utilized to directly break bonds at absorption sites. In the earliest models, researchers proposed a Beer-Lambert law based on the dependence of the ablated depth D on the energy density ED (J mm^{-2}) adopted, following a complete photolysis of the material:

$$D = \frac{1}{\alpha} \ln\left(\frac{ED}{ED_{th}} \right) \qquad [12.1]$$

α being the absorption coefficient (mm^{-1}) and ED_{th} the threshold energy density.

Alternatively, in the 'pure' photothermal process, this energy is subsequently converted into heat, which may produce high temperatures and high mechanical stresses in the absorbing region, leading to thermal degradation and causing massive ejection of the material. Pressure and mechanical stresses are dominant for pico-second pulses in the stress confinement regime, and can set off ablation by mechanical breakdown of the polymer matrix in the pure heating case. For longer pulse widths, however, the ejection process is predominantly thermally activated. Purely thermal models based on the thermally activated bond break process can work better in describing these events.

12.3.2 Parametric analysis

The laser used to create the step cavity for the repair process, is a Coherent® AVIA™ Q-switched solid state laser emitting in the third harmonic (355 nm). The pulse repetition rate (frequency) can vary between a single shot to 300 kHz, but the common workable range is 15–200 kHz. The pulse duration of the output beam ranges between 5 ns and 30 ns, depending on

Table 12.2 Laser characteristics of AVIA™ 355-23-300

M^2 (a.u.)	ϕ_{spot} (μm)	Rayleigh length (mm)	Angle of divergence (mrad)
1.08	23.28	1.114	21

the repetition rate. The beam is then focused by a 160 mm telecentric f-theta lens to a minimum spot diameter of 23 μm on the surface of the CFRP laminate.

The laser beam profile and power density distribution along the beam axis are measured according to ISO 11145/11146 with a Micro-Spot Monitor by PRIMES (Germany) to obtain the laser characteristics reported in Table 12.2.

For proper CFRP laser processing, it is necessary to know the absorption of the laser beam into the material, along with the ablation threshold of both carbon fibers and polymer matrixes. In this way it is possible to determine experimentally the range of variation of process parameters that should assure quick removal of fibers and reduced HAZ on the polymer matrix at the same time. Carbon fibers can remain totally absorptive of laser radiation irrespective of the wavelength adopted. Polymers reflectivity is also negligible, while energy lost due to transmission through the material can be significant since it depends on the chemical entangling of the macromolecular chains. The four polymers reported in Table 12.1 (polyamide PA6, polyphenylene-sulfide PPS, polyetheretherketone PEEK and epoxy resin) are analyzed by means of direct spectrophotometry. Figure 12.4 shows the absorption coefficients plotted for different wavelengths: all polymers are generally good absorbers of UV radiation while visible and IR spectra are mostly transmitted through. In the group tested, PEEK and PPS were more absorptive at the adopted wavelength, while PA6 and epoxy data reveal that absorption at 355 nm is reduced by more than 50%.

As a second main step, the *ED* threshold values for the ablation of the polymer matrix and carbon fibers are required to define a proper interaction between laser radiation and the whole composite material. These values can be derived experimentally using the same laser apparatus adopted for machining the cavity. In this case a 355 nm laser was used in the common working conditions: this machine set-up accounts a pulse repetition rate of 90 kHz, which determines pulse durations of about 30 ns. As an example of this application discussed in reference 18, experiments were carried out ablating grooves on a thin PEEK film and on a pre-preg PEEK lamina of carbon fibers (common Torayca® T300), perpendicularly to their orientation. Different polymer matrixes (e.g. the ones reported in Table 12.1) can also be analyzed with the same technique. Results evidenced that

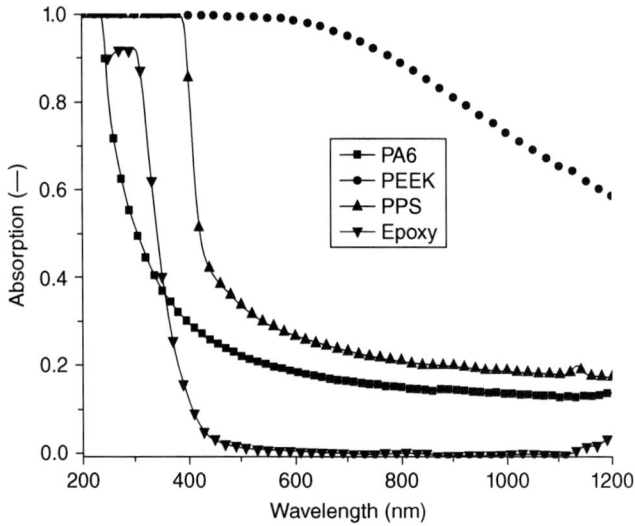

12.4 Transmission spectrum of PA6, PEEK, PPS and epoxy.

threshold values for polymers whose thermal properties are similar, are of the same order of magnitude. Data plotted in Fig. 12.5 show the dependence of the groove depth (measured by a 1 μm resolution optical microscope) on the increase of ED, calculated as follows:

$$ED = \frac{P_{average}}{V_{scan}\,\phi_{spot}} \quad (\text{J mm}^{-2}) \qquad [12.2]$$

$P_{average}$ being the output power given by the laser source at a fixed frequency, V_{scan} the scanning speed and ϕ_{spot} the diameter of the spot once the beam is focused on the specimen surface (23 μm for the experimental set-up used). $P_{average}$ is modulated in a range of 5–20 W combined with V_{scan} in the range of 1–2 m s^{-1} to obtain the values of energy density reported in Fig. 12.5. Three different grooves were ablated onto the specimens using different combinations of average power and scanning speed to obtain the same value of energy density and ensure repeatability of the obtained results. The low dispersion of experimental data allowed the use of energy density the as single parameter that includes process variables in energetic terms. This is because the pulsed laser emission produces a uniform removal rate along the scanned trajectory. This assumption appears reasonable for these types of UV sources which operates at very high frequencies.

For the above mentioned reasons, pulse repetition rate plays an important role in the ablation process and represents a further relevant process parameter since it determines the pulse overlap for a given scanning speed.

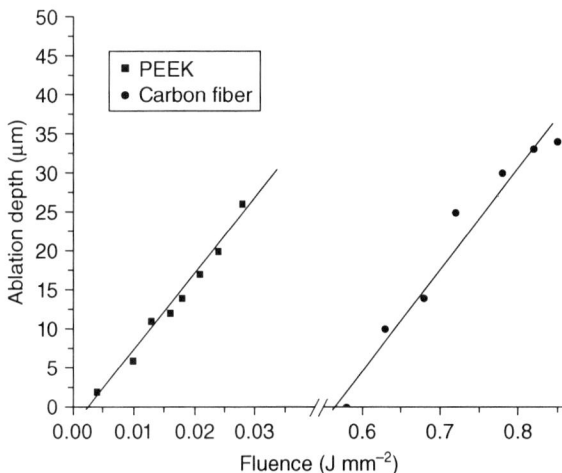

12.5 Groove depth on PEEK film and CF lamina versus energy density, at a fixed pulse repetition rate of 90 kHz.

For the laser source adopted, the maximum $P_{average}$ of 23 W is obtained with a frequency of 90 kHz. Values of V_{scan} of 1, 1.5 and 2 m s^{-1}, combined with a 90 kHz pulse repetition rate give, respectively, overlaps of 46%, 22% and 5%, while no overlap is obtained using 3 m s^{-1}. Especially for low conductivity materials such as the PEEK matrix (k = 0.25 W m^{-1} K^{-1}), the energy remains localized in the hit area and an insufficient pulse overlap results in a discontinuous groove or in an irregular geometry of the edges. CF-lamina behaves differently due to the extremely high heat conduction (up to 50 W m^{-1} K^{-1} along the fiber axis), and continuous grooves can be obtained, also with a nearly negligible pulse overlap.

Overlaps higher that 75% tend to induce temperature increase resulting in a melted zone for the polymer matrix or in swelling of carbon fibers. In the present study, a pulse frequency of 90 kHz was preferred in order to obtain the highest values of the average power. Using Eq 12.2, this choice allows one to increase the scanning speed for a given energy density, and consequently to reduce ablation time.

A pulse overlap between 25% and 50% was used as an adequate compromise for a good ablation of fibers and low thermal damage of the matrix. The scanning speed was then varied in the range of 0.5–1.5 mm s^{-1} and combined, using Eq. 12.2, with different values of the average power. It can be noted that the depth of the grooves ablated on PEEK showed a proportional behaviour in the range of low *ED*. In this case, a linear approximation could be adopted to predict the effects of the process parameters on the ablated area. A similar linear trend is common in polymer ablation: it was noticed also by Romoli *et al.* (2007)[19] in the case of polymethyl methacrylate (PMMA) vaporization, at least for low values of energy density. A threshold value of about 5 mJ mm^{-2} can be experimentally derived for PEEK, which is lower with respect to the one estimated for the fibers (0.58 J mm^{-2}) by two orders of magnitude. This phenomenon explains the complexity of the ablative process, which must ensure enough energy to remove the fibers and prevent heat damage to the matrix as well. The higher dispersion of the experimental data visible in the case of the fibers is due to the difficulties in measuring a discontinuous material like the CF lamina. Anyhow, the dependence of depth on energy density, described by the Beer-Lambert law in Eq. 12.1, fits the non-linear distribution of the experimental data, thus demonstrating that the thermal effects are negligible for the ablation of a single lamina. This assumption has to be verified again for the machining of through holes in the bulk composite material, since the heat conduction in the direction of the laminate thickness cannot be assumed to be negligible.

12.4   The laser-based repair process for CFRP

The current generation of diode-pumped solid state (DPSS) lasers with a high beam quality enables an efficient frequency doubling and frequency tripling using non-linear optical crystals. In this way, lasers can emit a doubling of the frequency in the green region (532 nm or 512 nm), or the so called second harmonic; and a frequency tripling in the UV region (355 nm), the third harmonic. With a doubling of the doubled frequency, the laser can even emit 266 nm, the fourth harmonic. This frequency-multiplied laser radiation can be generated with short pulses and ultra short pulses, and can be combined with high repetition rates of up to a few 100 kHz. Nevertheless, the radiation emits with an almost ideal Gaussian beam profile (see Fig. 12.6) and very high beam quality factor of $M^2 < 1.1$ (see Table 12.3).

12.4.1 Laser set-up

The presented investigations were performed with the previously mentioned DPSS laser from Coherent® (Deutschland) GmbH, Type AVIA™ 23-355-300. The laser generates, in the first step, a beam with a wavelength of $\lambda = 1064$ nm, and with the integrated optical crystal the laser emits in the third harmonic with a wavelength of $\lambda = 355$ nm. The Nd:YAG rod is pumped by diode lasers with an average output power of 23 W, obtainable at 90 kHz. Furthermore the Q-switch technique allows short pulses with pulse durations of 15–30 ns, a maximum repetition rate of up to 300 kHz, and a pulse-peak intensity in the MW cm^{-2} region. The main features of the used laser source are listed in the Table 12.3. The laser beam profile and the

(a)

(b)   Z (mm)

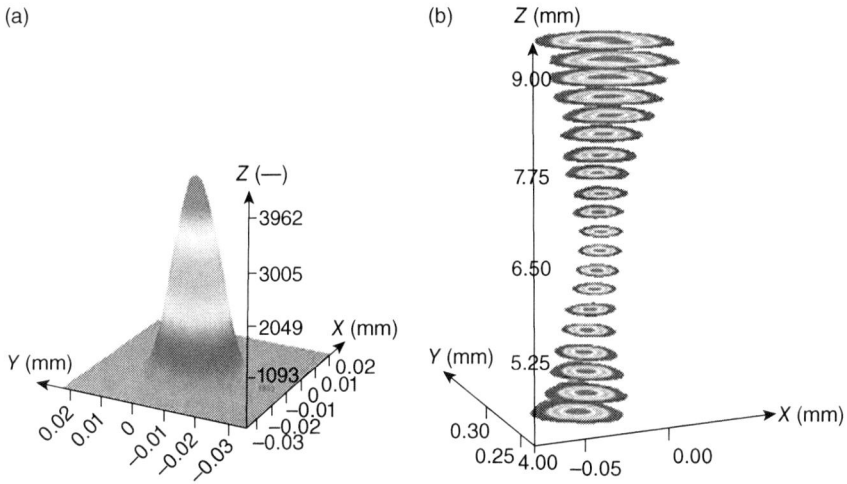

12.6 Measured laser characteristics: (a) beam profile, (b) envelope of profiles along beam axis.

Table 12.3 Main features of the Q-switched, frequency tripled DPSS laser (type: Coherent AVIA™ 355-23 300)

Pulse duration	t_P = 15–30 ns
Repetition rate	Up to 300 kHz
Wavelength	λ = 355 nm
Average power	P_A = 23 W at 90 kHz
Pulse energy	E_P = 255 μJ

power density distribution along the beam axis of the AVIA™ 355-23 300 were measured according to ISO 11145/11146 with a Micro-Spot Monitor by PRIMES (Germany) and are plotted in Fig. 12.6.

To emphasize the efficiency of the laser technique used in this investigation, experiments with layer removal of CF-PEEK were performed using the 355 nm laser. The laser beam was guided by mirrors into a galvometer driven scanner unit, which allowed a highly dynamic beam deflection. Figure 12.7 shows a draft of the adopted set-up: the laser beam is conveyed by mirrors into a galvo-driven scanning head which allows moving the beam on the workpiece surface. The very high pulse frequency of this DPSS laser combined with the high accuracy (± 1 μm) of the mirror equipped head, makes it possible to generate beam scanning speed up to 4.0 m s^{-1}. In addition, the described laser machining set-up allows various complex repair geometries (see Fig. 12.8a,b,c) and enables the designing engineer to obtain the optimal repair performance with a repair geometry that fits to the

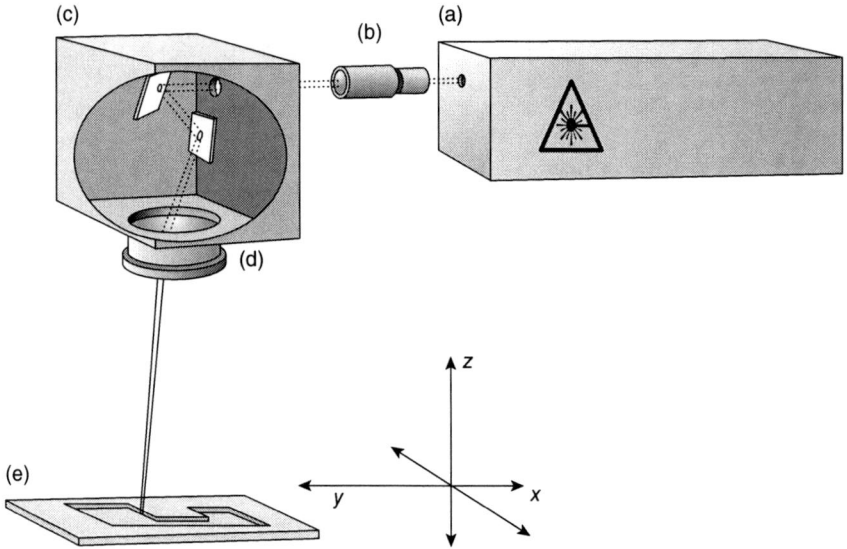

12.7 Draft of the experimental set-up: (a) UV laser source, (b) optics (beam expander, z-shift, etc.), (c) mirror beam guiding, (d) telecentric f-theta lens, (e) CFRP laminate on *xyz* positioning system.

damage of the laminate. Furthermore, the accuracy and the scanning speed retain the previously mentioned values of ±1 μm and up to 4.0 m s^{-1}, respectively.

12.4.2 Cavity laser hatching

The high accuracy of this laser machining process allows scarf ratios of a step depth to step width of 1:10 or even smaller. The motivation for the lower scarf ratio of 1:10 can be easily derived from Fig. 12.8d, which shows an obviously reduced total area needed for repair compared to that of the standard ratio of 1:20. This is relevant for composite parts with a curved or small-area surface to be repaired.

This experimental set-up can be complemented with a z-shift device to enable a 3D laser treatment of real CFRP parts (see Fig. 12.9b). Independent of a 2D or 3D geometry, the ablation of each layer will be realized by hatching the area with parallel scanning lines. In order to flatten the ablated surface, the hatching direction is alternated 0–90° for every couple of layers, thus negating the influence of fiber direction on the depth of the cavity as shown in Fig. 12.9a.

The hatching distance for material removal can be varied from a nearly complete overlap to a gap of multiples of the focus diameter between the

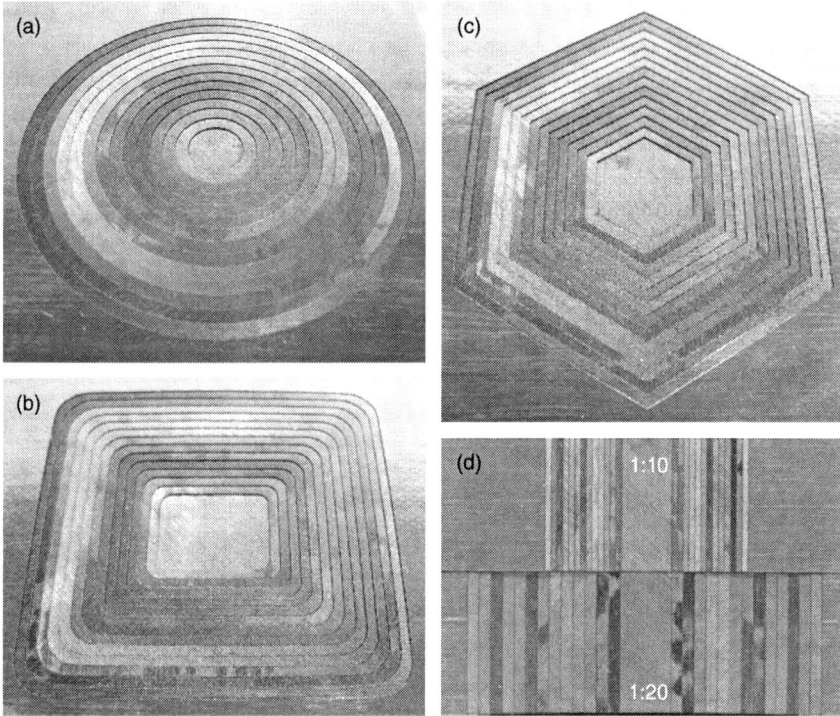

12.8 The laser machining set-up enables various complex repair geometries (a, b and c) and scarf ratios of 1:20 or 1:10 or smaller (d).

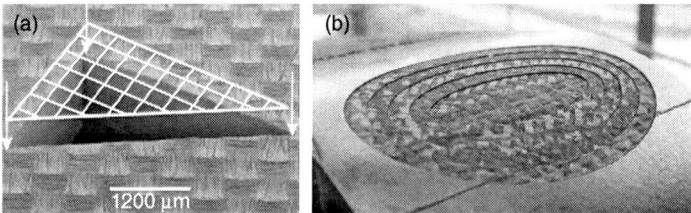

12.9 Laser material ablation by multiple scanned hatch cycles in alternated hatch directions: (a) the hatching principle, (b) a real 3D repair structure.

lines. Depending on the material and the area dimensions to be removed, a smaller hatch distance results in increased laser energy per area and will lead to a larger ablation depth. So the total ablation depth can be increased with the number of cycles and a shorter hatch distance. On the other hand, the investigations of Denkena *et al.*[20] show an increased ablation rate with a hatching distance in the range of the focus diameter. This effect can be explained by the fact that the ablation characteristic for larger hatch distances moves towards the expulsion of close-cropped fiber clusters instead of vaporizing the bulk material. This technique increases the material removal rate, but leads to a slightly rougher surfaces. However, from this relation an optimization of the process speed can be derived, up to a certain degree. According to Eq. 12.2, the depth per cycle, ranging from 1 μm to 25 μm, can be controlled by varying the laser parameters. Once determined, the depth per cycle (a cycle is defined as a couple of crossed layers) for a specified parameter set, the total depth of the cavity, can be then controlled only by the number of cycles. The depth per cycle normally increases with a reduction of hatch distance (Fig. 12.10).

The dependence of the depth per cycle from the hatch distance, as shown in Fig. 12.10, is non-linear and related to a thermal accumulation in the volume to be removed. These thermal effects are not negligible in the case of hatch distances lower than 30 μm.

The described laser machining set-up (see Fig. 12.7), considering the correlation of ablation rate and thermal accumulation, allows a high precise ply-by-ply material removal of CFRP laminates without any thermal damage or delamination, as illustrated by the microcomputed tomography (μCT) picture in Fig. 12.11.

It is, then, possible to obtain a HAZ-free material removal using the photochemical bond-breaking provided by a 355 nm wavelength laser. In addition it is possible to obtain a high resolution of the cavity depth by means of the low penetration of UV-laser radiation. A precise step-cavity geometry with a ratio of 1:20 is displayed in Fig. 12.11, right picture, which illustrates the high-resolution layer-by-layer removal.

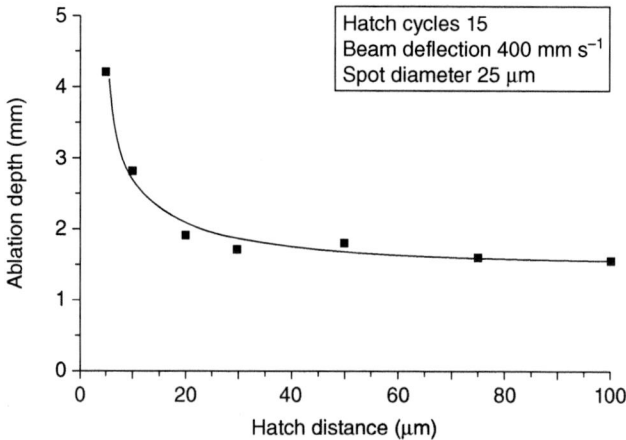

Hatch cycles 15
Beam deflection 400 mm s^{-1}
Spot diameter 25 μm

12.10 Non-linear decrease of the ablation depth in relation to the hatch distance.

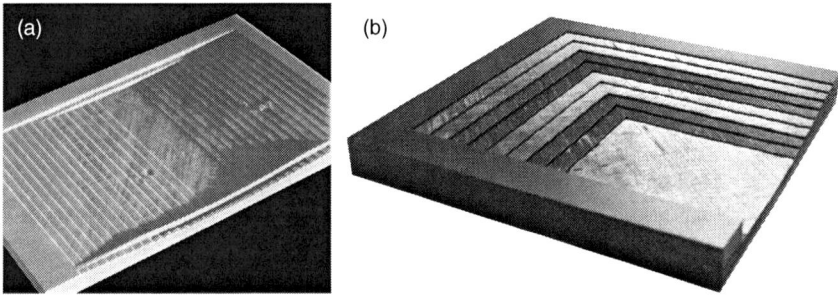

12.11 (a) μCT image of a step cavity geometry, (b) image of a repair structure. Both images with a ratio 1:20 and performed with: wavelength = 355 nm; repetition rate = 90 kHz; pulse duration = 15 ns; deflection speed = 0.4 m/s.

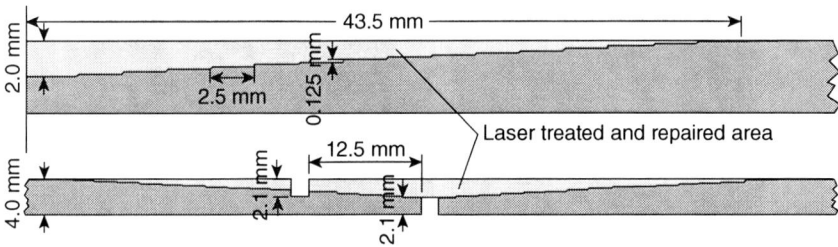

12.12 Specimen geometry and position of the laser treated repair area.

12.4.3 Mechanical tests on repaired structures

Based on the knowledge gained from the initial experiments, a full composite repair scenario has been developed to investigate the influence of laser ablation on the shear strength. For this purpose, stepped structures were fabricated using the described laser based technique. The material used for the mechanical testing was unidirectional CF-PEEK (Tape Toho-Tenax, P-Yarn/Vestakeep). The interlaminar shear strength of repaired laminates and reference specimens were measured and evaluated during the investigations according to DIN 65148 at the German Aerospace Center Institute of Structures and Design. The specimen geometry and the position of the laser-treated repair area are illustrated in Fig. 12.12.

As can be seen from this figure, the slots had been positioned in such a way that the entire load had to be carried by the repair zone. The specimens of one test series came from one laminate that was cut for references and specimens as illustrated by Fig. 12.13.

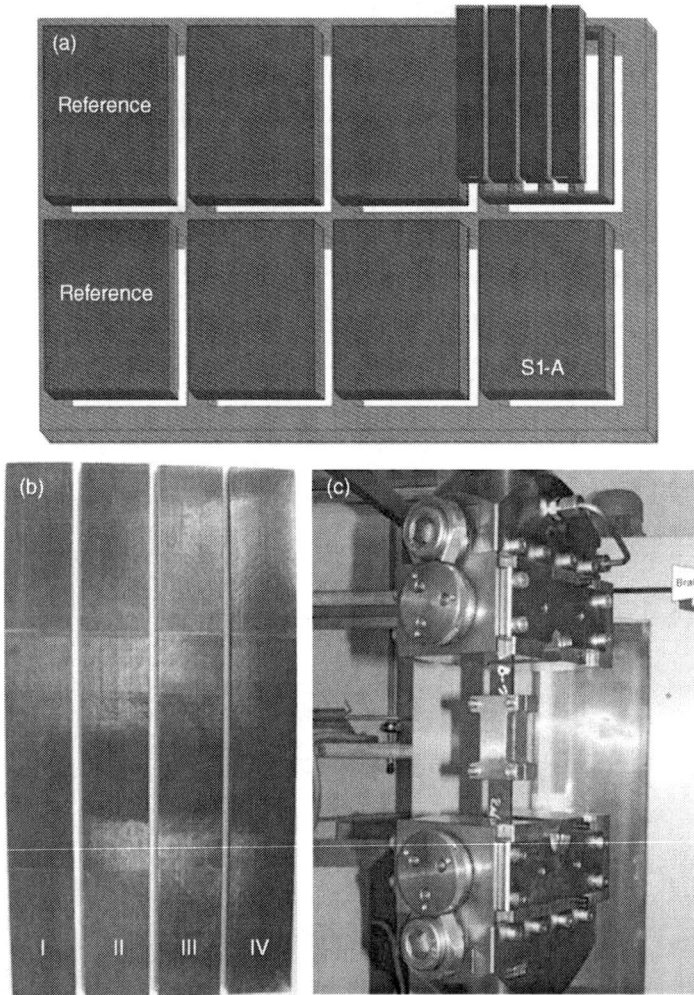

12.13 Interlaminar shear strength of repaired laminates and 20 reference specimens. (a) Cutting strategy, (b) results, (c) test machine (according to DIN 65148).

The first testing-series was used for analyzing the comparability of the repaired and reference specimens. Manufacturing of these specimens used the previously mentioned blanks with their laser-based mountings. The repair itself had been made by using the vacuum consolidation technique (Fig. 12.3) as for the laminates. Typically scarf angles from 1:20 to 1:50 were used for bonded repairs, which could cause the area of the scarf to be more than double the size of a bolted repair area. One reason to minimize the scarf angle is the possibility of repairing curved structures, to meet

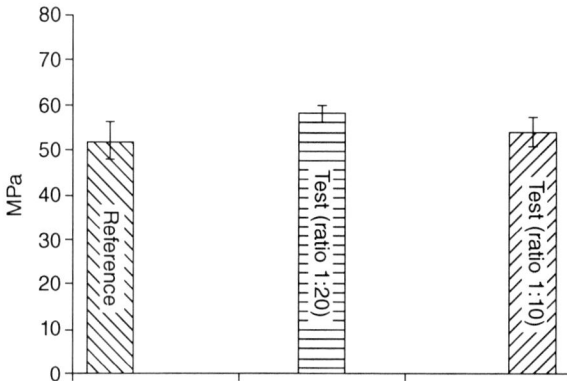

12.14 Results of interlaminar shear strength test for scarf ratios 1:20 and 1:10.

the stiffness and strength requirements. Additionally, the treated area will be smaller. The interlaminar shear strength of repaired laminates and 20 reference-specimens (according to DIN 65148) were measured and evaluated during the first investigations. According to the usual and current repair process, the examination started with a scarf angle of 1:20 (10 specimens).

Because of the above mentioned reasons for smaller scarf ratios, the interlaminar shear strength test continued at 1:10 (8 specimens). The results are shown in Fig. 12.14. The average values of all the specimens were in a range between 50 and 60 MPa. This leads to the conclusion that either the scarf angles of the repaired specimens have no negative bearing on the interlaminar shear strength or the failure was not caused by the stepped joining.

To be able to reach an exact conclusion, a polished micro-section analysis and an examination of the point of fracture were additionally carried out. Figure 12.15 shows the point of fracture of a 1:20 and a reference specimen. Several steps can be observed on the 1:20 specimen, which means that the failure was actually in the stepped joining (area of repair) and that the interlaminar shear strengths of the repaired specimens with the scarf angles were not decreased. Further investigations will focus on the strength behaviour of a repaired specimen under tension, according to DIN 2561. Additional scarf angles, dynamic tests and the effects of contamination, temperature and moisture will also be carried out.

12.5   Conclusions

A novel application concerning the repair process of laminates reinforced with carbon fibers was carried out in a stepped lap configuration by

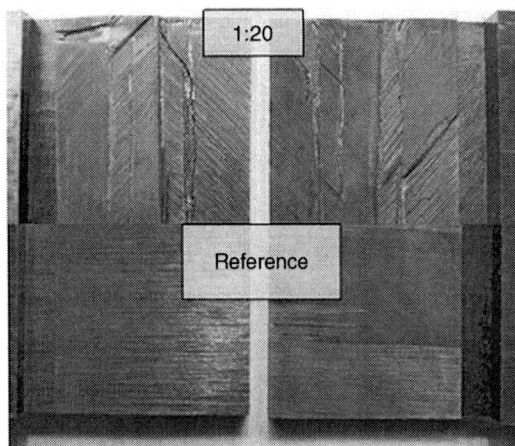

12.15 Point of fracture of a scarf ratio 1:20 and a reference specimen.

substituting the manual grinding with laser hatching. It was proved that a DPSS laser emitting in the UV region presents a valid alternative for processing CFRPs, especially in the field of precise machining. Experimental results have demonstrated that the proposed technique allows a removal process for CF-PEEK which is not affected by thermal damage. Moreover, this new process is fast enough to guarantee a successful economic implementation. The well-known drawbacks related to the extreme difference in thermal properties between fibers and matrix are overcome by a proper choice of repetition rate and pulse overlap, and a consequent modulation of the energy density on the hatched lines. The possibility of establishing a material removal mechanism based not only on direct vaporization but also in ejection of chopped fibers from the treated surface, was proved throughout the evaluation of this particular laser application.

The increase of scarf angle that characterizes the step cavity, provides a further possibility of process optimization. The ratio conventionally used in the manual process is limited to 1:20, due to the geometries of the grinding tools. This problem was overcome with scanning technology whose 5 μm-resolution allowed ratios of 1:10. This decreases the cavity dimensions and consequently the process time. The effective residual strength of laminates repaired with these ratios was investigated accordingly to the standard test method of DIN 65148. Results revealed that the laser hatching of the step cavity did not induce any embrittlement of the laminate. This final step represents a further effort towards the certification and standardization of the proposed technique for use in the aerospace sector.

12.6   References

1 Moutier J., Fois M., Picard C., 2009, Characterization of carbon/epoxy materials for structural repair of carbon/BMI structures, *Composites, Part B*, **40**, 1–6.

2 Flöck J., Friedrich K, Yuan Q., 1999, On the friction and wear behavior of PAN- and pitch-carbon fiber reinforced PEEK composites, *Wear*, 304–311.

3 Adams R.D., Comyn J., Wake W.C., 1997, *Structural Adhesive Joints in Engineering*. London: Chapman & Hall.

4 Matthews F.L., Kilty P.F., Godwin E.W., 1982, A review of the strength of joints in fiber-reinforced plastics. Pt. 2: Adhesively bonded joints. *Composites*, **13**(1), 29–37.

5 Tong L., Soutis C., 2003, *Recent Advances in Structural Joints and Repairs for Composite Materials*. Kluwer Academic Publishers, Dordrecht, The Netherlands.

6 Baker A., Dutton S., Kelly D., 2004, *Composite Materials for Aircraft Structures*. 2nd ed. Reston (VA): American Institute of Aeronautics, Inc.

7 Campilho R.D.S.G., de Moura M.F.S.F., Pinto A.M.G., Morais J.J.L., Domingues J.J.M.S., 2009, Modelling the tensile fracture behaviour of CFRP scarf repairs, *Composites, Part B*, **40**, 149–157.

8 Gunnion A.J., Herszberg I., 2006, Parametric study of scarf joints in composite structures, *Composite Structures*, **75**, 364–376.

9 Harman A.B., Wang C.H., 2006, Improved design methods for scarf repairs to highly strained composite aircraft structure, *Composite Structures*, **75**, 132–144.

10 Harman A.B., Wang C.H., 2005, Analytical and finite element stress predictions in two-dimensional scarf joints. In: *Eleventh Australian International Aerospace Congress (AIAC-11)*, Melbourne, Australia, March 13–17.

11 Johnson C.L., 1989, Effect of ply stacking sequence on stress in a scarf joint, *American Institute of Aeronautics and Astronautics Journal*, **2**(1), 79–86.

12 Wang C.H., Gunnion A.J., 2008, On the design methodology of scarf repairs to composite laminates, *Composites Science and Technology*, **68**, 35–46.

13 Armstrong K.B., Bevan L.G., Cole W.F., 2005, *Care and Repair of Advanced Composites*; SAE International, Warendale, PA, 2005, p. 548.

14 Whittingham B., Baker A.A., Harman A., Bitton D., 2009, Micrographic studies on adhesively bonded scarf repairs to thick composite aircraft structure; *Composites, Part A*, **40**, 1419–1432.

15 Goodfellow Catalogue, available on line at http://www.goodfellow.com (accessed 11th February 2011).

16 Li Z.L., Zheng H.Y., Lim G.C., Chu P.L., Li L., 2010, Study on UV laser machining quality of carbon fiber reinforced composites, *Composites, Part A*, **41**, 1403–1408.

17 Prasad M., Conforti P.F., Garrison B.J., 2007, On the role of chemical reactions in initiating ultraviolet laser ablation in poly(methyl methacrylate), *Journal of Applied Physics*, **101**(10), Article number 103113.

18 Fischer F., Romoli L., Kling R., 2010, Laser-based repair of carbon fiber reinforced plastics, *CIRP Annals – Manufacturing Technology*, **59**(1), 203–206.

19 Romoli L., Tantussi, G., Dini G., 2007, Layered laser vaporization of PMMA manufacturing 3D mould cavities, *CIRP Annals – Manufacturing Technology*, **56**(1), 209–212.

20 Denkena B., Völkermeyer F., Kling R., Hermsdorf J., 2007, Novel UV-laser applications for carbon fiber reinforced plastics. In: APT07, *International Conference on Applied Production Technology, Production of Aircraft Structures*, Bremer, Germany, 99–108.

Part III
Special topics in machining composite materials

13

High speed machining processes for fiber-reinforced composites

H. ATTIA, National Research Council of Canada, Canada,
and McGill University, Canada, A. SADEK, McGill University,
Canada and M. MESHREKI, National Research Council
of Canada, Canada

Abstract: Due to the lack of information on high speed drilling and
routing of fiber-reinforced polymers (FRPs), experimental investigations
were conducted by the authors. For the high speed drilling of FRPs, the
findings of the literature and the experiments are presented in detail and
cover the observed trends for the forces and temperatures, tool wear
mechanisms, tribological interaction, as well as the hole quality in terms
of surface roughness, delamination and geometrical errors. Similarly, the
high speed milling of composites is discussed and the dynamic aspects of
the process and the trends for the forces, temperatures, tool wear and
machined surface quality are presented.

Key words: high speed machining, drilling, milling, FRP, tool wear,
surface quality.

13.1 Introduction

The concept of high speed machining (HSM) was first introduced in the
early 1930s. The concept was based on some experimental observations
related to the machining of non-ferrous metals. The cutting temperatures
were found to increase in direct relationship with the increase in cutting
speed until a certain critical speed, above which the cutting temperatures
would decrease.[1] The range of cutting parameters corresponding to the
peak cutting temperature was defined as the 'non-workable' range. On the
other hand, the range of higher cutting speeds causing reduced temperatures opened the door for further research aiming at adopting high speed
machining in industry.

The HSM techniques are characterized by a substantial enhancement in
cycle-time reduction, waste minimization and high machine utilization.
High speed machining makes it possible for a manufacturing facility to
increase the volume of production with a fairly flexible range of product
diversity. This explains the harmony found between the principles of agile
and flexible manufacturing systems and the HSM techniques.

333

Several definitions have been stated for HSM, specific to the workpiece material and according to the process parameters including high spindle speed, high feed, high spindle speed and feed or high cutting speed.[1-3] The criteria commonly used to define HSM include the following:

(i)   The DN number, which is the product of the bearing bore diameter of the spindle (in mm) and the maximum spindle speed (in rev/min). For HSM, $DN > 5 \times 10^5 - 10^6$.

(ii)   The dynamic criterion, where the threshold of the HSM is reached when the tooth passing frequency of the cutting tool approaches the dominant natural frequency of the most flexible system mode.[4] This definition implies that the tool geometry and its unsupported length (overhang) play an important role in determining HSM. In practical terms, this criterion is linked to the spindle speed where practical sweet spots on the stability lobes are identifiable.

(iii) The thermal criterion, where, the cutting temperature reaches a maximum and then starts to decrease with the increase of the cutting speed.[5]

The peculiar difficulties associated with defining what is 'high speed machining' of composite materials are due to its dependence on the material composition and the type of machining process itself. In fact, defining a global range of process parameters that suits all the HSM processes, work materials, and cutting tools is almost impossible, especially for machining such highly complex materials as fiber reinforced polymers (FRPs). On the other hand, a categorization of low, medium, and high cutting speed ranges can be applicable for each workpiece material or family of materials. Then a sub-category can be defined based on the mechanics of different cutting processes.

For drilling and end milling/trimming operations, the first DN criterion can hardly be met unless very high rotational speeds >50 000 rpm are used. The application of the second dynamic criteria is quite difficult due to the sensitivity of the system natural frequency on the unsupported length of the tool and the design of the spindle and the machine tool. From a thermal perspective, it is the authors' opinion that the third criterion should rather be linked to the composite material glass transition limit and decomposition temperature.

Considerable research efforts have been exerted to define the range of HSM for different processes and work materials. For example, milling of carbon fiber-reinforced polymers (CFRPs) at surface cutting speeds of 1200 m/min to 8000 m/min was defined to be high speed milling; transitional speeds range from 600 m/min to 1200 m/min, and below that would be considered as low speed milling.[1] In terms of the cutting speed, the definition of the range of HSM parameters of non-ferrous materials varied from

(250–460 m/min) for high speed milling[3] to (9000–45000 m/min) for high speed turning.[1] In drilling and end milling, which are the most common machining processes used with composite materials, the tool is relatively small. Under these conditions, the surface cutting speed is not practically high as compared to the values for metals; nevertheless, the rotational speed can be very high with today's machine tool technology.

Considering the abrasive nature of chips of CFRP, another limiting factor is the tool life, which depends on its material and coating. A criterion that is quite relevant to machining of composites is the dependence of the quality attributes of the machined part (e.g. delamination and thermal damage) on the cutting speed. The tool life criterion seems to be, indeed, the practical measure of HSM.

Based on the above, the authors are proposing the scale shown in Fig. 13.1 to identify the ranges of rotational speeds for conventional, high speed and very high speed machining with small drilling and end milling tools.

However, the so far developed definitions for HSM of FRPs were based on general practices and expert observations and/or the available machine tool kinematic capabilities. The discussion of HSM of FRPs in this chapter is meant to provide a guideline for defining the ranges of HSM of FRPs based on the previously mentioned considerations but considering the impact of the process parameters on the produced part quality and tool life.

To better understand the dynamic, tribological, and thermal characteristics of high speed machining of composites, an extensive experimental study has been conducted by the authors at the Aerospace Manufacturing Technology Center (AMTC) of the National Research Council Canada (NRC) on the drilling and routing of CFRP. The machining was performed on a quasi-isotropic laminate comprised of 35 plies of 8-harness satin woven graphite epoxy prepreg with a final cured thickness of 6.35 ± 0.02 mm. The experiments were conducted on a Makino A88ε machining center. To achieve a spindle speed up to 40000 rpm, an IBAG spindle speed attachment, which has a 1 kW power, was used.

A system approach to the high speed machining of composites program at the National Research Council of Canada (NRC), that recognizes the nonlinear nature of the tool–workpiece tribological interaction, is schematically shown in Fig. 13.2. The system consists of three elements; namely, the

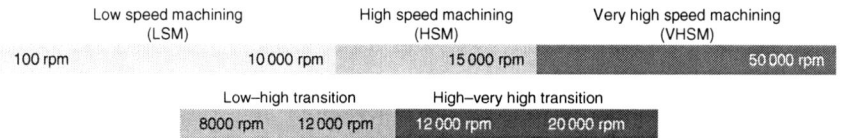

Low speed machining (LSM)		High speed machining (HSM)	Very high speed machining (VHSM)
100 rpm	10 000 rpm	15 000 rpm	50 000 rpm
	Low–high transition	High–very high transition	
	8000 rpm 12 000 rpm	12 000 rpm 20 000 rpm	

13.1 Scale of rotational speeds used in machining of FRPs.

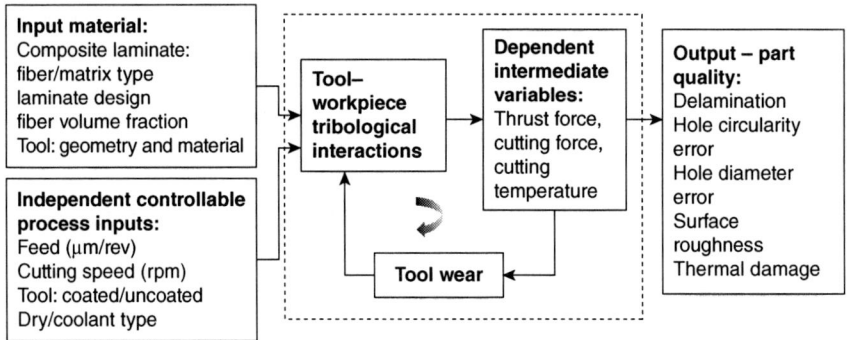

13.2 System approach for assessment of tool wear in drilling of composites.

independent input variables (materials and cutting conditions), the dependent intermediate variables (cutting forces, temperature and friction) and finally the outputs (desirable quality attributes of the hole). The nonlinear behavior and closed-loop interactions of this tribo-system are demonstrated by the fact that the tool wear process is both affecting and being affected by the intermediate process variables. While cutting forces and temperature control the mechanism and the kinetics of the wear process, the latter, in turn, alters the tool geometry, deteriorates the cutting capability of the tool and thus ultimately dictates the hole quality of the laminate.

13.2   Overview of high speed drilling (HSD) of fiber-reinforced polymers (FRPs)

The main objective for employing high speed drilling (HSD) is to facilitate higher productivity of the hole making process. The principle behind the enhancement of the drilling process performance with high speed is based on the relationship between the rotational speed and the axial feed of the drilling tool. The increase in rotational speed of the drilling tool for a fixed axial feed reduces the feed per tooth, which can significantly reduce the drilling forces within a certain range, and accordingly reduces the force-dependent material damage (e.g. interlaminar delamination). However, the high rotational speed is associated with undesirable tool dynamics and friction-induced temperature rise that leads to accelerated tool wear and thermal material damage. On the other hand, reducing the feed per tooth to extremely low values could result in a reversed force trend because of high cutting pressures, in this case depending on the tool–workpiece pair.

Except for the results reported by Lin and Chen,[6] all the very high speed drilling (VHSD) results (in the range of 16 000 to 40 000 rpm) presented in the sections to follow were generated by the authors at AMTC-NRC, using a solid carbide, two-flute, 6 mm diameter drill.

13.3 Thermal aspects and cutting forces in HSD of FRPs

13.3.1 Cutting temperatures

The thermal damage during high speed drilling is a critical aspect that has to be considered while studying the impact of different process parameters on hole quality attributes. The fiber reinforced polymers (FRPs) are sensitive to the temperature rise during machining because the matrix material holding the fibers deteriorates and decomposes at relatively low temperatures compared to metals. This can result in matrix burnout, causing material de-bonding on the machined surface and/or within the heat affected zones. Furthermore, the temperature rise during drilling of FRPs may accelerate the tool wear significantly, resulting in a further rise in the cutting temperature and forces. The severity of the thermal damage is controlled by the amount of friction-induced temperature rise, which is controlled, among other factors, by the rotational speed and the material exposure time. The latter depends on the feed and the depth of the hole.

Monitoring reliably the temperature rise profiles of the tool and/or the workpiece in the cutting zone during the drilling process has many implementation and accuracy challenges. Thermo-couples can be inserted in the coolant through-holes of the tool[7,8] and this will require a slip ring connection, as shown in Fig. 13.3a, to transfer the thermocouple signal while rotating. However, the capacity of most of the available slip ring devices remains limited, below the range of rotational speeds used for VHSD. Hence, using an IR camera for temperature monitoring with VHSD at a relevant location (e.g. the hole exit) becomes more practical

(a)   (b)

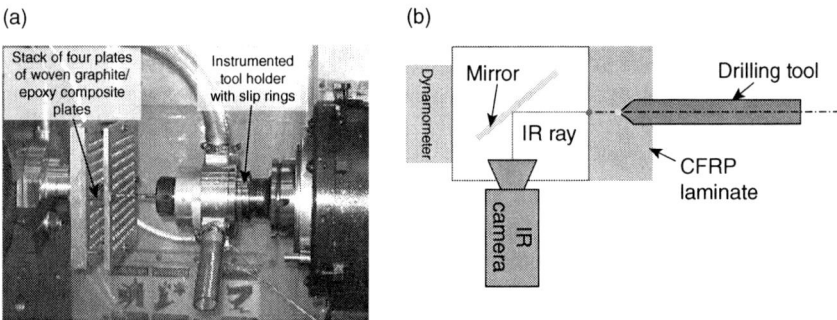

13.3 Temperature measurement using (a) slip ring and (b) IR camera.

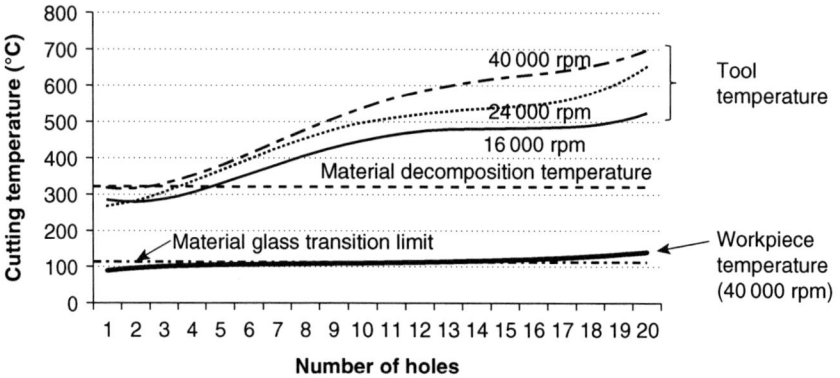

13.4 Effect of rotational speed and tool wear on the maximum cutting tool and workpiece temperatures.

and convenient. The large compartment required to accommodate the IR camera to monitor the temperature at the exit of the hole places the cutting zone at a position beyond the recommended measuring distance from the dynamometer face, which might significantly affect the accuracy of the measured forces. This problem is solved using a special reflection mirror to reflect the emitted IR rays towards the lens of the IR camera, as shown in Fig. 13.3b.

Figure 13.4 shows the trend of the maximum recorded tool and workpiece temperatures for VHSD of 20 holes, which was measured by the authors. The temperature increased with the increase of the rotational speed as a result of the corresponding increase of the cutting energy input to the cutting zone.

The maximum tool temperatures at the rotational speeds 40000 rpm, 24000 rpm and 16000 rpm exceeded the material decomposition temperature after the second, fourth, and fifth holes, respectively. This may also be attributed to the difference of the tool wear-rate associated with each rotational speed. After the ninth hole, a decrease can be seen in the slope of the temperature curves for the 16000 rpm and 24000 rpm rotational speeds, while only a slight decrease can be seen in the slope of the temperature for the 40000 rpm after the same drilled length. After the sixteenth hole, the slope of the temperature for all the rotational speeds starts increasing again, due to the severe tool wear. The maximum tool tip temperature for the 40000 rpm is almost five times the maximum workpiece temperature at the cutting area. The larger thermal conductivity of the solid carbide tool material compared to that of the CFRP workpiece material explains the higher heat flow towards the tool rather than towards the workpiece.[9] Figure 13.5 shows that the workpiece temperature remained below the glass transition

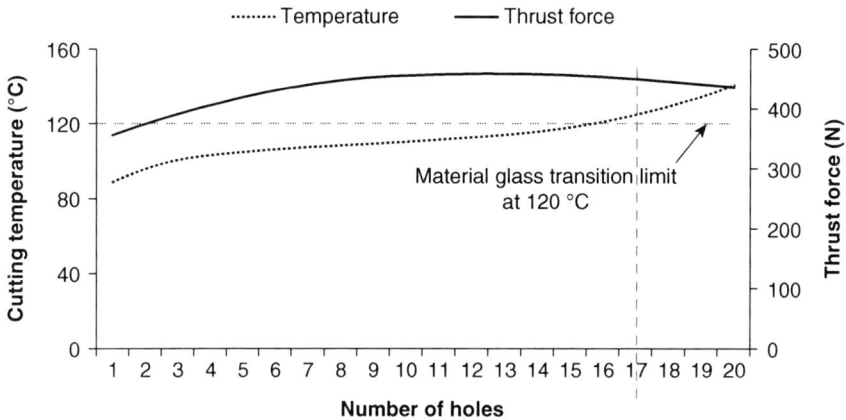

13.5 The effect of tool wear on the maximum temperature of the workpiece exit surface and on thrust force (drilling at 40 000 rpm and 8 m/min axial feed).

limit (~120 °C) until the sixteenth hole, after which the thrust forces started to decline due to, possibly, the matrix softening.

13.3.2 Cutting forces

Investigating the forces required for the hole making using different drilling parameters gives an effective indication of the quality of the drilling process, in terms of the produced hole quality, tool life, and energy consumption. As previously mentioned, one of the main objectives of employing high speed drilling is to achieve high productivity at relatively lower forces.

Figure 13.6 shows the effect of the rotational speed, within the low-to-high speed range, on the thrust and cutting forces.[8] It shows that the thrust and cutting forces decrease as the speed increases. It is evident that the benefits of high speed and high feed could be realized through the reduction of cutting forces during drilling.

The trends shown in Fig. 13.6 are in agreement with most of the work published in the open literature, e.g.[10–14] for low speed drilling (1000–4000 rpm) of different types of FRPs, using HSS and solid carbide tools. The force reduction with high cutting speeds was explained as being a result of the friction-induced temperature rise that causes polymer matrix softening.

The performance of VHSD (above 15 000 rpm) of CFRPs was investigated by the authors for high productivity axial feeds of 8 and 12 m/min (equivalent to drilling time of 60 and 40 ms, respectively, for an 8 mm thick plate). Figure 13.7 shows that for both axial feeds, the thrust forces decreased

(a)

Feed rate (mm/rev)

—◆—0.02 —■·0.1 --▲--0.4 ···×···0.8

(b)

Feed rate (mm/rev)

—◆—0.02 —■·0.1 --▲--0.4 ···×···0.8

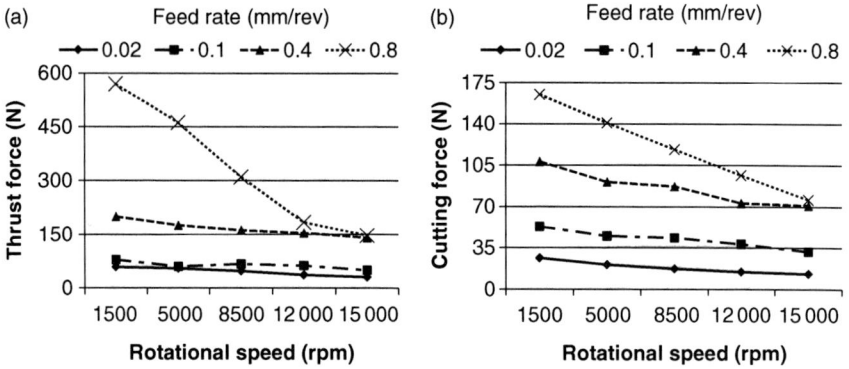

13.6 Effect of spindle speed on (a) thrust and (b) cutting forces in low and high ranges of cutting speed.[8]

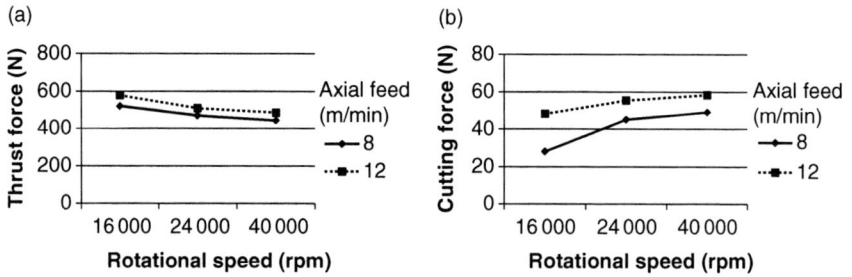

(a)

(b)

13.7 Very high speed drilling of woven CFRP laminate. (a) Thrust and (b) cutting forces.

with higher rotational speeds due to the decrease in the feed per tooth. This implies that HSD can make it possible to drill holes at high productivity axial feeds that are unworkable in the lower cutting speed range because of the defects (e.g. delamination) associated with the relatively high thrust force. One can also observe that the higher axial feed resulted in higher thrust forces for this range of rotational speeds. An increase in the axial feed leads, however, to an increase in cutting forces, as shown in Fig. 13.7b. This is attributed to the tool corner wear and the significant lateral friction between the tool and the workpiece at high rotational speeds.

Work carried out by Lin and Chen[6] on HSD of woven CFRPs investigated the drilling forces at high and very high speed ranges (210–850 m/min) in combinations with low feed rates. In Fig. 13.8, the thrust and cutting forces increase significantly at higher rotational speeds as the feed rate increases. It was explained that as the rotational speed increases, the effect

(a)

(b)

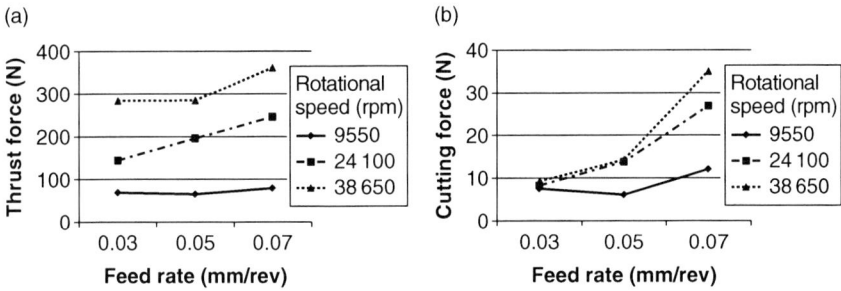

13.8 Effect of feed rate and ultra-high rotational speeds on (a) thrust forces and (b) cutting forces for drilling of woven CFRP.[6]

of lateral friction and tool wear become more significant, thus increasing the cutting forces for a fixed feed.

Aside from the effect of the feed per tooth on the cutting forces, the increase in cutting forces at higher cutting speeds could also be attributed to the strain rate, and the lateral friction between the tool and the workpiece.[15] Although the pressing of the cutting edge on the soft matrix at higher temperatures may result in force reduction, this effect is not expected to be significant due to the superior fiber strength compared with the matrix strength.

13.4   Tribological aspects in HSD of FRPs

13.4.1 Friction at the tool–workpiece interface

Friction between metallic components is governed by various mechanisms; namely, adhesion, deformation, abrasion, and mechanical interlocking. When one of the contacting bodies is a composite material, the additional visco-elastic effect of the flexible resin plays an important role. A model that describes this effect and its dependence on the sliding speed V has been proposed in:[16,17]

$$\mu = (a + bV) \times e^{(-cV)} + d \qquad [13.1]$$

where μ is the coefficient of friction and the constants a, b, c, and d are dependent on the material combination, the normal pressure, and the surrounding medium. It shows that, in principle μ increases with the sliding speed due to the initial predominance of the bracketed term, passing through a peak that is followed by an overall decrease as the exponential term becomes the dominant one. The influence of increasing the normal contact pressure was found to reduce μ and to shift the visco-elastic peak towards a lower speed. Other factors that complicate the friction between the cutting tool and the composite material include the peeling-off of the fibers from the matrix under the action of shear deformation,[18] as well as the fiber orientation and volume fraction, which affect the series of events leading to crack formation and fracture of the fibers and creation of chips.[19] This complex phenomenon explains the seemingly contradictory results on the effect of sliding speed on the friction of composite materials. While the results presented by Mondelin et al.[20] showed that sliding velocity has no significant influence on friction coefficient μ at the tool–work material interface when machining CFRP with monocrystalline diamond in the range 10–20 m/min, the results presented by Lancaster[21] demonstrated that in the range of high velocities, the material elastic behavior is prevalent in the contact zone and, as a result, the friction force decreases with increase in sliding velocity. The positive effect of the sliding speed was reported by Danaelan and Yousif,[22] who showed that the friction coefficient between CFRP composite and stainless steel increases as the sliding velocity increases in the range of 0.1–0.28 m/s.

13.4.2 Tool wear mechanisms and progression

The tool wear mechanisms encountered in the machining of composites are characterized by some unique features, due to the thermo-mechanical interactions of the tool–workpiece system.

(i)   *The thermal aspect.* As observed by Malhotra,[10] the temperature rise on the tool cutting edge during the machining of (CFRP) may exceed a threshold level of 300 °C[7,9] that causes a reduction of fiber-matrix interfacial shear strength[23] and ultimately the matrix burnout and the acceleration of the fiber pullout.

(ii)   *The mechanical fracture aspect.* The cutting process in CFRP is entirely based on fracture, as opposed to shearing phenomenon in metals,[23] due to the presence of the fibers that impair uniform plastic deformation. Due to the brittleness of the thermoset matrix, chips tend to fracture at earlier stages in the form of a powder,[7,24] particularly at high machining speeds due to the increase in strain rate and the reduced chain sliding.[25] Also, owing to the inhomogeneous structure and the brittleness of carbon fibers, the indenting chisel edge of the drill causes the fracture of hard fibers inside the soft epoxy matrix.[26]

These two aspects make the abrasive tool wear more dominant through the following two modes:

(i)   *Hard abrasion by fractured WC grains.* Dynamic stresses are generated on the WC hard grains due to the impacts from the reinforcement, the broken fibers and the powder-like chips.[26] These stresses result in crack initiation and propagation inside the WC grains, which eventually cause the WC grains to fall away partly or wholly by brittle fracture,[27] These particles contribute to the three-body abrasion wear of the tool as they slide over the rake and clearance faces.

(ii)   *Soft abrasion mode.* Due to the relatively low hardness of carbon fiber (CF) relative to tungsten carbide (WC) grains, the WC grains cannot be abraded by the carbon fibers pulled out of the matrix. These fibers can, however, damage the relatively soft Co binder,[28–30] through a three-body abrasion process. As the binder wears deeper, the exposed area of the WC grains increases and the fracture of these grains by fatigue is accelerated. The removal of a small amount of the cobalt binder by the abrasive particles results in a rapid process of crack nucleation. Under the cyclic loading imposed by the abrasive particles, these cracks propagate into the subsurface, following a tungsten carbide grain inter-granular path, until microscopic spalling takes place. The formulation of this theory was carried out by Bardetsky *et al.*[31] It is evident that this wear mechanism is promoted by the machining-induced matrix damage; namely, matrix burnout, fiber pullout and fiber fracture.

Another wear mechanism, which operates in parallel, is attributed to the anisotropy of the mechanical properties of WC, which results in material removal by the shearing of very thin platelets parallel to its prismatic plane. It was demonstrated by Jia and Fischer[32] that the prismatic plane hardness (H = 11 GPa) is much smaller than that of the perpendicular plane (H = 22 GPa). This explains why the relatively soft silicon nitride (H = 18 GPa) does not wear, but the harder WC–Co material is transferred to it.[32] Intensive adhesion wear of the WC tools in contact with the chip and the machined surface of fiber glass reinforced polymeric composite has also been observed under some cutting conditions.[33]

The mechanism and progression of solid carbide tool wear in HSD of CFRPs was studied thoroughly by Rawat and Attia.[34] It was shown that the hardness of the cobalt binder varies between 410 and 1225 HV. The overall hardness of the composite material H_{CFRP} is 80 Barcol, which is equivalent to 418 HV. The measured hardness H_{epoxy} of the epoxy matrix is 73 HV. With a fiber volume fraction of $V_f = 60\%$, and by applying the rule of mixture, the hardness of the carbon fiber is estimated as $H_{CF} = 648$ HV. The twist drill is subjected to aggressive abrasive action in the presence of hard graphite fibers embedded inside the soft epoxy matrix as well, as fractured tungsten carbide (WC) grains. The fluctuating chip load acting on the cutting edge leads to completely different wear characteristics of the WC drills when used for drilling CFRP composites, as compared to metals.

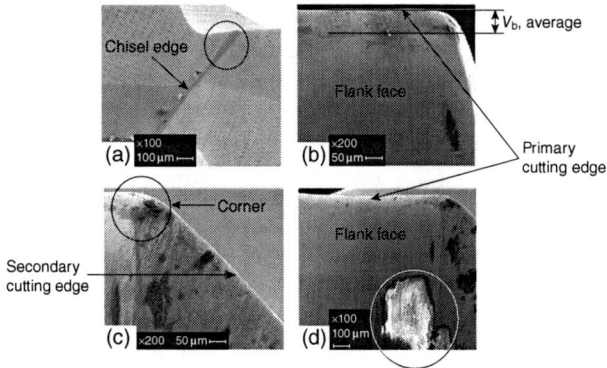

13.9 SEM images of wear damage observed during drilling at a spindle speed of 15 000 rpm and feed rate of 0.1 mm/rev: (a) chisel edge wear, (b) flank wear on the primary cutting edge showing abrasive wear, (c) rounding of corner of drill, and (d) adhesion of carbon on the flank face and the corner of the drill.[34]

Rawat and Attia's analysis[34] showed that fracture (chipping) and abrasion were the main mechanisms observed in HSD of deep holes in CFRP at 12 000 and 15 000 rpm and feed rate of 0.1 mm/rev. At the beginning of the deep hole drilling process, chipping was observed at the rake faces of the chisel edge and the primary cutting edge, the corner of the drill, and the secondary cutting edges. This was attributed to the brittleness of the WC tools that makes them unable to sustain the high stress concentration on the sharp edge at the beginning of the drilling pass.

Figure 13.9 depicts the abrasive wear observed at the chisel edge, the flank face of the primary cutting edge, and the secondary cutting edge.[34] Tool wear by abrasion was observed on the rake and flank faces of the primary cutting edge of the WC drill as a result of hard abrasion by fracture of the WC grains and soft abrasion mode.

The abrasive wear was found to be more severe on the flank face compared with that on the rake face of the primary cutting edges of the WC tool used for HSD of FRPs. The fractured WC grains and the powdery chips formed at low feed rates and high spindle speeds rub against the flank face, resulting in a 'three-body' abrasive wear mechanism. On the other hand, the FRP powdery chips on the rake face of the primary cutting edge are free to escape through the flutes of the drill, causing less severe abrasion, unlike the famous chip sliding mechanism in the case of metals.

As shown in Fig. 13.9, an evidence of the adhesion of carbon residues on the flank face can be observed. However, the adhesion wear was not found to be as dominant as the abrasive wear of the WC tools in the study of Rawat and Attia.[34]

(a)

(b)

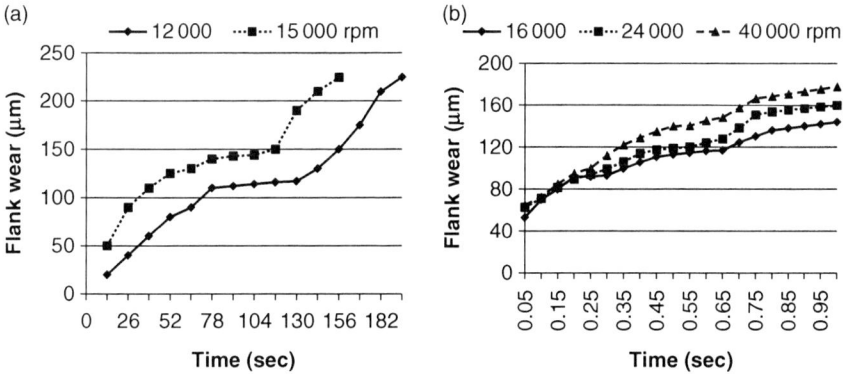

13.10 The effect of rotational speed and number of drilled holes on the flank tool wear: (a) 12 000 and 15 000 rpm, (b) 16 000, 24 000, and 40 000 rpm.

For the HSD of CFRPs, the progression in the flank wear is not uniform and can be divided into three distinct regions, as shown in Fig. 13.10a.

(i) *Initial (or primary) wear region.* Wear in this region is caused by chipping or micro-cracking. The high range of temperatures encountered in HSD catalyses the surface oxidation process that can significantly reduce the hardness and the abrasive wear resistance of WC–Co. At the start of drilling, the new cutting edges, having sharp corner radii, carry cutting forces over relatively small chip contact areas. Consequently, the extremely high contact pressure results in bulk sub-surface flow and causes the tool–workpiece system to behave as a heavily loaded system,[26,34] resulting in high wear rate.

(ii) *Steady wear (secondary) region.* After the initial wear or cutting edge rounding, the increase in the area of contact between the tool and workpiece results in lower contact stresses. As the tribo-system becomes a lightly loaded system, the wear rate is reduced and becomes nearly constant with the increase in the sliding distance (or time). This is also accompanied with improvement in the surface micro-roughness.

(iii) *Severe (ultimate, catastrophic or tertiary) wear region.* As the flank wear reaches a second critical value, the cutting force and temperature increase rapidly. The combined effect of thermal softening of the workpiece and tool materials and the increase in applied contact pressure cause the tribo-system to behave again as a highly loaded system, exhibiting a sharp increase in wear rate. These self-induced changes in the tribological system under consideration underline the very nature of this nonlinear dynamic system. Compared with the case of high rotational drilling speeds, where flank face wear can progress through initial, steady, and ultimate wear stages over around 500 holes, the VHSD drilling tool can reach severe levels of flank wear after 20 holes, as shown in Fig. 13.10b, which was obtained in the experiments conducted by the authors.

The axial feed used for the tool wear plot in Fig. 13.10 is considered to be in the ultra-high range of axial feeds, which was meant in order to test the performance of VHSD with highly productive feeds. The drilling tool experienced a rapid evolution of flank tool wear under such conditions. The direct relationship between the rotational speed and flank wear is clear from Fig. 13.10. The tool wear trend reported by Lin and Chen[6] was considered, in order to compare the effect of the axial feed on the tool wear, which can be projected on the balance of productivity and tool cost. Lin and Chen[6] reported the tool flank wear after 95.4 mm of drilled length (16 holes) at 38650 rpm to be close to that shown in Fig. 13.10 for 40000 rpm after the same number of holes. Thus, it could be tentatively concluded that the effect of axial feed on tool wear could be minor compared to the effect of the rotational speed.

13.4.3 Tribological interactions

The analysis conducted in Rawat and Attia[34] on the effect of tool wear on HSD forces showed that both the thrust force and cutting force increased with the increase in flank wear, as shown in Fig. 13.11. The entry and exit delamination factors were observed to be increasing as a consequence of the increase in the cutting and thrust forces, respectively, with the increase in the flank tool wear. For VHSD, Fig. 13.12 shows a significant increase in the thrust and cutting forces with the increasing drilling length, because of the accelerated tool wear associated with high speeds.

13.5   Hole quality

Sanjay et al.[8] developed machinability maps, shown in Fig. 13.13, for drilling of woven CFRP laminates using a wide range of low to high rotational speeds and feed rates. The produced maps can be used as a process optimization tool that relates the hole quality attributes (namely entry delamination, hole size accuracy, hole circularity, and surface quality) to the drilling parameters. Sanjay et al.[8] demonstrated a case of process parameter selection that satisfies certain hole quality and productivity requirements using machinability maps to illustrate the proposed approach. It is worth noting that due to the heterogeneous nature of composites, the variability of the test results was reported to be acceptable and to fall within 10%.

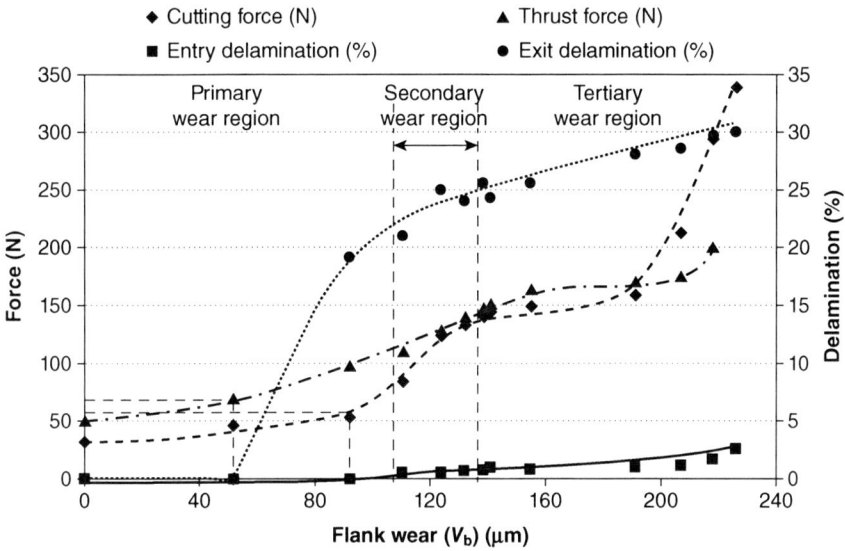

13.11 Effect of flank wear on thrust force, cutting force, entry delamination and exit delamination at spindle speed of 15 000 rpm and feed rate of 100 μm/rev.[34]

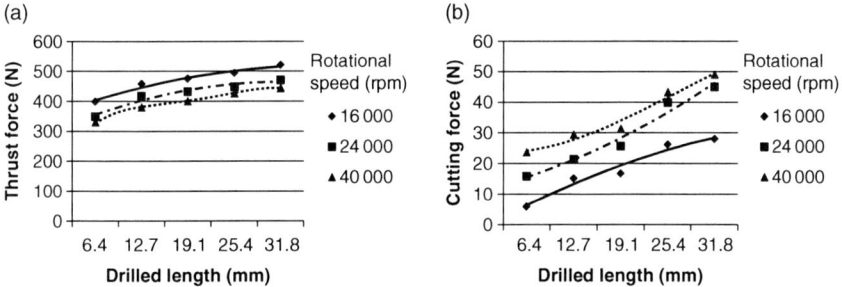

13.12 Effect of rotational speed and tool wear on (a) thrust and (b) cutting forces.

13.5.1 Delamination: exit, entrance, internal

Delamination is the most common and most critical defect associated with machining of FRP laminate composites. It can severely deteriorate the mechanical performance of FRP materials in service.[35,36] Delamination is a major challenge in the machining of laminate and this explains the extensive trials that have been done to quantify, correlate or to model delamination in relation to the machining parameters.[12,25,37,38] Entry and exit

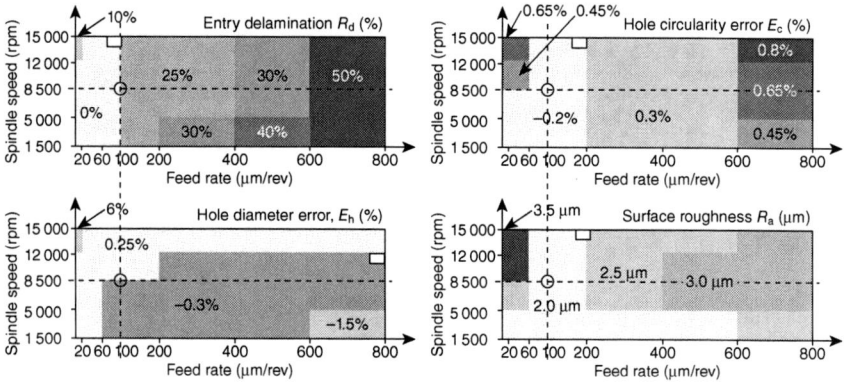

13.13 Machinability maps for drilling of CFRP at low to high rotational speeds and feeds.[8]

delamination are damage types that take place in the transient sections during drilling and are controlled by various process parameters (e.g. rotational speed, feed, and drilling tool point angle).[39,40] The delamination damage could be quantified by a delamination factor ($\phi_d = D_{max}/D_{nominal}$), where D_{max} is the maximum diameter that contains the observed delamination zone, and $D_{nominal}$ is the nominal hole diameter.

The study conducted by Tagliaferri et al.[41] focused on finding a relationship between the rotational speed to feed ratio and the size of delamination during drilling of GFRPs. Di Paola et al.[42] monitored the crack growth propagation as the drilling tool exited the hole. Enemuoh et al.[43] developed multi-objective function optimization, one of which was to obtain delamination-free holes in CFRPs. Their results recommended using high rotational speeds and low feeds for producing delamination free holes. The trends of delamination size in references 41 to 46 were found to agree with each other, as well as with the trend reported in reference 8. They attributed this increase in delamination size to the thrust forces that increased with the increase in feed rate.

Figure 13.14 shows the entry and exit delamination factors associated with VHSD of a woven CFRP laminate, conducted by the authors, at rotational speeds of 16 000 rpm, 24 000 rpm, and 40 000 rpm and an axial feed of 8 m/min. A direct relationship between thrust and cutting forces and the exit and entry delamination has been defined, respectively, for a wide range of low speed and HSD parameters. However, the VHSD with high axial feed exhibited a similar trend of exit delamination and a reversed trend of entry delamination when compared with low speed and HSD. Furthermore, the delamination factor values shown in Fig. 13.14 for VHSD represent an average value of a considerably large variance.

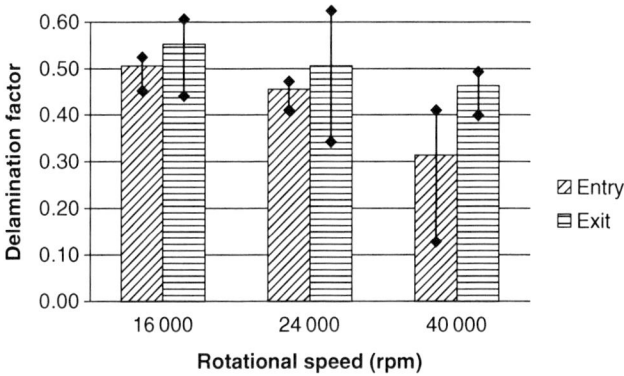

13.14 Effect of rotational speed with axial feed rate 8 m/min, on the entry and exit delamination factors.

Gaitonde *et al.*[39] developed a response surface methodology (RSM) based mathematical model in order to analyze the main and interaction effects of cutting speed (4000 rpm to 40 000 rpm), feed rate, and drilling tool point angle on the delamination factor. The delamination factors were shown to be more sensitive to the change in feed rate, and exhibited a nonlinear trend with the change in cutting speeds. Rubio *et al.*[47] investigated the delamination of CFRP laminates through the same range of cutting speeds with high axial feeds (1000 mm/min to 9000 mm/min) and point angles 85° and 115°. The analysis in references 39 and 47 support the trend depicted in Fig. 13.14 in the sense that the delamination size was found to decrease with the increase in rotational speed for a fixed axial feed.

In the case of entry delamination, the peeled-up layer in the vicinity of the chisel edge is very thin and narrow because of the brittleness of the thermoset resin, and has a considerable thickness of backing material underneath. This allows the primary cutting edges of the drilling tool, rotating at high speed, to instantaneously cut the narrow peeled-up layers before the delamination propagates beyond the diameter of the final hole. On the other hand, in the case of low rotational speeds, the entry delamination has enough time to propagate beyond the diameter of the final hole.

In the case of exit delamination, the thin bottom layer acts as its own backing material, in which the fibers can be free to bend outwards and break at a position beyond the edge of the final hole rather than being cut. For the same axial feed, higher rotational speeds reduce the thrust forces, which accordingly reduce the severity of exit delamination. Figures 13.15a, b, and c show an internal section along the depth of the holes drilled at the axial feed rate of 8 m/min and rotational speeds 16 000 rpm, 24 000 rpm, and 40 000 rpm, respectively. In Fig. 13.15a, a distribution of voids that represents

(a)   (b)   (c)

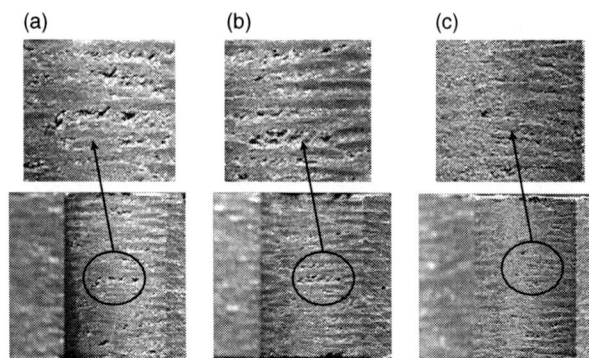

13.15 Internal delamination for holes drilled at feed rate of
8 m/min and rotational speeds of (a) 16000 rpm, (b) 24000 rpm
and (c) 40000 rpm.

internal delamination at different levels along the hole depth can be clearly
seen for the case of 16000 rpm. These voids almost disappeared in Fig.
13.15c, for the case of 40000 rpm where the thrust forces are lower than
the case of 16000 rpm.

13.5.2 Surface quality

The surface roughness of CFRP drilled holes could give some indication on
the level of damage of the produced hole in terms of fiber pullout voids,
fiber fuzziness, and internal delamination voids. However, this requires
extensive characterization in order to develop a model that relates a certain
type of damage relative to the measured surface hole quality. The machinability maps developed in reference 8 showed the average surface roughness 'R_a' relative to cutting speed and feed rate. As shown in Fig. 13.13, the
increase in surface roughness with the increase in feed was more significant
than the change in surface roughness with the cutting speed. The surface
roughness values obtained for VHSD, determined by the authors and shown
in Fig. 13.16, exhibit a similar trend to that of the holes produced by HSD
in Fig. 13.13. It is worth mentioning that the extreme variability of the material structure along the depth of the hole results in inconsistent trends in
measured surface roughness for some cases.

13.5.3 Hole size error

The machinability maps developed by Rawat and Attia[8] for HSD of CFRPs
showed that the hole size error turned from a positive error (oversize) to
a negative error (undersize) as the feed rate increased and the speed

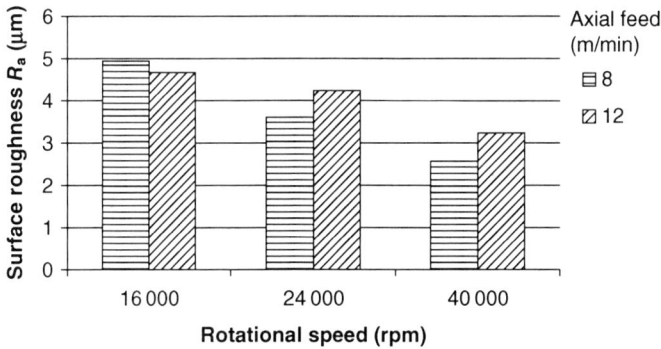

13.16 Effect of rotational speed and axial feed on the surface roughness of CFRP holes performed by VHSD.

decreased. The rotational speed showed insignificant effect on the hole circularity at low and medium feed rates, while the hole circularity error increased at high feed rates with high speeds. The best surface quality was shown to be achieved at low and medium feeds with the entire range of speeds.

13.6   Overview of high speed milling of FRPs

Little is known about the high speed slotting/routing of composites. Most of the work has focused on the drilling of carbon fiber-reinforced plastics (CFRPs). A preliminary study has been presented for the milling of CFRPs.[48] The focus of the study was on the chip formation mechanism and the effect of the cutting direction with respect to the fiber orientation for unidirectional continuous carbon fiber-reinforced epoxy with a fiber volume fraction of 60%. In this study, the maximum spindle speed was only 3000 rpm using a single square carbide insert tool. Similarly, a study was conducted to characterize the chip formation mechanism in orthogonal edge trimming of unidirectional and multidirectional graphite epoxy laminate panels of 4 mm thickness with 3501–6 resin and IM-6 fibers composites.[49] The trimming was performed using polycrystalline diamond (PCD) tool inserts at surface cutting speeds in the range of 4 to 14 m/min. Experimental investigations were also presented for the milling of glass fiber reinforced plastics (GFRP) whereby a 5 mm, two-flute cemented carbide (K10) tool was used to mill polyester matrix with glass fibers.[50] The maximum speed and feed covered in this work were 7000 rpm and 0.06 mm/tooth, respectively. It was shown that the cutting forces and the surface roughness increase with increasing feed and decreasing speed, while the delamination increases with the increase of both the feeds and speeds. In a different study on the milling of CFRP,[51] the design of experiments was used to characterize the effect of the process parameters on the final part quality. The maximum speed and feed were limited to 2500 rpm and 0.29 mm/tooth, respectively. Similar trends as the previous studies were found in terms of the effect of the feed and speed on the forces, the surface roughness, and the delamination factor. In all the previous studies, the maximum feed rate was 850 mm/min.

In a study by Lopez De Lacalle *et al.*[52] the use of multi-tooth routers for the high speed trimming of Carbon and Kevlar fiber reinforced plastics was investigated. The main advantage of these cutters over helical end mills is that they tend to reduce the axial forces that are critical for the machining of composite structures and plates with lower rigidity in the transverse direction. The multi-tooth routers, made with micro-grain carbide, were tested at high spindle speeds of 16 000 and 18 000 rpm and a feed of 0.2 mm/rev. It was found that the wear was mainly due to abrasion; thus, thicker coatings contributed to increasing the tool life. It was also found that new coatings based on nanostructures (TiAlN and SiC) did not provide better results than AlTiN coating. For the lay-up material that was used in these experiments, the surface finish was better when up-milling was used as opposed to down-milling. The use of a multi-tooth tool with micro-grain carbide and 6% Co, coated with a 4 μm-thick monolayer of TiAlN increased the tool life to 50 m of machined length after which a poor machined surface was detected. PCD tools, which are substantially more expensive than carbide tools, were shown to improve the tool life as compared to solid carbide end mills, but they were outperformed by multi-tooth cutters for high speed milling conditions.

Due to the lack of information on high speed milling of composites, all the results presented in the sections to follow were generated by the authors at the Aerospace Manufacturing Technology Centre (AMTC) of the National Research Council of Canada (NRC). A 6.35 mm, four-flute, solid carbide end mill tool was used. The slotting was performed along the full thickness of the composite material. For the same cutting speed and feed, the slotting was performed using the following tool overhang lengths (TL): TL1 = 38 mm, TL2 = 31 mm, and TL3 = 24 mm. The tool wear was investigated on the flank and the rake faces after each 32 mm of cutting distance, while the total cutting distance was kept at 96 mm. The following test matrix was used:

- spindle speed (rpm): 10 000, 20 000, 30 000, and 40 000;
- feed (mm/min): 250, 500, and 1000.

For the size of the tool used in these experiments, the above spindle speeds translate into surface speeds of 200 to 800 m/min. In addition, to compare the effect of the feed per tooth rather than the feed rate, additional experiments were conducted for the same speed and the following feeds

(μm/tooth): 6.25, 12.5, and 25. For the rotational speed range of 10000 to 40000 rpm, these feeds are equivalent to 250 to 4000 mm/min. During slotting, the forces were measured using a Kistler dynamometer 9255B and the temperatures were measured using an infra-red camera. After machining, the slots were characterized in terms of the straightness errors, the dimensional errors, the surface roughness, and the delamination. To assess the repeatability, the cuts were repeated twice, using sharp tools each time.

13.7   Dynamic characteristics in high speed milling of FRPs

For high speed machining, the effect of the tool dynamics plays an important role in terms of the quality of the machined surfaces. In this study, it was noticed that a slight increase in the tool over-hang could induce larger vibrations due to the loss of rigidity. To determine the frequency response function (FRF) of the tool, impact tests were performed whereby accelerometers were placed at the tip of the tools. The hammer impacts were applied at the same locations of the accelerometers. By comparing the FRFs in the x- and y-directions, shown in Fig. 13.17, one can notice that increasing the tool length by only 7 mm (equivalent to the tool diameter) resulted in a decrease in the magnitude of the dominant natural frequencies by more than 30%, representing a substantial loss of rigidity. This would be expected to induce large variations in the cutting forces and temperatures for large tool lengths (TL1: 38 mm and TL2: 31 mm), which would certainly affect the surface roughness, the slot dimensional errors, and the straightness. In general, for large tool lengths and small feeds, chatter was observed due to the cyclic engagement and disengagement between the tool and the

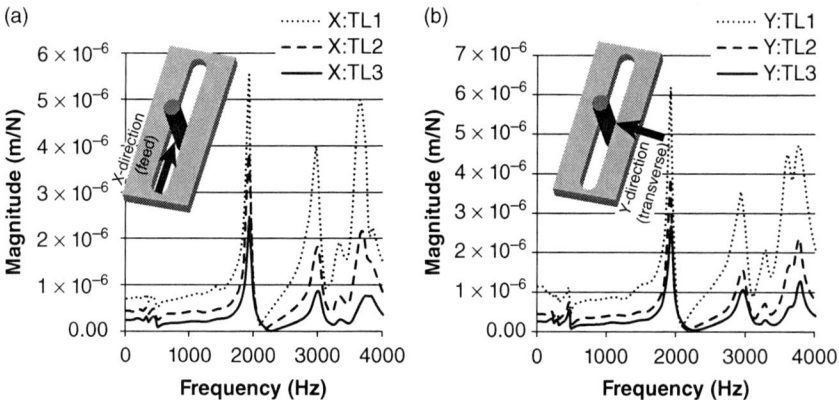

13.17 FRF for end mills in x- and y-directions for different tool lengths.

workpiece. From the FRF analysis, it was found that the dominant natural frequencies of the tools were 2000 Hz, 3000 Hz, and 4800 Hz. By comparing the speeds at which the tools were running (10 000–40 000 rpm), and taking into account the effect of the tooth passing frequency, the frequencies of the machining forces fall in the range of 166 Hz to 2600 Hz. For a speed of 30 000 rpm, the tooth passing frequency is 2000 Hz, which coincides with the most dominant natural frequency of the tool (see insets in Fig. 13.7). As will be shown in the following sections, this results in higher geometrical errors and surface roughness as compared to the other parameters.

13.8 Cutting forces and thermal aspect in high speed milling of FRPs

13.8.1 Cutting forces

The three measured components of the cutting forces were the feed force along the slot direction, the transverse force perpendicular to the slot, and the axial force along the axis of the tool (see the insets of Fig. 13.17). It was found that as the speed increased, the forces decreased until a speed of 30 000 rpm, and then they started to slightly increase, as shown in Fig. 13.18. It is expected that as the speed increases for the same feed rate, the feed per tooth will decrease, thus leading to a decrease in the cutting forces. At the same time, the small feed per tooth could lead to higher cutting pressures that could lead to an increase of the cutting forces. In addition, the increase in speed could induce higher frictional forces between the tool and the workpiece. These opposing effects could explain the change of the forces above 30 000 rpm. It was found that the axial forces followed

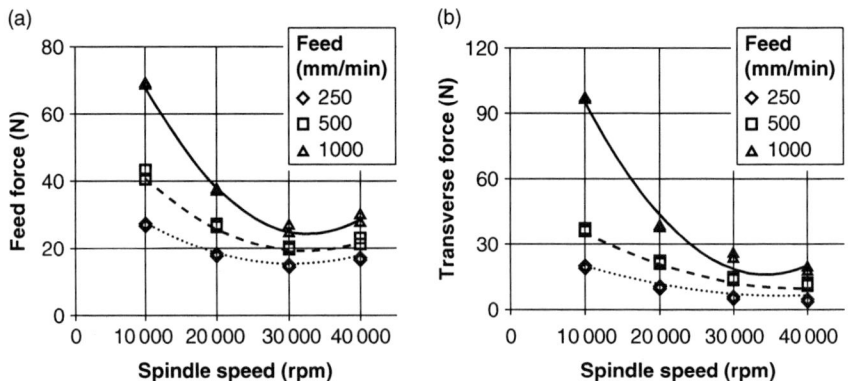

13.18 Effect of speed on the forces for different feed rates, using a sharp tool.

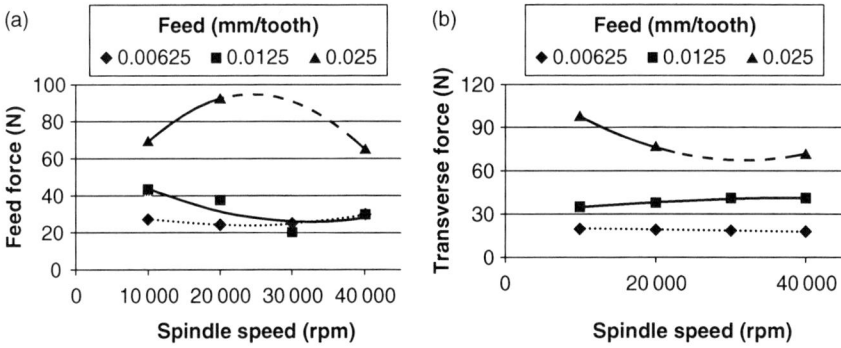

13.19 Effect of speed on the forces for different feed rates per tooth, using a sharp tool.

the same trend; however, their values were lower than the feed and the transverse forces. The repeatability of the measured forces was found to be less than 4% which is shown in Fig 13.18 by having two points for each test. For all the cases, a direct linear proportionality was found between the feed and the forces for a given speed.

To focus on the effect of the feed per tooth rather than the feed rate, the forces were plotted for fixed values of feeds per tooth as shown in Fig. 13.19. For low values of feed per tooth, the effect of the speed on the feed and transverse forces was minor. For higher values of feeds per tooth, the feed force tended to increase with speed, and then it gradually decreased, which can be attributed to the thermal softening that could occur to the workpiece material. The transverse forces tended, however, to decrease throughout the full range of spindle speeds. It is worth noting that at 30 000 rpm and 0.025 mm/tooth feed, the tool broke, while at lower and higher speeds, the cutting proceeded smoothly. This could be attributed to the dynamics of the tool at 30 000 whereby, its natural frequency coincided with the tooth-passing frequency.

Due to the high rotational speeds of the tool, the dynamics of the tool play an important role in the process. It was found that a small change in the stiffness of the tool had a significant impact on the cutting forces. The large vibrations of the tools were the source of the large force variations. These vibrations were more pronounced for the tools with larger over-hangs due to the increased flexibility. In addition, the damping of the vibration was small for the small values of feeds per tooth due to the relatively small engagements of the tool with the workpiece. An example of the measured forces is shown in Fig. 13.20, whereby, the slotting was conducted at 40 000 rpm and 2000 mm/min feed. As can be seen, for the tool length of 38 mm, the variations of the forces exceeded 30 N in the feed and transverse

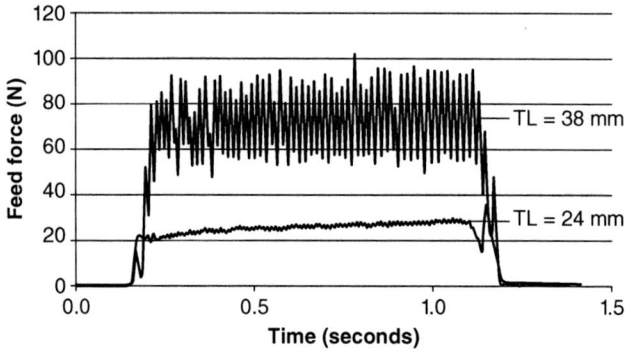

13.20 Force signal for slotting at 40 000 rpm and 2000 mm/min using different tool lengths.

directions. As the cutting action took place with smaller tool lengths, the variations were considerably reduced. The large variations of the forces resembled the cases of vibration-assisted drilling, where the tool is having intermittent cutting actions. This was found to be advantageous in the drilling process if properly controlled. In this case, however, it was found that the forces tended to increase with higher over-hangs in most of the cases.

13.8.2 Cutting temperature

The temperature of the tool was measured at its outer surface (at the top plane of the workpiece), as shown in the inset in Fig. 13.21a. For the slotting operation with a tool length of 24 mm, the temperatures are shown in Fig. 13.21. Similar to the trends observed with the cutting forces, the temperature decreased as the spindle speed increased until 30 000 rpm, and then it started to increase again. This could be explained as follows: As the speed increases, the feed per tooth decreases, which is expected to generate lower forces and temperatures. At the same time, very small feeds per tooth could result in a higher specific cutting pressure which will tend to increase the cutting forces and temperatures. The effect of the feed per tooth on the temperature is shown in Fig. 13.21b, whereby for a given feed per tooth, the temperature tended to increase with increasing speed.

The tool length and the feed per tooth also affect the cutting and tool temperature variation. For large over-hangs and small feeds per tooth, the variations in temperatures were higher. This is mainly attributed to the dynamic engagements of the tools as it is progressing in the slotting operation due to the vibrations of the tool. As shown in Fig. 13.22, the higher feeds and shorter tools tended to generate more stable cuts since the flutes are more restricted during cutting.

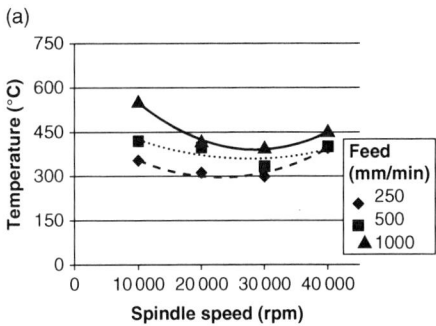

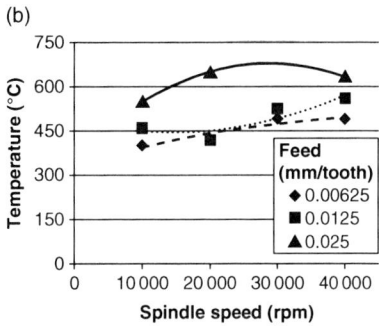

13.21 Effect of spindle speed and (a) feed rate, and (b) feed per tooth, on the cutting temperature using a sharp tool.

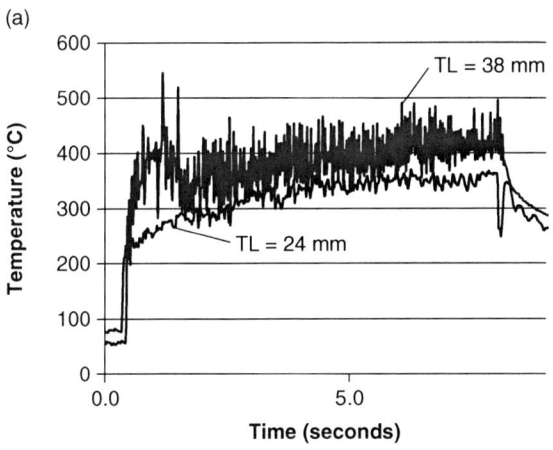

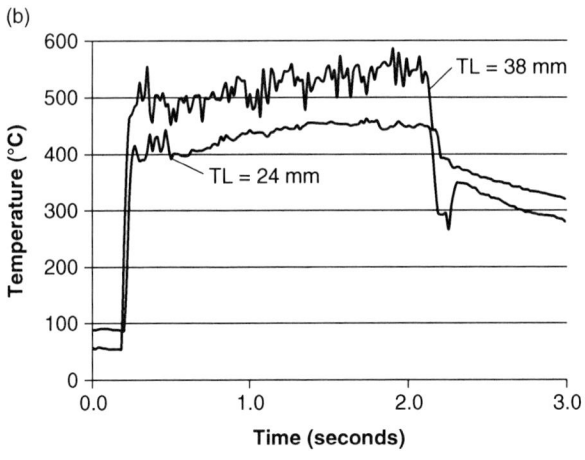

13.22 Temperature signals for slotting at 30 000 rpm, for different tool lengths. (a) Feed rate: 250 mm/min, (b) feed rate: 1000 mm/min.

Tribological aspect

Due to the relatively short length of the cut, a small amount of wear was noticed on the flank face with a maximum magnitude of 40 μm. On the rake face, isolated cases of edge chipping and abrasion were observed. This observation cannot, however, be generalized as a trend. For the feeds of 250 and 500 mm/min, the wear was highest at 40000 rpm while for the 1000 mm/min feed the highest wear was at 30000 rpm. For all the cutting conditions that were tested, the effect of the cutting length on the milling forces and the temperatures was analyzed and plotted as shown in Figures 13.23 and 13.24. As expected, even if the tool wear is small, an increase in the forces and the temperatures with the progression of the cutting process was noticed.

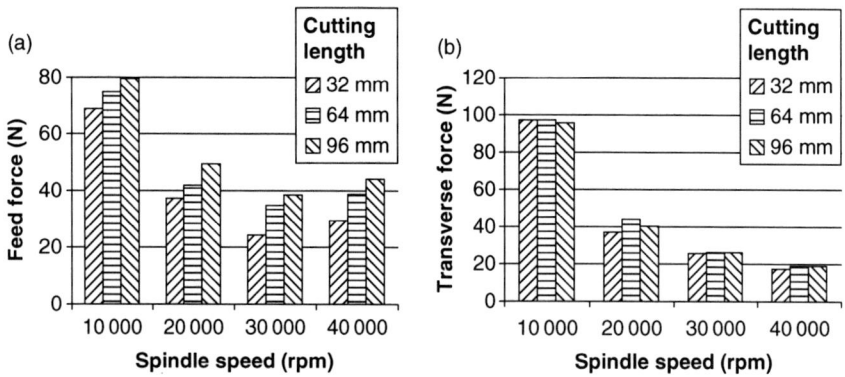

13.23 Effect of the tool wear for a feed rate of 1000 mm/min on: (a) feed force, (b) transverse force.

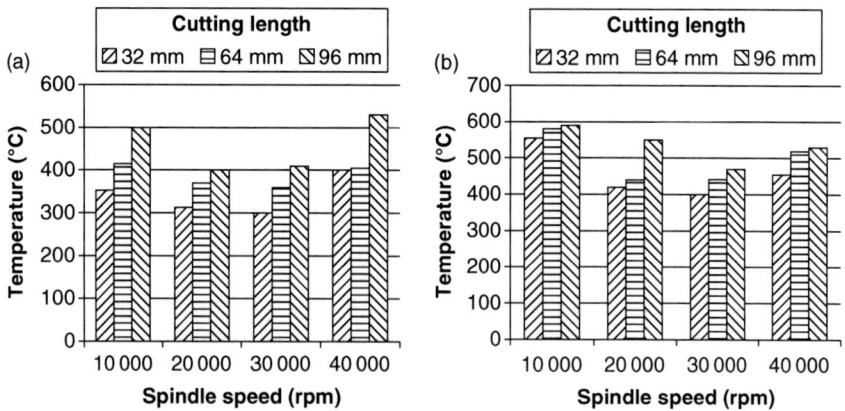

13.24 Effect of the tool wear on temperature for feed rates of (a) 250 and (b) 1000 mm/min.

13.9   Surface quality and geometrical errors

In the high speed milling of composites, the surface can be subjected to high forces and temperatures. These will tend to affect the quality of the machined surfaces as well as the dimensional and geometrical accuracy of the machined features.

13.9.1 Dimensional and geometrical errors

The dimensional and geometrical errors are governed by the deflection and vibrations of the tool relative to the workpiece, the tool wear, and the thermal deformations. As expected, it was shown in this study that for large tool lengths, the machined surfaces tended to lose straightness as compared to the short tool length, as shown in Fig. 13.25. In general, it was found that the straightness errors increased with increasing feeds. This is mainly attributed to the increase in the cutting load which leads to higher tool deflections. For large tool lengths (TL1), the straightness errors increased until 30000 rpm, and then it starts to decrease. Similar trends could be detected for the slot dimensional errors, as shown in Fig. 13.26, whereby the small tool over-hangs tended to generate precise slots with errors less than 0.4%, while for TL1 and TL2, the dimensional errors could reach up to 8.0%. One can notice that the increase of the tool length from 24 mm to 31 mm led to a relatively higher increase in errors as compared to the change from 31 mm to 38 mm. For the all the geometrical measurements, the repeatability errors were found to be around 20%.

13.9.2 Surface quality

For a given feed rate, the decrease in the feed per tooth with increasing speed, which generates lower forces, temperatures, and less tool deflections, results in a better surface finish. This can be noticed in the trends in

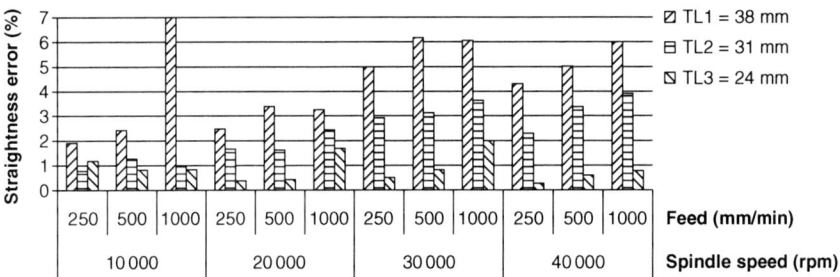

13.25 Effect of speeds on slot straightness for different tool lengths.

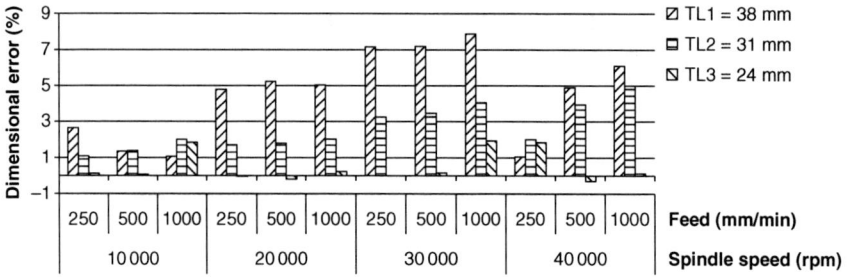

13.26 Effect of speeds on dimensional errors for different tool lengths.

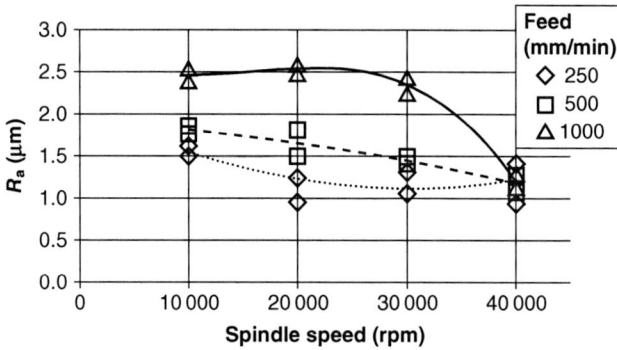

13.27 Effect of the speed on the surface roughness for different feeds, using a sharp tool.

Fig. 13.27. In addition, the effect of increasing the feed rate for the same speed resulted in an increase in the surface roughness. One can note that it was possible to achieve a better surface finish while increasing the speed more than four times and maintaining the same feed. Large vibrations were induced for large tool lengths which were detrimental for the surface finish. Fig. 13.28 shows the difference in the surface roughness when the tool length is decreased. For the long tools (TL1 and TL2), the lower the feeds and speeds, the better was the surface finish.

13.9.3 Delamination

In the routing process, the axial forces are relatively lower than the feed and transverse forces. Consequently, the surface delaminations were lower than in the cases of drilling. For this study, surface delamination and fiber tearing were noticed only when the direction of the fibers in the top laminate was perpendicular to the feeding direction of the tool. The vibrations of the tools, the feed rate, and the speed played a major role when it came

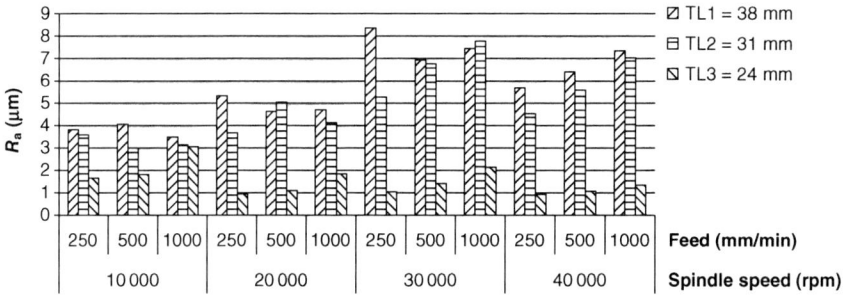

13.28 Effect of the speeds on the surface roughness for different tool lengths.

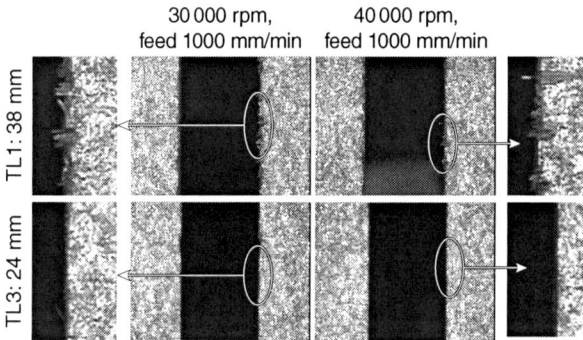

13.29 Images for the top surfaces of the slots for a feed rate of 1000 mm/min. TL1: 38 mm and TL3: 24 mm.

to the delamination and fiber tearing. Fig. 13.29 shows the images of the top surface of the slots for the tool lengths 38 and 24 mm, speeds of 30 000 and 40 000 rpm and feed of 1000 mm/min, respectively. One can see that for the large tool length, pronounced fiber tearing and delamination were generated at the edges of the slots. In general it was noticed that more damage was generated at the speed of 30 000 rpm. This is mainly attributed to the high flexibility of the tool and the dynamics of the tool, whereby the natural frequency of the tool coincided with the tooth-passing frequency of the cutting forces. For the short tool lengths, the quality of machined slots was very comparable.

13.10 References

1 King, R. I., 1985, *Handbook of High Speed Machining Technology*, Chapman and Hall, New York.

2 Erdel, B. P., 2003, *High-speed Machining*, Society of Manufacturing Engineers, Dearborn, Michigan.

3 Schulz, H. and Moriwaki, T., 1992, High-speed machining, *CIRP Annals – Manufacturing Technology*, **41**(2), pp. 637–643.

4 Smith, S. and Tlusty, J., 1987, Update on High-speed Milling Dynamics, *Symporium on Intelligent Manufacturing, Proc. American Society of Mechanical Engineers, Production Engineering Division (PED)*, Vol 25, pp. 153–165.

5 Longbottom, J. M. and Lanham, J. D., 2006, A review of research related to Salomon's hypothesis on cutting speeds and temperatures, *International Journal of Machine Tools and Manufacture*, **46**(14), pp. 1740–1747.

6 Lin, S. C. and Chen, I. K., 1996, Drilling carbon fiber-reinforced composite material at high speed, *Wear*, **194**(1–2), pp. 156–162.

7 Chen, W.-C., 1997, Some experimental investigations in the drilling of carbon fiber-reinforced plastic (CFRP) composite laminates, *International Journal of Machine Tools and Manufacture*, **37**(8), pp. 1097–1108.

8 Rawat, S. and Attia, H., 2009, Characterization of the dry high speed drilling process of woven composites using machinability maps approach, *CIRP Annals – Manufacturing Technology*, **58**(1), pp. 105–108.

9 Weinert, K. and Kempmann, C., 2004, Cutting temperatures and their effects on the machining behaviour in drilling reinforced plastic composites, *Advanced Engineering Materials*, **6**(8), pp. 684–689.

10 Malhotra, S. K., 1990, Some studies on drilling of fibrous composites, *Journal of Materials Processing Technology*, **24**, pp. 291–300.

11 Davim, J. P. and Reis, P., 2003, Drilling carbon fiber reinforced plastics manufactured by autoclave Experimental and statistical study, *Materials & Design*, **24**(5), pp. 315–324.

12 El-Sonbaty, I., Khashaba, U. A. and Machaly, T., Factors affecting the machinability of GFR/epoxy composites, *Composite Structures*, **63**(3–4), pp. 329–338.

13 Tsao, C. C., 2008, Investigation into the effects of drilling parameters on delamination by various step-core drills, *Journal of Materials Processing Technology*, **206**(1–3), pp. 405–411.

14 Tsao, C. C., 2008, Experimental study of drilling composite materials with step-core drill, *Materials & Design*, **29**(9), pp. 1740–1744.

15 Ahmad, J., 2009, *Machining of Polymer Composites*, Springer, New York.

16 Moore, D. F., 1975, *Principles and Applications of Tribology*, Pergamon Press, Oxford; New York.

17 Kragelsky, I. V., Dobychin, M. N. and Kombalov, V. S., 1982, *Friction and Wear-Calculation Methods*, Pergamon Press, Oxford.

18 Tsukizoe, T. and Ohmae, N., 1983, Friction and wear of advanced composite materials, *Fibre Science and Technology*, **18**(4), pp. 265–286.

19 Koplev, A., Lystrup, A. and Vorm, T., 1983, The cutting process, chips, and cutting forces in machining CFRP, *Composites*, **14**(4), pp. 371–376.

20 Mondelin, A., Furet, B. and Rech, J., 2010, Characterisation of friction properties between a laminated carbon fibres reinforced polymer and a monocrystalline diamond under dry or lubricated conditions, *Tribology International*, **43**(9), pp. 1665–1673.

21 Lancaster, J. K., 1972, Lubrication of carbon fibre-reinforced polymers. Part I – Water and aqueous solutions, *Wear*, **20**(3), pp. 315–333.

22 Danaelan, D. and Yousif, B. F., 2008, Adhesive wear performance of cfrp multi-layered polyester composites under dry/wet contact conditions, *Surface Review and Letters*, **15**(6), pp. 919–925.

23 Yoda, S., Takahashi, R., Wakashima, K. and Umekawa, S., 1979, Fiber/matrix interface porosity formation in tungsten fiber/copper composites on thermal cycling, *Metallurgical Transactions A*, **10**(11), pp. 1796–1798.

24 Dharan, C. K. H., 1978, Fracture mechanics of composite materials, *Journal of Engineering Materials and Technology, Transactions of the ASME*, **100**(3), pp. 233–247.

25 Hocheng, H. and Puw, H. Y., 1992, On drilling characteristics of fiber-reinforced thermoset and thermoplastics, *International Journal of Machine tools and manufacture*, **32**(4), pp. 583–592.

26 Teti, R., 2002, Machining of composite materials, *CIRP Annals – Manufacturing Technology*, **51**(2), pp. 611–634.

27 Masuda, M., Kuroshima, Y. and Chujo, Y., 1993, Failure of tungsten carbide–cobalt alloy tools in machining of carbon materials, *Wear*, **169**(2), pp. 135–140.

28 Blombery, R. I., Perrot, C. M. and Robinson, P. M., 1974, Abrasive wear of tungsten carbide-cobalt composites. I. Wear mechanisms, *Materials Science and Engineering*, **13**(2), pp. 93–100.

29 Larsen-Basse, J., 1978, Abrasion mechanisms – delamination to machining, *Proceeding of International Conference on The Fundamentals of Tribology*, MIT press, Cambridge, MA, pp. 679–689.

30 Larsen-Basse, J. and Koyanagi, E. T., 1979, Abrasion of WC-Co alloys by quartz, *Journal of Lubrication Technology Trans ASME*, **101**(2), pp. 208–211.

31 Bardetsky, A., Attia, H. and Elbestawi, M., 2007, A fracture mechanics approach to the prediction of tool wear in dry high-speed machining of aluminum cast alloys – Part 1: Model development, *Journal of Tribology*, **129**(1), pp. 23–30.

32 Jia, K. and Fischer, T. E., 1997, Sliding wear of conventional and nanostructured cemented carbides, *Wear*, **203–204**, pp. 310–318.

33 Velayudham, A., Krishnamurthy, R. and Soundarapandian, T., 2005, Evaluation of drilling characteristics of high volume fraction fibre glass reinforced polymeric composite, *International Journal of Machine Tools and Manufacture*, **45**(4–5), pp. 399–406.

34 Rawat, S. and Attia, H., 2009, Wear mechanisms and tool life management of WC-Co drills during dry high speed drilling of woven carbon fibre composites, *Wear*, **267**(5–8), pp. 1022–1030.

35 De Albuquerque, V. H. C., Tavares, J. M. R. S. and Durão, L. M. P., 2010, Evaluation of delamination damage on composite plates using an artificial neural network for the radiographic image analysis, *Journal of Composite Materials*, **44**(9), pp. 1139–1159.

36 Persson, E., Eriksson, I. and Zackrisson, L., 1997, Effects of hole machining defects on strength and fatigue life of composite laminates, *Composites Part A: Applied Science and Manufacturing*, **28**(2), pp. 141–151.

37 Hocheng, H. and Tsao, C. C., 2005, The path towards delamination-free drilling of composite materials, *Journal of Materials Processing Technology*, **167**(2–3), pp. 251–264.

38 Hocheng, H. and Tsao, C. C., 2006, Effects of special drill bits on drilling-induced delamination of composite materials, *International Journal of Machine Tools and Manufacture*, **46**(12–13), pp. 1403–1416.

39 Gaitonde, V. N., Karnik, S. R., Rubio, J. C., Correia, A. E., Abrão, A. M. and Davim, J. P., 2008, Analysis of parametric influence on delamination in high-speed drilling of carbon fiber reinforced plastic composites, *Journal of Materials Processing Technology*, **203**(1–3), pp. 431–438.

40 Karnik, S. R., Gaitonde, V. N., Rubio, J. C., Correia, A. E., Abrão, A. M. and Davim, J. P., 2008, Delamination analysis in high speed drilling of carbon fiber reinforced plastics (CFRP) using artificial neural network model, *Materials & Design*, **29**(9), pp. 1768–1776.

41 Tagliaferri, V., Caprino, G. and Diterlizzi, A., 1990, Effect of drilling parameters on the finish and mechanical properties of GFRP composites, *International Journal of Machine Tools and Manufacture*, **30**(1), pp. 77–84.

42 Di Paola, G., Kapoor, S. G. and Devor, R. E., 1996, An experimental investigation of the crack growth phenomenon for drilling of FRP composites, *ASME Journal of Engineering Industry*, **118**, p. 6.

43 Enemuoh, E. U., El-Gizawy, A. S. and Chukwujekwu Okafor, A., 2001, An approach for development of damage-free drilling of carbon fiber reinforced thermosets, *International Journal of Machine Tools and Manufacture*, **41**(12), pp. 1795–1814.

44 Caprino, G. and Tagliaferri, V., 1995, Damage development in drilling glass fibre reinforced plastics, *International Journal of Machine Tools and Manufacture*, **35**(6), pp. 817–829.

45 Davim, J. P. and Reis, P., 2003, Study of delamination in drilling carbon fiber reinforced plastics (CFRP) using design experiments, *Composite Structures*, **59**(4), pp. 481–487.

46 Khashaba, U. A., 2004, Delamination in drilling GFR-thermoset composites, *Composite Structures*, **63**(3–4), pp. 313–327.

47 Campos Rubio, J. C., Abrão, A. M., Eustáquio Faria, P., Correia, A. E. and Davim, J. P., 2008, Delamination in high speed drilling of carbon fiber reinforced plastic (CFRP), *Journal of Composite Materials*, **42**(15), pp. 1523–1532.

48 Hocheng, H., Puw, H. Y. and Huang, Y., 1993, Preliminary study on milling of unidirectional carbon fibre-reinforced plastics, *Composites Manufacturing*, **4**(2), pp. 103–108.

49 Arola, D., Ramulu, M. and Wang, D. H., 1996, Chip formation in orthogonal trimming of graphite/epoxy composite, *Composites Part A: Applied Science and Manufacturing*, **27**(2), pp. 121–133.

50 Davim, J. P., Reis, P. and António, C. C., 2004, A study on milling of glass fiber reinforced plastics manufactured by hand-lay up using statistical analysis (ANOVA), *Composite Structures*, **64**(3–4), pp. 493–500.

51 Davim, J. P. and Reis, P., 2005, Damage and dimensional precision on milling carbon fiber-reinforced plastics using design experiments, *Journal of Materials Processing Technology*, **160**(2), pp. 160–167.

52 Lopez De Lacalle, N., Lamikiz, A., Campa, F. J., Valdivielso, A. F. and Etxeberria, I., 2009, Design and test of a multitooth tool for CFRP milling, *Journal of Composite Materials*, **43**(26), pp. 3275–3290.

14
Cryogenic machining of composites

Y. YILDIZ, Dumlupinar University, Turkey
and M. M. SUNDARAM, University of Cincinnati, USA

Abstract: This chapter presents the progress made in the cryogenic machining of composite materials. Difficulties in machining of composite materials are outlined in the introductory section. This is followed by an examination of the historical development of cryogenic science and its applications in industry. The section on cryogenic treatment includes the following topics: characteristics of cryogenic treatment, and low temperature characteristics and cryogenic treatment of composites. Subsequently, state-of-the-art cryogenic machining, especially with cryogenic workpiece/chip cooling, cryogenic cutting tool cooling, and cryogenic machinability of some composite materials, is discussed. Finally, a general evaluation of cryogenic machining of composite materials is presented.

Key words: cryogenic machining, composites, machinability, cryogenic treatment.

14.1 Introduction

There has been a rapid growth in the use of composite materials in many engineering applications, – such as aerospace, automotive, electrical and sport industries, because of their superior properties, especially high specific modulus (modulus per unit weight) and specific strength (strength per unit weight) and thus low weight (Hull and Clyne, 1996). Specific examples include wings, landing gear, helicopter rotor blades, drive shafts, transmission units, roof panels, engine pistons (Funatani, 2004), insulators, and canoes. Because of their varying anisotropic, inhomogeneous laminate structures and material properties, the machining of composite materials is significantly different from the machining of metals and alloys in terms of chip formation, cutting forces and heat transfer. However, conventional metal-cutting tools and techniques are still used for the most part in the cutting of composites. Difficulties in the machining of fiber reinforced polymer (FRP) composite laminates have been summarized as follows (Ho-Cheng and Dharan, 1990; Gordon and Hillery, 2003):

- delamination of the composite material, due to local dynamic loading caused by different stiffness of the fiber and matrix;
- spalling, chipping and delamination of the material on exit from cutting;

365

- pulled out and crushed fibers causing fuzzing;
- heat build-up during cutting of FRP composites – a serious problem because the matrix material has a low thermal conductivity compared with metals and many inorganic materials.
- cutting tools damaged by abrasive fibers rounding the cutting edges prematurely during machining. In addition, the difference in hardness between the fiber and matrix may lead to edge chipping of the tool and also the tool may become clogged by melted matrix material.

The drilling performance of glass and carbon fiber reinforced composites (GFRP and CFRP) was evaluated from machined hole quality, with emphasis on surface delamination depending on cutting parameters and tool material/geometry (Abrao et al., 2007). A similar study was also performed to determine the machinability characteristics of carbon fiber reinforced plastic composites in terms of tool life, cutting forces and delamination factors (Shyha et al., 2010). The effect of micro structural parameters, such as carbon nanotube (CNT) orientation regarding the cutting direction, CNT loading, and level of dispersion within the matrix, on the machinability of aligned carbon nanotube composites was investigated by a finite element machining model depending on chip morphology, cutting forces, surface roughness and surface/subsurface damage (Samuel et al., 2010). There are also numerous studies reported on the machinability of several kinds of metal matrix composites (MMCs) in terms of surface finish, tool life, cutting forces and chip formation. Machinability has been investigated in those studies, depending on some parameters such as cutting speed, feed rate, depth of cut and type of cutting tools (PCD, CBN, TiC, Si_3N_4, Al_2O_3, and WC) and depending on composite characteristics such as particulate size, volume fraction and type of reinforcement (Basavarajappa et al., 2006). Severe, quick and premature tool wear and a poor machined surface are reported as the major problems in machining of SiC-particle-reinforced aluminum matrix composites (Quan et al., 1999) due to the fact that the hard abrasive ceramic components increase the mechanical characteristics of these MMCs and thus considerable difficulties appear when machining the materials using conventional methods such as turning, drilling, milling and sawing (Luliano et al., 1998). Since composite materials are generally abrasive and have very low thermal conductivity, it can be inferred that one of the biggest problems in machining of composite materials is excessive heat and consequently increased tool wear, as in metal cutting.

Conventional cooling methods and cutting fluids have been used as a supplementary to overcome and improve the machinability difficulties of composite materials. The purpose of the application of cutting fluids in machining operations has been stated as reducing cutting temperature by cooling, and reducing friction between the tool, chip and workpiece by lubrication (Adler *et al.*, 2006). However, there are some incompatibilities in results related to conventional cooling in machining composite materials. The effect of cutting fluid on the machinability of aluminum-based metal matrix composites reinforced with SiC or Al_2O_3 particles were investigated by Hung *et al.* (1997) and they reported that cutting fluid did not affect the machining performance significantly in terms of tool life, surface finish and cutting forces. Kannan and Kishawy (2008) studied cutting forces, tool wear, surface integrity and chip formation for dry and wet turning of A356 silicon carbide particulate reinforced aluminum metal matrix composites. Small differences were reported between the cutting forces in dry and wet cutting conditions and this result was correlated with rapid abrasion of the cutting tool (coated tungsten carbide) by the particles, even under wet cutting conditions, and incapability of the cutting fluid to create any protective film to reduce the frictional conditions on the tool flank face. The only pronounced results were seen at higher cutting speeds (240 m/min) in terms of cutting forces and reduced tool flank wear due to the effective cooling. Surface quality was also deteriorated by using coolant and the micro-hardness of the machined surface increased under wet cutting conditions. In another study (Shetty *et al.*, 2009), the machining performance of discontinuously SiC (15 vol. %) reinforced aluminum composites (DRAC)-(6061 Al/15% SiC 25p) were investigated under dry, oil oblique water emulsion, and steam lubricated conditions. High pressure steam conditions gave the best results in the turning of DRAC composites in terms of cutting force and cutting temperature reductions.

Cryogenic temperatures or cryogenic cooling has become an alternative to conventional cooling methods because of the health and environmental problems of conventional cutting fluids and the superior effects of cryogenic cooling on machining performance. Several advantages of cryogenic machining over the conventional machining methods were stated as follows (Hong, 1991): Cryogenic cooling is able to reduce tool wear and increase tool life; reduce friction force and power consumption; increase cutting efficiency; reduce workpiece distortion; improve surface quality, chip breakability and chip removal for precision machining; reduce manufacturing time and manufacturing cost by providing machining with higher cutting speeds; maintain a clean operation with no environmental hazard; and it is more economical than conventional machining. Cryogenic science has been applied to almost all kinds of traditional and non-traditional manufacturing processes including turning (Wang and Rajurkar, 1997), milling (Hong, 1999), drilling (Bhattacharyya and Horrigan, 1998), grinding (Paul and Chattopadhyay, 2006) and electro discharge machining (Sundaram *et al.*, 2009) of almost all kind of materials, including composites (Yildiz and Nalbant, 2008).

14.2   Key aspects of cryogenic science

Cryogenic science is a branch of low-temperature physics concerned with the effects of very low temperatures (less than about 123 K (−150°C)) and it extends down to absolute zero −273°C (−459°F). Low temperature applications were limited to the use of natural ice for many centuries because of the difficulty in producing lower temperatures. Reaching temperatures below 0°C were not realized until the development of thermodynamics, in spite of the idea of absolute zero at −273°C being put forward in the mid-1700s. So, the development of thermodynamic fundamentals was the primary contributor to advances in cryogenics between about 1850 and 1900. The only significant applications of cryogenics involved the use of cryogenic liquids around 1950. The liquefaction technology for these cryogens was developed primarily in the years between 1850 and 1900, and the transfer of the liquefaction technology to industry and the rapid scale-up of liquefaction rates occurred primarily in the years from 1900 to 1950 (Timmerhaus and Reed, 2007). These historical progressions are summarized in Fig. 14.1 in detail. The normal boiling points of the permanent gases such as helium, hydrogen, neon, nitrogen, oxygen, and air, lie below −150°C (123 K) or (−240°F) (Barron, 1999). Nitrogen is the most common cryogen used in cryogenic science applications since it is the major constituent of air, accounting for 78% volume, and it is a colorless, odorless, tasteless and non-toxic gas. Applications of cryogenics vary in industry, such as aerospace, electronics, chemistry, biology and food transportation (Sitting and Kidd, 1963). Cold/sub-zero and cryogenic treatments have also been used in industry for material processing techniques because of their favorable effects on mechanical, thermal and electrical properties of most engineering materials, as discussed by Barron (1985).

14.2.1 Cryogenic treatment

From the definition of cryogenics, cryogenic treatment is a heat treatment process and can be characterized by its applications whereby temperatures

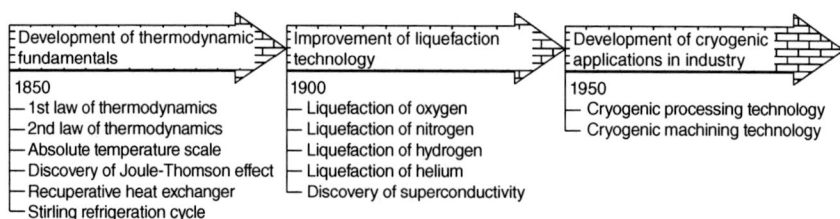

14.1 Development of cryogenic science from 1850.

below −150°C, often at liquid nitrogen (LN$_2$) temperature (−196°C), are used to cause beneficial changes in the subjected material properties. A variant of this process is 'shallow cryogenic treatment', where cold or sub-zero temperatures (but higher than cryogenic temperatures) are used; this can extend down to about −80°C. Cryogenic treatment has been applied particularly in automotive and manufacturing industry for a wide variety of tools such as drills, taps, reamers, end mills, broaches, blades, and gear cutters on many parts, such as crankshaft, castings and bearings to increase their strength, hardness and wear resistance. Remarkable savings in industry by this method have been recorded (Sweeney, 1986; Vaccari, 1986). Interest of industry in the application of cryogenic treatment began with the early space program and its stimulation to manufacturers during 1950s and 1960s. Companies involved first tried treating their tool materials or parts by immersing them in liquid nitrogen. However, this attempt resulted in damaging subjected materials by thermal shock. For reducing the probability of thermal shock, direct contact of liquid nitrogen with the specimens needs to be avoided. Some techniques used in cryogenic treatment systems for this purpose include programmable temperature controllers, solenoid valves to control the liquid nitrogen (LN$_2$) flow and thermocouples to ensure cooling of components at a controlled rate and to a controlled temperature (Carlson, 1991; Bowes, 1974; Levine, 2001).

Characteristics of cryogenic treatment

Generally, cryogenic treatment is performed in three main stages, namely a slow cooling stage (the cool-down cycle/period) in which the parts are cooled from ambient temperature to cold/cryogenic temperatures during a time period (degrees per hour or minute); a soaking stage in which the parts are maintained at cold/cryogenic temperatures for a given duration (hour); and a tempering/warming stage (warm-up cycle/period) in which the parts are heated from cold/cryogenic temperatures to tempering temperatures during another time period (degrees per hour or minute). These stages and their conditions depend on the desired properties, time–cost and the shape and size of the parts (Carlson, 1991; Chillar and Agrawal, 2006). Figure 14.2 shows a typical cryogenic treatment cycle. Since the cooling stage had little effect on the final properties of the material being treated, it is recommended that the materials be cooled as rapidly as possible to the treatment temperature without causing thermal shocks in order to minimize the treatment time thus the cost. The soaking time, in which the material stays at the cryogenic temperatures is important for the final properties; this soaking time is required for the atoms in the material to disperse to new locations (Reitz and Pendray, 2001). Barron (1982) investigated the effect of cold

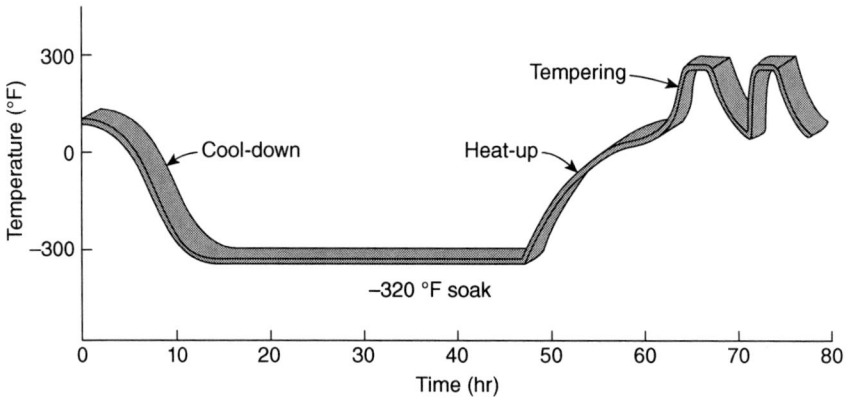

14.2 A typical cryogenic treatment cycle (Vaccari, 1986).

treatment (189 K) and cryogenic treatment (77 K) on the wear resistance of many tool steels and stainless steels by using an abrasive wear test apparatus. He used same cooling rate (3°C min^{-1}) and soaking time (24 h) for both treatments. He found that wear resistance was increased by both treatment methods and the results were close to each other for stainless steels. However, in most of the tool steels, the cryogenic treatment showed better results. It has been shown that wear resistance in the cryogenic treatment of tool steel materials increases with longer soaking time at cryogenic temperatures by improvement of the finer carbide distribution in the microstructure of the subjected materials (Collins, 1996). A similar study (Mohan Lal et al., 2001) also proved that soaking time is more important than lowering the temperature, by performing a comparison between two different cryogenic treatment characteristics (163 K for 24 hours and 93 K for 6 hours) on similar kinds of tool steel materials. Tempering can be performed by single, double or triple cycles, and it is generally performed after the cryogenic treatment to improve impact resistance of the treated materials (Carlson, 1991). Related studies have suggested that the tempering process should be performed after a proper cryogenic treatment cycle for the ultimate effect (Nordquist, 1953; Molinari et al., 2001). One of the reasons for the effects of cryogenic treatment has been found to be transformation of retained austenite into martensite, and another was precipitation of large numbers of very fine carbides for steel materials (Carlson, 1991; Collins, 1996). In the case of copper alloys, it has been claimed that the cryogenic process increases the homogeneity of the crystal structure, resolving gaps and dislocations of the alloying elements and consequently improving structural compactness and thus thermal and electrical conductivity (Trucks, 1983). This claim has been proved by more recent researchers (Isaak and

Reitz, 2008). The effects of low temperatures and cryogenic treatments on composite materials are examined in the following section.

14.2.2 Low temperature characteristics and cryogenic treatment of composites

The determination of low temperature characteristics of composite structures will be beneficial for manufacturing and machining industries. This subject has been investigated over a wide range of materials by researchers. Hence, this section has been divided into two parts: metal matrix composites and polymer matrix composites.

Metal matrix composites

Gayda and Ebert (1979) investigated the effect of cryogenic cooling on the tensile properties of metal–matrix (6061 aluminum alloy-stainless steel) composites. They first heated the specimens to 260°C and held them at this temperature for 30 minutes, subsequently cooling them by air. Thereafter, the specimens were refrigerated at −196°C by immersion in liquid nitrogen for 15 min. and subsequently warmed by air. They reported that the strength of the composites was increased, depending on the compressive residual stress state in the matrix.

Cryogenic temperature properties of some metal matrix composite (MMC) materials, (P100/6061, 25%-SiC/2124, 25%-B_4C/6061), were investigated for space applications to determine their ability to withstand alternate heating and cooling conditions (Sherman, 1990). For this purpose, MMC materials were exposed to temperatures between −320°F and +250°F, thermally cycled for 2000 and 5000 cycles, and tested for mechanical properties (ultimate strength, modulus of elasticity, strain-to-failure and proportional limit). It was reported that the tensile modulus of P100/6061 decreased by 10% from room temperature to −315°F. This result was explained by inhibition of aluminum contraction during the 390°F temperature change caused by aluminum and graphite bonding, thereby producing a stress greater than the yield strength of the aluminum. However, the tensile strength increased due to yield strength of 6061 increasing significantly during the treatment. Both B_4C/6061 and SiC/2124 showed excellent cryogenic properties and increased notch sensitivity at −315°F. The tensile strength of B_4C/6061 was reduced by thermal cycling between −315°F and 250°F for 2000 and 5000 cycles due to overaging.

The influence of cryogenic treatment on the wear properties of an aluminum matrix composite material reinforced by Al_2O_3 particles and fibers was examined (Chen *et al.*, 2006) and it was shown that treatment improved the wear property of composite materials effectively.

It was reported that cryogenic treatment caused a martensitic phase transformation in Cu-Zr-Al bulk metallic glass (BMG) composite and this transformation caused remarkable improvements of the microhardness and the ultimate compression fracture strength at 72 h cryogenic treatment time (Ma *et al.*, 2010a). The effect of cryogenic treatment on the microstructure and mechanical properties of bulk metallic glass (BMG) matrix composites was investigated (Ma *et al.*, 2010b). For the cryogenic treatment, the samples were immersed in liquid nitrogen for several hours (2, 4, 8, 12, 24, 48, 60 and 72 hours). It was reported that the microstructure and morphology of the test samples were changed after the cryogenic treatment and this result contributed to an increase of the fracture strength and micro-hardness of the composite materials; up to 30% and 18%, respectively, depending on treatment time.

Polymer matrix composites

The effects of cryogenic processing on various kinds of composite materials have been investigated widely and show better strength, hardness and wear resistance than the pure matrix alloy (Kalia, 2010). However, there are also some problems related to the low temperature characteristics of composite materials. The main trouble, especially with carbon fiber reinforced polymeric composites, is micro-crack formation due to the thermal stresses caused by different coefficients of thermal expansion of the components at their interfaces. This negativity also causes damages such as potholing and delamination (Timmerman *et al.*, 2003).

The mechanical behavior of Al_2O_3-fiber-reinforced lead-doped BPSCCO ($Bi_2Sr_2Ca_2Cu_3O_x$) high-temperature superconducting (HTS) composites was investigated at cryogenic temperature (77 K) by Miyase *et al.* (1995). The elastic modulus of that composite obtained at 77 K was found to be significantly lower than that at room temperature, which was correlated with the appreciable amount of matrix cracking observed. It was reported that the large difference in coefficient of thermal expansion between the matrices and the fiber materials during cooling from room to cryogenic test temperatures introduced complex microscopic thermal stresses in the composite, leading to initial cracking. Kim and Choi (2003) determined some thermo-acoustic emission (thermo-AE) characteristics of unidirectional carbon fiber/epoxy composite laminates at cryogenic temperatures. Strong AE amplitude with low and high frequency bands during the initial stage of cryogenic cooling showed that the development of large cracks and/or micro failures accompanying fiber breakages existed. In another study (Shindo *et al.*, 2006), the fatigue behavior of woven-fabric glass/epoxy composite laminates comprised of bisphenol-A epoxy resin with E-glass reinforcement at 47% fiber volume fraction was investigated at temperatures of 77 and 4 K by immersing the test specimens in liquid nitrogen and liquid helium for around 15 minutes. The low-temperature ultimate strength of the laminates was almost twice those of the room-temperature specimens. While the fatigue limit was about 88 MPa at room temperature, it was 167 MPa at 77 K and 141 MPa at 4 K. However, delaminations and several matrix microcracks were problems at low temperatures.

A 3D micromechanical finite element damage model, confirmed by experimental observations, was introduced for the prediction of matrix crack development in a unidirectional carbon fiber-reinforced composite (T300/RS-7) at low temperatures, beginning from 121°C to 25°C, and to −50°C (Peterson et al., 2008). The model predictions indicated that longitudinal tensile stress in the matrix exceeds the strength of the epoxy for most fiber volume fractions (35–75%) and causes matrix cracking at low temperatures.

The crack resistance of composite materials under cryogenic temperatures was improved by incorporating nano/micro-sized fillers such as Al_2O_3, and MWNTs (Multi-walled carbon nanotubes) in the resin formulation (Kim et al., 2007).

In another study (Whitley and Gates, 2003), the thermal and mechanical behaviors of a carbon fiber reinforced polymeric-matrix composite (PMC) in various laminate configurations were examined at cryogenic temperatures (−196°C and −269°C). These cryogenic test conditions were achieved by immersing the test specimens in liquid nitrogen and liquid helium, respectively. While shear modulus and shear strength increased up to 35% as the temperature decreased, tensile modulus and tensile strength decreased up to 35%.

So, it seems that results from changes in temperature are not always smooth and consistent. In addition, it has also been reported (Stevens et al., 2006) that there was not seen any change in the microstructure of carbon fiber-reinforced magnesium composites (C/Mg) and cryogenic treatment did not cause any cracking in the specimens when the cryogenic treatment was applied at −196°C for 5 min.

Fiber/matrix adhesion strength has an important role to play for the improvement of mechanical properties of fiber reinforced polymer composites, and cryogenic treatment has been used as an approach to improve interaction and bonding properties between fibers and resin. Rashkovan and Korabel'nikov (1996) investigated the effect of cryogenic treatment on carbon fiber strength and adhesion to the matrix. It was reported that the mean strength of the fiber increased after cryogenic treatment by the removal of amorphous pyrolyzate deposits from the carbon fiber surface. There have been some remarkable studies investigating the effect of cryogenic treatment on mechanical (Zhang et al., 2004) and tribological (Zhang and Zhang, 2004) properties of carbon fiber reinforced composites. In these

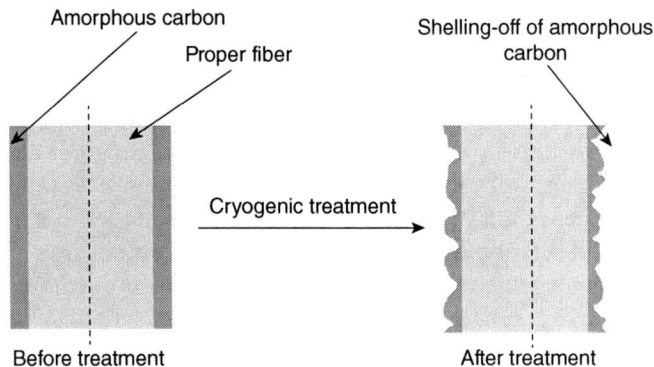

14.3 Diagram of the mechanism of the cryogenic treatment on carbon fibers (Zhang *et al.*, 2004).

researches, M-2007S carbon fibers were cryogenically treated by immersing them in liquid nitrogen for various times (1, 5, 10 and 20 min), and afterwards these fibers were combined into a bisphenol-A type epoxy resin matrix (DER331). Strength tests for mechanical properties and sliding wear tests were performed for the wear resistance investigations. The effect of cryogenic treatment on the carbon fiber structure is schematically illustrated in Fig. 14.3. The carbon fiber is composed of two parts and is subject to a contraction of the amorphous layer and an axial expansion of the proper fiber at cryogenic temperatures, due to the difference in the coefficients of thermal expansion between those two parts. This situation causes an increase of shear stress exceeding the shear strength between the proper fiber and the amorphous layer. The result improves the fiber/matrix interfacial bonding due to increased surface roughness by tiny fragments and some striations along the fiber axis. Hence, flexural modulus and strength of these composites were increased effectively by the optimum cryogenic treatment conditions. In addition, cryogenic treatment caused a reduction in the fiber diameter to some degree. It was shown that wear resistance of those composites was increased at high sliding pressures after cryogenic treatment, due to improved fiber and interface strength.

Mechanical properties of a carbon fiber (65% by volume) reinforced epoxy polymer composite (Cytec IM7/977) were investigated at cryogenic temperatures (Kim and Donaldson, 2006). Microcrack and delamination within the laminates under thermal and mechanical loadings were determined using acoustic emission. It was reported that thermo-mechanical properties, such as transverse modulus, shear modulus, transverse shear modulus and transverse strength, were increased by a temperature reduction of 23°C to −196°C.

The effect of cryogenic treatment on the structure and mechanical properties of polyamide (PA-6) reinforced by glass threads (30%) was investigated by Novik *et al.* (2008). The following cycle was used for cryogenic treatment by liquid nitrogen, using a 10 minute exposure time for each temperature: Heating to 100°C and cooling to 21°C (Step 1); cooling to −196°C and heating to 21°C (Step 2); cooling to −196°C and heating to 100°C (Step 3). It was reported that the strength of the composite increased by 7–10%, and the elongation increased to 140% after Step 3 of the cycle. In another study, pitch-based, short carbon fibers (15 wt%) were treated cryogenically and after, were incorporated into a polyimide matrix to form a composite. The cryogenic process was performed by immersing the carbon fibers in liquid nitrogen for 10 minutes. These composites were examined by XPS and SEM analysis. The flexural strength and tribological properties of those composites were evaluated and it was reported that the surface of the treated fibers had become rougher and thus mechanical and tribological properties were improved due to the enhanced fiber–matrix interfacial bonding (Zhang *et al.*, 2008). A similar study was performed for hybrid PTFE/Kevlar fabric/phenolic composite (Zhang *et al.*, 2009). Hybrid PTFE/Kevlar fabric was immersed in liquid nitrogen for various minutes for the cryogenic treatment. This treatment increased the roughness of the PTFE and Kevlar fiber surfaces. The tribological properties (friction coefficient, wear rate) of the fabric/phenolic composite were influenced to an extent depending on cryo-treatment time. While the lowest wear rate was achieved at 10 minutes treatment time, wear was increased at 20 minutes, which means that excessive cryo-treatment is unfavorable for wear resistance of those composites.

A polymeric composite material, polyimide (PI) and polyetherimide (PEI) was cryogenically treated in a study by Indumathi *et al.* (1999) for wear resistance. Cryogenic treatment was performed at liquid nitrogen temperature (77 K) for 24 hours, the samples being cooled to that temperature in 8 hours by controlled cooling in a cryogenic chamber. After that, these samples were heated back to room temperature with a controlled heating rate of 35°C/h. Pin-on-disk tests showed that abrasive wear performance (increases of between 2% and 35%) and hardness (increases of between 4% and 13%) of the treated composites improved significantly as compared to untreated samples. These improvements varied depending on the configuration of the composite material.

A general evaluation

The cryogenic temperature characteristics of several fiber (alumina, glass, carbon) reinforced composite materials were investigated in detail in relation to their mechanical and thermal aspects. Some of those outcomes are summarized in Table 14.1 and in Fig. 14.4, respectively. Generally, the

Table 14.1 Mechanical properties of some fiber-reinforced composites at different temperatures

Fiber	Resin system	Fiber vol. (%)	Temperature (K)	Young's modulus E (GPa)	Tensile strength σ_u (GPa)	Fatigue strength $\times 10^4$ (GPa)
Alumina	PEEK	—	295	96	0.52	0.43
			76	102	0.80	0.56
			4	101	0.88	0.68
E-glass	Epoxy	75	295	43.3	1.41	—
			4	—	1.68	—
S-glass	Epoxy	60	293	59.5	1.11	0.60
			77	63.2	1.45	0.79
T-300 graphite	Epoxy	61-N/A	293–300	—	0.82–163	0.69–1.14
			77–76	—	0.77–1.52	0.69–1.46
T-700 graphite	Epoxy	—	295	144	1.53	—
			76	157	2.11	—
T-1000 graphite	Epoxy	—	295	175	2.34	—
			76	195	2.04	—

From Reed and Golda (1997).

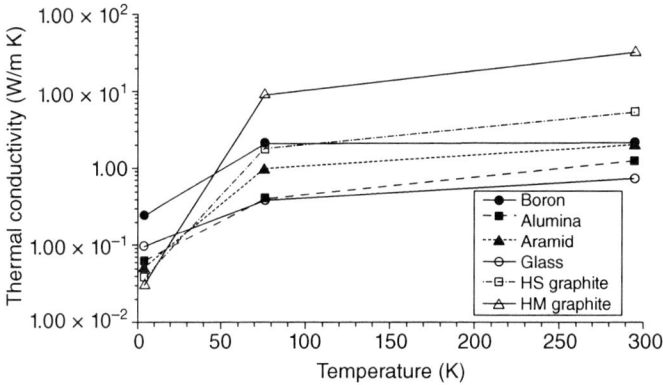

14.4 Thermal conductivity of unidirectional composites (Schutz, 1998). HS, high strength; HM, high modulus.

Young's modulus and tensile strength of alumina, glass and carbon/graphite fiber reinforced composites increase as the temperature decreases. Reductions of carbon fiber composites' tensile strengths are attributed to brittleness of the fibers at low temperatures and strain mismatch between the fibers and the epoxy at low temperatures. Additionally, transverse microcracking and axial delaminations within the epoxy resin were observed during the fatigue tests. While the glass fiber composites had the highest thermal conductivity at very low temperatures, carbon fiber composites had the highest thermal conductivity at temperatures above 25 K. As a consequence, it can be concluded that the effect of low temperatures and cryogenic treatment on composite materials varies, depending on composite material type and properties. Before processing a composite material, the literature should be examined in detail for the ultimate effects. In the following sections, cryogenic machining and cryogenic machining of composite materials are discussed in detail.

14.3   State-of-the-art cryogenic machining

Cryogenic machining technology has been developed by the cooperation of companies in the aerospace, automobile, machine tool, cutting tool, and cryogenic industries to improve material machinability and machining efficiency. This effort has been especially targeted towards high speed cutting of hard-to-cut materials, such as sticky titanium alloy, soft aluminum alloy, ductile carbon steel and abrasive composite materials (Hong, 1991). Cryogenic cooling employing liquid nitrogen as a cryogenic coolant has been explored since the 1950s in the metal cutting industry. However, it could not be examined over a wide range at the beginning, due to the high costs associated with early cryogenic technology. Uehara and Kumagai (1969,

1970) attempted the initial efforts towards studying the fundamentals of the effects of cryogenic machining, and researchers have applied this cutting method almost exclusively for all manufacturing processes over the last few decades. Some important functions of cryogenic machining were defined by Hong and Zhao (1999) as removing heat effectively from the cutting zone and then lowering cutting temperatures, moderating the frictional charac- teristics at the tool–chip interface, and modifying the properties of the workpiece and the cutting tool material, hence to improve the machinability of the materials. It was pointed out that the hot-strength and hot-hardness of the cutting tool remain high with cryogenic cooling and thus tempera- ture-dependent tool wear reduces significantly under all machining condi- tions (Wang and Rajurkar, 2000). It was also pointed out that cryogenic cooling had remarkable advantages over conventional emulsion cooling in terms of higher productivity, lower productivity cost and safer and healthier conditions for workers and the environment (Hong, 2001). Cryogenic machining strategies have been applied to the cutting operations in differ- ent ways. These strategies include cryogenic cooling of workpiece/chip, indirect cryogenic cooling of the tool, and direct cryogenic cooling of the tool by delivering the cryogen to the tool-chip or tool/work interfaces. Liquid nitrogen (LN_2) has been commonly used as the cryogenic coolant in cryogenic machining applications because of its nonflammability.

14.3.1 Cryogenic workpiece/chip cooling

A cryogenic cooling flow can be directed toward the workpiece or chip to improve machinability and chip breakability by transformation of the mate- rials from ductile to brittle. Figure 14.5 shows sample applications of these methods for turning operations. In the workpiece cooling method, the work- piece is pre-cooled cryogenically ahead of the cutting tool edge with a nozzle. In this design, thermocouples were used to monitor workpiece tem- perature down to LN_2 temperature before starting the cutting process. Cryogenic workpiece cooling has also been performed as an enclosed bath.

In the chip cooling method, the size, shape and position of the nozzle, mounted close to the cutting edge, were designed to cover the chip arc and chip faces with the LN_2 flow. Significant chip breakability improvement of AISI 1008 low carbon steel was demonstrated by this method (Ding and Hong, 1997). In addition, Hong and Ding (2001a) achieved a 62% reduction in cutting temperatures by pre-cooling the workpiece during the machining of Ti-6Al-4V. Improvements in tool flank wear growth in the machining of low carbon steel by these methods were reported (Hong and Ding, 2001b). However, pre-cooling the workpiece or enclosing the workpiece in a cryo- genic bath has practical difficulties in the production line. It is necessary to change configurations of the machine parts for this application; it may cause

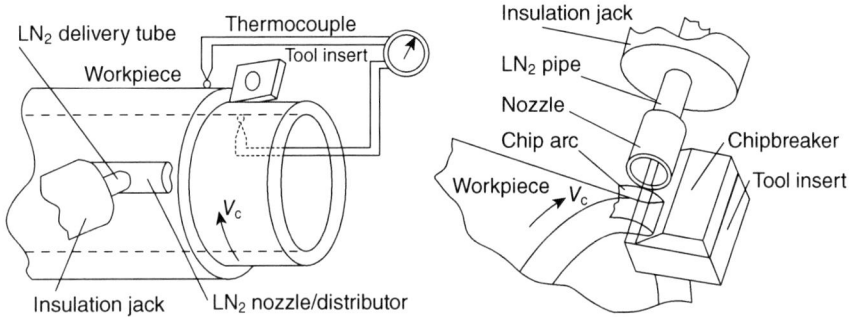

14.5 Sample cryogenic workpiece and chip cooling applications (Hong *et al.*, 1999; Hong and Ding, 2001a).

dimensional change of the workpiece; high liquid nitrogen consumption can make the process economically unfeasible; and these approaches negatively increase the cutting forces and abrasion to the tool (Hong *et al.*, 2001a).

14.3.2 Cryogenic cutting tool cooling

Cutting tools in machining operations have been cryogenically cooled in two ways – direct and indirect cryogenic cooling methods.

Direct cryogenic cooling

As an approach, cutting tools have been cryogenically cooled directly by a general flooding of the cryogen into the cutting zone, as in conventional emulsion cooling. However, this approach can cause very high liquid nitrogen consumption and cooling of unwanted areas causing a negative effect such as pre-cooling the workpiece (Hong *et al.*, 2001a). Alternatively, jet cooling approaches, featuring an LN_2 delivery micro-nozzle system have become more striking methods of direct cryogenic cooling. In these approaches, cryogens are sprayed, at a minimized coolant flow rate, to a localized zone of the tool rake and/or the tool flank faces, where the material is cut and maximum temperature and major areas of tool flank and crater wears are formed. Figure 14.6 shows an example design of the device developed for that method of turning operations. In this design, LN_2 can be injected parallel to the spindle axis (-Z direction) and/or perpendicular to the spindle axis (-X direction). In addition, nozzle and chipbreaker integration makes it easy to lift the removed chip and assists penetration of the gas/fluid into the tool–chip interfaces. In this design with two small sized nozzles (1.59 and 0.3 mm), the cryogen (LN_2) consumption is more economic in comparison with a general flooding method. LN_2

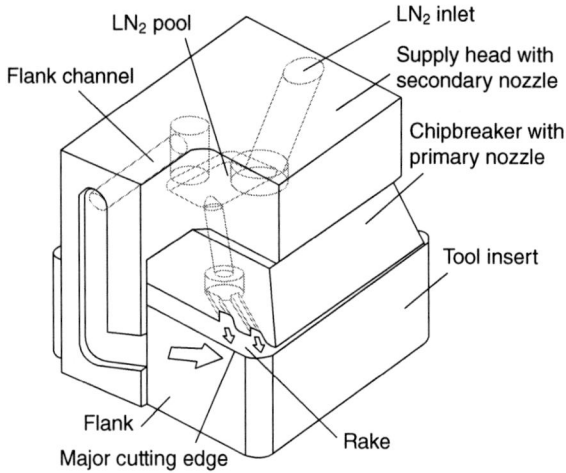

14.6 Assembly of two-nozzle LN₂ delivery system (Hong *et al.*, 2001a).

cannot be circulated inside the machine like conventional cooling fluids, and quickly evaporates as it is released at normal atmospheric pressure. By this approach, Hong and Broomer (2000) yielded tool life improvements from 43% to 67%, depending on the cutting parameters, in the machining of AISI 304 stainless steel; and Hong and Ding (2001a) reported a 75% reduction in cutting temperatures in the machining of Ti-6Al-4V. Hong (2001) also stated that this cryogenic machining approach eliminates the built-up edge (BUE) problem on tools because the cold temperatures reduce the possibility of chips welding onto the tool. The high pressure cryogenic jet also helps to remove possible BUE; therefore, it can produce better surface quality. On the other hand, it was also reported in machining of Ti-6Al-4V (Hong *et al.*, 2001b) that cryogenic cooling by this method increased the main cutting force and thrust force due to the fact that lower cutting temperatures caused stronger and harder workpiece materials. However, it was also determined that LN₂ formed a fluid cushion at the tool–chip interface and provided a lubrication effect by absorbing heat and evaporating quickly, and hence the feed force decreased at cold temperatures because of the lower friction between the chip and tool face.

Indirect cryogenic cooling

This method is also called 'cryogenic tool back cooling' or 'conductive remote cooling'. In this approach, the cutting zone on the tool is cooled by heat conduction from a cryogen-LN₂ chamber located at the tool face or the tool holder. So, LN₂ is not sprayed onto the cutting tool or work-piece.

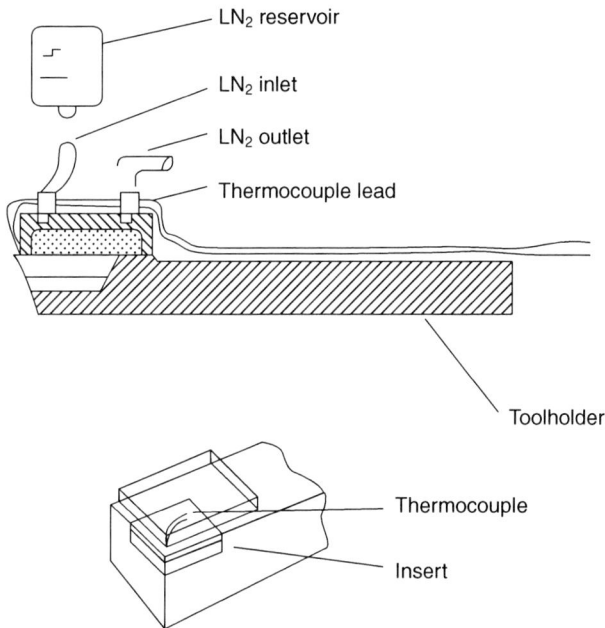

14.7 Tool cap and LN₂ reservoir (Wang and Rajurkar, 2000).

However, the success of this approach depends on tool geometry, the size of the surface area with which the cryogen will be in contact, the thermal conductivity of the tool material, the cutting insert thickness and the distance from the cryogen-LN_2 source to the highest temperature point at the cutting edge. Figure 14.7 shows an example application of this method for turning operations. In this system, liquid nitrogen is circulated through a reservoir built on top of the cutting tool to reduce cutting temperatures. It has been claimed that this system is more efficient than LN_2 sprays in terms of a stronger and much more stable cooling effect on the cutting insert. In addition, longer tool life can be achieved because of the maintenance of the hardness and strength of the tool material, and the reduction of chemical reactivity of the workpiece since there is no contact between the cryogen and workpiece and so there are no changes in properties of the workpiece. Wang and Rajurkar (1997, 2000) tested this system by machining various advanced materials such as RBSN (reaction bonded silicon nitride-98% Si_3N_4) (Wang et al., 1996), Ti-6Al-4V, Inconel 718 and tantalum by CBN 50 (cubic boron nitride) and H13A. For all of the materials tested, lower flank wear (V_B) with longer cutting distances was obtained by cryogenic cooling. Although there were no significant changes in the cutting forces in three directions (F_x, F_y and F_z) with and without cryogenic cooling, the reductions in cutting temperatures (more than 28% in RBSN machining with CBN) and surface roughness values by indirect cryogenic cooling were considerable. This system was noticeably and distinctly improved later by addition of a new concept called the 'hybrid machining method' (Wang *et al.*, 2003). In this new method, additional plasma heating is provided to soften the workpiece material for easier machining and the liquid nitrogen keeps the temperature of the tool insert low by indirect cooling. Incredible improvements in tool life (156%), surface roughness (250%) and cutting forces (30–50%) were reported by this technique when machining Inconel 718 material.

14.4   Cryogenic machinability of composite materials

The effect of low-temperature embrittlement on the machinability of bone, as a natural composite material, was investigated by Wiggins and Malkin (1974) for orthogonal drilling, because heat generation was a serious problem in the clinical machining of bone. In that study, a sample bone was immersed in substances of different temperatures (air, ice water, dry ice, and liquid nitrogen) during cutting, and specific cutting energy versus temperature was determined by measuring the forces. It was reported that the specific cutting energies required for cutting both longitudinal and transverse to the bone axis were reduced almost three times by lowering the temperature. Localized cooling was also explored for drilling characteristics in the study. For this purpose, a special drill was designed to allow injection of LN_2 to the drill tip during cutting. Torque and feed rate measurements showed that while the LN_2 application increased the feed rate, it decreased the specific cutting energy. A striking study (Bhattacharyya *et al.*, 1993) was performed in a turning operation to determine the cryogenic machinability of Kevlar aramid fiber reinforced plastics (KFRP) with longitudinal continuous fiber orientation and 40% volume fraction. A fuzzy surface finish was suggested as the problem for this kind of composite (KFRP), due to its low thermal conductivity, high toughness and flexibility. In that study, cryogenic cooling was achieved either by soaking the test pieces in liquid nitrogen for a predetermined period or by directly applying the liquid nitrogen onto the test pieces during machining. Cutting tests were performed depending on workpiece temperatures (Case 1 from 195°C to –185°C, Case 2 from 135°C to –105°C, and Case 3 from –75°C to –60°C), cutting speeds (50 m/min and 100 m/min), tool geometries (TPUX 160308 FR18 and TPUN 160304 F), constant depth of cut (1.0 mm) and feed rate (0.05 mm/rev). For the workpiece temperature, while Case 2 (1–2 minutes) and Case 3 (3–4 minutes) were achieved by dipping the test rod in liquid nitrogen, Case 1 was achieved by continual application of liquid nitrogen flow (0.4–0.5 L/min) onto the workpiece. Interestingly, sufficient pressure was provided by controlled heating of the LN_2 in a Dewar flask with an immersion heater. However, there were some practical difficulties related to Cases 2 and 3 in that cryogenic machining was restricted to a short period of a very limited time unless the specimen was re-cooled. Workpiece temperatures were measured by imbedding a thermocouple below the surface, 5 mm deep. Machinability of KFRPs was evaluated in terms of tool flank wear/tool life, surface roughness and cutting force improvements. It was shown that the magnitudes of tool flank wears, cutting forces and surface roughness (R_a) were quite low in all cases, notably with continual flooding of LN_2 for long continuous machining. Continual rubbing of loosened fibers, chip notching, edge rounding and higher temperatures were problems in Cases 2 and 3. These problems also increased the cutting forces (50% higher compared with those obtained in Case 1) and caused a severe deterioration of the surface finish. For these reasons, the approach of directly applying the liquid nitrogen onto the workpiece has been recommended for good machinability of KFRP composites. It was assumed that a compressive stress producing a clamping force occurred on the fiber due to the difference of thermal expansion coefficients of the resin ($80–100 \times 10^{-6}$/K) and the fiber (about -4×10^{-6}/K and 50×10^{-6}/K in longitudinal and transverse directions, respectively). Additionally, the epoxy matrix becomes stiffer at the cryogenic temperatures and thus the fibers are held more rigidly causing a reduction of fiber deflection. Interestingly, the failure condition changed from a bending rupture to shear fracture. These outcomes were concluded from microscopic investigation of the cutting zone during micro-machining at ambient and cryogenic temperatures. Bhattacharyya and Horrigan (1998) also investigated the machinability of continuous Kevlar-epoxy composites (KA-060/ADR 240) by drilling with modified high speed steel (HSS) drill bits under cryogenic conditions. Cryogenic cooling was obtained through the application of liquid nitrogen at the drill site at a flow rate of 0.2–0.3 L/ min through a 3 mm nozzle. This modification to standard HSS drills was carried out by producing a negative point angle ($-20°$). Drilling tests were performed by varying drill helix geometry and cutting speeds. In spite of the increase in drilling thrust forces under cryogenic conditions, improved delamination, surface finish, hole quality/dimensional accuracy and superior tool life were reported in comparison with ambient temperature tests. Cryogenic conditions produced a better surface regardless of the drill geometry and type. This result was associated with increased stiffness of the matrix at cryogenic temperatures. Additionally, compressive stress due to the difference between thermal-expansion coefficients of the matrix and fiber provided a better bounding of fibers, and thus cleaner shear fractures were obtained rather than bending and tearing. Negligible tool wear under cryogenic conditions was assumed to be a consequence of improved fiber cutting and a reduction in tool temperature. This improved fiber cutting also reduced torque magnitudes under the cryogenic conditions. Deficiencies such as fuzzy, uncut and protruding fibers at the entry and exit faces of the holes in drilling the KFRP laminates were minimized to a large extent by using a backing plate, leaving thin (<0.2 mm) resin-rich layers on the laminate surfaces around the drilling location. While severe delamination was observed under cryogenic conditions with modified drills and without backing plate drilling, this delamination was eliminated by employing standard drills. However, by employing the backing plate, minimum delaminations were achieved under cryogenic conditions for both standard and modified drills. The delamination results conformed with predictions made by a linear-elastic fracture mechanics model and also by a three-dimensional, orthotropic, finite element model. However, the possibility crack propogation was also reported. Kim and Ramulu (2004) investigated the effect of cryogenically treated C2 grade standard twist carbide (~6% cobalt) drill performance on the machinability of fiber reinforced thermoplastic (FRP) – multidirectional graphite/PIXA-M composite (combining IM-6 graphite fibers with a PIXA-M thermoplastic matrix). This composite material was obtained from the Boeing Company and is used in NASA's high speed research programs to make titanium–graphite panels and wing skins. The machinability of that composite was investigated in terms of the drilled hole quality, cutting forces and tool flank wear. Hole quality was evaluated in terms of material integrity, surface finish, delamination, and area of uncut fibers in the holes. An approximately 5% to 10% reduction (depending on depth) was found in the thrust force and drilling torque when using cryogenically treated drills. There was a remarkable reduction (30%) in R_a values obtained by cryogenically treated drills for the entire feed and speed range. In addition, large improvements in the area of fiber protrusion (30%) and delamination length (60%) were achieved by the cryogenically treated drills. It was claimed that cryogenically treated tools could minimize fiber pullout because of their superior abrasion resistance, and this high abrasion resistance could improve the cutting capability in drilling of those composites by leading a complete cut on the fibers at the last ply rather than push them down. The abrasion resistance of carbide tools was associated with increased micro-hardness after the cryogenic treatment. However, it was shown that the cryogenically-treated carbide tools wore approximately 10% faster than the untreated carbide tools. It was claimed that pitting took place on the cutting lips of the cryogenically-treated drills and the treatment process could have increased the brittleness of the carbide drills since both carbide and binder cobalt generally became brittle at cryogenic temperatures. Large amounts of residual stresses induced on the tools after the cryogenic treatment were also proposed as an explanation.

The effect of cryogenic cooling on the drilling-machinability of a Kevlar 49 fiber reinforced composite material was investigated by Ahmed (2004).

The machinability was evaluated depending on cutting forces, specific energy and the hole quality in terms of delaminating and fiber breakage of the machined surface. TiN coated HSS drills at four different temperature ranges (room temperature, 0°C, –60°C and –120°C) were used. Feed rate and speed were the variable parameters for the drilling experiments. Cryogenic cooling was performed by using liquid nitrogen through a specially designed fixture which held the workpiece in place during drilling. In this system, the workpiece laminate was cooled to a minimum temperature of –160°C then allowed to warm up to the desired drilling temperature – these temperatures were recorded by thermocouples inserted in the laminate specimen. It was observed that the thrust force (F_z) and the torque (M_z) increased remarkably as the laminate temperature decreased and this was attributed to an increase of strength and stiffness of the fiber and resin under cooling. The calculated specific cutting energy also increased according to the increased cutting force and torque. On the other hand, about a 400% improvement in the delaminating damage factor (F_d) was reported when drilling was performed at the lower temperatures, compared with that at room temperature and this result was associated with the difference of thermal expansion coefficients of the resin and the fiber at low temperatures. Therefore, this change caused a compressive stress on the fiber, this stress caused the epoxy to become stiffer at cryogenic temperatures, and thus the fibers were maintained in a more rigid condition. Stewart (2004) showed that tool wear and tool forces were reduced when using cryogenically-treated C2 tungsten carbide tools for the turning of medium density fiberboard (MDF). It was claimed that cryogenic treatment (–306°F) affected retention of the cobalt binder by changing the crystal structure of the tool. Similarly, in the machining of MDF four double-flute, solid, tungsten carbide router bits were used (Gisip et al., 2009). Cryogenic treatment was performed to below –149°C and it was shown that tool wear was reduced from 76.2% to 46% when the cryogenically treated cutters were used together with refrigerated air. This result was connected to the microstructure of the cryogenically treated tools indicating a tight grain structure. It was also reported that the supportive cooling also provided retention of the cobalt binder in the cryogenically-treated tools during machining, which resulted with less tool wear.

A remarkable study was carried out for improving the grindability of ceramic matrix composites (CMC-AlSiTi, Al_2O_3-SiC-TiC, 46.1 vol% Al_2O_3, 30.9 vol% SiC whiskers and 23 vol% TiC powder) with cryogenic cooling (Singh et al., 2010). The effect of the cryogenic cooling on the surface and subsurface quality, grinding forces and specific grinding energy were determined according to four process parameters, namely grinding wheel speed, table speed, grain size and depth of cut. Cryogenic cooling was applied on the grinding zone by a spray method. Tangential and normal grinding forces,

Table 14.2 General evaluation of the cryogenic machining of composites

Application of cryogenics	Composite material property	Machining method/ cutting tool	Major findings	Brief explanation	Reference
Cryogenic workpiece cooling and liquid nitrogen spraying	Bone	Drilling	Reduction in specific cutting energy and feed rate enhancement	Improvement in the low temperature embrittlement	Wiggins and Malkin, 1974
Cryogenic workpiece cooling	Kevlar aramid fiber reinforced plastics, Kevlar 49/epoxy (Sicomin SR 1600), longitudinal	Turning/K20 type uncoated carbide insert (TPUX 160300 FR18, TPUN 160304 F)	Poor surface finish and increase in tool wear and cutting forces	Continual rubbing of loosened fibers, chip notching, edge rounding and overheating	Bhattacharyya et al., 1993
Liquid nitrogen spraying	continuous fiber orientation with 40% volume fraction		Improvement in tool wear, surface finish and cutting forces	Effective cooling, minimal bending and fiber pullout	Bhattacharyya et al., 1993
Liquid nitrogen spraying	Kevlar/epoxy composite (KA-060/ADR 240) continuous, unidirectional, 55% volume fraction	Drilling/HSS twist drills with 29° and 57° helix angles	Improved surface finish, negligible tool wear, reduced torque against increased thrust force and minimal delamination	Cleaner shear fracture rather than bending and tearing, due to better clamping of fibers at low temperatures	Bhattacharyya and Horrigan, 1998

Cryogenic treatment	Graphite/PIXA-M multidirectional FRP composite	Drilling /C2 Grade twist carbide (6% cobalt) drills	Reduction in thrust force, torque, surface roughness, fiber protrusion and delamination length, faster tool wear	High abrasion resistance induced by cryogenic treatment on tools, effective cutting of fibers rather than pushing them	Kim and Ramulu, 2004
Cryogenic workpiece cooling	Kevlar 49 fiber reinforced composite	Drilling/TiN coated HSS drills	Increased thrust force and torque, improvement in delamination factor	Increased strength and stiffness of fiber and resin under cooling	Ahmed, 2004
Cryogenic treatment	Medium density fiberboard (MDF)	Turning/C2 tungsten carbide	Reduction in tool wear and tool forces	Alteration in microstructure of cryogenically treated tools	Stewart, 2004
Cryogenic treatment	Medium density fiberboard (MDF)	Router/tungsten carbide	Reduction in tool wear	Retaining of the cobalt binder of cryogenically treated tools during machining	Gisip et al., 2009
Liquid nitrogen spraying	Ceramic matrix composites (AlSiTi)	Surface grinding/ metal bonded diamond grinding wheel	Improvement in the surface quality and reduction in grinding forces	Lubrication effect of the cryogen and sustained grit sharpness with cryogenic cooling	Singh et al., 2010

specific grinding energy, sub-surface damage and surface roughness were found to be lower in the case of cryogenic grinding when compared with dry grinding. Improvements in the surface quality of the ground surface with cryogenic cooling were attributed to a lubrication effect of the cryogenic mist at the grinding zone under high pressure application of the cryogen. It was also assumed that cryogenic cooling maintained grit sharpness during grinding, which reduced rubbing phenomena and thus reduced grinding forces. Meanwhile, the abrasive grits were worn and could not maintain their edge sharpness in the case of dry grinding, which increased the forces due to the rubbing, along with the shearing. It was also stated that specific grinding energy was reduced by cryogenic cooling, depending on the reduction in cutting forces.

14.5   Conclusions

In this chapter, the development of cryogenic science and cryogenic machining of composite materials has been discussed in detail. In addition, the effects of lower temperatures and cryogenic treatment on metal matrix and polymer matrix composites have also been presented. Cryogenic cooling in machining operations has been studied for more than six decades. However, cryogenic machining of composite materials still remains an under-explored topic in comparison with various kind of steels and advanced engineering materials. Table 14.2 gives a summary of progress made in the cryogenic machining of composite materials.

14.6   Acknowledgments

Financial support provided by the University of Cincinnati under the URC Faculty Research Grant program is acknowledged. The authors are grateful to Professor Kamlakar P. Rajurkar from the University of Nebraska-Lincoln (UNL) for his invaluable help and support.

14.7   References

Abrao AM, Faria PE, Rubio JCC, Reis P and Davim JP (2007), 'Drilling of fiber reinforced plastics: A review', *Journal of Materials Processing Technology*, **186**, 1–7.
Adler DP, Hii WS, Michalek DJ and Sutherland JW (2006), 'Examining the role of cutting fluids in machining and efforts to address associated environmental/health concerns', *Machining Science and Technology*, **10**, 23–58.
Ahmed M (2004), *Cryogenic Drilling of Kevlar Composite Laminates*, MSc thesis, King Fahd University of Petroleum & Minerals, Dahran, Saudi Arabia.
Barron RF (1982), 'Cryogenic treatment of metals to improve wear resistance', *Cryogenics*, **22**, 409–413.

Barron RF (1985), *Cryogenic Systems*, Second Edition, Oxford University Press, New York Clarendon Press, Oxford, 13–38.

Barron RF (1999), *Cryogenic Heat Transfer*, Taylor & Francis, Philadelphia.

Basavarajappa S, Chandramohan G, Rao KVN, Radhakrishanan R and Krishnaraj V (2006), 'Turning of particulate metal matrix composites – review and discussion', *Proc. IMechE Part B: J. Engineering Manufacture*, **220**, 1189–1204.

Bhattacharyya D, Allen MN and Mander SJ (1993), 'Cryogenic machining of Kevlar composites', *Materials and Manufacturing Processes*, **8**(6), 631–651.

Bhattacharyya D and Horrigan DPW (1998), 'A Study of hole drilling in Kevlar composites', *Composites Science and Technology*, **58**, 267–283.

Bowes RG (1974), 'The theory and practice of sub-zero treatment of metals', *Heat Treatment of Metals*, **1**(1), 29–32.

Carlson EA (1991), *Cold Treating and Cryogenic Treatment of Steel*, ASM Handbook, Volume 4, 203–206, ASM International, Novelty, OH, USA.

Chen WG, Zhou X, Zhang Q and Gu CQ (2006), 'Influence of cryogenic treatment on wear properties of aluminum matrix composite material reinforced by Al_2O_3 particle and fibre hybrid', *Transactions of Materials and Heat Treatment*, **27**(5), 10–12.

Chillar R and Agrawal SC (2006), 'Cryogenic treatment of metal parts', Advances in Cryogenic Engineering, AIP Conference Proceedings (*Transactions of the International Cryogenic Materials Conference – ICMC*, Keystone, Colorado, 29 August – 2 September, 2005), Volume 52A, Melville, New York, 77–82.

Collins DN (1996), 'Deep cryogenic treatment of tool steels: A review', *Heat Treatment of Metals*, **23**, 40–42.

Ding Y and Hong SY (1997), 'Improvement in chip breaking in machining a low carbon steel by cryogenically precooling the workpiece', *Journal of Manufacturing Science and Engineering, Trans. of ASME*, **120**(1), 76–83.

Funatani K (2004), 'Heat treatment of automotive components: Current status and future trends', *Transactions of the Indian Institute of Metalurgy*, **57**(4), 381–396.

Gayda J and Ebert LJ (1979), 'The effect of cryogenic cooling on the tensile properties of metal-matrix composites', *Metallurgical Transactions A*, **10A**, 349–353.

Gisip J, Gazo R and Stewart HA (2009), 'Effects of cryogenic treatment and refrigerated air on tool wear when machining medium density fiberboard', *Journal of Materials Processing Technology*, **209**, 5117–5122.

Gordon S and Hillery MT (2003), 'A review of the cutting of composite materials', *Proceedings of the Instruction of Mechanical Engineers, Part L: Journal of Materials: Design and Applications*, **217**, 35–45.

Ho-Cheng H and Dharan CKH (1990), 'Delamination during drilling in composite laminates', *Journal of Engineering for Industry (Transactions of the ASME)*, **112**(3), 236–239.

Hong SY (1991), 'Economical cryogenic machining for high speed cutting of difficult-to-machine materials', *Proceedings of First International Conference on Manufacturing Technology*, Hong Kong, December 27–29.

Hong SY (1999), *Milling Tool with Rotary Cryogenic Coolant Coupling*. Patent Cooperation Treaty (PCT), WO 99/60079.

Hong SY and Zhao Z (1999), 'Thermal aspects, material considerations and cooling strategies in cryogenic machining', *Clean Products and Processes*, **1**, 107–116.

Hong SY, Ding Y and Ekkens RG (1999), 'Improving low carbon steel chip breakability by cryogenic chip cooling', *International Journal of Machine Tools & Manufacture*, **39**, 1065–1085.

Hong SY and Broomer M (2000), 'Economical and ecological cryogenic machining of AISI 304 austenitic stainless steel', *Clean Products and Processes*, **2**, 157–166.

Hong SY and Ding Y (2001a), 'Cooling approaches and cutting temperatures in cryogenic machining of Ti-6Al-4V', *International Journal of Machine Tools & Manufacture*, **41**, 1417–1437.

Hong SY, Ding Y and Jeong WC (2001a), 'Friction and cutting forces in cryogenic machining', *International Journal of Machine Tools & Manufacture*, **41**, 2271–2285.

Hong SY, Markus I and Jeong W (2001b), 'New cooling approach and tool life improvement in cryogenic machining of titanium alloy Ti-6Al-4V', *International Journal of Machine Tools & Manufacture*, **41**, 2245–2260.

Hong SY and Ding Y (2001b), 'Micro-temperature manipulation in cryogenic machining of low carbon steel', *Journal of Materials Processing Technology*, **116**, 22–30.

Hong SY (2001), 'Economical and ecological cryogenic machining', *Journal of Manufacturing Science and Engineering*, **123**(2), 331–338.

Hull D and Clyne TW (1996), *An Introduction to Composite Materials*, Cambridge University Press, New York, USA.

Hung NP, Yeo SH and Oon BE (1997), 'Effect of cutting fluid on the machinability of metal matrix composites', *Journal of Materials Processing Technology*, **67**, 157–161.

Indumathi J, Bijwe J, Ghosh AK, Fahim M and Krishnaraj N (1999), 'Wear of cryo-treated engineering polymers and composites', *Wear*, **225–229**, 343–353.

Isaak CJ and Reitz W (2008), 'The effects of cryogenic treatment on the thermal conductivity of GRCop-84', *Materials and Manufacturing Processes*, **23**, 82–91.

Kalia S (2010), 'Cryogenic processing: A study of materials at low temperatures', *Journal of Low Temperature Physics*, **158**, 934–945.

Kannan S and Kishawy HA (2008), 'Tribological aspects of machining aluminium metal matrix composites', *Journal of Materials Processing Technology*, **198**, 399–406.

Kim D and Ramulu M (2004), 'Cryogenically treated carbide tool performance in drilling thermoplastic composites', *Transactions of NAMRI/SME*, **32**, 79–85.

Kim MG, Hong JS, Kang SG, Kim CG and Kong CW (2007), 'Improvement of the crack resistance of a carbon/epoxy composite at cryogenic temperature', *Key Engineering Materials*, **334–335**, 365–368.

Kim RY and Donaldson SL (2006), 'Experimental and analytical studies on the damage initiation in composite laminates at cryogenic temperatures', *Composite Structures*, **76**, 62–66.

Kim YB, Choi NS (2003), 'Characteristics of thermo-acoustic emission from composite laminates during thermal load cycles', *KSME International Journal*, **17**(3), 391–399.

Levine J (2001), 'Cryoprocessing equipment', *Heat Treating Progress*, **2**(1), 42–52.

Luliano L, Settineri L and Gatto A (1998), 'High-speed turning experiments on metal matrix composites', *Composites Part A*, **29A**, 1501–1509.

Ma G, Chen D, Chen Z and Li W (2010a), 'Effect of cryogenic treatment on microstructure and mechanical behaviors of Cu-based bulk metallic glass matrix composite', *Journal of Alloys and Compounds*, **505**, 319–323.

Ma GZ, Chen D, Chen Z, Liu JW and Li W (2010b), 'The effect of cryogenic treatment on the microstructure and mechanical properties of $Cu_{46}Zr_{46}Al_8$ bulk metallic glass matrix composites', *Materials Science-Poland*, **28**(2), 595–601.

Miyase A, Yuan YS, Wong MS, Schön J and Wang SS (1995), 'Cryogenic and room-temperature mechanical behaviour of an Al_2O_3-fibre-reinforced high-temperature superconducting $(Bi, Pb)_2Sr_2Ca_2Cu_3O_x$ ceramic matrix composite', *Superconductor Science & Technology*, **8**, 626–637.

Mohan Lal D, Renganarayanan S and Kalanidhi A (2001), 'Cryogenic treatment to augment wear resistance of tool and die steels', *Cryogenics*, **41**, 149–155.

Molinari A, Pellizzari M, Gialanella S, Straffelini G and Stiasny KH (2001), 'Effect of deep cryogenic treatment on the mechanical properties of tool steels', *Journal of Materials Processing Technology*, **118**, 350–355.

Nordquist WN (1953), 'Low temperature treatment of metals', *Tooling and Production Magazine*, **7**, 72–100.

Novik IG, Sechko AE, Sviridenok AI and Voina VV (2008), 'Influence of heat treatment regimes on the structure and mechanical properties of polyamide composites', *Materials Science*, **44**(6), 844–849.

Paul S and Chattopadhyay AB (2006), 'Environmentally conscious machining and grinding with cryogenic cooling', *Machining Science and Technology*, **10**, 87–131.

Peterson EC, Patil RR, Kallmeyer AR and Kellogg KG (2008), 'A micromechanical damage model for carbon fiber composites at reduced temperatures', *Journal of Composite Materials*, **42**(19), 2063–2082.

Quan YM, Zhou ZH and Ye BY (1999), 'Cutting process and chip appearance of aluminum matrix composites reinforced by SiC particles', *Journal of Materials Processing Technology*, **91**, 231–235.

Rashkovan IA and Korabel'nikov YG (1996), 'The effect of fiber surface treatment on its strength and adhesion to the matrix', *Composite Science and Technology*, **57**, 1017–1022.

Reed RP and Golda M (1997), 'Cryogenic composite supports: A review of strap and strut properties', *Cryogenics*, **37**, 233–250.

Reitz W and Pendray J (2001), 'Cryoprocessing of materials: A review of current status', *Materials and Manufacturing Processes*, **16**(6), 829–840.

Samuel J, Kapoor SG, DeVor RE and Hsia KJ (2010), 'Effect of microstructural parameters on the machinability of aligned carbon nanotube composites', *Journal of Manufacturing Science and Engineering*, **132**, 051012-1-9.

Schutz JB (1998), 'Properties of composite materials for cryogenic applications', *Cryogenics*, **38**, 3–12.

Sherman RG (1990), *Behavior of Metal Matrix Composite Materials at Cryogenic Temperatures*, Nevada Engineering & Technology Corporation, NAVSWC TR 90–272.

Shetty R, Pai R and Rao SS (2009), 'Experimental studies on turning of discontinuously reinforced aluminium composite under dry, oil water emulsion and steam lubricated conditions using Taguchi's Technique', *G.U. Journal of Science*, **22**(1), 21–32.

Shindo Y, Takano S, Horiguchi K and Sato T (2006), 'Cryogenic fatigue behavior of plain weave glass/epoxy composite laminates under tension-tension cycling', *Cryogenics*, **46**, 794–798.

Shyha I, Soo SL, Aspinwall D and Bradley S (2010), 'Effect of laminate configuration and feed rate on cutting performance when drilling holes in carbon fibre reinforced plastic composites', *Journal of Materials Processing Technology*, **210**, 1023–1034.

Singh V, Ghosh S and Rao PV (2010), 'Grindability improvement of composite ceramic with cryogenic coolant', *Proceedings of the World Congress on Engineering*, Vol 11 WCE, June 30–July 2, London, UK.

Sitting M and Kidd S (1963), *Cryogenics, Research and Applications*, D. Van Nostrand Company Inc., Princeton, New Jersey, New York, 4–6, 62–73, 203–209.

Stevens MR, Todd RI and Papakyriacou M (2006), 'Thermal expansion behaviour of ultra-high modulus carbon fibre reinforced magnesium composite during thermal cycling', *Journal of Materials Science*, **41**, 6228–6236.

Stewart HA (2004), 'Cryogenic treatment of tungsten carbide reduces tool wear when machining medium density fibreboard', *Forest Products Journal*, **54**(2), 53–56.

Sundaram MM, Yildiz Y and Rajurkar KP (2009), 'Experimental study of the effect of cryogenic treatment on the performance of electro discharge machining', *Proceedings of the ASME International Manufacturing Science and Engineering Conference (MSEC 2009)*, West Lafayette, Indiana, USA.

Sweeney TP (1986), 'Deep cryogenics: The great cold debate', *Cryogenics*, **18**(2), 28–32.

Timmerhaus KD and Reed RP (2007), *Cryogenic Engineering, Fifty Years of Progress*, Springer, New York, USA.

Timmerman JF, Hayes BS and Seferis JC (2003), 'Cure temperature effects on cryogenic microcracking of polymeric composite materials', *Polymer Composites*, **24**(1), 132–139.

Trucks HE (1983), 'How cryogenics is used for the treatment of metals', *Manufacturing Engineering*, **91**(6), 54–55.

Uehara K and Kumagai S (1969), 'Chip formation, surface roughness and cutting force in cryogenic machining', *Annals of CIRP*, **17**(1), 409–416.

Uehara K and Kumagai S (1970), 'Characteristics of tool wear in cryogenic machining', *Annals of CIRP*, **18**(1), 273–277.

Vaccari JA (1986), 'Deep freeze improves products', *American Machinist & Automated Manufacturing*, **130**(3), 90–92.

Wang ZY, Rajurkar KP and Murugappan M (1996), 'Cryogenic PCBN turning of ceramic (Si₃N₄)', *Wear*, **195**(1–2), 1–6.

Wang ZY and Rajurkar KP (1997), 'Wear of CBN tool in turning of silicon nitride with cryogenic cooling', *International Journal of Machine Tools & Manufacture*, **37**(3), 319–326.

Wang ZY and Rajurkar KP (2000), 'Cryogenic machining of hard-to-cut materials', *Wear*, **239**, 168–175.

Wang ZY, Rajurkar KP, Fan J, Lei S, Shin YC and Petrescu G (2003), 'Hybrid machining of Inconel 718', *International Journal of Machine Tools & Manufacture*, **43**, 1391–1396.

Whitley KS and Gates TS (2003), *Thermal/mechanical response of a polymer matrix composite at cryogenic temperatures*, NASA Center for AeroSpace Information (CASI), NASA/TM-2003-212171.

Wiggins KL and Malkin S (1974), 'Cryogenic enhancement of machinability of bone', *Advances in the Astronautical Sciences*, 169–170.

Yildiz Y and Nalbant M (2008), 'A review of cryogenic cooling in machining processes', *International Journal of Machine Tools & Manufacture*, **48**, 947–964.

Zhang H, Zhang Z and Breidt C (2004), 'Comparison of short carbon fibre surface treatments on epoxy composites. I. Enhancement of the mechanical properties', *Composites Science and Technology*, **64**, 2021–2029.

Zhang H, Zhang Z (2004), 'Comparison of short carbon fibre surface treatments on epoxy composites. II. Enhancement of the wear resistance', *Composites Sciences and Technology*, **64**, 2031–2038.

Zhang X, Pei X, Mu B and Wang Q (2008), 'Effect of carbon fiber surface treatments on the flexural strength and tribological properties of short carbon fiber/polyimide composites', *Surface and Interface Analysis*, **40**, 961–965.

Zhang ZZ, Zhang HJ, Guo F, Wang K and Jiang W (2009), 'Enhanced wear resistance of hybrid PTFE/Kevlar fabric/phenolic composite by cryogenic treatment', *Journal of Materials Science*, **44**, 6199–6205.

15
Analyzing the machinability of metal matrix composites

M. BALAZINSKI, École Polytechnique de Montréal, Canada,
V. SONGMENE, Université du Québec, Canada and
H. A. KISHAWY, University of Ontario Institute of Technology
(UOIT), Canada

Abstract: The strength and the physical properties of composite materials are no doubt improved by the presence of the reinforcing particles. However, the higher energy requirement, the rapid tool wear and the associated poor surface finish during machining of metal matrix composites (MMCs) are leading drawbacks of the presence of the reinforcement particles. As a result, the applications of MMCs are still limited in certain areas. Understanding the deformation behavior and the effects of the reinforcement on the chip formation and the machinability is therefore crucial for proper tool selection, machining cost minimization and the expansion of MMCs to higher-volume applications.

Several factors of the particulate, such as nature, shape, size, volume fraction and distribution, are responsible for the mechanical behavior of MMCs, and thereby the cutting forces and the tool wear mechanisms. Some of these process indicators are still difficult to predict not only because of the nature, size and distribution of the reinforcements but also due to role that these reinforcements play during chip formation and the way in which the particles are removed or cut. This chapter presents the effects of reinforcement on chip formation, tool wear and part quality during the machining of MMCs.

Key words: MMCs, machining, forces, tool wear, chip formation.

15.1 Effect of the nature of the particle: soft and hard particles

Improving the machinability of particulate MMCs and developing machining data for popular cutting tools are the most promising ways to convince designers and manufacturers to use particulate MMCs in their applications. Developing composites containing both soft lubricating particles and hard particles, tailoring the volume percentage of each type of reinforcement, and developing efficient machining strategies can do this. The machinability of composite materials depends on the matrix alloy, on heat treatment, and on the nature, size and amount of the reinforcing particles (Lane, 1992).

394

Ceramic-reinforced metal matrix composites (CMMCs) are known for their high wear resistance, but also for the difficulties encountered during their machining. The hard and abrasive reinforcing ceramic particles that improve the wear resistance of these composites abrade the tool flank face, causing excessive wear. In order to improve the machinability and the friction properties of particulate composites, aluminum MMC reinforced with both soft lubricating graphite particles and hard silicon carbide particles has been developed. In fact, Inco J. Roy Research Center developed a new family of composite materials, named GrA-Ni®, containing both hard particles (SiC or Al_2O_3) for wear resistance and soft particles (graphite) for improved friction and machinability (Rohatgi et al., 1993; Bell et al., 1997; Songmene and Balazinski, 1999). The first composite of this family was named GrA-Ni 10S-4G and consisted of an A356 aluminum matrix reinforced with 10 vol% SiC and 4 vol% nickel-coated graphite. The microstructure of GrA-Ni® consists of a matrix of aluminum-silicon (white) with a dispersion of either SiC particles (small black phase, Figure 15.1b, c and d), or Al_2O_3 (small black phase, Fig. 15.1a, and coarser graphite particles

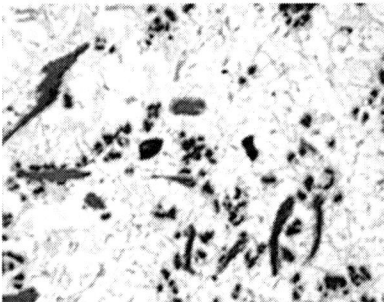

(a) GrA-Ni 5A.4G 5 vol% Al_2O_3,
4 vol% Ni-Gr

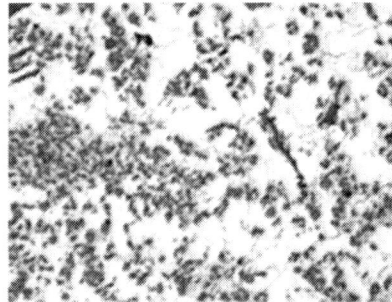

(b) GrA-Ni 10S.4G 10 vol% SiC,
4 vol% Ni-Gr

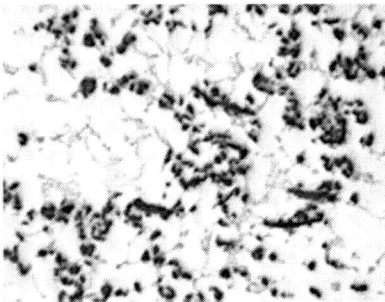

(c) GrA-Ni 6S.2.5G 6 vol% SiC,
2.5 vol% Ni-Gr

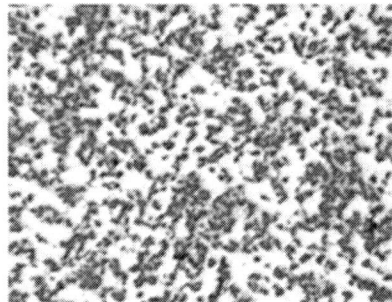

(d) Al-9Si.20SiC 20 vol% SiC

15.1 SEM images of MMCs studied (Songmene and Balazinski, 1999).

(large black phase, Fig. 15.1a, b and c). Also found in the matrix are nickel-based intermetallic precipitates (Al_3Ni), which are formed on solidification of the alloy as a result of the dissolution of the nickel coating in molten aluminum. By coating the graphite with nickel, the wettability of the coated particles is improved and this facilitates their incorporation into the aluminum alloy (Bell et al., 1997). Researchers (Rohatgi et al., 1993; Rohatgi and Narendranath, 1993) demonstrated the beneficial effects of a dispersion of graphite within the composite, which acts as a solid lubricant in the aluminum silicon alloys. The introduction of both soft graphite particles and hard silicon carbide particles into the aluminum matrix improves its wear resistance while the graphite lowers the coefficient of friction of the resultant composite.

15.1.1 Machining performance of graphitic particle reinforced MMCs when milling with carbide tools

Songmene and Balazinski (1999) studied the machinability of a non-reinforced aluminum alloy (Al380) and GrA-Ni® composites during milling with TiCN-coated carbide inserts and drilling with 10 mm uncoated high-speed steel (HSS) twist drills. The particle reinforced MMCs (MMCp) tested (Fig. 15.1) contained nickel-coated graphite particles as seen in Fig. 15.2.

The wear observed (Fig. 15.2) when milling these composites varies from no measurable wear on the insert that cut the alumina composite to heavier wear on those that cut aluminum reinforced with SiC particles only. As can be seen in Fig. 15.2b, c and d, the hard SiC particles (2700–3500 HV) grind the flank of the tool in a similar way to a grinding wheel.

For the case of Al_2O_3-reinforced GrA-Ni®, the hardness of the reinforcing alumina particles (2100–2300 HV) is lower than that of the TiCN coating (~3000 HV) of the tool. Alumina particles cannot abrade the face of the tool, which fails by microcracking and fatigue (Tomac et al., 1992).

As the density of the harder particle increases, so does the abrasive load on cutting tools and thus the wear rate. This explains why the tool that cut the aluminum composite containing 20 vol% SiC (Fig. 15.2d) wears quicker than both of those that cut the composite containing 6 vol% SiC (Fig. 15.2b) and the one containing alumina (Fig. 15.2a).

Figure 15.3 displays the wear curves observed when milling at a cutting speed of 61 m/min. GrA-Ni 5A.4G showed a lower wear rate when compared with all the other composites. When the reinforcing particles are alumina and nickel-coated graphite, the wear rate is quite similar to that observed during the machining of a high silicon content aluminum such as Al380, but the wear progresses quicker at the beginning when cutting alumina graphitic MMC (Fig. 15.3).

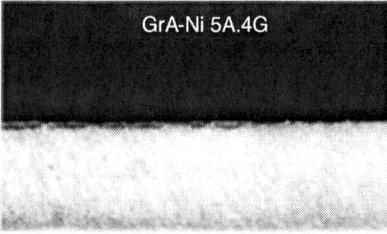

(a) Alumina and graphite

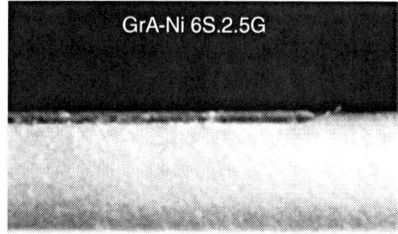

(b) SiC and Gr (low content)

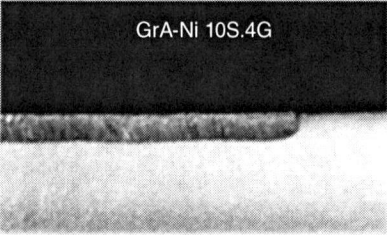

(c) SiC and Gr (high content)

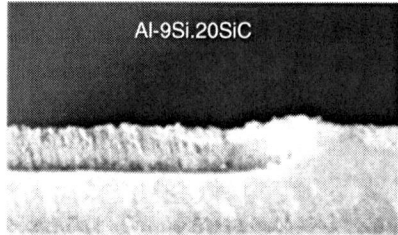

(d) SiC only

15.2 Wear on TiCN-coated carbide milling inserts after 10 minutes of cut. Cutting speed: 61 m/min; feed rate: 0.254 mm/rev (Songmene and Balazinski, 1999).

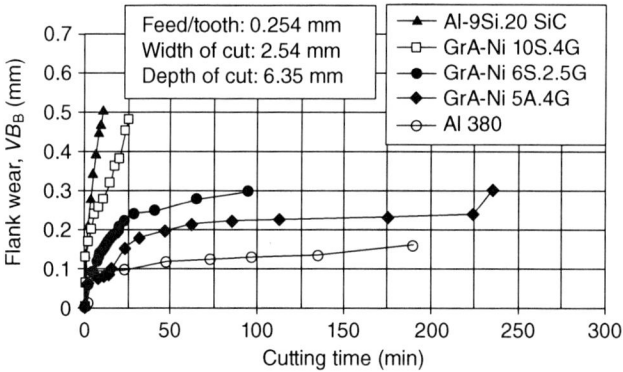

15.3 Wear progression at 61 m/min cutting speed when machining with TiCN-coated carbide milling inserts (Songmene and Balazinski, 1999).

A material with a high wear rate, such as Al-9Si.20SiC (Fig. 15.3), is more abrasive since the wear mechanism is pure abrasion due to the harder particles that grind the flank face of the cutting tool. The higher the content of harder particles (e.g. SiC), the higher the wear rate and thus the poorer the machinability. On the other hand, the higher the content of graphite, the better the machinability since the graphite lubricates the cutting.

A study of the cutting speed–tool life relationship for each material (Fig. 15.4) showed that the slope of the Taylor relation is quite comparable and all these composites have a higher Taylor exponent ($0.69 < n < 0.95$), which denotes the lower influence of cutting speeds on the tool life. It was shown in previous studies (Songmene et al., 1997) that the feed rate has a greater impact on tool life than the cutting speed when machining graphitic SiC-reinforced MMCs.

Songmene and Balazinski (1999) performed dry drilling tests to evaluate the cutting force when drilling 12.7 mm deep holes in various composites using 10 mm diameter uncoated high-speed steel (HSS) twist drills at a cutting speed of 7 m/min and a feed rate of 0.254 mm/rev. The study revealed a 50% increase in the torque when drilling Al-9Si-20S compared with that of the GrA-Ni 10S.4G material (Fig. 15.5) but forces were similar to those for aluminum alloy.

Figure 15.5 illustrates the relative machinability indicators of the composites expressed in terms of speeds for a 60-minute tool life and drilling torque. The better machinability of GrA-Ni® over existing SiC-reinforced MMCs such as Al9Si.20SiC was explained by the presence of graphite flakes in the aluminum matrix, which act as a solid lubricant, as reported earlier by Rohatgi and Narendranath (1993) and Ames and Alpas (1993). Also, the hardness of GrA-Ni® alloy is lower (~70 HRB) than that of Al-9%Si-20 vol% SiC (~75 HRB) (Bell et al., 1997). Low values of hardness and

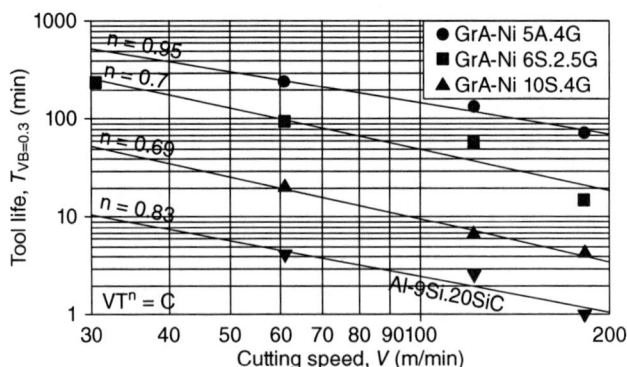

15.4 Tool life as a function of cutting speed when milling with TiCN-coated carbide inserts (Songmene and Balazinski, 1999).

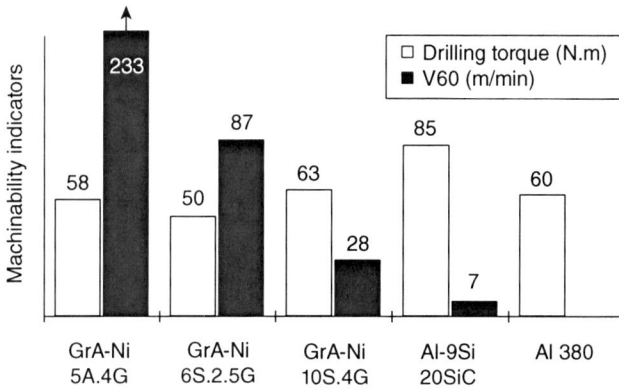

15.5 Comparison of machinability indicators: drilling torque and cutting speed for 60 minutes tool life – V60 (Songmene and Balazinski, 1999).

strength are favorable for machinability, except in very ductile materials, which may generate built-up edge, burrs, and poor finish.

15.1.2 Machining performance of graphitic MMCp when drilling with high speed steel tools

The main goal of the work (Songmene *et al.*, 2002) was to study the chip formation mechanisms and use the findings to further explain whether it is the matrix or the reinforcements that influence the cutting forces. Drilling torque and thrust force were analyzed together with shear angle and chip form to explain the machining behavior of composites containing hard particles (SiC or Al_2O_3) and soft lubricating particles (graphite). The spindle speed used varied from 56 to 710 rpm.

The workpiece materials studied were the aluminum alloy A356 and the composites (a, b and c of Fig. 15.1) consisting of the A356 aluminum alloy matrix with different reinforcements.

Figures 15.6 and 15.7 summarize the thrust force recorded as a function of spindle speed and feed rates when drilling graphitic MMCp using 4 mm HHS drills (118° drill point angle) under dry machining conditions. The measured data showed that the normal load did not vary with the cutting speed but increased at higher feed rates. This result was already known for a number of metals, such as steels, aluminum alloys, bronze, copper and MMCs (Al-SiC). The thrust force showed no distinct sensitivity to the rein-forced materials as compared to the unreinforced A356 aluminum alloy. The evolution of the thrust force as a function of feed rate for all the tested materials was comparable (Fig. 15.6). This result, which applies only to the

Feed rate: 0.08 mm/rev
Drill diameter: 4 mm

◆ A356
■ GrA-Ni 5A.4G
▲ GrA-Ni 10S.4G
✕ GrA-Ni 6S.2.5G

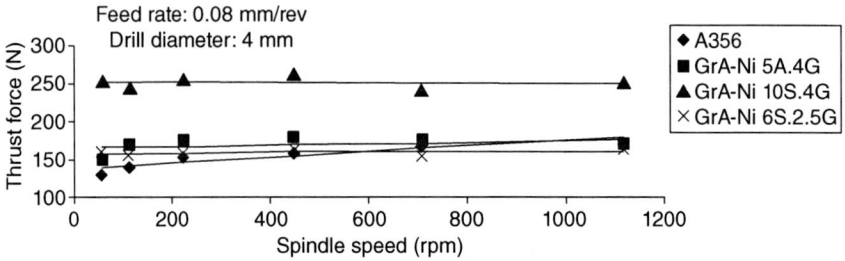

15.6 Thrust force for drilling GrA-Ni composites and its matrix alloy A356 as a function of spindle speed (Songmene *et al.*, 2002).

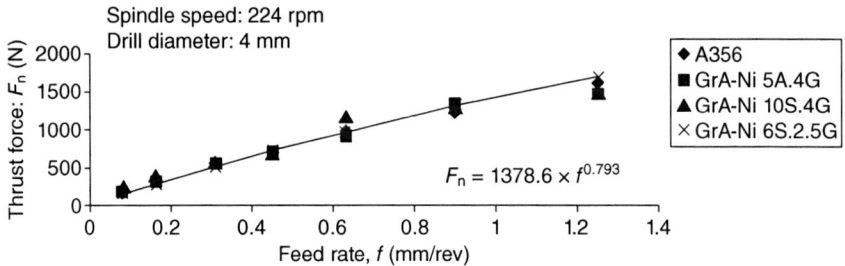

Spindle speed: 224 rpm
Drill diameter: 4 mm

◆ A356
■ GrA-Ni 5A.4G
▲ GrA-Ni 10S.4G
✕ GrA-Ni 6S.2.5G

$F_n = 1378.6 \times f^{0.793}$

15.7 Evolution of thrust force for drilling GrA-Ni composites and its matrix alloy A356 with feed rate (Songmene *et al.*, 2002).

materials tested, should be taken with caution because of the limited number of materials examined.

It is obvious that a small amount of graphite tends to reduce the negative effect of the SiC particulates and a fine layer of graphite was observed on the surfaces of the drill, which acts as good lubricant that annihilates the effect of the hard reinforcing particulates. Therefore it can be concluded that small amount of graphite was sufficient to provide an effective lubrication.

All observations converge into the fact that the thrust force is mainly due to the friction between tool and chip. The main consequence of this observation is that the cutting forces increase with increase in feed rate due to an increase in the contact surface and therefore an increase in the friction. Therefore, taking into account the results presented in Fig. 15.7, the influence of the feed rate on thrust force for the A356 aluminum alloy and the composites tested is given by:

$$F_n = A \cdot f^{0.8} \qquad\qquad [15.1]$$

where F_n is the thrust force (N), A is a constant and f is the feed rate (mm/rev)

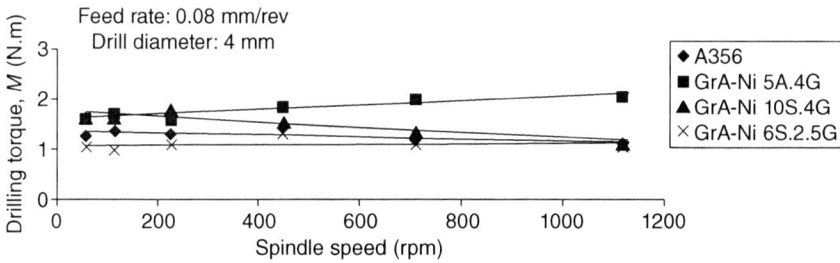

15.8 Comparative drilling torques for A356 matrix and GrA-Ni® composites (Songmene *et al.*, 2002).

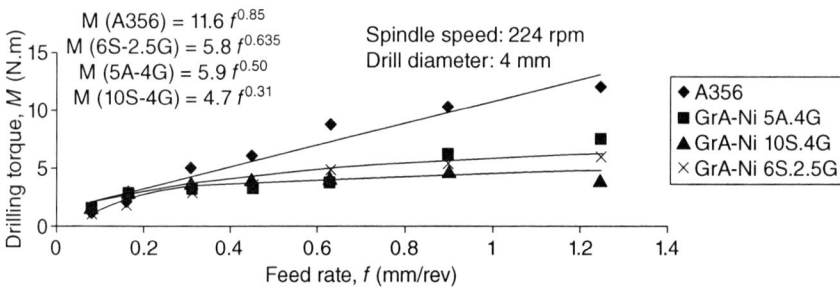

15.9 Drilling torque as a function of workpiece material and cutting speed (Songmene *et al.*, 2002).

This law was obtained for numerous materials (Masounave *et al.*, 1998).

Figures 15.8 and 15.9 show the torque obtained for the same materials at various drilling speeds and feed rates. As shown, the presence of soft and hard particles decreases the drilling torque and the higher the content of reinforcements the lower is the torque. Built-up edge (BUE) was observed on the cutting tool edge for all the materials analyzed that led to a stable drilling process. In particular, BUE led to the formation of a rough surface finish on the workpiece. The torque does not change with the cutting speed (Fig. 15.8).

15.2   Chip formation

The presence of the reinforcement helps to improve the chip breakability (Figures 15.10 and 15.11). The chip form obtained with graphitic MMCs appeared to be determined firstly by the volume content of Ni-coated graphite particles and secondly by the total ceramic particle content within the composite. A reduction in ceramic particle content reduces the possibility of forming serrated chips and thus generated long chips.

	GrA-Ni 10S.4G	GrA-Ni 6S.2.5G	GrA-Ni 5A.4G	Al-9Si 20SiC	Al 380
Milling					
Drilling					

15.10 Chip forms (Songmene and Balazinski, 1999).

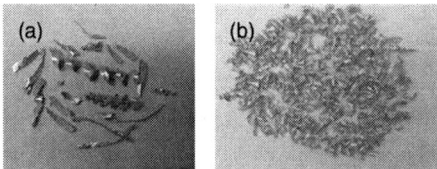

15.11 Image of chip forms obtained when drilling (a) A356 aluminum alloy and (b) GrA-Ni 10S-4G composite (Songmene *et al.*, 2002).

There were no significant differences in chip forms obtained during the milling process; this can be explained by the discontinuous nature of the milling. Conversely, the chip segmentation when drilling GrA-Ni composites (Fig. 15.11b) is quite different when compared to the chips obtained on non-reinforced aluminum alloys such as A356 and A380 (Figures 15.10 and 15.11a). The segmentation of the chip of the composite is due to the reinforcing particles.

The results of the measurements of the chip length (Fig. 15.12) and chip thickness (Fig. 15.13) confirm this observation. The dimensions (length and thickness) of the chips obtained on composites were very small compared to those obtained on the A356 alloy. As expected, the chip thickness followed the same trend as the one observed previously on the shear angle. It increases with the feed rate, and the non-reinforced alloy generated thicker chips (Fig. 15.13).

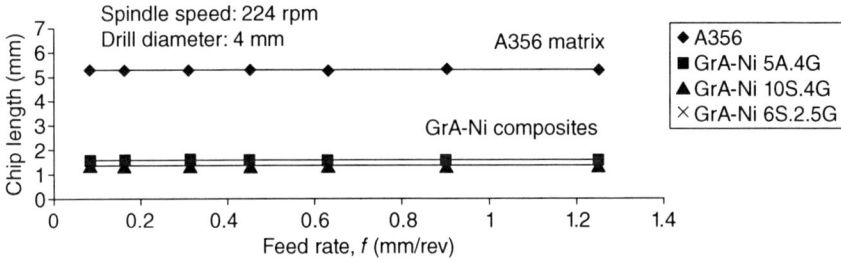

15.12 Length of chip as a function of feed rate for A356 and GrA-Ni®
composites (Songmene *et al.*, 2002).

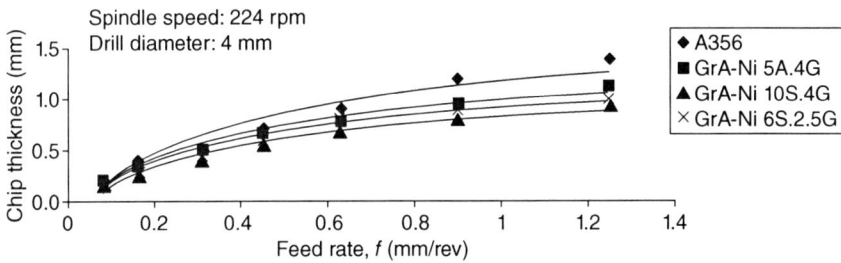

15.13 Deformed chip thickness of the GrA-Ni® composites and its
matrix.

Figures 15.14a, b, c and d display the results of quick-stop tests. The drill-press spindle was stopped suddenly at the same time that the feed motion was reversed. In this way, the drill edge disengaged leaving the chip formed on the workpiece. Figure 15.14a highlights the chip formation mechanism when drilling the A356 alloy: the tool motion generates a work-hardened zone (I) on the workpiece near the tool tip; there is a narrow and localized shear zone (II) materialized by the shear angle. The chip bends and comes into contact with the outer surface of the part being machined (Zone III). The reacting force applied to the chip contributes to breaking it (Zone IV).

Figure 15.14b shows the shear zones on a A356 chip being formed. The cracks initiate and progress in the euteutic zone (grey phase), which is different from what is observed on the chip formation of the composite.

The tool action shears the graphite pocket, liberating the graphite that acts as a solid lubricant on the tool rake and flank faces (Figures 15.14c and d). This reduces the friction and the tool wear. A magnified image of the chip section is presented in Figure 15.14c. It shows shearing bands on the outer surface of the chip (Fig. 15.14d). These shear bands denote the initiation of partial chip segmentation.

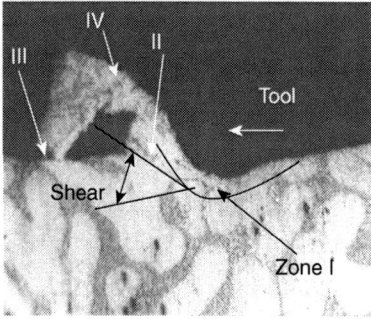

(a) Section of A356

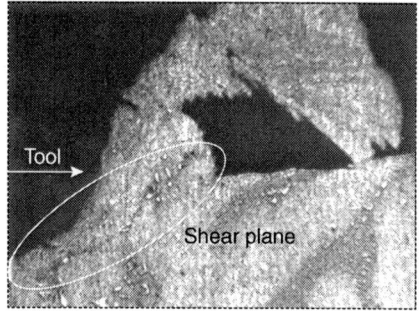

(b) Shearing plan on a section of A356

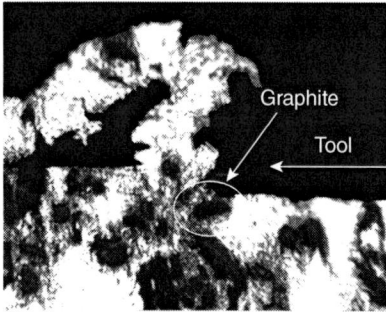

(c) Section of GrA-Ni 5A-4G

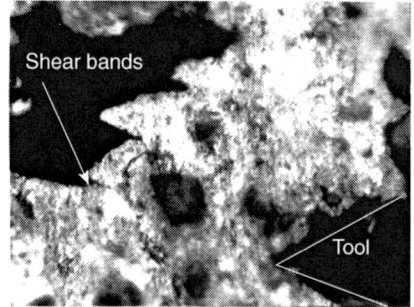

(d) Shear bands on GrA-Ni 5A-4G

15.14 Sections of unreinforced A356 alloy and a composite GrA-Ni 5A-4G showing the chip formation as observed in quick-stop tests (Songmene *et al.*, 2002).

15.3   Effect of particle shape

Most particulate reinforced MMCs contain irregular shaped particles with sharp corners, which imposes real difficulties on the study of the mechanical behaviors of the composites. During the past decades, researchers have simplified these irregular shapes to more simple geometries such as spherical, cylindrical, or oval, and have carried out extensive studies on the effect of these regular shapes on the behavior of the MMC. Bao *et al.* (1991) studied the role of the particles in reinforcing ductile matrix materials using finite element analysis and considering cylindrical-shape reinforcements. The reinforcements were assumed to be rigid and perfectly aligned, and the authors pointed out that the orientation and the distribution of high aspect ratio particles may also have a profound effect on the strength of MMCs. They suggested that the high strengthening due to aligned high aspect ratio particles would not be achieved if the orientation of the particles was random. Li and Castaneda (1993) proposed a variational method and investigated the effect of particle shape on the effective behavior of ductile metals reinforced by aligned spheroid inclusions with linear-elastic properties. The overall stress and strain relations were obtained in terms of the three distinct loading modes: axisymmetric tension (relative to the symmetry axis of the inclusions), longitudinal shear (along the symmetry axis) and transverse shear (perpendicular to the symmetry axis). In this work, it was found that the effective plastic behavior of the composites was strongly affected by the particle shape, and the significance of the effects of different particle shapes significantly depended on the loading mode.

Later, Li and Ramesh (1998) modeled the high strain rate mechanical behavior of particulate reinforced MMCs with the unit cell model approach. The overall response of the composite under uniaxial compression was computed numerically using finite element analysis. In their work, the effect of particle shape was investigated under various strain rates. The strengthening effectiveness of the particles was found to be not only a function of the shape, but also to be dependent on the aspect ratio and the strain rate. Cylindrical particles gave a more effective reinforcing function than spherical particles, especially at high strain rates. Compared to unit aspect ratio particles, particles with high aspect ratios were found to be more effective as reinforcements no matter whether the shape was spheroidal or cylindrical. The studies of these researchers all suggest the significant role that the particle shape plays in the strengthening of the MMCs. Cylindrical-shaped particles are commonly recognized as a more effective reinforcement under uniaxial loading condition. However, varying loading mode and strain rate may change the reinforcing effect caused by the particle geometry. These suggestions shed light on the understanding of the deformation behavior of MMCs during machining, in which extreme shearing and high strain rates are always encountered. Unfortunately, in the community of metal cutting, serious studies of these effects do not seem to be carried out yet.

15.4   Effect of particle size

It has been commonly recognized that the SiC particle strength is inversely proportional to its size. Since the probability of a strength-limiting flaw being present also increases when the particle volume increases, a relatively large amount of particle cracking occurs during extrusion before further machining. The load-carrying ability of these cracked particles is almost zero and therefore the monotonic strength of MMCs with larger particles is lower than that of those with smaller particle sizes, for the same volume fraction. Chawla and Shen (2001) studied the effect of particle size on the tensile behavior of Al 2080/SiCp composites, with volume fractions from 10–30 vol%. For the composites with particle size below 20 μm, the tensile strength of the composite was improved by the decrease in particle size, but particle cracking was observed for the case of particle sizes above 20 µm, leading to lower strengths of the composites than that of unreinforced matrix alloy. Nan and Clarke (1996) developed a hybrid analytical methodology for calculating the deformation response of particulate reinforced MMCs over the full range of particle sizes. The effect of particle size, distribution and volume fraction on the deformation response of T4-Al with 15 vol% SiC MMCs was investigated, based on the continuum mechanics based model. The predicted results followed the trend of the previous work by Chawla and Shen (2001). While the effect of the size was found to be significant when the particle size was smaller than 0.5 µm, only a slight effect of the particle size on the tensile behavior of the composite was observed when the particle size increased (larger than 10 µm). When the particle size was in the intermediate range, the mechanical behavior of the composites showed a significant dependence on both mechanisms.

During machining, the effect of particle size plays an important role in the chip formation. Kannan et al. (2008) conducted an experimental investigation on the generated forces and the consequent change of microstructure of the MMCs. The effects of particle size and volume fraction were also studied. The machining forces were found to increase with the average particle size. In fact, a larger particle size indicated the larger average spacing between the particles. This spacing between the particles significantly affects the flow strength of the composite. For the composites with larger particles, the increased spacing and the corresponding dislocation density make a stronger barrier to the plastic flow. This causes an increased dynamic shear stress during cutting and leads to an increase in the cutting forces. Furthermore, higher stress intensity near the particle takes place and may cause particle fracture, which will further accelerate tool wear and deteriorate the surface finish. Two-body and three-body abrasive wear have been reported as the dominant wear mechanisms during machining particulate reinforced MMCs (Xiaoping and Seah, 2001; Ibrahim et al., 2004; Kannan and Kishawy, 2006b). The cracked particles are normally responsible for the two-body abrasion with the cutting tool flank face. On the other hand, three-body abrasion is attributed to the particles debonded from the matrix due to interfacial cracking. As seen in Fig. 15.15, the micrograph of the uncoated carbide cutting tools used for machining 6061/10% MMCs (Kannan et al., 2006) shows the flank wear associated with the uniform abrasion wear tracks. The presence of grooves parallel to the cutting direction along the tool flank face indicates the dominance of the two-body abrasion mechanism. The micro cracks and pits on the flank face are mainly caused by three-body abrasion, due to the indentations made on the tool material by the hard particles debonded from the matrix.

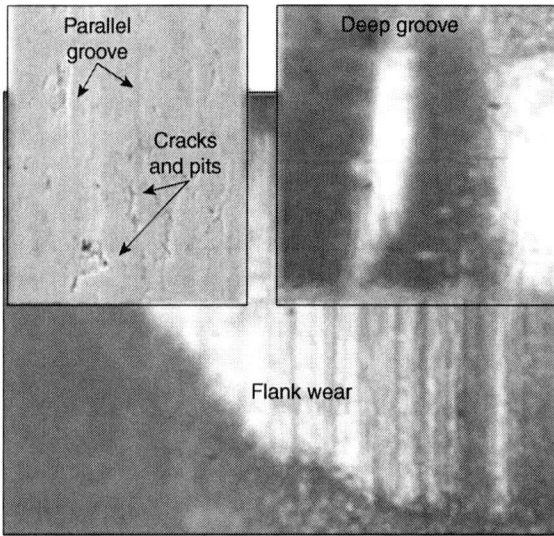

15.15 Flank wear on uncoated carbide cutting tool.

During machining, the plastic grooves formed by the two-body abrasion on the tool flank face mainly depend on the number and size of the abrasive particles. As previously mentioned, the bigger the particle size, the higher the possibility for the particles to break and therefore more and larger cracked hard particles will come into contact with the cutting tool. This results in a higher volume loss of the tool material due to the two-body abrasion mechanism.

Dai *et al.* (2004) studied the effect of particle size on the formation of an adiabatic shear band in 2024 Al matrix composites reinforced with 15% volume fraction of 3.5, 10 and 20 mm SiC particles; they demonstrated that the onset of adiabatic shear banding in the composites strongly depends on the particle size and that the adiabatic shear banding is more observed in the composite reinforced with small particles than in the composite with large particles. It is therefore expected that a composite with fine particles will lead to more broken chip.

According to Masounave *et al.* (1998), the influence of the reinforcing particles' type and size on machinability of MMCs can be illustrated as shown in Table 15.1. Composites of Group IV are almost impossible to machine while those of Group I are easy to machine. MMCs containing fine and hard particles (Group III) can be machined with coated tungsten carbides but a high chip cross-section (feed rate and depth of cut) must be used (Lane, 1992). When the particles are hard, abrasive and large, machining of the composite is very difficult.

Table 15.1 Classification of reinforcing particles for MMCs

Nature of particles	Particle size	
	Fine	Large
Soft: MoS$_2$, graphite and talc	Group I	Group II
Abrasive: SiC, SiO$_2$, Al$_2$O$_3$ and diamond	Group III	Group IV

15.5   Effect of particle volume fraction on tool wear and cutting forces

Machining forces were found to increase with increasing volume fractions of particles. A strong relation has been found to exist between the micro structural changes taking place in the matrix and the corresponding changes in the yielding behavior of the composites. One of the common features of MMCs is that the coefficient of thermal expansion of the metal matrix is much higher than that of the reinforcement particulate. This thermal mismatch may cause dislocation around the interface between the matrix and the particles, which will lead to failure. As a result, the plastic deformation of the matrix material occurs and the whole composite is strengthened. These strengthening particles serve as nucleation sites for the precipitates leading to the second phase precipitates. Therefore, it is obvious that the average dislocation density in the matrix increases with increasing particle volume fractions. Furthermore, the plastic flow of the matrix is locally constrained due to an increase in the number of particles attaching the matrix boundaries. For the reasons discussed above, the monotonic yield strength and the work hardening rate of the composite are commonly observed to be increased by the addition of discontinuous particulates to the aluminum matrix. On the other hand, the micro-yielding around the particles is usually initiated with stress concentrations before the macro-yielding occurs. The existence of more particles indicates more stress concentration points and therefore the stress needed to initiate the micro-yielding decreases with the increase in the particle volume fraction. However, this lower micro-yielding stress caused by the increase in the particle numbers does not seem to influence the overall macro-yielding behavior. From the measured cutting forces and TEM observations of the line defects in particulate reinforced Al 6061 MMCs with various volume fractions (Kannan *et al.*, 2008), the role that the particle volume fraction plays in the matrix strengthening mechanism can be clearly seen and explained in terms of dislocation density, as discussed above.

   The particle volume fraction is also found to affect the MMC machinability. For the same sized particles, an increase in the volume fraction increases the number of cracked hard particles coming into contact with the cutting tool. The number of particles debonded from the matrix will be increased by the increase in the volume fraction as well. Moreover, a higher volume fraction of particles may cause an increase in the hardness of the matrix composite and lead to the more active process of two-body and three-body abrasions. Normally, the percentage increase in the tool volumetric wear rate due to a three-body abrasion mechanism is higher than that due to a two-body abrasion mechanism. As shown in the work by Kannan *et al.* (2006), during cutting Al 6061 MMCs, an increase in the volume fraction of the alumina reinforcement from 10% to 20% resulted in the volumetric wear rate due to the two-body wear process increasing by 40%, compared with the 350% due to the three-body abrasion process. However, since the cutting tool material lost along the flank face is caused mainly by the plowing and micro-cutting action of the cracked particles, the two-body abrasion process dominates the tool flank wear. From the above analysis, it seems that the number of cracked particles, instead of the total number, is the determinant factor for the cutting tool flank wear rate. This may serve as the basic concept for the effect of particle shape and particle distribution on machinability, and explain the phenomenon that the coarser the particle, the more severe the tool wear. Therefore, decreased tool wear rate and improved machinability can be expected during machining matrix composites reinforced with fine and uniformly distributed particles.

15.6   Conclusions

- The incorporation of graphite particles into aluminum MMCs improves the machinability of the composites. Graphitic aluminum MMC reinforced with alumina was found to be easier to machine than that reinforced with both SiC and graphite and that containing SiC particles only. The GrA-Ni® composites can be classified in order of decreasing machinability as: (i) GrA-Ni 5A.4G; (ii) GrA-Ni 6S-2.5G; and (iii) GrA-Ni 10S.4G.
- When machining MMC, the cutting force is governed by the feed rate used and the nature of the reinforcing particles. The presence of the graphite particles in the matrix helps to lubricate the tool and thus reduce the cutting forces. A thin layer of graphite was found on the tool surfaces. There was no significant difference in forces recorded on the composites whether they contain SiC or Al_2O_3 particles. The graphite, even in small amounts, lessens the negative effect of the SiC particulates.
- Classical relations relating the thrust force and the drilling torque to the feed rate, as established by Shaw (1955) and confirmed by Masounave *et al.* (1998), were found to be applicable to the GrA-Ni® composites

and the A356 aluminum alloy used as the matrix for the GrA-Ni® composites.

- The tool wear rate depends on the nature and the percentage of ceramic particles within the composite. GrA-Ni® containing 5 vol% Al_2O_3 and 4 vol% graphite showed the lowest wear rate compared with other composites.
- The chip form obtained on graphitic MMCs appeared to be determined firstly by the volume content of Ni-coated graphite particles and secondly by the total ceramic particle content within the composite. Reduction in ceramic particle content reduces the possibility of forming serrated chips and thus generated long chips similar to those observed when drilling pure aluminum alloy such as Al380 and A356. As noticed by Tomac *et al.* (1992) and Songmene and Balazinski (1999), the reinforcing particles acted as chip breakers and helped to generate serrated chips. The chips break easily and liberate the graphite that acts as solid lubricant at the tool–chip interface.

15.7   References

Ames, W. and Alpas, A.T., 1993, Sliding wear of an Al-Si alloy reinforced with silicon carbide particles and graphite flakes, *Proceedings of ASM Materials Congress*, Pittsburgh, Pennsylvania, Oct. 17–21:27–35.

Bao, G., Hutchinson, J.W. and McMeeking, R.M., 1991, Particle reinforcement of ductile matrices against plastic flow and creep, *Acta Metallurgica et Materialia*, **39**, 1871–1882.

Bell, J.A.E., Stephenson, T.F., Warner A.E.M. and Songmene, V., 1997, physical properties of graphitic silicon carbide aluminum metal matrix composite, SAE Paper No. 970788, SAE International Congress and exposition, Detroit, Michigan, Feb. 24–27.

Chawla, N. and Shen, Y., 2001, Mechanical behavior of particle reinforced metal matrix composites, *Advanced Engineering Materials*, **3**, 357–370.

Dai L.H., Liu, L.F and Bai Y.L., 2004, Effect of particle size on the formation of adiabatic shear band in particle reinforced metal matrix composites, *Materials Letters*, **58** 1773–1776.

Ibrahim, C., Turker, M. and Seker, U., 2004, Evaluation of tool wear when machining SiC reinforced Al-2014 alloy matrix composites, *Materials and Design*, **25**, 251–255.

Kannan, S. and Kishawy, H.A., 2000a, On the role of reinforcements on tool performance during cutting of metal matrix composites, *Journal of Manufacturing Processes*, **8**, 67–74.

Kannan, S. and Kishawy, H.A., 2006b, Surface characteristics of machined aluminium metal matrix composites, *International Journal of Machine Tools and Manufacture*, **46**, 2017–2025.

Kannan, S., Kishawy, H.A. and Balazinski, M., 2006, Flank wear progression during machining metal matrix composites, *Transactions of the ASME, Journal of Manufacturing Science and Engineering*, August **128**, 787–791.

Kannan, S., Kishawy, H.A. and Deiab, I., 2008, Cutting forces and TEM analysis of the generated surface during machining metal matrix composites, *Journal of Materials Processing Technology*, **209**, 2260–2269.

Kishawy, H.A., Kannan, S. and Balazinski, M., 2005, Analytical modeling of tool wear progression during turning particulate reinforced metal matrix composites, *CIRP Annals – Manufacturing Technology*, **54**, 55–58.

Lane, C., 1992, *Machining of Composite Materials* (A95-15178 02-37), ASM International, Materials Park, OH, pp. 17–27.

Li, G. and Castaneda, P.P., 1993, The effect of particle shape and stiffness on the constitutive behavior of metal-matrix composites, *International Journal of Solids Structures*, **30**, 3189–3290.

Li, Y. and Ramesh, K.T., 1998, Influence of particle volume fraction, shape, and aspect ratio on the behavior of particle-reinforced metal-matrix composites at high rates of strain, *Acta Material*, **46**, 5633–5646.

Masounave, J., Maugendre S. and Scheed L., 1998, Prédiction des efforts de perçage des métaux, *Matériaux & Techniques*, **9–10**, 7–16.

Nan, C.W. and Clarke, D.R., 1996, Influence of particle size and particle fracture on the elastic/plastic deformation of metal matrix composites, *Acta Metallurgica et Materialia*, **44**, 3801–3811.

Rohatgi, P.K. and Narendranath, C.S., 1993, Tribological properties of Al-Si-Gr-SiC hybrid composite, *Proc. of ASM Materials Congress*, Pittsburgh, Pennsylvania, Oct. 17–21: 21–25.

Rohatgi, P.K., Bell J.A. and Stephenson T.F., 1993, *Aluminum-base metal matrix composite*, European Patent Ep0567284A2, Inco, April 20 and October 27.

Shaw, M.C., 1955, Principle of cutting, *Journal of the American Society of Mechanical Engineers (ASME)*, **77**, 103–114.

Songmene V. and Balazinski M., 1999, Machinability of graphitic MMCs as a function of reinforcing particles, *Annals of CIRP*, **48/1**, 77–80.

Songmene, V., Stephenson, T.F. and Waner A.E.M., 1997, Machinability of graphitic silicon carbide aluminum metal matrix composite GrA-NiTM, *ASME Int. Mech. Eng. Congress and Expo.*, Dallas, Texas, Nov. 19–21:193–200.

Songmene, V., Balout, B. and Masounave, J., 2002, Drilling of metal matrix composites: Cutting forces and chip formation, in T. Lewis, *Light Metals, Proceedings of the International Symposium on Enabling Technologies for Light Metal and Composite Materials and their End-products, 41th Conference of Metallurgists of CIM*, Montréal, QC, 11–14 August, 633–651.

Tomac, N., Tonnessen, K.. and Rasch, F.O. 1992, Machinability of particulate aluminium matrix composites, *Annals of CIRP*, **41/1**, 55–58.

Xiaoping, L. and Seah, W.K.H., 2001, Tool wear acceleration in relation to workpiece reinforcement percentage in cutting of metal matrix composites, *Wear*, **247**, 161–171.

16

Machining processes for wood-based composite materials

G. KOWALUK, Warsaw University of Life Sciences – SGGW, Poland

Abstract: Wood is probably the best known and widely recognized natural composite material: the cellulose fibers are arranged in complex bunches and are bonded with lignin. The decrease in 'fresh' wood resources, combined with the need to make use of the waste products resulting from wood processing, have made further developments in wood-based composite materials highly desirable. Combining wood with other materials allows composites to be developed with specific desirable characteristics, which are often better than the properties of wood alone. To achieve these properties, special techniques and tools for machining are required. Developments in improving the materials for tool production, as well as new construction methods for the machines, help to process wood-based composite materials efficiently while still ensuring a high quality product. Such developments are discussed in this chapter.

Although advanced tools and machines have been developed and are in use, there is a need for a better basic understanding of the characteristics of wood-based materials and of the factors influencing the machining process. In addition to these matters, this chapter discusses the current state of the art and research trends in the development of wood-based composite materials.

Key words: wood, composite, machining, particleboard, laminate.

16.1   Introduction

Composite materials, which are combinations of two or more components with different properties, have been used by mankind for thousands of years; for example, the traditional Chinese finishes made from compressed sheets of paper moistened with a clear or coloured self-hardening varnish from *Rhus verniciflua*, usually called the varnish tree. Wood is probably the best known and most widely recognized natural microcomposite, whose cellulose fibers are arranged in complex bunches and bonded with lignin. Wood offers many advantages, such as high strength-to-weight ratio and relatively easy processing. However, there are also some disadvantages associated with the use of wood. Wood has different mechanical parameters along three major directions – longitudinal, tangential and radial. This leads

412

to special requirements regarding the use of wooden elements in construction or in any load-bearing application. Other problems connected with the use of wood are its ever decreasing availability and increasing cost, and the consequent need to use lower quality wooden raw materials and/or wood waste products. Therefore, the use of wooden composite materials has become popular and is on the rise. Although the properties of solid wood vary among species, between trees of the same species, and between pieces from the same tree, the properties of reconstituted wood can be easily controlled and changed if needed.

The development of wood-based composites has resulted in changes to the tools and parameters for machining such materials. The machining of laminated particleboards or flooring panels, the surfaces of which are finished by hard laminate films, requires durable tools, mostly made from tungsten carbide or polycrystalline diamonds. The quality of brittle laminate particleboard milling is kept high by increasing the cutting speed. The positive aspect of the development of the wood-based composites, from a processing perspective, is that the number of processing operations and machines required can be lower than for solid wood, and could even be reduced to just two or three: sawing, drilling and edge finishing. Many modern one-stand CNC routers are able to realize all these operations and more.

16.2   Wood-based composite materials

Maloney (1986), in his proposal of a classification for wood-based composites, mentions among others: plywood, hardboard and particleboard, as well as wood–nonwood and wood fiber–agricultural fiber composites. These materials, which are now well known and frequently used, are good representatives of wood-based composites as a whole: they differ in the form (scale) of the wooden elements.

Plywood is a flat panel built-up of sheets (plies) of veneer, bonded with glue under pressure. Plywood can be constructed from either softwood or hardwood. Typically, plywood is constructed with an odd number of layers, with the direction of the grain in adjacent layers oriented perpendicular to one another. Thanks to the alternation of grain direction in adjacent plies, plywood panels have dimensional stability across their width. This grain direction alternation also results in fairly similar axial strength and stiffness properties in perpendicular directions within the panel plane, and helps to give plywood high strength-to-weight and strength-to-thickness ratios. Plywood is produced in two different forms for different usages: construction plywood and decorative plywood. An example of the use of plywood as a structural engineered material is the production of I-joists. The predominant technique used in the machining of plywood is sawing.

A need to utilize large quantities of lower quality wood particles, such as mill residues, sawdust, planer shavings and other relatively homogeneous materials from various wood industries has resulted in the growth of the particleboard industry. As well as wood, other lignocellulosic raw materials can be used in particleboard production, including rape, willow and black locust (Kowaluk 2009a, b; Kowaluk *et al.* 2010). The production stages for the particle and fiber panels are generally the same. The wood particles, e.g. oriented strand board (OSB) strands, particles or fibers, are blended with glue and pressed under elevated temperature and pressure. OSB is a panel for structural use, produced from thin wood (usually softwood) strands bonded with resin. Because of the typical aspect ratio of the strands (strand length divided by width) of about three, and thanks to the special orientation of the strands in the layers of the panel, OSB panels have a greater bending strength in the oriented directions. OSB panels are used mostly for roof, wall, and floor sheathing in wooden and prefabricated constructions, as well as for the members in I-beams. When used as flat panels, OSB panels are mainly machined by sawing.

Typical particle panels are produced in three layers. The surface layers are made from fine particles to provide a smooth surface for laminating, overlaying or painting. The core layer is made of a coarser material, allowing a more efficient use of raw materials. The particles are bonded with an amine resin, or, to a much lesser extent, with phenol-formaldehyde and isocyanates. As well as resin, paraffin or a microcrystalline wax emulsion is added to improve the short-term moisture resistance of the panels. The blended particles are pressed under elevated temperature and pressure. After pressing, the surfaces of the panels are sanded and can be finished, veneered or overlaid with other materials to provide a decorative surface. Today, approximately 85% of particleboard is used in the production of furniture and cabinets. The machining of particleboards when used in furniture production is conducted by sawing, milling and drilling.

The term 'fiberboard' is often used for both hardboard and medium-density fiberboard (MDF). However, the fundamental difference between the two types is the method of mat formation: for hardboard, the mat is formed using the wet-method, whereby the fibers are moved in a water suspension and pressed without the addition of a bonding agent. MDF fibers are blended with a resin (most often an amine resin), dried and formed into a mat through use of pneumatic-mechanical techniques. In hardboards, the natural tendency of the lignocellulosic fibers to create larger conglomerates is utilized to bond them together. Both hardboard and MDF mats are then pressed under elevated pressure and temperature. There are three types of dry-formed fiber panels: high-density fiberboard (HDF), medium-density fiberboard (MDF) and low-density fiberboard (LDF). The main application of HDF panels is in flooring materials, while

LDF panels are used as wall covering. The furniture industry is by far the largest user of MDF, where it is frequently used in place of solid wood. The more regular structure of the MDF panels across the thickness (compared to particleboards, for example) allows deep routing (milling) of the faces and shaping of the edges for use in furniture frontages or doors. Sawing, milling and drilling, along with deep routing, are the most common methods used for the machining of fiberboards.

16.3   Major machining techniques

The most common method used for machining wood-based composite materials is sawing on circular saws. In the case of some materials, and for certain purposes, the sawing process can be the only machining process, for example, when machining plywood, OSB or hardboard, and also in the production of furniture elements from laminated particleboards or MDF. Because modern circular saws ensure high quality and accuracy of machining, the finishing of elements processed in such a way covers the narrow surfaces of the panels with plastic or paper decorative bands. The sawing parameters depend on the machined material, the tool used, the construction of the machine, the assumed efficiency, and so on. The rotational speed of a 350 mm diameter 96 tooth circular saw with manual feed is in the range of 3000–6000 rotations per minute, where the feed speed is less than 20 m per minute. Saws with a mechanical feed can achieve higher feed speeds and should then be operated at a higher rotational speed. The cutting speed when sawing can be calculated from the equation:

$$v = \pi \cdot D \cdot n \qquad\qquad [16.1]$$

where: v is cutting speed, D is cutting diameter, n is rotational speed.

Finished wood-based composites, especially those with hard laminated surfaces, cause intensive tool wear. To solve this problem, new types of saw blades have been developed (Fig. 16.1). The cutting edges of the saw blades now consist of 3–5 mm thick tips made from tungsten carbide or polycrystalline diamonds (PCD), fixed to the saw body. The blades have special laser-made notches to reduce thermal stress (see Fig. 16.1). The noise generated by the saw blades when operated at high rotational speed is reduced by producing the body of the blade from two parts, which are bonded together.

Depending on the thickness of the wood-based panel and the tool diameter, an industrial sawing machine can cut as many as eight boards simultaneously. Modern machines of this type are automated and computer-controlled. The cuts are conducted according to a cutting plan, which can be prepared outside of the machine. Even material loading and manipulation are automated, ensuring high efficiency and quality of machining.

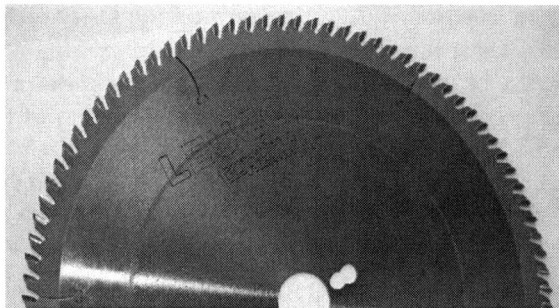

16.1 Saw blade for laminated particleboard machining.

After sawing, milling is the next most commonly conducted process in the machining of wood-based composites. Milling machines work with milling heads, where the cutting edges are replaced when worn-out, or with tools with fixed blades made from tungsten carbide or PCD (see Fig. 16.2). Depending on the purpose, the diameter of the tools can be from 80 to 250 mm. Milling heads with a small diameter have from two to four edges and can work at rotational speeds of over 10 000 rotations per minute. Generally, the rotational speed should increase when the tool diameter decreases. This is due to the need to ensure an optimal, high cutting speed, which is proportional to the tool diameter according to Eq. 16.1.

The milling of wood-based composite materials is principally carried out by machines, in which the milling process is one of several different processes. Laminated particleboards, for example, are machined in this way. Panels with 'gross' dimensions, prepared previously by other processes, undergo milling, after which the machined surfaces are covered with a decorative band made of plastic or other material, which is fixed to the surface by a hot-melt glue. The band edges are equalized to the panel surface by additional milling units. Depending on the type of decorative band, machines such as this can be operated at a feed speed of up to 30 m per minute. HDF panels for flooring are another type of wood-based composite machined by milling. In this case, special profiles are created on the narrow surfaces of the panels, allowing the panels to be connected when mounted on the floor. The efficiency of the processing lines of HDF flooring panels arises from the high feed speed, which is often in excess of 200 m per minute.

As mentioned previously, MDF panels have a more regular structure than particleboards, providing the opportunity to machine the wide surfaces of the panels. Such operations are conducted on computer-controlled

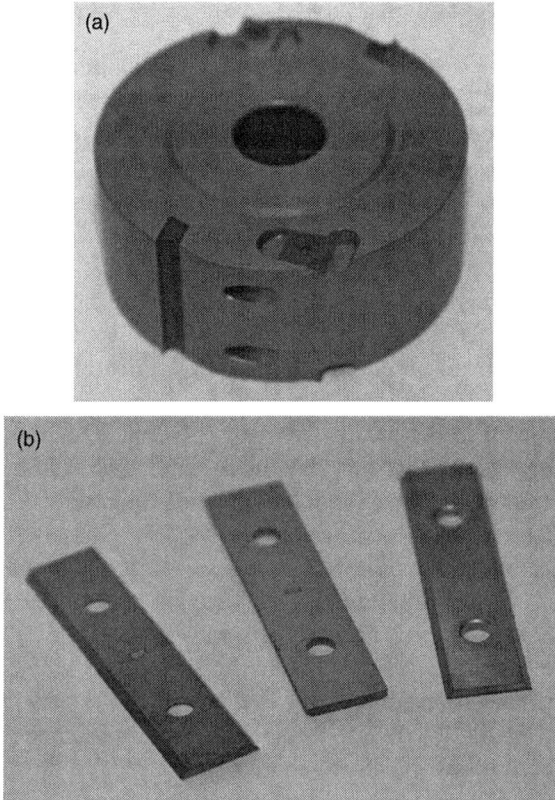

16.2 (a) Typical two edge milling head and (b) replaceable tips.

machining centers, CNC (Computerized Numerical Control) routers, which are usually equipped with several different tools/aggregates, including saws, milling heads and drills, as well as gluing units for the edge bands. Depending on the complexity of the router, the machine can be operated with one or more units working simultaneously: For example, when one is running the milling process, the second can prepare the set of drills from the tool pod. The working area can also allow the machining of one or more elements: while one element is being processed, a second can be manually or automatically replaced. The machined material is fixed on the machine stand by vacuum. The CNC routers, depending on the type of unit, work with various tools. The milling tools are constructed as pin tools with a diameter (relating to purpose) of less than 50 mm. The engines of the routers, constructed as electro spindles with ceramic bearings, ensure a spindle rotational speed of over 20 000 rotations per minute.

16.4   Selected machining problems

Although the machining of some wood-based composite materials, such as plywood or OSB, does not pose problems, and requires similar processing operations to solid wood, laminated particleboards or MDF (HDF, LDF) require special treatment. The laminate layer, which fulfils the protective and decorative functions, is hard and brittle, while the supporting layer is much less hard. This leads to fundamental differences in crack propagation between the laminate and supporting layers, which must be taken into account to ensure the high quality of the machined material.

As mentioned above, sawing on circular saws is the most common method of machining wood-based composites. The indicator of the quality of the machining of the laminated panels is the condition of the panel edge. It is important to underline that the saw blade creates panel edges when both entering and leaving the machined panel. Two different problems arise as a result. As the saw blade enters the panel, the cutting edge presses against the laminate layer. The continuity of the laminate is broken when the maximal stresses exceed the laminate strength. According to Palubicki *et al.* (2008), the modulus of elasticity of a laminate is about 12 GPa, while the modulus of elasticity of the surface layer of particle board is about 3.8 GPa. The most problematic scenario is when the part under the laminate is a particle of bark. The modulus of elasticity of this type of particle is only 0.022 GPa. As a result, a situation can arise in which the laminate bends, because it is not rigidly supported by the soft bark underneath. Cutting this sort of material can be compared to cutting a hard, rigid chocolate icing on a soft cake: the icing never breaks directly under the cutting edge, but some distance away. To prevent uncontrolled breaking of the laminate, the following rules should always be fulfilled: the edge radius (wear of the tool) (Fig. 16.3) should be as small as possible, and a high cutting speed should be maintained (Palubicki *et al.*, 2007). A small edge radius helps to increase the local stresses on a relatively small area. A high cutting speed also favours this phenomenon. To extend the

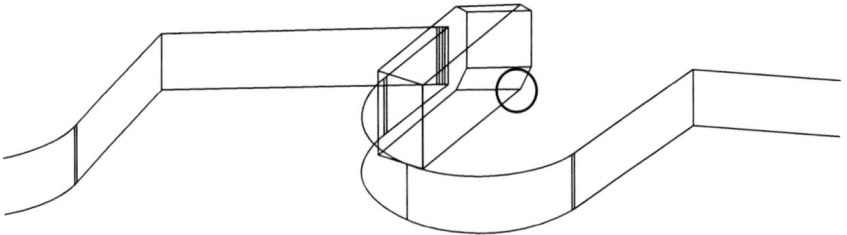

16.3 The quality of the panel's edge is generated by the circular saw tooth's side cutting edge.

length of time before the tool needs to be replaced and/or sharpened, Palubicki (2006) has proposed increasing the cutting speed as tool wear progresses.

Another critical situation occurs when the saw blade leaves the machined panel. The forces acting on the laminate layer now try to tear off the material. Although the moved material is supported by a table, the construction of the sawing machine never allows the material close to the rotating tool to be supported. This unsupported area could be improperly divided, and damage to the laminate edge could occur. The solution to this problem is the use of an additional pre-cut saw blade, which is mounted directly before the main tool (Fig. 16.4). The pre-cutter rotates in the opposite direction to the main tool, so the material machined by this additional saw blade is pressed into the core of the panel. The height of the cutting layer machined by the pre-cutter is less than 2 mm. The role of the pre-cutter is to make a low-depth cut, to divide the material, which could otherwise be damaged by the main tool. The width of the cut mark from the pre-cutter is slightly wider than the width of the main tool cut. This solution has recently been applied in almost all sawing machines used for machining of laminated panels.

During typical milling of wood-based composite materials, especially laminated panels, which generate by far the most problems with regard to achieving proper machining quality, the plane of the tool rotation is parallel to the machined material (except in special processing). The forces working on the outer layers of the panel try to tear off it, rather than bend it, as in the case of sawing. Beer *et al.* (2002) proved that laminated particleboard is characterized by higher hardness than other wood-based composites, and that crack propagation is rapid. Because of the high hardness of the laminates, the wear of tools machining such materials is also intensive. Due to this, the edges of the tools used for processing such panels are produced from hard materials, such as tungsten carbide or PCD. The high performance level of PCD tools used in the processing of wood-based composites was confirmed by Philbin and Gordon (2005). The difficulty is that these

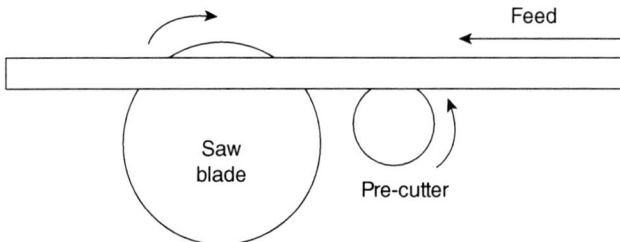

10.4 The application of the pre-cutter in sawing machines.

materials are hard and brittle (see Fig. 16.5). Because of the extremely dynamic machining process and high cutting speed, a brittle edge such as this can be catastrophically damaged (see Fig. 16.6). Thus, the geometry of such tools cannot be the same as that of typical tools made from high speed steel (HSS). The edge angle is higher, and depending on the producer, can be about 55° for edges made from tungsten carbide. As was shown by

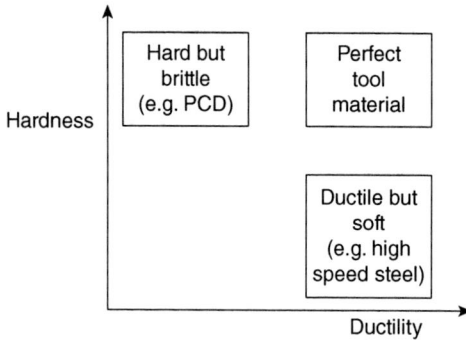

16.5 The dependence between the cutting edge material parameters.

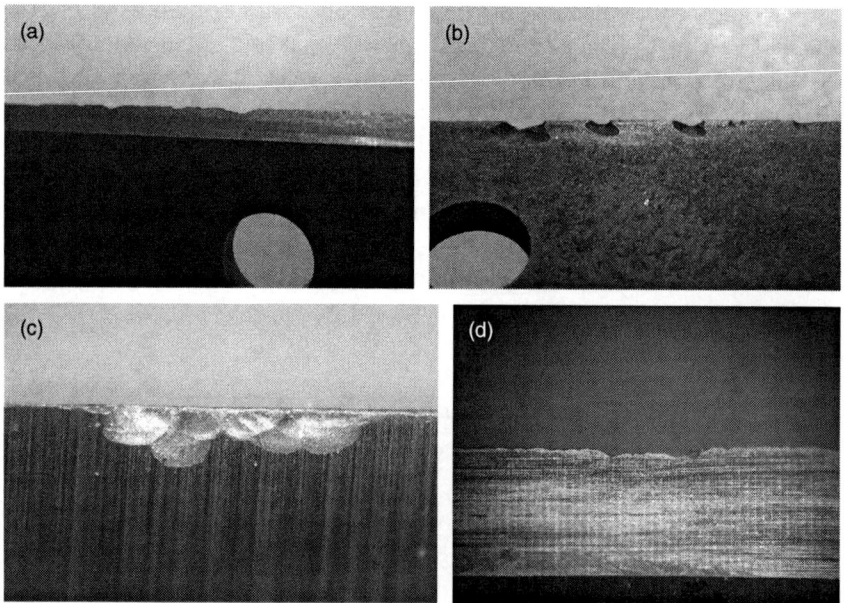

16.6 Examples of cutting edge catastrophic damage (a, b, c, d).

Kowaluk *et al.* (2009 b), the edge recession decreases when the edge angle increases.

Such a large edge angle causes extended stresses in the machined material. As long as the edge radius is low, the material can be cut before uncontrolled breaking occurs. With an increase in radius, when the edge recession reaches a certain level, the machined material can break before cutting takes places because of the 'overtaking crack' (Kowaluk *et al.*, 2004). This is confirmation of the importance of monitoring tool wear, especially when the machining is carried out at a high feed speed: if the critical wear of the tool is judged incorrectly, unacceptable damage can be caused by the worn-out tool during the rest of the production process (see Fig. 16.7).

The development of a new composite based on lignocellulosic raw materials offers promising results in the machining field. Panels produced from alternative raw materials, such as willow, black locust or rape, have a lower friction coefficient than panels produced from industrial chips. Because tool wear is partly caused by friction between the machined material and the tool, the wear intensity of the tools used for machining panels produced from alternative raw materials can be reduced. It was proved (Kowaluk *et al.*, 2007) that panels produced from rape particles caused a lesser degree of wear of the tool, after the same time period, compared with panels produced from industrial particles.

Wood–plastic composite machining trials were conducted by Buehlmann *et al.* (2001). The materials investigated were five different commercially-available woodfiber–plastic composites. Solid wood (white pine) was also tested for comparison purposes. The general conclusion drawn from these investigations is that the lowest degree of edge wear is found in solid wood machining. The reason for this is hypothesized to be the contamination content in the wood–plastic composites, as well as pigments used for plastic coloring.

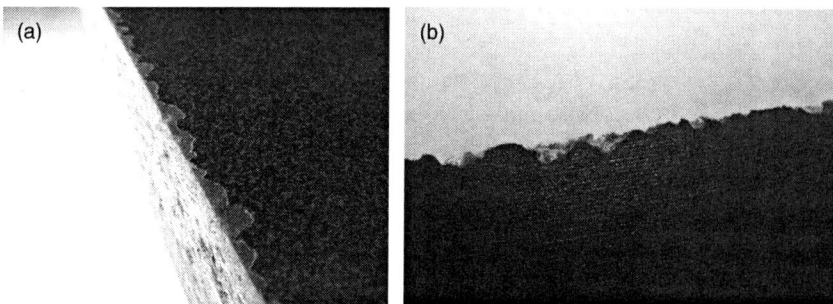

16.7 Effects of laminated particleboard machining with worn tool (a, b).

16.5   Future trends

The intensive development of wood-based composite materials, which is being driven forward by the need to utilize wood resources in a rational way, also requires that progress be made in the machining of such materials. If we analyze the wood-based composite and tool as a cooperative system, then changes to one side of the system causes (or should cause) a change on the opposite side. On the basis of contemporary research in the field of wood-based composites and machining tools, probable future developments can be hypothesized.

Research into wood-based composites that is currently underway can directly and significantly influence the machining processes adopted, and is focused on:

- reducing the density of the material,
- increasing the share of the recycled materials used in the production of wood-based composites,
- changing the structure of the surface layers of the panels in order to improve the quality of machining.

The investigations into reducing the density of wood-based composites, especially particleboards, based on a desire to make saving in raw materials and to improve logistical aspects, are conducted principally by changing the density of the core layers. Because of the need to ensure the correct panel strength, the parameters can be manipulated only to a certain extent. Typical three-layer particleboards, where the surface layers are responsible for the bending strength and the core layer is responsible for the internal bond, are produced in one process on continuous presses. One possible means of controlling the parameters of the layers more effectively is the production of three-layer panels in separate processes (Kowaluk *et al.*, 2011). The lowering of the density of the panels has a positive effect on machining processes. Most of the machining parameters, such as cutting force, feed force, edge recession and noise decrease along with the density of the machined material. There is no need to use special tools or machines to process these materials.

The increasing use of recycled lignocellulose materials in the production of wood-based composite materials could have a significant effect on the machining of such composites. The problems with proper selection and cleaning of raw material from second-hand furniture, flooring materials, doors and windows, etc., which have not yet been resolved, must be taken into account during machining. Any remaining metal or plastic parts or other unexpected bodies from the recycled raw material, present as a result of poor or incomplete separation, can cause intensive edge wear or catastrophic tool damage. Although current edge materials are more resistant to abrasion, they are still not resistant to wear caused by large mineral or metal parts within the machined product. Until new tool materials are developed, the most important aspect has been the implementation of a correct sensitive system to monitor tool wear.

As mentioned previously, the high hardness gradient through the depth of the laminated particleboard often causes damage to the machined element. According to Beer (2009), investigations in the field of laminated wood-based panels should focus on changing/improving the supporting layer under the laminate. One potential method is the addition of wood fibers to the surface layer of particleboards. Thus, the empty spaces that can occur in the surface layer structure could be filled by comparable soft and plastic wood fibers. The other potential means of modifying the surface layer of panels is the introduction of an additional transitory layer between the laminate and the surface of the panel. This layer should reduce the high hardness gradient and improve the support of the laminate.

Research concentrating on the creation of the perfect edge creation and the modification of tools is carried out by taking advantage of the possibilities offered by HSS and tungsten carbide/PCD tools. Tools made from HSS have a small edge angle and high ductility, whereas tungsten carbide/PCD tools have high hardness. The perfect tool should have an abrasion-resistant surface and a small edge angle with a ductile body, resistant to both dynamic stresses and catastrophic damage. To achieve this, modification of the surface of the steel tools is carried out. Selected chemical components, such as boron nitride or titanium compounds, are placed in the tool surface during a special deposition process. The surface-modified tools should be sharpened from one side only, to secure/extend the surface modification effects.

Newly developed cutting tools can be also used in the cutting of wood-based composites. Laser cutting technology, which works well in the processing of other materials, has also been introduced for wood cutting and can be applied to wood-based material cutting. Because of limitations caused by the burned surface after laser cutting, this method could be used in, for example, pre-cutting of laminate layers, or could alternatively be used to produce items with complicated shapes, where the burned narrow surface of the panel is not a critical point.

The rapid progress that has been made in tool material engineering, including nanotechnology, can help with the development of so-called 'intelligent tools'. This visionary idea involves tools composed of layers, each of which is responsible for a chosen parameter, e.g. hardness, bending strength, thermal conductivity, etc. The special crystallized structure of metal could be used to create a self-sharpening edge, through breaking in a predicted direction.

16.8 Different wood-based composite materials (from the top: particleboard, MDF, plywood).

16.6   Conclusions

The development of wood-based composite materials (see Fig. 16.8), brought about by the need for both rational and sustainable use of wooden raw materials, should lead to similar developments in machining technologies for these materials. It is difficult to predict the future of wood-based composite materials and methods of processing them. It should therefore be borne in mind that the potential developments discussed in this chapter, although prepared on the basis of the current state of the art, are by no means certain.

16.7   References

Beer, P. 'The mechanism of machining of composite materials with gradient of hardness' (in Polish). *Proc. of Seminary 'Wood Technology – New Threads in Technique and Technology'*. Poznan, Poland, 25.06.2009.

Beer, P., Ede, Ch., Gindl, M., Stanzl-Tschegg, S. 'Cutting tests of laminated particle board regarding work of fracture and chips plastic deformation. *Proc. Of Wood Science and Engineering in the Third Millennium*, Transylvania, University Brasov, Romania, 2002.

Buehlmann, U., Saloni, D., Lemaster, R.L. 'Wood fiber–plastic composites: Machining and surface quality'. *15th Int. Wood Machining Seminar*, Anaheim, CA, 30.07.–01.08.2001.

Kowaluk, G. (a) 'Influence of the method of milling on the geometry of fibrous chips and bending strength of produced particleboards'. *Proc. of 3rd Int. Sc. Conf. Woodworking Techniques*, Zalesina, Croatia, 02.–05.09.2009.

Kowaluk, G. (b) 'Influence of the density on the mechanical properties of the particleboards produced from fibrous chips'. *Ann. WULS-SGGW, For. and Wood Technol.* **68**, 2009.

Kowaluk, G., Borysiuk, P., Boruszewski, P., Maminski, M., Fuczek, D. 'Particleboards engineered through separate layer bonding'. In preparation (2011).

Kowaluk, G., Frackowiak, I., Beer P., Palubicki, B., Szymanski, W. 'Comparison of the tool wear in milling of the particleboards produced from wood and rape straw'. *Proc. of 3rd Int. Symp. on Wood Machining, Properties of Wood and Wood Composites Related to Machining*, COST Action E35, Lausanne, Switzerland, 21.–23.05.2007.

Kowaluk, G., Dziurka, D., Beer, P., Sinn, G., Tschegg, S. 'Influence of particleboard production parameters on work of fracture and work of chips formation during cutting'. *EJPAU* **7**(1), #01, 2004.

Kowaluk, G., Palubicki, B., Frackowiak, I., Marchal, R., Beer, P. (a) 'Influence of ligno-cellulosic particles on tribological properties of boards'. *Eur. J. Wood Prod.* DOI 10.1007/s00107-0090362-9, 2009.

Kowaluk, G., Palubicki, B., Fuczek, D. 'Influence of the raw materials parameters on the properties of the fibrous chips and particleboards'. *Proceedings of 7th International Scientific Conference Chip and Chipless Woodworking Processes*, Terchova, Slovakia, 08.–12.09.2010.

Kowaluk, G., Szymanski, W., Palubicki, B., Beer, P. (b) 'Examination of tools of different materials edge geometry for MDF milling'. *Eur. J. Wood Prod.* **67**, 2009.

Maloney, T.M. 'Terminology and products definitions – A suggested approach to uniformity worldwide'. *Proc. of 18th Int. Union of Forest Research Organization World Congress*, 1986 September; Ljubljana, Slovenia. IUFRO World Congress Organizing Committee.

Palubicki, B., Kowaluk, G., Beer, P. 'FEM analysis of stress distribution during cutting in laminated particleboard'. *Ann. WULS-SGGW, For. and Wood Technol.* **63**, 2008.

Palubicki, B. *Researches on Machining Quality of Furniture Elements Made out of Laminated Particleboard*. Doctoral thesis. Typescript of Dept. of Woodworking Machinery and Basis of Machine Construction, Faculty of Wood Technology, Poznan University of Life Sciences, Poznan, Poland, 2006.

Palubicki, B., Olejniczak, K., Kowaluk, G., Hryc, J., Beer, P. 'The change of laminated particleboard edge quality while sawing with progressing teeth wear'. *Proc. of 3rd Int. Symp. on Wood Machining, Properties of Wood and Wood Composites Related to Machining*, COST Action E35, Lausanne, Switzerland, 21.–23.05.2007.

Philbin, P., Gordon, S. 'Characterization of the wear behavior of polycrystalline diamond (PCD) tools when machining wood-based composites'. *Journal of Materials Processing Technology* **162–163**, 2005.

17
Machining metal matrix composites using diamond tools

S. S. J O S H I, Indian Institute of Technology Bombay, India

Abstract: Among the various cutting tools, polycrystalline diamond (PCD) tools cause the least damage to machined surfaces, besides providing the longest tool life in machining of metal matrix composites. The focus of this chapter is on assessing the performance of diamond tools in the machining of the composites by analyzing tool wear, quality and integrity of machined surfaces, and fundamental aspects of their machining mechanics.

Key words: machining, metal matrix composites, PCD, tool wear, failure mechanisms, machined surface finish, integrity, MMCs.

17.1   Introduction

Modern composite materials have evolved to provide 'tailor-made' high performance materials for the advanced engineering and technology fields. The composite material evolution began with the development of fiber reinforced plastic (FRP) composites in the early 1930s and graduated to metal matrix composites (MMCs), and ceramic matrix composites (CMCs) in the early 1990s.

MMCs are a class of composite materials in which light weight, relatively low strength alloys of aluminum, magnesium or titanium are given additional strength and stiffness by adding second-phase particles, whiskers, fibers, wires or filaments. During the 1990s, numerous research and development activities aimed at improving their processing technology, properties and effectiveness have led to wider applications of MMCs.[1–4] Examples of their applications in the automotive industry include engine-connecting rods, propeller shafts and brake disks; and in the leisure industry, items such as bicycle frames and tennis racquets. In most of the applications, the efficiency in use was improved by a reduction in weight, and acceleration improvements brought about by a reduction in the inertia of the moving masses.[5]

Among MMCs, the focus is on composites made of an aluminum matrix reinforced with particles or fibers of silicon carbide or alumina, because of their attractive mechanical properties, such as high strength-to-weight ratio, high wear and temperature resistance and low density. In

426

the recent past, much progress has been made both in composite material formulations and in the routes for their manufacture. Various processing routes that give near net-shape components have been developed; these include powder metallurgy, co-spraying, low-pressure liquid metal infiltration and squeeze-casting.

Despite these developments, machining is still often necessary and it is difficult to obtain close tolerances and high surface finish in the machining of components made of MMCs, primarily due to their extremely abrasive nature.[5] Consequently, over the last two decades, the area of machining MMCs has received widespread attention.[6-12] In general, the focus of these studies has been on:

(i)   identifying suitable tool materials for machining of MMCs,
(ii)   arriving at parameters for their precise machining and,
(iii)   achieving desired quality and properties on the newly machined surfaces.

In the 1980s, Brun et al.[6] investigated the performance of PCDs and other tools in the machining of high volume fraction Al/SiC particulate composites. They realized that PCD tools are superior to PCBN, ceramics, coated ceramics and cemented carbide tools in machining MMCs. They attributed this to the fact that the hardness of PCDs is greater than that of most of the reinforcements used in MMCs.

PCDs have also found application in machining of many other materials, such as hypereutectic aluminum–silicon alloy components,[13] non-ferrous alloys of copper,[14] abrasive plastics and plastic composites such as glass fiber,[15] wood composites such as chipboard and MDF,[16] and so on. In the latter part of 90s, many researchers recommended PCD as the most suitable cutting tool material for machining MMCs.[4-12,17-20] Recently, ultra-precision turning of MMCs at very low feed rates and depths of cut has also been carried out using single point diamond tools (SPDT); and PCD tools with various geometries have also been used.[21-23] It has been demonstrated that a nanometric level of surface finish is attainable on MMCs.

One of the main features of the PCD tools is their edge sharpness. Therefore, when PCD tools are used, the reinforcements in the subsurface zone are usually cut cleanly, leaving them nearly undamaged. Use of other tool materials tends to fracture or pull the reinforcements out instead of cutting them cleanly.[18]

In previous studies the performance of PCD tools in machining MMCs has been assessed based on three major criteria, as depicted in Fig. 17.1. This chapter focuses on understanding the role and capabilities of PCD tools in the area of machining of MMCs, covering the current state of the art, and including inputs from the author's research work. It is divided into the three major sections shown in Fig. 17.1.

```
┌─────────────────────────────┐
│ Performance of PCD tools in  │
│      machining of MMCs       │
└─────────────────────────────┘
```

⇓   ⇓   ⇓

| Tool life, tool wear and productivity | Machined surface roughness and integrity | Chip formation and machining mechanics |

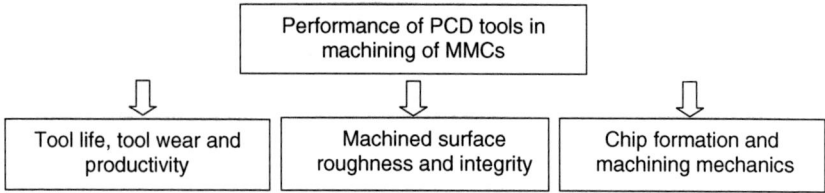

17.1 Machinability assessment of MMCs using PCD cutting tools.

17.2   Tool life, productivity and tool failure/wear mechanisms

Tool life and tool wear of PCD tools in the machining of MMCs has been the single most researched topic as it has direct relevance to the applications of MMCs in the real-life environment. The fundamentals of material machining indicate that tool life is associated with cutting speed, whereas productivity is governed by cutting parameters such as feed and depth of cut, along with cutting speed. The tool failure and wear mechanisms are dependent on tool material composition, grain size, geometry of the tools and the machining environment. In the following sections, the performance of PCD tools in machining MMCs as a function of tool life, productivity and tool failure/wear mechanisms is discussed in detail.

17.2.1 Tool life and productivity

Studies on the machining of MMCs using PCD tools report either the evaluation of Taylor's exponent in machining or the progression of flank wear land width as a function of machining time and cutting speed.[4,5,9,11,24-36]

The Taylor's tool life exponent in machining MMCs using PCD tools is as high as 0.6, indicating an excessive dependence of the tool life on the cutting speed, as given by $-V = v^{1/n}/C^{1/n}$. Other studies on the progression of flank wear on PCD tools show that these tools have a much higher tool life than cemented carbides at cutting speeds in excess of 1000 m/min (see Fig. 17.2a). Nevertheless, a reasonable tool life of 10–50 minutes for carbide tools can be obtained by reducing the cutting speed to 50–200 m/min (Fig. 17.2a). The performance of cemented carbide tools is poorer on hybrid composites compared with on SiC reinforced MMCs (Fig. 17.2b). The life of cemented carbides on hybrid composites is of the order of 1 minute at cutting speeds of less than 6 or 7 m/min (Fig. 17.2b).[5]

Apart from the cutting speed, the composition of the PCD tool influences its life during the machining of MMCs. An increase in the volume of reinforcement by 50% causes a decrease in the tool life of about 21%. This is attributed mainly to the increased rate of abrasion due to the greater

(a)

Cutting time to 0.2 mm
flank wear (min)

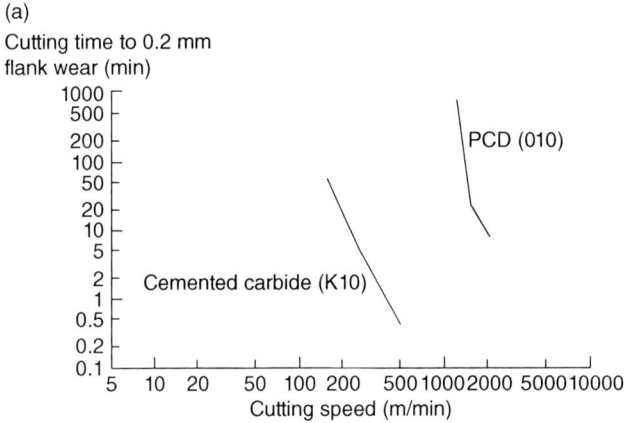

PCD (010)

Cemented carbide (K10)

Cutting speed (m/min)

(b)

Cutting time to 0.2 mm
flank wear (min)

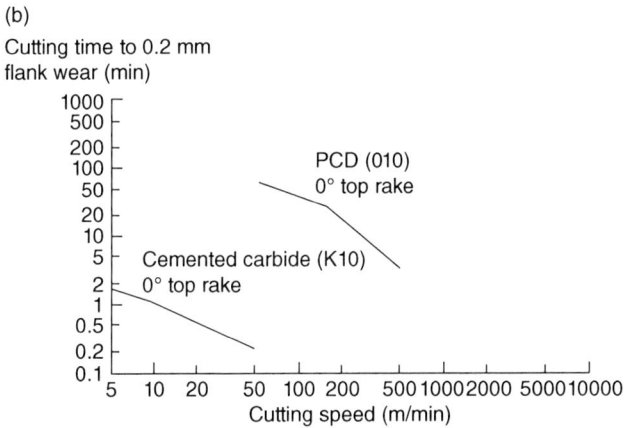

PCD (010)
0° top rake

Cemented carbide (K10)
0° top rake

Cutting speed (m/min)

17.2 (a) Taylor lines for cemented carbide and PCD (010) on Al-20%
Si alloy.[5] (b) Taylor lines for cemented carbide and PCD (010) on
Al-hybrid 15 vol% SiC/5 vol% short alumina fiber.[5]

number of abrasive particles per unit area in the chip. Further, an increase
in the size of the reinforcement particles from 9.3 µm to 12.8 µm makes the
wear 160% faster by tool abrasion and reduces the tool life by about 60%.[24]

When machining MMCs, the PCD flanks show the presence of deep wear
marks (Fig. 17.3a). Also, the band-width of the marks increases with increasing cutting speed. EDX analysis of the tool surfaces reveals that the wear
marks are covered with a thin film of matrix material (aluminum).[22–27]
However, the band-width of marks reduces by 30% after etching as this
dissolves the matrix material (see Fig. 17.3b). It appears that the flank surfaces during the machining are subjected to very high plastic deformation
(stress) due to intense rubbing between the work surface and tool flank,
and also due to the high ductility of the aluminum matrix.[28]

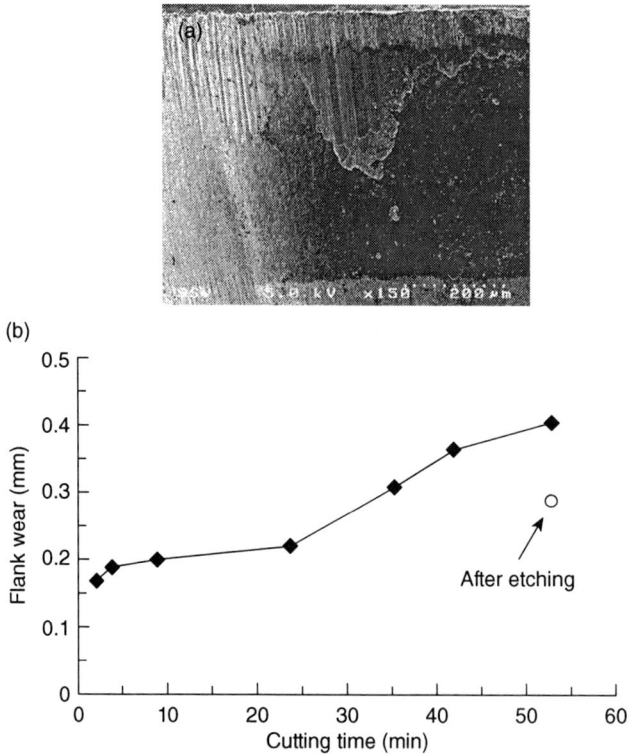

17.3 (a) SEM micrograph of the PCD flank face after 53 min of cutting.[28] (b) Variation of band-width of wear marks on PCD flank face with cutting time.[28]

Table 17.1 summarizes comparative performance of PCD tools in machining MMCs. It clearly shows that PCD tools have significantly higher tool life when measured in terms of machining time or progression of flank wear land width, at extremely high cutting speeds of the order of 500 to 1000 m/min. At such high speeds, no other tool material except polycrystalline cubic boron nitride (PCBN) is able to perform as well as PCD tools.

It is also observed from Table 17.1 that flank wear is doubled if the PCD grain size is reduced from 25 μm to 10 μm (Table 17.1, row 2). This indicates that PCD tools with larger grain sizes have better tool life. Tools with zero rake angle appear to give better performance than the tools with positive as well as negative rake angles (Table 17.1, row 3). The PCD tools show more chipping and larger flank wear when their nose radius is smaller (see Fig. 17.4a) for a chipped PCD tool.[9] Figure 17.4b shows that tools with a larger tool nose radius have more resistance to wear by abrasion.

Table 17.1 Comparative evaluation of tool life of PCD tools in machining of MMCs

MMC material	Tool life criteria	Tool life (cutting speed, m/min)	Remarks
Al/SiC/10p/20p/30p (Ref. 4 and 11)	Cutting time	Infinite: (8 to 100 m/min) for PCD; 5–8 s for WC (8 m/min); 60–80 s for self-propelling WC (88 m/min)	Rotary WC tools have longer life than stationary
AlMgSi1Mn/SiC-20% vol. (Ref. 7 and 24)	Cutting time and V_b	6000 s V_b = 0.2 mm, 10 μm size) 6000 s (V_b = 0.1 mm, 25 μm size) (at 500 m/min)	Wear doubled for 10 μm grain size PCD. Similar results in Ref. 13, 37–38
Al356-20%SiC (Ref. 9)	Flank wear	V_b = 0.02, 0.06 and 0.09 mm for 0°, –5° and +5° rake angles, respectively (at 894 m/min and feed: 0.35 mm/rev)	0° rakes angles are better than positive and negative rake tools
Al356-20%SiC (Ref. 9)	Flank wear	V_b = 0.02 and 0.09 mm for 1.6 and 0.8 mm TNR, respectively (at 894 m/min, feed: 0.35 mm/rev)	More chipping for 0.8 mm TNR
Al356-20%SiC (Ref. 9)	Flank wear (V_b = 0.18 mm)	V_b = 0.18 at 670 m/min and V_b > 0.18 mm at 894 m/min (at 670 m/min and 894 m/min)	
Al-20%Si alloy and hybrid 15% SiCp and 5% short Al_2O_3 fibre (vol.) (Ref. 5)	Flank wear (V_b = 0.2 mm)	10–50 min (at 50–1500 m/min); 1–10 min (at 5–100 m/min)	

Continued

Table 17.1 Continued

MMC material	Tool life criteria	Tool life (cutting speed, m/min)	Remarks
A356/SiC/20p (Ref. 18)	Flank wear ($V_b = 0.24$ mm)	45 min (at 250 m/min)	
Al alloy + 10% Al_2O_3 (Ref. 31)	Flank wear ($V_b = 0.24$ mm)	2 min (at 700 m/min) 15 min for PCD; 7 min for WC + TiN coated (at 500 m/min)	
Al alloy + 20% Al_2O_3 (Ref. 18)	Flank wear ($V_b = 0.3$ mm)	8 min for PCD; 3.5 min for WC + TiN coated (at 500 m/min)	
Al alloy + 20% SiC (Ref. 18)	Flank wear ($V_b = 0.3$ mm)	9 min for PCD (at 500 m/min), 0.5 min for WC + TiN coated (at 150 m/min)	
359/SiC/20p (Ref. 18)	$V_b = 0.2$ mm (CVD diamond coated), $V_b = 0.29$ mm for PCD	18 min (at 500 m/min), 18 min for CVD diamond coated (at 500 m/min)	CVD diamond coated tools are competitor for PCD
Al/SiC/20p (Ref. 33)	Cutting distance = 1800 m	$V_b = 0.1$ mm (at 50 m/min); $V_{bmax} = 0.5$ mm for PCBN (at 50 m/min)	
Al/SiC/20p (Ref. 33)	Cutting distance = 900 m	$V_b = 0.1$ mm (at 400 m/min); $V_{bmax} = 0.5$ mm) for PCBN (at 400 m/min)	
A359/SiC/20p (Ref. 34)	Cutting time, $V_b = 0.25$ mm	350 min (at 300 m/min); 100 min (at 500 m/min); 35 min (700 m/min)	Taylor's equation developed

TNR, tool nose radius.

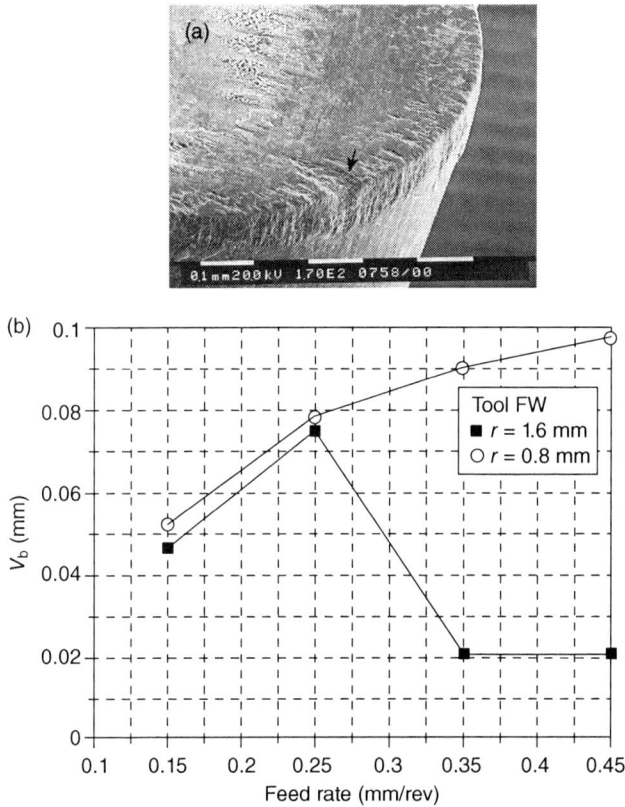

17.4 (a) SEM image illustrating PCD tool wear by chipping
(v = 894 m min^{-1}, depth of cut = 1.5 mm, f = 0.35 mm rev^{-1},
r = 0.8 mm, α = 0°); (b) effect of tool nose radius, r, on the tool flank
wear (FW; v = 894 m min^{-1}, depth of cut = 2.5 mm, α = 0°).[9]

A comparison between the performance of PCD tools and tungsten
carbide (WC) TiN coated and CVD diamond coated tools was made in row
9 of Table 17.1.[10–12] The cutting time of the coated insert is about half that
of the PCD inserts; the life of CVD diamond-coated carbides lies some-
where in between.[31]

17.2.2 Tool wear and failure mechanisms of PCD tools in machining MMCs

Studies have found that the predominant PCD tool wear mechanism is
abrasion by the hard reinforcement particles in the MMCs.[9,19,35–42] The
tool is subjected to mechanical loads due to cutting and frictional forces
(Fig. 17.5). Also, the reinforcements in the composites cause impacts on the

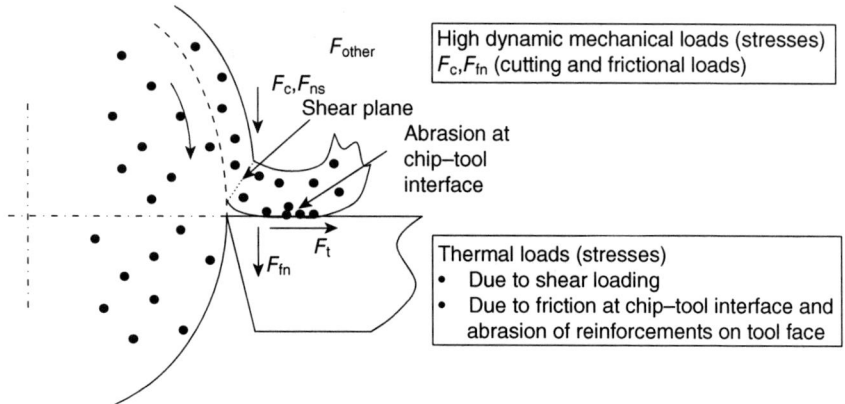

High dynamic mechanical loads (stresses)
F_c,F_{fn} (cutting and frictional loads)

Abrasion at chip–tool interface

Thermal loads (stresses)
• Due to shear loading
• Due to friction at chip–tool interface and abrasion of reinforcements on tool face

F_{other}: cutting, shear, frictional forces

17.5 Tribological system in the machining of MMCs material. F_c, cutting force; F_{fn}, frictional normal force; F_t, thrust force.

tool surface during machining. Finally, the system has thermal loads due to intense heating along the shear plane, and localized friction between reinforcements and tool surface (see the tribological system in Fig. 17.5).[43] The mechanical and thermal loads give rise to a number of wear phenomena in the PCD tool–MMC machining system. They are as follows:

(i)   abrasion wear,
(ii)   adhesion wear,
(iii) chipping, wear scars and tool breakage.

Abrasive wear

The dominant wear mechanism in machining MMCs by PCD tools is abrasion, primarily occurring by impacts of reinforcements on the cutting edge, and by the sliding motion of reinforcement particles on the rake and clearance faces.[42] These mechanisms have been investigated by a number of researchers.

The wear pattern and wear mechanisms of single crystal diamond (SCD) and polycrystalline diamond (PCD) tools were investigated during ultra-precision turning of Al/SiCp MMCs.[23] The authors used wet machining conditions. The results showed that micro-wear, chipping, cleavage, abrasive wear and chemical wear were the dominant wear mechanism on the SCD tools. The PCD tools mainly suffered from abrasive wear on the face as well as on the flank surfaces (see Fig. 17.6a and b). The magnitude of wear at both locations increased gradually with an increase in the cutting distance.[23] The wear on the rake face was more significant than that on the flank face

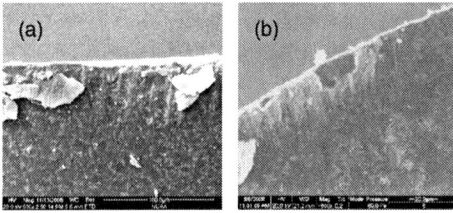

17.6 Abrasive wear of PCD after cutting 15 vol% SiCp/2009Al composite for 3.6 km: (a) flank, (b) rake face.[23]

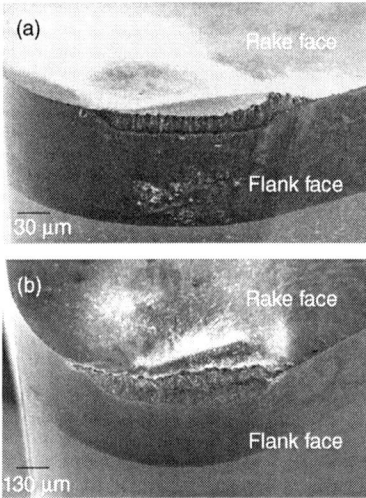

17.7 SEM of the PCD tools used to machine Al-SiC MMC without coolant (before etching): (a) at 50 m/min for a distance of 1760 m; (b) at 400 m/min for a distance of 880 m.[33]

of the tool. Grooves on the flank face (Fig. 17.6a) were formed by a combination of two-body and three-body abrasion between the workpiece material and the tool. The multiple bodies for the abrasion were available from the irregularly shaped SiC particulate reinforcements and the detached particles from the matrix during machining.[23]

The grooves on the PCD tool face and flank surface were due to formation of hard Al_2O_3 at the tool edge, which is hard enough to cause this grooving wear on PCD tools (Fig. 17.7a and b). In addition, the soft aluminum matrix causes seizure, leading to the pull-out of PCD grains. A third possible reason for the groove formation is abrasion of the tool faces by the SiC reinforcement particles.

Another parameter that influences PCD tool wear in machining MMCs is the size of grains in the tool. PCD tools with coarse grains possess high abrasion resistance. However, the coarser grains may lead to a substantial drop in the fracture resistance of the tools, which in turn might have a negative influence on their overall performance during machining.[9,42]

As far as the processing parameters are concerned, the feed rate affects the wear of PCD tools during machining of MMCs. With an increase in the feed rate, for a fixed volume of material removed, the tool surfaces will have less contact with the abrasive composites, thus increasing the tool life.[9] Also, with an increase in the feed rate, thermal softening of the material takes place. It has been suggested that as the workpiece becomes soft and reinforcement particles get pressed into the workpiece, less abrasion occurs on the tool tip.[8]

In yet another study,[44] on the turning of magnesium-based MMC (ZC71 reinforced with 20 vol% SiC particles) under dry conditions, three types of tools were used – TiN-coated tungsten carbides, PCD-coated and PCD inserts. During machining, the TiN-coated tools were destroyed instantly, while the PCD coatings showed good resistance against abrasion until the PCD coated film was torn. The PCD inserts gave the longest tool life.

Adhesion and adhesive wear

The softer matrix metals used in MMCs usually cause sticking at the tool cutting edges and surfaces. In the initial stages of machining, the rake face comes into direct contact with the underside of the newly formed chip. The adhered chip material protects the tool from further abrasive wear for a short period of time. However, with the continuous growth and breaking of the chip material, and the built-up edge (BUE), the stuck material on the rake face gets removed leading to severe adhesive wear on the rake face.

In machining Al/SiC MMC at a cutting speed of 400 m/min without a coolant, severe adhering of work material on the edges, face and flank of the tools took place for a PCBN tool (see Fig. 17.8a). This could have been due to an increase in thermal softening of the chip material. The adhering tendency was reduced considerably when a coolant was used (see Fig. 17.8b). On the other hand, the severity of adherence on the tip of the PCD tools was significantly lower than that observed on the PCBN tools (Fig. 17.7a and b). It shows that the PCD tools have a lower propensity for work-material adhesion than the PCBN tools.

Chipping, wear scar and breakage

Apart from abrasive and adhesive types of wear, PCD tools also show some other wear phenomena. El-Gallab and Sklad[9] found that unstable built-up edge may induce tool chipping and adversely affect the surface finish.[9] It

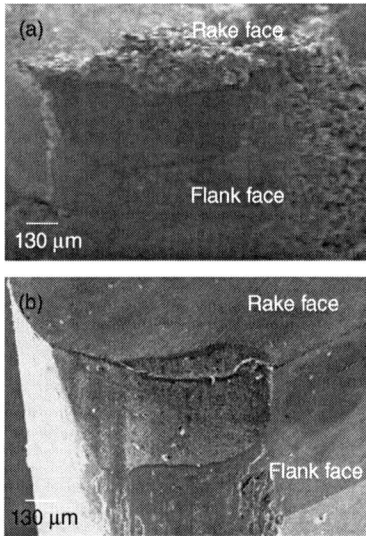

17.8 SEM of the BN300-PCBN tools used to machine Al-SiC MMC at 400 m/min for a distance of 880 m (before etching): (a) without coolant; (b) with coolant.[33]

was further reported that with negative rake tools, chips are trapped between the tool face and the workpiece causing more damage to the tool face. While, the positive rake angle tools showed excessive pitting on the tool face,[24] a decrease in the tool nose radius from 1.6 mm to 0.8 mm caused excessive chipping and crater wear of the PCD tools.[24]

The PCD tools also showed the presence of two types of wear scars. The first one was similar to a lightly ground, fractured surface and the second type involved grooves over a large number of grains and gave the tools a rough polished appearance.[6]

In another interaction of PCD tools with MMCs during machining, the tools broke easily due to shock by the coarse SiC particles. However, with a negative edge on the tool and with an adequate corner radius, the tools offered better resistance against breaking. The coarse-grained PCDs may have lost some of the grains due to fatigue stress, which in turn scraped the tool.[9]

The formation of BUE and its subsequent breakage may cause fracture of the PCD tools. In the turning experiments at depths of cut from 1.5 to 2.5 mm, a large BUE was formed, which when broken, was found to fracture the PCD tips.[9] Lane[45] suggested that the use of cutting fluid helps flush the chips away and detach the BUE. This can avoid breakage of the tool tip.

In the ultra-precision machining of 2009Al/SiC/15p MMC, PCD tools show lesser chipping on the edges than SCD tools. This is because, though the former have high hardness, their toughness and bending strength are higher than that of the single crystal tools.[23]

17.3   Machined surface and sub-surface integrity

Besides tool wear, another critical factor that it is necessary to evaluate in machining MMCs is the machined surface and sub-surface quality and integrity. It is well known that any damage to the reinforcements in the sub-surface zone can lead to deterioration in the performance of the products. When PCDs are used for machining of MMCs, the reinforcements in the subsurface zone are usually cut cleanly and remain virtually undamaged because of the sharpness of the cutting edges. On the other hand, with carbide tools, the widespread reinforcement fracture prevails on the machined subsurface regions.[46]

The machined surface integrity is a function of surface topography and sub-surface integrity, which is governed by work, tool and process related parameters during machining of MMCs (see Fig. 17.9).[47] In the following sections, various factors influencing surface roughness and integrity of the machined surfaces as shown in Fig. 17.9[47] are discussed in detail.

17.3.1 Surface roughness in machining of MMCs using PCD cutting tools

Effect of work material related parameters

Work material parameters, such as the composition of the composite and the morphology of the reinforcement, influence the roughness of machined surfaces on MMCs. The effect of the MMCs' composition was investigated

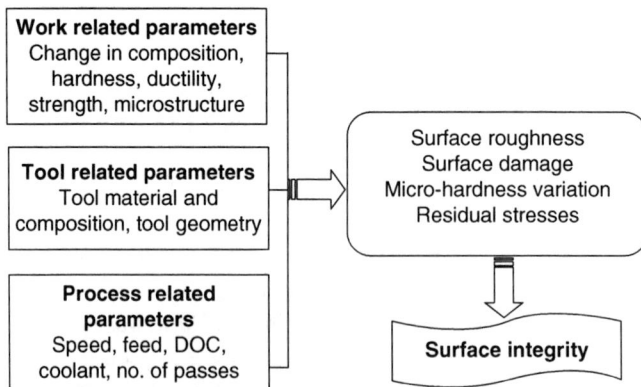

17.9 Surface integrity as a function of work, tool and process related parameters.[47] DOC, depth of cut.

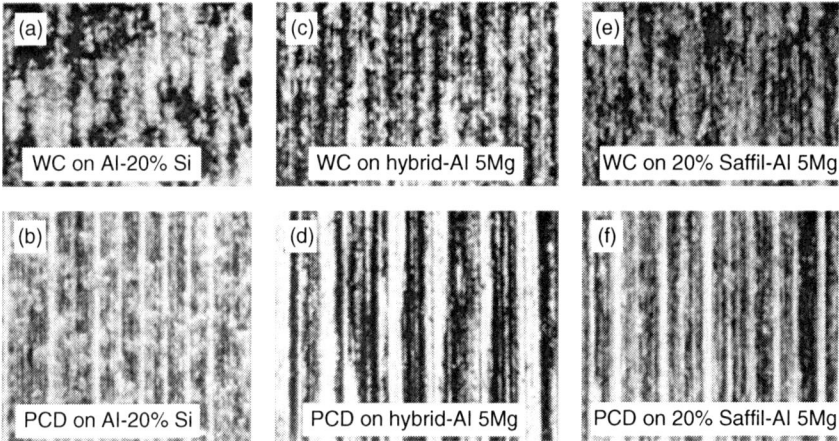

17.10 Appearance of surfaces cut with cemented carbide and PCD tools.[5]

by machining Al-20% Si alloy, the Al-hybrid 15% SiC/5% Al$_2$O$_3$ MMCs and the Al-20% alumina fiber MMCs, using cemented carbides and PCD tools (see Fig. 17.10a–f). The PCD-turned surfaces show distinct feed marks on all the composites (Fig. 17.10b, d and f) and considerable burnishing is evident on the cemented-carbide machined surfaces (Fig. 17.10a, c and e).[5]

In order to understand the effect of tool nose radius and the size of the reinforcement in the MMC on the surface roughness, turning experiments were performed on Al–2124/SiCp composites with 20 and 30 vol% of SiCp reinforcement of 220 and 600 mesh sizes. The feed rates (10, 60 and 110 μm/rev) and depths of cut (50, 100 and 150 μm) were chosen such that they were close to the size of reinforcement (15 and 65 μm). The tool-nose radius was 0.2, 0.4 and 0.8 mm and the PCD tools were with a rake angle of 5°.[48] The lowest surface roughness (the best surface finish) was obtained at the lowest value of feed-rate, the smaller particle size and the largest tool-nose radius (Fig. 17.11a). The highest surface roughness was obtained with the highest feed-rate, the larger particle size and the smallest tool-nose radius (see Fig. 17.11b).

Effect of tool related parameters

A number of studies have shown that the PCD tool geometry and compositional parameters cause various effects on the quality of the surface generated. For instance, it is observed that an increase in PCD grain size results in a significant deterioration in surface finish, since the PCD grains with size >25 μm can easily get pulled out of the cutting edge.[7]

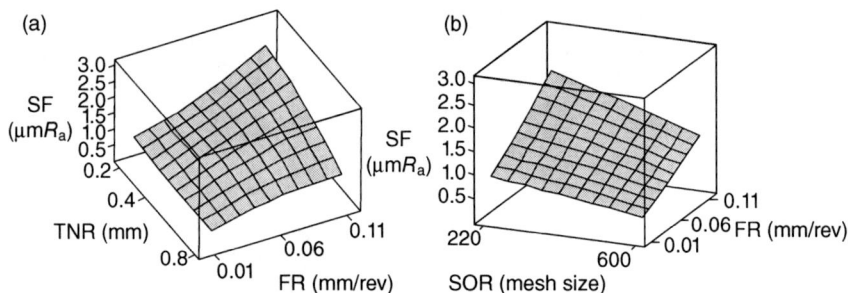

17.11 Response surface analysis. Relative influence of the three factors on the surface roughness: (a) surface finish (SF) vs. feed-rate (FR) and tool-nose radius (TNR), (b) surface finish vs. feed-rate and size of reinforcement (SOR).[48]

Tool geometry also affects the surface roughness on the machined components. A negative rake angle led to a greater surface roughness and this was attributed to clogging of the hot chips between the tool and the machined surface causing severe damage to the machined work surface. PCD tools with a small tool nose radius (r = 0.8 mm) produced a better surface finish than those with nose radii of 1.6 mm (cutting speed = 894 m/min, feed rate = 0.25 mm/rev, depth of cut = 2.5 mm and α = 0°).[6] However, at higher cutting rates and/or cutting speeds, tools with a nose radius of 1.6 mm out-performed those with a nose radius of 0.8 mm. This is attributed to increased edge chipping of the tools with small nose radii.[9]

In an investigation of ultra-precision machining of SiCp/2024Al and SiCp/ZL101A composites with PCD tools, the surface roughness obtained on them was of 26.65–49.70 nmR_a and 24.41–135.50 nmR_a, respectively. Keeping the machining conditions unchanged, single point diamond turning of SiCp/2024Al composite resulted in a surface finish of 23.66–111.23 nmR_a when machining with a feed rate in the range 1–10 μm/rev and depth of cut in the range 1–10 μm. These results suggest that a nanometric surface finish can be achieved as long as the cutting tools and machining parameters are appropriately chosen.[21]

Figure 17.12 shows the variation of surface roughness (nmR_a) with cutting distance for the three types of tools in the ultra-precision turning of a 15 vol% SiCp/2009Al composite.[23] The round edged SCD tools gave a good cutting performance in the initial cutting stages. However, surface roughness increased remarkably after severe flank wear on machining for a distance of 2 km. The PCD tool showed steady and favorable cutting performance up to a cutting distance of 6 km. Also, the surface roughness was less than 45 nmR_a and varied in the range of 12–15 nm. This is because the adhesive wear on the rake face and abrasive wear on the flank gradually

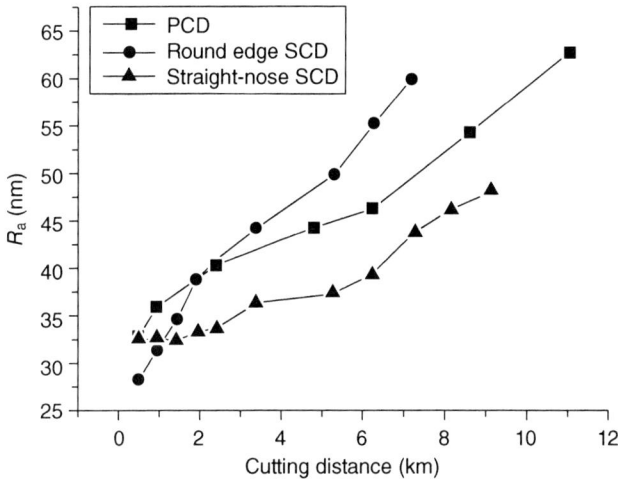

17.12 Surface roughness R_a vs. cutting distance when ultra-precision turning of 15 vol% SiCp/2009Al (n = 1000 rpm, f = 3 μm/rev, depth of cut = 10 μm).[23]

increased with an increase in the cutting distance. The straight-nosed SCD tool gave the best cutting performance among the three types (Fig. 17.12). The machined surface roughness R_a was less than 49 nm after the tool had cut over 9 km. This indicates that single crystal diamond (SCD) tools offer much better machined surfaces on composites.

Effect of process related parameters

The fundamentals of metal machining indicate that to achieve a high surface finish, a low feed rate and a low tool nose radius should be used. In the case of machining MMCs, however, the situation is complex and this principle is not directly applicable.

The least and most uniform surface roughness (R_a) was obtained in a facing operation with a completely new PCD tool.[49] The R_a value reduced significantly when the flank wear on the PCD tool had reached 300 μm.

In another study, evaluation of the maximum peak-to-valley height and arithmetic mean roughness versus cutting time when turning A356/SiC/20p with PCD inserts showed that a surface finish of R_a < 0.8 μm was possible when adequate cutting speed and feed rate were employed.

In a study of the machinability of an Al alloy reinforced with SiC particles (V_f = 10%) and Ni coated graphite particles (V_f = 5%), two types of diamond tool materials were used: PCD and diamond coated carbides (DCC). In this work, the feed rate was found to be a significant factor influencing surface roughness, while machining with both the PCD and the DCC tools. The cutting speed did not influence the R_a significantly. At higher cutting speeds, unless chatter or vibrations occured, the surface finish was independent of the speed. In most tests, PCD tools generated a better finish than DCC tools.[50] At 400 m/min, the surface roughness could be improved with the application of a coolant. This was attributed to a reduction in the severity of material transfer onto the machined surface.

Tomac and Tonnessen[8] found that an excellent surface finish, comparable to a ground surface, was obtained with a mixture of 1 part mineral oil and 20 parts of water as the cutting fluid in turning Al-Si/SiC/14p with a PCD-coated carbide tool.

Pramanik and co-workers[51] found that, at low feeds, the surface roughness of an MMC was controlled by particle fracture or pull-out. However, at higher feeds, it was controlled by the feed. (The surface roughness of the non-reinforced alloy is known to be controlled by the feed).

In the ultra-precision machining of Al6061/15SiCp MMCs using single crystal diamond tools (with rake angle of −25°, front clearance angle of 10° and tool nose radius of 0.762 mm), it was observed that the surface roughness and surface integrity could be significantly improved by using high spindle speed and low tool feed rate.[52] Here, the depth of cut did not affect the surface roughness except under low spindle speed conditions.

The hole surfaces produced in drilling Al 6061/10 and 20 vol% (Al_2O_3)p MMCs using PCD drills were smooth.[53] Lower feed rates gave lower surface roughness with a better bearing ratio. A negatively skewed distribution indicated that, generally, the surface had blunt peaks. Such surfaces are good for a sliding bearing, especially since the deep valleys will hold lubricant, while the blunt peaks will not penetrate into the mating surface.

17.3.2 Sub-surface integrity in machining of MMCs using PCD cutting tools

The integrity of the machined sub-surface is a common problem in machining of MMCs. The reduced integrity is due to various reasons, such as: debonding, and pull-out and fracture of reinforcements with consequent generation of voids and pits. Such effects are known to influence the physical and mechanical properties of components made of MMCs.

Typical scanning electron micrographs of machined surfaces on MMCs showing the most common defects are depicted in Fig. 17.13a–d. Usually, voids are created around the reinforcement particles on the machined surfaces (Fig. 17.13b). When a reinforcement particle is pulled out, it creates a groove on the machined surface (Fig. 17.13c). Similarly, when a reinforcement particle is crushed (Fig. 17.13d) during the machining process, it leaves a number of pits around and its fragments spread over the machined surface.

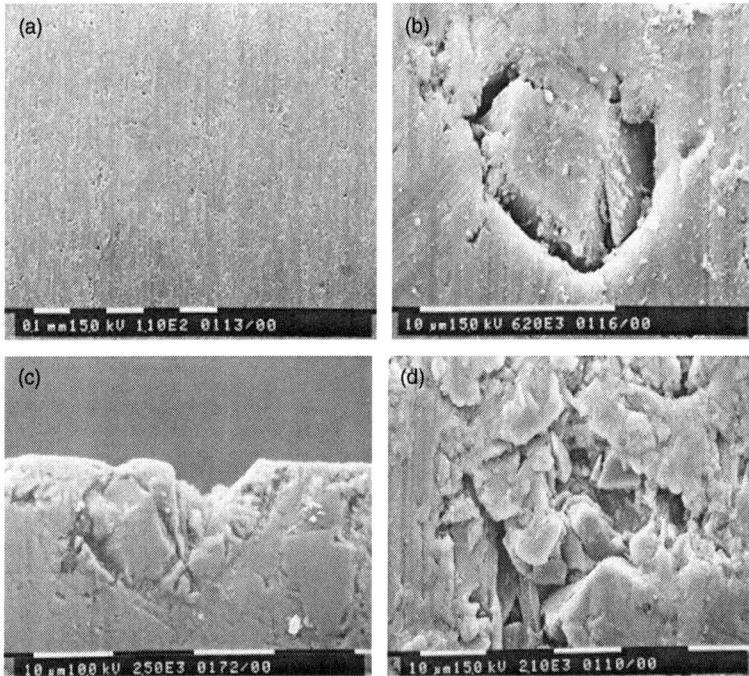

17.13 Typical SEM images of machined surfaces on MMCs at cutting speed of 670 m/min, a depth of cut = 1.5 mm and feed rate of 0.45 mm rev^{-1}, for r = 1.6 mm and α = 0°: (a) typical topography, (b) voids around the SiC particles, (c) pulled-out SiC particles and (d) fractured or crushed SiC particles.[53]

The interaction of PCD tools during the turning of MMCs is illustrated pictorially in Figs 17.14 and 17.15. When the particle of whisker is directly cut by the diamond tool, neither pits nor cracks are left on the cut surfaces (Figs 17.14a[52] and 17.15a.[22]). On the other hand, when the particles or whiskers are pulled out, pits or cracks are formed on the machined surfaces, (Figs 17.14b[52] and 17.15b[22]).

The machined surfaces after using PCDs on Al/SiC particulate composites under all kinds of cutting conditions showed that surface defects such as pits, voids, micro-cracks, fine grooves (scratches), protuberances and matrix material tearing were present. All the defects had an intimate relationship with the process of SiC particle removal.[21] As a result of de-coherence between the SiC particles and the aluminum matrix, some SiC particles were pulled out of the machined surface, which left behind many pits on the surfaces (see Fig. 17.16a). During the course of SiC particles being pulled out or crushed, the matrix around them also became torn. Needless

(a)

(b)

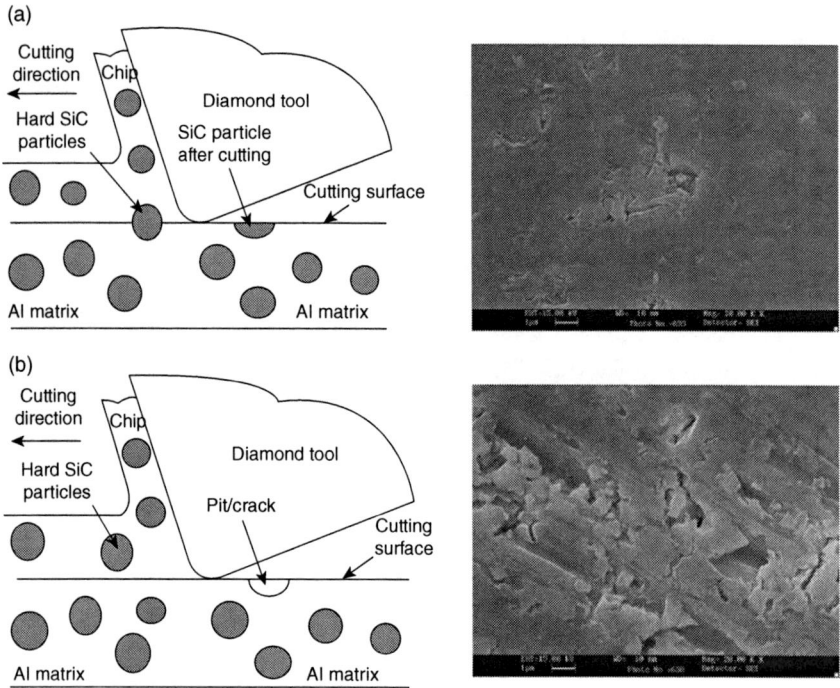

17.14 Illustration of (a) cut-through and (b) pulled-out mechanisms in cutting SiC particles.[22]

to say, the depth of the pits depended on the size of the SiC particles removed. For the SiCp/2024Al used in the tests, the pits could be as deep as several hundred nm (see the AFM image of a pit Fig. 17.16b).

It was also observed that some of the voids and microcracks were formed either due to part of the SiC reinforcement detached from the matrix or due to the orientation of the SiC reinforcement, so as to accommodate deformation of the matrix around it (see Fig. 17.17).

In yet another study, it was observed that the effect of speed and feed on the residual stress induced in machined surfaces of non-reinforced alloy was different from that with the MMC. Both longitudinal and transverse residual stresses on the non-reinforced alloy surface were tensile and increased with increase of speed and feed. On the other hand, the presence of reinforcement particles induced compressive residual stresses on the machined surfaces on MMCs. An increase in feed reduced the longitudinal compressive residual stress, but had negligible influence on the transverse stress on the MMC. The influence of speed on the residual stress of the MMC was not significant.[54]

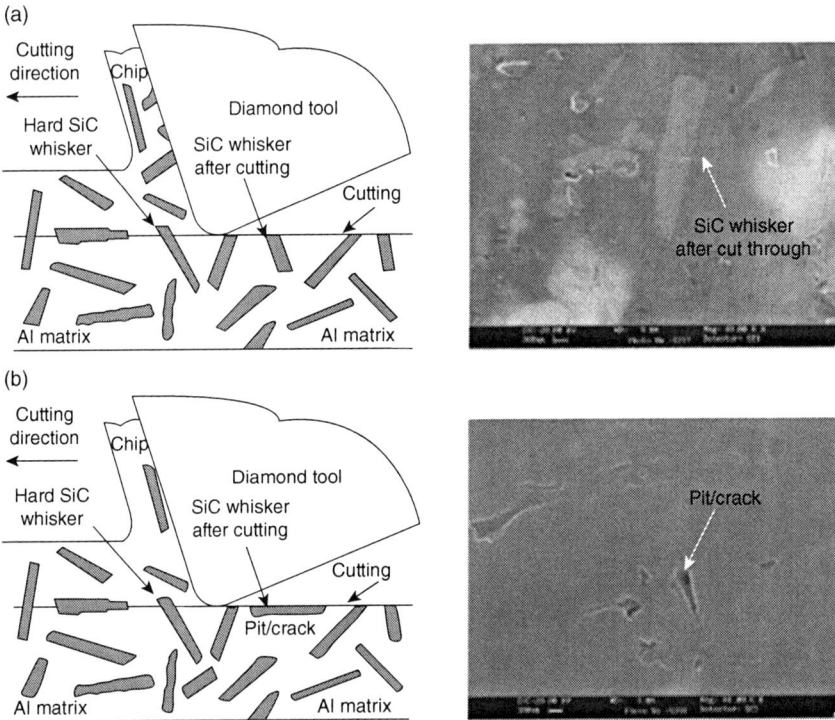

17.15 Illustration of the (a) cut-through and (b) pulled-out mechanisms in cutting SiC whiskers.[22]

Micro-hardness measurements of machined cross-sections indicate that a few micro meters below the cut surface, the hardness had increased beyond the bulk hardness of the material (Fig. 17.18a–c). This is attributed to an increase in the dislocation density due to plastic deformation in this region. Moving further away from the machined surface, the hardness started to decrease and reached the bulk hardness.[53] As observed from Fig. 17.18a–c, the depth of the machining-affected layer is influenced by the cutting parameters. As the cutting speed, feed rate and/or depth of cut increase, the depth of the affected layer increased. Moreover, hardening increased when any of the cutting parameters were increased. An increase in cutting temperature, which leads to thermal deformation in the matrix material, could be one of the reasons behind hardening in the machined sub-surface. It is probably worth mentioning that the measurements of hardness at the machined surface and to a depth of 10 μm had a large scatter. This is due to particle pull-out, particle fracture and crack growth preferentially around the SiC particles in this region. However, no cracks were observed at depths below 10 μm.[53]

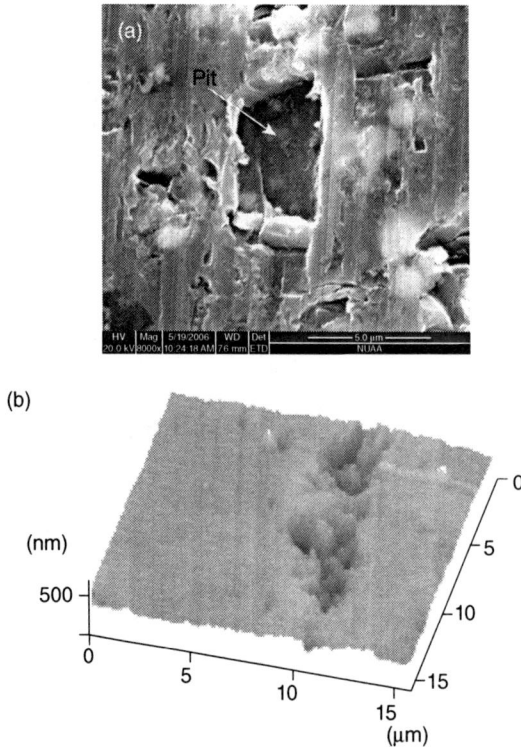

17.16 Pit left on the surface when SiC particle was pulled out (for 15 vol% SiCp/2024Al): (a) SEM image, (b) AFM image.[21]

17.4   Chip formation and mechanics of machining

Studies on chip formation help us to understand the mechanisms of machining MMCs and ways to improve chip breakability, which in turn leads to an improvement in the productivity of the machining process. At the same time, a knowledge of the mechanics of machining help us to understand the dynamic characteristics of the machining process, which in turn helps selection of appropriate equipment and toolings. In the following sections, the relevant studies are described.

17.4.1 Chip formation in machining MMCs

Fundamental experiments on orthogonal machining of Al/SiC particulate MMCs, including chip freezing trials to understand the chip formation mechanism in machining, were done by Joshi and co-workers.[10,12] PCD tools were used during these experiments.

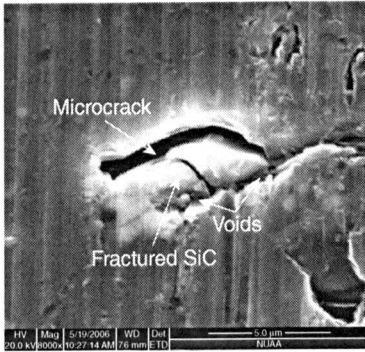

17.17 Voids and micro-cracks formed around the SiC particle (15 vol% SiCp/2024Al).[21]

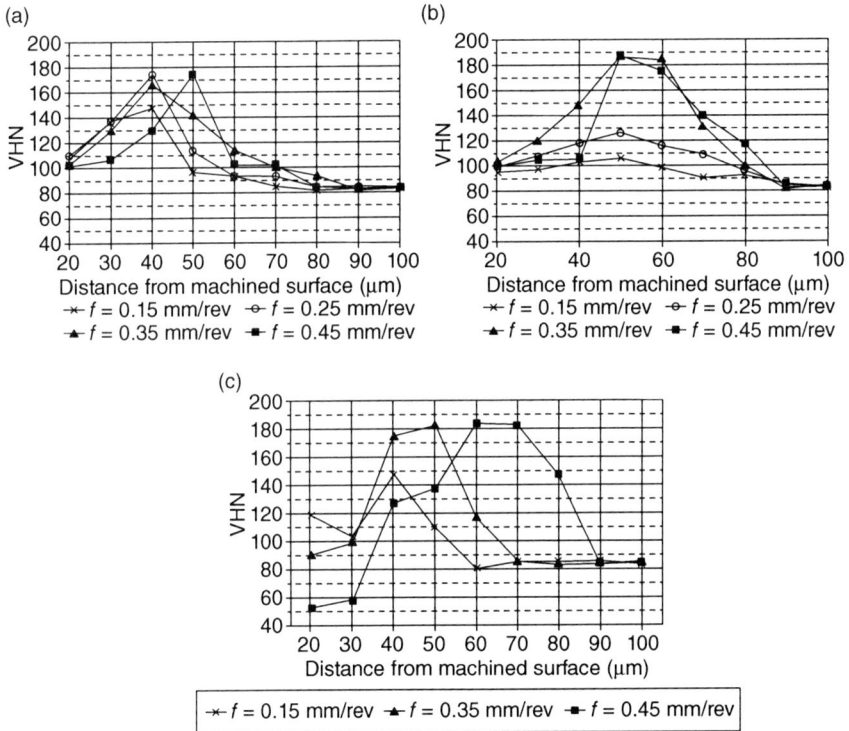

17.18 Micro-hardness-depth profiles (VHN) on cross-sections of machined surfaces: (a) at cutting speed = 670 m min^{-1}, depth of cut = 2.5 mm; (b) at cutting speed = 894 m min^{-1}, depth of cut = 1.5 mm; and (c) at cutting speed = 670 m min^{-1}, depth of cut = 2.5 mm (r = 1.6 mm, $\alpha = 0°$).[53]

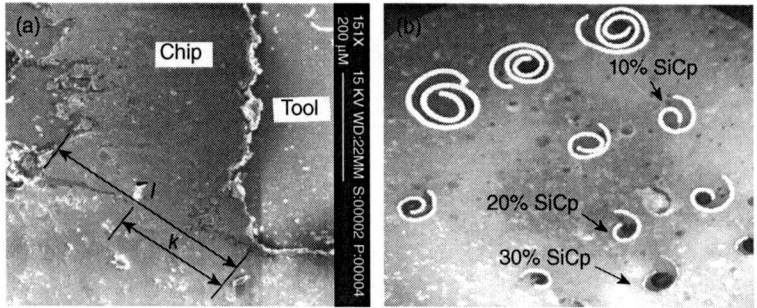

% Vol. of SiCp	00	10	20	30
Material failure strain in tension	0.30	0.049	0.027	0.019
Number of chip curls	2–3	1–1$^{1}/_{2}$	$^{3}/_{4}$–1	$^{3}/_{4}$–1

17.19 (a) Typical deformation on the shear plane in machining of Al/SiCp MMCs showing the extent of flow type (*k*) and fracture (*l* − *k*) deformation,[10] (b) typical chips in Al/SiCp composites,[12] (c) number of chip curls as a function of failure strain.[12]

The chip freezing trials showed that the chip formation mechanism in machining of MMCs involves initiation of a gross fracture from the chip-free surface and its propagation towards the tool nose (Fig. 17.19a).[10] The extent of the propagation of the gross fracture depends upon the cutting speed and volume of reinforcement in the composites.[10] The higher the cutting speed, the lower the contribution of gross fracture in the deformation of material along the shear plane. At the same time, the higher the volume of reinforcement, the larger is the contribution of gross fracture in the deformation of material along the shear plane (see Fig. 17.19a). A ratio (*k*/*l*) has been derived to quantify the extent of ductile and brittle (fracture) deformation along the shear plane in machining an Al/SiC particulate composite (Fig. 17.19a). Here, *l* represents the total length of the shear plane and *k* represents the length of ductile deformation from the tool nose. The ratio *k*/*l* increases with an increase in the cutting speed, indicating an increase in flow-type deformation (Fig. 17.19a).

Analysis of chip formation mechanism carried out by Nakayama[55] also showed that in MMC machining, the chips are formed by the initiation of cracks from the chip-free surface. The application of shear stress by the tool causes shear concentration around the edges of particles, leading to the generation of voids. With further shearing, coalescence of voids accelerates crack growth and its propagation. This eventually causes formation of chips bearing 'saw-tooth' profiles while machining.

In another case, the chip formation in orthogonal machining of Al/SiCp MMCs showed a systematic pattern of chip breaking that was proportional to the strain at failure that the MMC material could withstand (Fig. 17.19b and c).[12] As observed from the geometry of the chips those in aluminum matrix alloy curl by 2–3 circles; however, as the volume of SiCp reinforcement in the MMC increases, the number of chip circles reduces to the strain that the material can withstand (Fig. 17.19b and c). The chip-breaking strain in composites can be predicted using two classical models.

(i)   *Nakayama's model.*[55] This model predicts chip failure strain based on the geometry of the chips:

$$\varepsilon_b < \frac{t_2}{2}\left[\frac{1}{R_0} - \frac{1}{R_L}\right]$$ [17.1]

where ε_b = strain in the chip material, t_2 = chip thickness (mm), R_0 = radius of initial chip curl (mm), R_L = radius of final chip curl prior to fracture (mm).

(ii) *Zang and Peklenik's model.*[56] This model predicts chip failure strain based on the geometry and mechanical properties of the chip material:

$$\frac{2\sigma_{sc}}{E_c} + \varepsilon_b = \left[\frac{h_c}{r_g}\right]\left[1 - \frac{1}{2k}\right]$$ [17.2]

where, $k = r_g/r_c$ and σ_{SC} = yield strength of the chip material (MPa), E_c = modulus of elasticity of chip material (GPa), h_c = chip thickness (m), r_g = radius of chip curvature at the breaking point (m), r_c = radius of chip curvature when chip flows out of the chip breaker groove (m).

It was observed that at the various parametric conditions, the chip failure strain in machining Al/SiCp/10p composites predicted by Nakayama's model (Eq. 17.1) was closer to the experimental values (Fig. 17.20) than those predicted using Zang and Peklenik's model (Eq. 17.2). It is therefore thought that the simpler of the two equations, i.e Nakayama's model (Eq. 17.1), which gives chip failure strains based on their geometry, is sufficient to describe the chip breaking process in machining MMCs.

In general, it is found that the chip breakability in machining MMCs improves due to the presence of the reinforcement particles. Short chips are formed under almost all conditions in MMCs. However, with the non-reinforced alloy, chips of almost similar shape (but long and unbroken) are formed at all the cutting conditions.[51] In another experiment, on machining

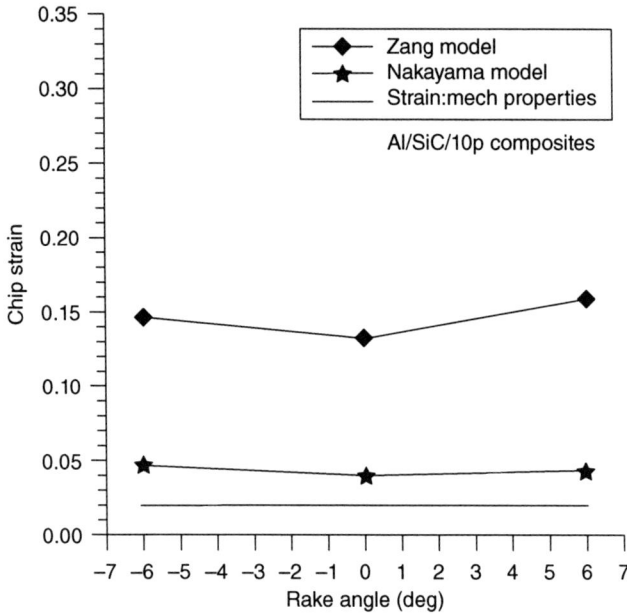

17.20 Comparison of chip failure strains evaluated using the models.[12]

of A359/SiC/208 MMC, it was evident that the reinforcements help produce semi-continuous chips. Hence, the machining process had better control.[57]

It is also reported that the reinforcement particles do not influence shear and friction angles significantly with variation of the feed rate. In the case of MMCs, shear and friction angles increase very little with an increase in speed. However, in the case of non-reinforced alloy, initially the shear angle decreases and the friction angle increases. However, at low speed but after a certain speed, the shear angle increases and the friction angle remains constant with further speed increase.[51]

An increase in the feed rate causes the chip forms to change from continuous to discontinuous. This facilitates easy chip disposal and provides a brief relief to the tool from abrasion by chips under its surface. Also, an increase in feed causes the work to become soft, thereby reducing the abrasion effect of reinforcement particles.[8]

In ultra-precision turning of MMCs, chips formed by a PCD tool were more discontinuous and fragmented (Fig. 17.21a) than those formed by a straight-nose SCD tool, Fig. 17.21b.[23] This can be explained by the fact that the cutting edge radius of PCD is about ten (or more) times larger than that of SCD. The PCD tools therefore merely smeared the surface when the undeformed chip thickness was of the same order as the cutting edge radius. Hence, they produced fragmented chips.

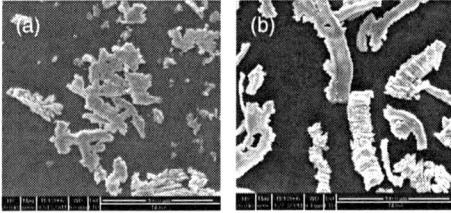

17.21 Chips obtained by machining using (a) PCD tools, (b) SCD tools.[23]

In another study,[57] it was observed that the nature of chips formed during machining Al/SiC particulate composites changed with the extent of tool wear. When the tool was sharp, long or small, washer-type helical chips were formed owing to the constraint on chip flow by the tool holder. However, once the tool started to become blunt, the chip type changed to a loose arch type. This was because of unstable built-up edge on the blunt tool tip which began operating as a chip breaker.

17.4.2 Mechanics of machining

Examination of the mechanics of machining MMCs usually involves measurement of the cutting forces during machining. It is evident that machining MMCs may not require cutting forces that are way higher than their un-reinforced counterparts. Indeed, in some instances the forces for machining of un-reinforced alloys are more. This complex variation is due to:[51]

(i)   differences in their work hardening and thermal softening characteristics,
(ii)   fracture that occurs at the shear plane and tool–chip interface, and
(iii)   differences in their strain and strain rate responses.

When dry machining MMCs, the tool forces were measured using a Kistler three-component dynamometer. It was observed that the cutting forces decreased with an increase in cutting speed and/or depth of cut (Fig. 17.22). This could be attributed to the thermal softening of the work material. The cutting forces, however, increased with an increase in feed rate. An increase in the feed rate increased the chip load on the tool without any increase in the thermal softening (Fig. 17.23).

Evaluation of the average power and specific cutting pressure during the turning of A356/SiC composites with PCD tools is shown in Figs 17.24 and 17.25, respectively. It was observed that with cutting speeds of 250 to 350 m/min, a feed of 0.1 mm/rev and a depth of cut of 1 mm, the machining power required was around 1 kW. Under similar machining conditions, the cutting

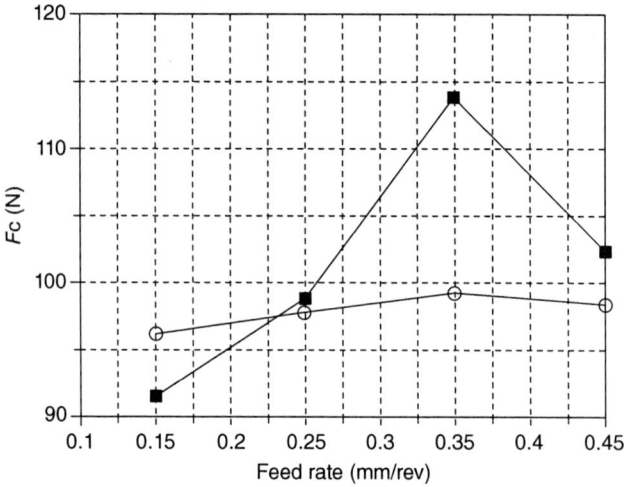

17.22 Effect of cutting speed on the cutting forces (PCD tool: $r =$ 1.6 mm, $\alpha = 0°$. Filled squares: $v = 670$ m min^{-1}, depth of cut = 1.5 mm. Open circles: $v = 894$ m min^{-1}, depth of cut = 1.5 mm).[9]

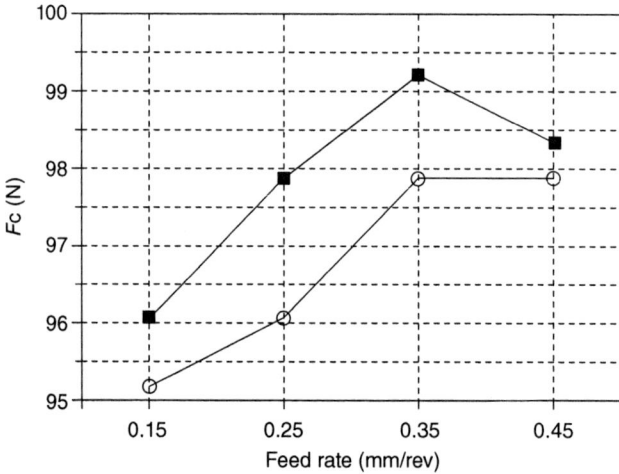

17.23 Effect of depth of cut on the cutting forces (PCD tool: $r =$ 1.6 mm, $\alpha = 0°$. Filled squares: $v = 894$ m min^{-1}, depth of cut = 1.5 mm. Open circles: $v = 894$ m min^{-1}, depth of cut = 2.5 mm).[9]

pressure varied between 1500 to 2500 N/mm^2 (Fig. 17.25).[18] An increase in the cutting speed caused a reduction in the specific cutting pressure. However, this conclusion was not valid when the cutting speed was increased excessively, as rapid tool wear then occurred.[18]

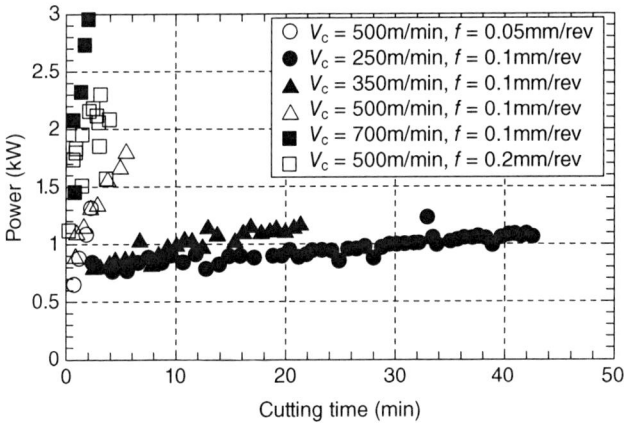

17.24 Power vs. cutting time in turning A356/SiC/20p as function of cutting speed and feed using PCD inserts.[18]

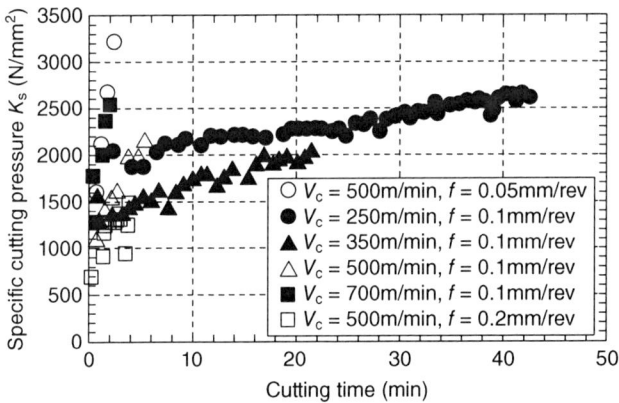

17.25 Specific cutting pressure vs. cutting time in turning A356/SiC/20p, with PCD inserts.[18]

As far as the mechanics of machining of MMCs during other processes such as drilling and milling are concerned, very limited studies are available. This is primarily because cutting tool materials with longer tool lives are not available for performing these operations. Also, due to the intermittent nature of such machining operations, the highly impact sensitive PCD tools cannot be used to perform them. Nevertheless, some studies on drilling of MMCs with PCD drills have been done. The measurement of thrust and torque in drilling in 6061A/Al$_2$O$_3$/20p composites using PCD and carbide tipped tools showed that under all the parametric conditions, the thrust as well as the torque are significantly lower with the PCD tools (Fig. 17.26a–d).

(a)

(b)

(c)

(d)

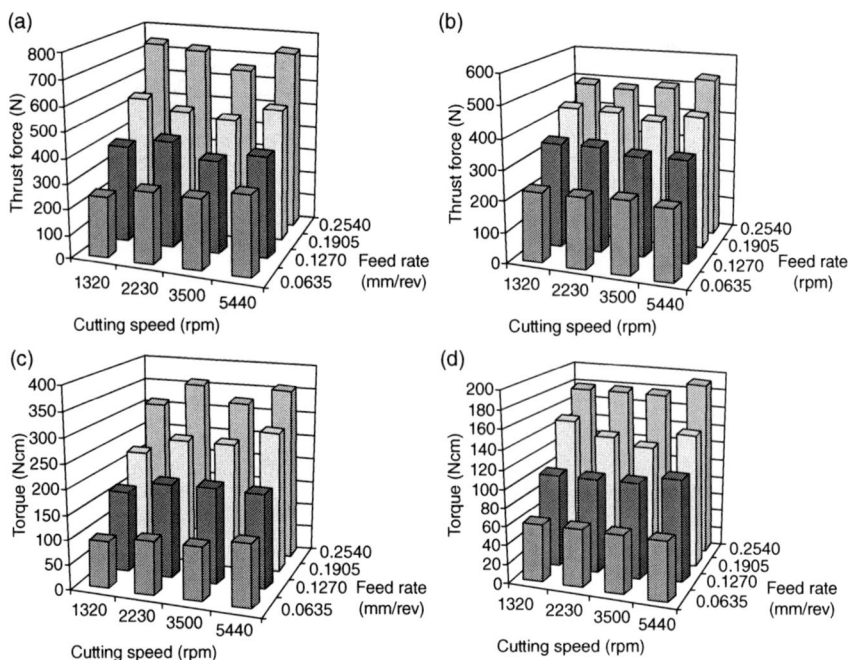

17.26 Drilling forces for (10 vol% (Al_2O_3)p/6061). (a) Thrust force carbide-tipped drills, (b) thrust force PCD drills, (c) torque carbide-tipped drills (10 vol% (Al_2O_3)p/6061), (d) torque PCD drills.[58]

Both the thrust and torque significantly increased with an increase in feed rate; however, such increase was not observed with an increase in the cutting speed (Fig. 17.26a–d). The thrust and torque also increased with an increase in the volume of reinforcement in the composites.[58]

17.5   Conclusions and future trends

- It is clear that PCD is the most suitable cutting tool material for the machining of MMCs (with volume of reinforcement ~20%) at cutting speeds in excess of 500 m/min. Other cutting tools materials, such as carbides, coated carbides and ceramics, have very poor tool life at these cutting speeds, so their use is not economical for machining MMCs.
- Besides the PCD tools, CVD diamond coated and PCBN tools have comparable tool life in machining MMCs.
- In the ultra-precision (extremely low feed rates and depths of cut) machining of MMCs, PCD tools demonstrate comparable performance to that of single crystal diamond tools.

- Multiple parameters influence the performance of PCD tools in machining MMCs: they include the volume of reinforcement in the MMCs, cutting speed and size of grains in the PCD tool.
- The prominent mechanism of wear of PCD tools is abrasion. However, sticking of soft matrix in MMCs coupled with the reinforcement fragments cause initiation of numerous failure mechanisms on PCDs, e.g. adhesive wear, chipping, wear scar and tool breaking.
- In MMC machining, the PCD tools, owing to their edge sharpness, cause significantly lower surface and sub-surfaces damage on the machined surfaces than carbide tools.
- The machined surface roughness on MMCs is influenced by several parameters, such as composition of the MMCs, tool nose radius, tool geometry and magnitude of flank wear on the tools. Low feed rate, large tool nose radius and use of cutting fluids are recommended to improve the machined surface on MMCs.
- In ultra-precision machining of MMCs using PCD and SCD tools, a very fine finish, of the order of $10\text{--}30\ \mathrm{nm} R_a$ was achieved on the machined surfaces.
- Poor machined surface integrity on MMCs is due to de-bonding, pull-out and fracture of reinforcements, and consequent generation of voids, pits and micro-cracks on the surfaces. However, no cracks are observed at a depth of 10 μm beneath the machined surfaces.
- Chip formation mechanisms in machining MMCs show both gross fracture and flow type deformations. The chip breakability in machining is improved by the presence of reinforcements and can be predicted reasonably well using simple chip geometry-based strain evaluation.
- The MMCs require machining forces that are comparable to their unreinforced counterparts. Thus, the presence of reinforcements does not add to the difficulty in their machining from a cutting forces point of view.
- Therefore, saving PCD tools from severe abrasion on tool faces caused by reinforcements in MMCs remains the main challenge in machining these materials.
- There is not enough knowledge available on the use of other machining processes on MMCs such as drilling, milling and grinding.

17.6   Acknowledgments

The author would like to express his thanks to Dr Uday Dabade, Assistant Professor, Walchand College of Engineering, Sangli, India, for providing immense help in the compilation and assimilation of literature towards the preparation of this manuscript. The author would also like to thank Harshad Sonawane, PhD candidate working with the author for his help in drafting this manuscript and preparing art work.

17.7   References

1   A. B. Pandey, K. L. Kendig, J. Lewandowski, and S. R. Shah (Eds), *Affordable Metal-matrix Composites for High-performance Applications*, TMS Publications, USA, 2003.

2   V. M. Kervorkijan, Commercial viability of Al-based MMCs in the automotive segment, *Materials and Manufacturing Processes*, **14**(5) (1999), pp. 639–645.

3   J. E. Allison, G. S. Cole, Metal matrix composites in the automotive industry: Opportunities and Challenges, *Journal of Minerals, Metals and Materials Society*, Jan (1993), pp. 19–24.

4   S. S. Joshi, N. Ramakrishnan, P. Ramakrishnan, PCD in the machining of DRA composites, *Industrial Diamond Review*, **3** (2001), pp. 177–182.

5   P. J. Heath, Developments in applications of PCD tooling, *Journal of Materials Processing Technology*, **116** (2001), pp. 31–38.

6   M. K. Brun and M. Lee, Wear characteristics of various hard materials for machining of SiC reinforced aluminium alloy, *Wear*, **104** (1985), pp. 21–29.

7   K. Weinert, and W. Konig, A consideration of tool wear mechanisms when machining metal matrix composites, *Annals of CIRP*, **42**(1) (1993), pp. 95–98.

8   N. Tomac, and K. Tonnessen, Machinability of particulate aluminium matrix composites, *Annals of CIRP*, **41**(1) (1992), pp. 55–58.

9   M. El-Gallab, M. Sklad, Machining of Al/SiC particulate metal-matrix composites. Part I: Tool performance, *Journal of Materials Processing Technology*, **83** (1998), pp. 151–158.

10   S. S. Joshi, N. Ramakrishnan, and P. Ramakrishnan, Microstructural analysis of chip formation during orthogonal machining of Al/SiCp composites, *Trans. ASME, Journal of Engineering Materials and Technology*, **123** (2001), pp. 315–321.

11   S. S. Joshi, N. Ramakrishnan, H. E. Nagarwalla, and P. Ramakrishnan, Wear of rotary carbide tools in machining of Al/SiCp composites, *Wear*, **230** (1999), pp. 124–132.

12   S. S. Joshi, N. Ramakrishnan, and P. Ramakrishnan, Analysis of chip breaking during orthogonal machining of Al/SiCp composites, *Journal of Materials Processing Technology*, **88** (1999), pp. 90–96.

13   M. Jennings, Fast tools for fast cars, *Industrial Diamond Review*, **49**(4) (1989), pp. 150–153.

14   W. Stief, Trumpets depend on precision, *Industrial Diamond Review*, **47**(3) (1987), pp. 112–115.

15   G. Spur, U. E. Wunsch, Turning FRP with Syndite Ð test results, *Industrial Diamond Review*, **45**(1) (1985), pp. 195–199.

16   H. Lach, PCD tools for woodworking, *Industrial Diamond Review*, **45**(4) (1985), pp. 166–167.

17   J. T. Lin, D. Bhattacharyya, C. Lane, Machinability of silicon carbide reinforced aluminium metal matrix composite, *Wear*, **181–183** (1995), pp. 883–888.

18   J. P. Davim, Diamond tool performance in machining metal-matrix composites, *Journal of Materials Processing Technology*, **128** (2002), pp. 100–105.

19   J. P. Davim, A. M. Baptista, Relationship between cutting force and PCD cutting tool wear in machining silicon carbide reinforced aluminium, *Journal of Materials Processing Technology*, **103** (2000), pp. 417–423.

20 M. El-Gallab, M. Sklad, Machining of Al/SiC particulate–metal matrix composites. Part III. Comprehensive tool wear models, *Journal of Materials Processing Technology*, **101** (2000), pp. 10–20.

21 Y. F. Ge, J. H. Xu, H. Yang, S. B. Luo, Y. C. Fu, Workpiece surface quality when ultra-precision turning of SiC$_p$/Al composites, *Journal of Materials Processing Technology*, **203** (2008), pp. 166–175.

22 C. F. Cheung, K. C. Chan, S. To, W. B. Lee, Effect of reinforcement in ultra-precision machining of Al6061/SiC metal matrix composites, *Scripta Materialia*, **47** (2002), pp. 77–82.

23 G. Yingfei, X. Jiuhua, Y. Hui, Diamond tools wear and their applicability when ultra-precision turning of SiC$_p$/2009Al matrix composite, *Wear*, **269** (2010), pp. 699–708.

24 C. Lane, The effect of different reinforcements on PCD tool life for aluminium composites, *Machining of Composite Materials, Proc. of ASM Materials Congress*, Nov. 1–5, 1992, Pennsylvania, USA, pp. 17–27.

25 S. Barnes, I. R. Pashby, D. K. Mok, The effect of workpiece temperature on the machinability of an aluminum/SiC MMC, *Trans. ASME, Journal of Manufacturing Science and Engineering*, **118** (1996), pp. 422–427.

26 C. Divakar, S. K. Bhaumik, A. K. Singh, Machining Al/SiCp MMC with wBN, cBN composite tools, in: *Proceedings of the Fourth International Conference on Composites Engineering*, Hawaii, USA, July 1997, pp. 913–914.

27 S. S. Cho, K. Komvopoulos, Wear mechanisms of multi-layer coated cemented carbide cutting tools, *Trans. ASME, Journal of Tribology*, **119** (1997), pp. 8–17.

28 J. E. Caroline, H. F. Andrewes, W. M. Laub, Machining of an aluminum/SiC composite using diamond inserts, *Journal of Materials Processing Technology*, **102** (2000), pp. 25–29.

29 K. Winert, A consideration of tool wear mechanism when machining metal matrix composites (MMC), *Annals of CIRP*, **42**(1) (1993), pp. 95–98.

30 C. Lane, Machining discontinuously-reinforced aluminum composites, in: *Tool and Manufacturing Engineers Handbook*, Vol. 7, Society of Manufacturing Engineers, 1983.

31 S. Durante, G. Rutelli, F. Rabezzana, Aluminum-based MMC machining with diamond-coated cutting tools, *Surface and Coatings Technology*, **94–95** (1997), pp. 632–640.

32 G. E. D'Errico, R. Calzavarini, Turning of metal matrix composites, *Journal of Materials Processing Technology*, **119** (2001), pp. 257–260.

33 X. Ding, W. Y. H. Liew, X. D. Liu, Evaluation of machining performance of MMC with PCBN and PCD tools, *Wear*, **259** (2005), pp. 1225–1234.

34 J. T. Lin, D. Bhattacharyya, C. Lane, Machinability of a silicon carbide reinforced aluminium metal matrix composite, *Wear*, **181–183** (1995), pp. 883–888.

35 N. P. Hung, C. H. Zhong, Cumulative tool wear in machining metal matrix composites. Part II: Machinability, *Journal of Materials Processing Technology*, **58** (1996), pp. 114–120.

36 L. Luliano, L. Settineri, A. Gatto, High-speed turning experiments on metal matrix composites, *Composites Part A*, **29A** (1998), pp. 1501– 1509.

37 X. Li, W. K. H. Seah, Tool wear acceleration in relation to workpiece reinforcement percentage in cutting of metal matrix composites, *Wear*, **247** (2001), pp. 161–171.

38 R. T. Coelho, S. Yamada, D. K. Aspinwall, M. L. H. Wise, The application of polycrystalline diamond (PCD) tool materials when drilling and reaming aluminium based alloys including MMC, *International Journal of Machine Tools & Manufacture*, **35**(5) (1995), pp. 761–774.

39 N. P. Hung, T. C. Tan, Z. W. Zhong, G. W. Yeow, Ductile-regime machining of particle-reinforced metal matrix composites, *Machining Science and Technology*, **3**(2) (1999), pp. 255–271.

40 G. E. D'Erico, R. Calzavarini, Turning of metal matrix composites, *Journal of Materials Processing Technology*, **119** (2001), pp. 257–260.

41 Z. J. Yuan, L. Geng, S. Dong, Ultraprecision machining of SiCw/Al composites, *Annals of CIRP*, **42**(1) (1993), pp. 107–109.

42 K. Weinert, A Consideration of tool wear mechanism when machining metal matrix composites (MMC), *Annals of CIRP*, **42**(1) (1993), pp. 95–98.

43 K. Weinert, *Machining of light-metal matrix composites*, STC-C Technical Presentation, January CIRP Meeting, Paris, 2000.

44 H. K. Tonshoff, J. Winkler, The influence of tool coatings in machining of magnesium, *Surface Coating Technology*, **94–95** (1997), pp. 610–616.

45 C. T. Lane, Machining discontinuously-reinforced aluminum composites, *American Machinist*, USA, Nov (1993), pp. 56–60.

46 Y. M. Quan, Q. X. Yu, L. J. Xie, Study on the adaptability of thick film diamond tool to cutting composites, *International Journal of Machine Tools and Manufacture*, **42** (2002), pp. 501–504.

47 U. A. Dabade, *Characteristics of Machined Surfaces on Al/SiCp Metal Matrix Composites Produced During Turning*, Ph. D. thesis, Indian Institute of Technology, Bombay, 2008.

48 A. C. Basheer, U. A. Dabade, S. S. Joshi, V. V. Bhanuprasad, V. M. Gadre, Modeling of surface roughness in precision machining of metal matrix composites using ANN, *Journal of Materials Processing Technology*, **197** (2008), pp. 439–444.

49 N. P. Hung, S. H. Yeo, B. E. Oon, Effect of cutting fluid on the machinability of metal matrix composites, *Journal of Materials Processing Technology*, **67** (1997), pp. 157–161.

50 R. Teti, Machining of composite materials, *Annals of CIRP*, **51**(2) (2002), pp. 611–634.

51 A. Pramanik, L. C. Zhang, J. A. Arsecularatne, Machining of metal matrix composites: Effect of ceramic particles on residual stress, surface roughness and chip formation, *International Journal of Machine Tools & Manufacture*, **48** (2008), pp. 1613–1625.

52 K. C. Chan, C. F. Cheung, M. V. Ramesh, W. B. Lee, S. To, A theoretical and experimental investigation of surface generation in diamond turning of an Al6061/SiC$_p$ metal matrix composite, *International Journal of Mechanical Sciences*, **43** (2001), pp. 2047–2068.

53 M. El-Gallab, M. Sklad, Machining of Al/SiC particulate–metal matrix composites. Part II. Workpiece surface integrity, *Journal of Materials Processing Technology*, **83** (1998), pp. 277–285.

54 J. P. Davim, J. Silva, A. M. Baptista, Experimental cutting model of metal matrix composites (MMCs), *Journal of Materials Processing Technology*, **183** (2007), pp. 358–362.

55 K. Nakayama, A study of chip breaker, *Bulletin of Japanese Society of Mechanical Engineers*, **5**(17) (1962), pp. 142–150.

56 Y. Z. Zang, J. Peklenik, Chip curl, breaking and chip control of difficult-to-machine materials, *Annals of CIRP*, **29**(1) (1980), pp. 79–83.

57 J. T. Lin, D. Bhattacharyya, W. G. Ferguson, Chip formation in the machining of SiC-particle-reinforced aluminium-matrix composites, *Composites Science and Technology*, **58** (1998), pp. 285–291.

58 M. Ramulua, P. N. Rao, H. Kao, Drilling of (Al$_2$O$_3$)p/6061 metal matrix composites, *Journal of Materials Processing Technology*, **124** (2002), pp. 244–254.

Index